# Composting of Farm Residues by Nature-Based Solutions

Rupesh Kumar Singh • Henrique Trindade
João Ricardo Sousa • Arne Sæbø
Vishnu D. Rajput • Tao Zhang
Editors

# Composting of Farm Residues by Nature-Based Solutions

## A Potential Key Towards Sustainable Agricultural Practices

Springer

*Editors*
Rupesh Kumar Singh
Centre for the Research and Technology of Agro-Environmental and Biological Sciences (CITAB), Inov4Agro
University of Trás-Os-Montes and Alto Douro (UTAD)
Vila Real, Portugal

Henrique Trindade
Centre for the Research and Technology of Agro-Environmental and Biological Sciences (CITAB), Inov4Agro
University of Trás-Os-Montes and Alto Douro (UTAD)
Vila Real, Portugal

João Ricardo Sousa
Centre for the Research and Technology of Agro-Environmental and Biological Sciences (CITAB), Inov4Agro
University of Trás-Os-Montes and Alto Douro (UTAD)
Vila Real, Portugal

Arne Sæbø
Særheim Research Station
Norwegian Institute of Bioeconomy Research
Klepp, Norway

Vishnu D. Rajput
Academy of Biology and Biotechnology
Southern Federal University
Rostov-on-Don, Russia

Tao Zhang
State Key Laboratory of Nutrient Use and Management, Beijing Key Laboratory of Farmland Soil Pollution Prevention and Remediation, Key Laboratory of Plant-Soil Interactions of Ministry of Education
College of Resources and Environmental Sciences, China Agricultural University
Beijing, China

ISBN 978-3-032-12254-4 ISBN 978-3-032-12255-1 (eBook)
https://doi.org/10.1007/978-3-032-12255-1



This Springer imprint is published by the registered company Springer Nature Switzerland AG
The registered company address is: Gewerbestrasse 11, 6330 Cham, Switzerland

# Preface

The present book entitled *Composting of Farm Residues by Nature-Based Solutions: A Potential Key Towards Sustainable Agricultural Practices* focuses on sustainable management of residues produced by various agriculture establishments including chicken farms, cattle farms, and horticulture/crops farms. Composting of organic residues involves a complex microbial dynamic and their interaction to decompose the organic matter. During this process, not only large amounts of greenhouse gases and ammonia are released to environment but the nutrient losses have also been recorded. Several new models have been evolved in the past decade to address these issues. Additionally, the nature-based solutions have been major attraction in recent European green deal to develop the sustainable practice toward improved environment health. This book focuses on a comprehensive exploration of evolution of composting techniques from traditional to modern methodologies.

The primary objective of this book is to disseminate the recent advances made in composting process to reduce the greenhouse gas emissions and restore the available nutrients under field conditions. The end product compost may reduce heavily the chemicals used in agriculture sector and restore the physio-chemical properties of soil by improving the natural microbial community toward normal state. Developing such practices by including nature-based solutions may foster the agriculture units/farms in a self-sustainable model while creating a big impact to environment health in a bigger picture. The book aims to fill the gap between scientific community, agriculture farmers, stakeholders, common peoples, and policymakers, for better understanding of risks associated with composting and their sustainable mitigation for maximum benefits in agriculture to create a better tomorrow.

This book emphasizes to explore the nature-based solutions to improve the composting process instead of synthetic chemical-based components. Unlike the other works that focus mainly on compost applications in field, this book includes the mechanisms of composting to application in a more environment-friendly manner to improve the sustainability and restore the soil and environmental health. The book also focuses to maximize the nutrient availability in the compost by reducing the losses during the process. Several new bio-based materials and microorganisms have been tested for this purpose and may potentially transform the composting science. The book also dedicates specific chapter to assess the potential risks in depth and provides guidelines for safer implementation in agriculture to minimize the negative effects for human as well as environmental health. Although significant improvements have been made in composting, monitoring, management, and application to the field since the past decade, the book addresses the potential challenges and opportunities and further encourages researchers to overcome and achieve the maximum positive impacts.

Overall, the book presents a holistic approach to sustainably manage the residues from agriculture farming sector to reduce greenhouse gases, restore nutrient loss, improve physiochemical properties of soil by application of compost to the fields, and improve the crop physiology for better production and reduced chemicals use. The book brings the key factors to develop self-reliance by reducing the dependency on chemical-based fertilizers and overall improvement of total environment.

Vila Real, Portugal Rupesh Kumar Singh
Vila Real, Portugal Henrique Trindade
Vila Real, Portugal João Ricardo Sousa
Klepp, Norway Arne Sæbø
Rostov-on-Don, Russia Vishnu D. Rajput
Beijing, China Tao Zhang

# Acknowledgments

This work is supported by National Funds by FCT – Portuguese Foundation for Science and Technology, under the projects UID/04033/2025: Centre for the Research and Technology of Agro-Environmental and Biological Sciences (https://doi.org/10.54499/UID/04033/2025) and LA/P/0126/2020 (https://doi.org/10.54499/LA/P/0126/2020) and ECONUTRI project, the European Union Horizon Europe Innovation program under the Grant Agreement No. 101081858.

# About the Book

The book entitled *Composting of Farm Residues by Nature-Based Solutions: A Potential Key Towards Sustainable Agricultural Practices* brings a wider understanding of problems and solutions regarding agriculture waste produce and their sustainable management. High demand of agriculture products forced this industry to meet the requirement: thus, high amount of organic residues is being generated. Management of this organic residue is of utmost importance to avoid environmental hazards, but also to recycle this material into compost to boost the circular economy. The composting is a complex process, during which a large amount of greenhouse gases and ammonia is released as pollutants, which also means the nutrients lost in the final compost. The new nature-based solutions to combat and address this issue is major need in recent European green deal toward improved environment health. Thus, this book intends to explore the various stages of complex composting process including traditional and modern technologies.

# Contents

**Composting: Historical Background to Modern Prototypes of the Composting Process** . . . . . . . . . . 1
Mudassar Khan, Sana Bashir, Ahmad Zia Khan, Francisco Saraiva, Rupesh Kumar Singh, Henrique Trindade, and João Ricardo Sousa

**Overview of Different Nature-Based Ingredients in the Composting Process and Their Potential Value Addition** . . . . . . . . . . 29
Miguel Baltazar, Sandra Pereira, Lia Dinis, Sandra Martins, and Cátia Brito

**Earthworm-Mediated Bio-composting and Basic Transformation Mechanisms in Organic Matter** . . . . . . . . . . 59
Elisabete Nascimento-Gonçalves, Tiago Azevedo, Catarina Medeiros, Maria Cristina Morais, Anabela R. Reis, João Ricardo Sousa, Paula A. Oliveira, Ana M. Coimbra, and Marta Roboredo

**Greenhouse Gas Emissions During Composting Under Field Conditions and Their Impact on Environmental Health** . . . . . . . . . . 89
Pooja L. R., Renu Singh, Mamta Bisht, Manoj Shrivastava, and H. M. Akshay Kumar

**Composting Techniques for Poultry Industry–Based Organic Residues and Their Sustainable Application to Improve Soil Nitrogen Content** . . . . . . . . . . 115
Catarina Medeiros, Tiago Azevedo, Elisabete Nascimento-Gonçalves, Maria Cristina Morais, João Ricardo Sousa, Paula A. Oliveira, Marta Roboredo, and Ana M. Coimbra

**Exploring the Organic Residual Matter from Cattle Industries and Microbial Dynamics in Composting** . . . . . . . . . . 137
Oyindamola John Samson, Doyinsola Amos Samson, Jamiu Oluwatosin Adeyemi, Hafeez Aderinsayo Adekola, and Ismail Babatunde Onajobi

**Evolution of Modern Methods to Understand the Microbial Community and Metabolic Dynamics of Compost** . . . . . . . . . . . . . . . . . 153
Alexandros Phokas, Panayiota Pissaridou, Johana Rodosthenous, Christiana Hadjimichael, and Michalis Omirou

**Composting: Improved Crop Performance and Resilience for Environmentally Friendly Practices** . . . . . . . . . . . . . . . . . . . . . . . . . 177
Sana Bashir, Tasmia Bashir, Ali Abdur Rehman, and Mudassar Khan

**Reducing the Chemical Fertilizer in Agriculture by Composting and Improving the Self-Reliance Perspective for Long-Term Impact** . . . . . . . . . . . . . . . . . . . . . . . . . . . . . . . . . . . . . . . . . 229
Yenda Damodhara Rao and Shweta Meshram

**Compost Application for Soil Health and Nutrient Restoration: Benefits and Limitations** . . . . . . . . . . . . . . . . . . . . . . . . . . . . . . . . . 249
Mithlesh Kumar, S. N. Pandey, Sayyada Shakeel, and Jitendra Kumar

**Improving the Nutrient Uptake in Crops by Compost Application in Diverse Agricultural Contexts** . . . . . . . . . . . . . . . . . . . . . . . . . . . . . 269
Kshirod Chandra Sahoo, Dewali Roy, Pooja Tamuk, Aditya Pratap Singh, and Arkaprava Roy

**Nanotechnology Advances in Composting** . . . . . . . . . . . . . . . . . . . . . . . 307
Faria Fatima, Deepti Srivastava, Abdul Mazeed, and P. Smriti Rao

**Composting: Future Perspectives in Sustainable Agricultural in Different Geographical and Climatic Conditions** . . . . . . . . . . . . . . . . 335
Arne Sæbø, Lidia Sas-Paszt, Luisella B. Celi, Henrique Trindade, Francisco Saraiva, Joao Ricardo Sousa, and Rupesh Kumar Singh

**Existing Challenges and Opportunities in Composting on Different Scales** . . . . . . . . . . . . . . . . . . . . . . . . . . . . . . . . . . . . . . . . . 359
Zhuoya Lv and Tao Zhang

**Composting: Safety and Regulatory Considerations in Sustainable Agriculture** . . . . . . . . . . . . . . . . . . . . . . . . . . . . . . . . . . . . . . . . . . . 379
Sudhir Kumar Upadhyay

**Index** . . . . . . . . . . . . . . . . . . . . . . . . . . . . . . . . . . . . . . . . . . . . . . . . 411

# About the Editors

**Rupesh Kumar Singh** Rupesh obtained his PhD in Biotechnology from Kumaun University, Nainital, India, and did his first postdoc in the plant biology department, University of Minho, Braga, Portugal. He joined University of Tras-O-Montes e Alto Douro (UTAD), Vila Real, Portugal, to do his second postdoc and worked to develop sustainable management practices in grapevine toward qualitative improvement. He then worked in industrial research as Principal Investigator and tested microbial cell-based formulation in compost to control the major diseases in mushrooms. Currently he is working as a researcher in the ECONUTRI project in the Department of Agronomy, UTAD, Vila Real, Portugal, where his major objectives are to develop innovative technology in composting to reduce greenhouse gas emissions. He has over 10 years of post-PhD experience in various roles in different scientific organizations nationally and internationally.

https://scholar.google.pt/citations?hl=en&user=vPIqviYAAAAJ

**Henrique Trindade** holds 5-year BSc degree in Agricultural at the University of Trás-os-Montes e Alto Douro (UTAD), and an MSc and PhD in Agrarian Sciences at UTAD. His main area of expertise is the study of the nitrogen cycle in intensive agricultural systems, namely, the application of techniques to improve utilization efficiency and reduce losses of this nutrient. He is serving as a Full Professor in the Department of Agronomy, UTAD. His teaching experience comprises lecturing curricular units to BSc, MSc, and PhD degrees and several specialized courses. He supervised 5 post-docs, 13 PhDs, 12 MSc, and 46 research scholarships and tutorships. He has also been a member of >40 academic referees (MSc and PhD). Regarding R&D units, he has been an integrated member of the Centre for the Research and Technology of Agroenvironmental and Biological Sciences (CITAB) since its creation in 2007, and currently a Vice-Director. From the very beginning, he was part of its Scientific Council and also the Coordinator of the Sustainable Agrofood Chains Group (2009–2012). In terms of scientific indicators, he was a member of 42 R&D projects/contracts, 12 of which were European/international projects, and has been the coordinator of the UTAD team in 8. He is author of 80 SCOPUS-indexed publications (h-index 28, >190 coauthors), 10 books/book chapters, >170 proceedings in scientific meetings, and invited speaker in >40 meetings. He has also been an editor/guest editor in eight journals and a reviewer in national and international journals. He has been a member of organizational/scientific committees of >40 scientific meetings.

**João Ricardo Sousa** has a PhD in soil science and currently working as a Professor in the Department of Soil Science, UTAD, Portugal. He has expertise in soil science, composting process, organic matter management, soil amendments, and process control. He has wider experience of teaching in UTAD and immense research experience in the area. Additionally, he is specialized in the method developments for characterization of organic matters before and after the treatments and final quality. He is author of several peer-reviewed scientific communications and has been part of supervision and jury of several master's and graduation thesis. The immense and diversified specialization of Prof. Ricardo is utmost necessary to complete this book in successful manner.

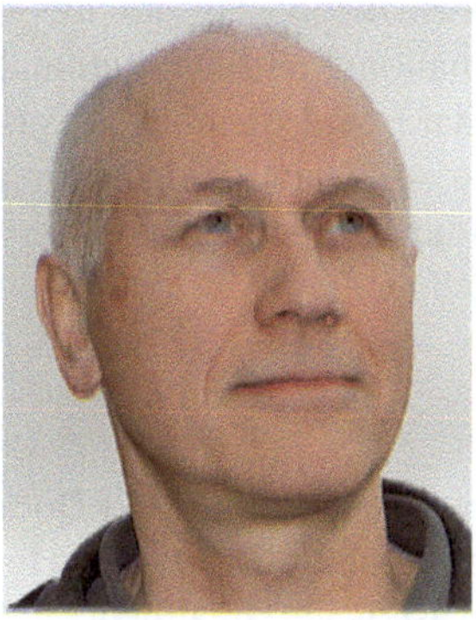

**Arne Sæbø** Dr Arne research professor, have a PhD in horticulture and presently working as leader of NIBIO Særheim Research station. His research is in the department of Urban Greening and in department of horticulture in Norwegian Institute of Bioeconomy Research (NIBIO) for vegetation's ability to collect pollution and the role of urban vegetation in storm water management. The last 5 years, topics related to waste treatment has been a major working area. The international engagement has been in several COST actions, Nordic networks, bilateral projects, for example with Poland and coordinator of the FACCE SURPLUS project "INTENSE", with seven European countries participating. Involved in EFI project "Biocities" as lead author of book chapter "Nature as pillar of Biocities" (2021). The work on waste treatment has been in a national project (MAFIGOLD) and two EU-projects (ECONUTRI and Hort2thefuture).

**Vishnu D. Rajput** is working as Principal Scientist and Head of International Laboratory on Nanobiotechnology at Southern Federal University, Rostov-on-Don, Russia. He comprehensively detailed the state of research in environmental science about "how nanoparticles/heavy metals interact with plants, soil, microbial community, and the larger environment" as well as possible remediation technology using nanoparticles/nano-biochar. His recent work is on nano-bioengineering for soil rhizospheric microbes and enhanced remediation using nanotechnology, CRISPR, and metagenomics. He is included in the world's top 2%

scientists (Stanford University, 2023, 2024), has obtained 2nd top rank in Environmental and Earth Science in Russian Federation, and has received the status of Highly Qualified Specialist. He has published more 500+ papers and scored 65 H-Index with **15,281** citations. In the last five years, he received appreciation letters (Letter of Gratitude, Certificate of Honor, Diploma Awards) from the Rector of Southern Federal University for breakthrough research in environmental and soil science. He is a leader of BRICS project Priority 2030 (Bioengineering of the rhizosphere), and mega projects and participating 18 national and international research projects. He is editorial board member of various international journals. He is currently supervising four postdoctoral fellow and four PhD students.

**Tao Zhang** Dr Tao is an Associate Professor and Doctoral Supervisor of China Agricultural University. His research area is mainly on wastewater treatment and recycling and utilization of agricultural waste. His research focuses to optimize, validate, and demonstrate nature-based novel solutions adapted into a holistic concept, which contributes to reduction of NO3-N and P leaching, control of N losses through ammonia volatilization, and mitigation of GHG emissions originating from the agricultural sector, including both plant and animal production. He has been a dynamic scientist and contributed through training and education to implement nature-based nutrient management tools that would improve air, soil, and water quality in China, and contribute to mitigation of global climate change.

# Composting: Historical Background to Modern Prototypes of the Composting Process

**Mudassar Khan, Sana Bashir, Ahmad Zia Khan, Francisco Saraiva, Rupesh Kumar Singh, Henrique Trindade, and João Ricardo Sousa**

## 1 Introduction to the Composting Process

Composting is the process of biologically degrading macromolecules such as proteins, amino acids, lipids, carbohydrates, cellulose, and lignin (Zhang et al. 2024). This degradation occurs in the presence of oxygen and an adequate supply of nutrients (N, P, and K) for a microorganism. The microorganism in question produces heat, $CO_2$, $H_2O$, $NO_3$, and $SO_4$ as by-products via exothermic reaction. Composting involves the controlled decomposition of organic materials by microbes (Flomo et al. 2023), and the final product of this process is a stable soil-like substance called compost that is essential for agriculture and the stabilization of the soil (Sayara et al. 2020). Different stages of the composting process are shown in Fig. 1.

The pH level (ranging from 6.5 to 9), moisture content, carbon nitrogen ratio based on the composition of the waste (typically 30:1), microbes, temperature, and particle size distribution are the major factors in the process. Bulk density is a pivotal factor in determining porosity and aeration for optimal microbial interaction. Beyond optimizing the moisture content, these bulking agents also enhance aeration between the particles in the compost, thereby preventing anaerobic conditions (Pardo et al.

M. Khan · F. Saraiva · R. K. Singh (✉) · H. Trindade · J. R. Sousa
Centre for the Research and Technology of Agro-Environmental and Biological Sciences (CITAB), Inov4agro, Department of Agronomy, University of Trás-os-Montes and Alto Douro (UTAD), Vila Real, Portugal
e-mail: rupesh@utad.pt

S. Bashir
CIMO, LA SusTEC, Instituto Politécnico de Bragança, Campus de Santa Apolónia, Bragança, Portugal

A. Z. Khan
Mediterranean Institute for Agriculture, Environment and Development Faculty of Science and Technology, University of Algarve, Faro, Portugal



R. K. Singh et al. (eds.), *Composting of Farm Residues by Nature-Based Solutions*,
https://doi.org/10.1007/978-3-032-12255-1_1

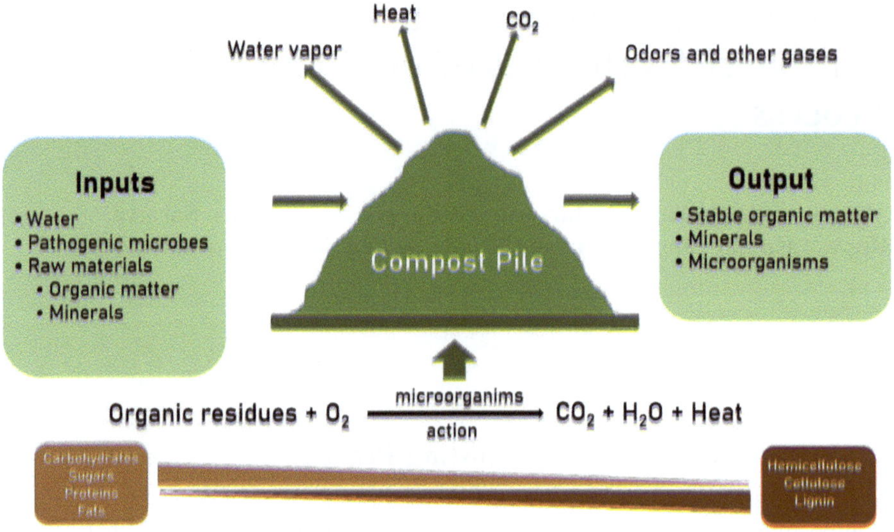


**Fig. 1** Physiological changes during the composting process

2021). Ensuring adequate aeration and periodic repositioning of the compost bed are crucial for the efficient execution of composting processes. The presence of adequate moisture content (40–60%) and the maintenance of an optimal carbon-to-nitrogen ratio (C/N) are important factors for ensuring optimal microbial activity during composting (Liaqat et al. 2023).

## *1.1 Objective of Composting*

The overuse of pesticides and artificial fertilizers in agriculture has neglected compost and other organic fertilizers, which were derived from waste. The consequences of this oversight have been manifold, including soil degradation, nutritional imbalances, insect outbreaks, and soil compaction (Ali 2025). However, composting organic waste, particularly food waste, has been demonstrated to be an effective strategy for restoring soil fertility. A substantial portion of the world's 1.3 billion tons of agricultural output is discarded each year, with food waste going to landfills, which are contributing to methane emissions. Composting organic waste has been shown to effectively reduce greenhouse gas emissions, eliminate environmental concerns such as odor, and enhance soil quality. By replacing synthetic fertilizers, compost functions as an important soil conditioner, hence promoting sustainable agricultural practices. The process of composting involves treating organic waste, neutralizing hazardous chemicals, and producing organic fertilizers, thereby enhancing soil quality and nutrient content (Khan et al. 2024).

## 2 Early History of Composting

The history of composting serves as a chronicle of human advancement and the escalating awareness of natural processes. During the transition from hunting to an agriculture-based society, there was a pivotal moment in human history, which coincided with the onset of better agricultural practices and the development of a more sophisticated understanding of the relationship between soil fertility and plant growth. Archaeological evidence from approximately 5000 B.C. from ancient Scotland indicates that the ancient Scots cultivated crops in heaps of rotting manure (Smout 2019). These observations likely emerged from the realization that plant growth appeared to be enhanced in areas where manure had been applied. Researchers have revealed clay tablets from the Akkadian Empire (circa 2350 B.C.) that are among the earliest known references to the making of compost for agricultural use. The Akkadian Empire, which was centered in the Mesopotamian Valley, modern-day Iraq, marked a significant development in agricultural practices (Mondal and Rao 2025). The practice of composting with worms can also be traced back to the early Egyptian era. Moreover, Cleopatra even went so far as to designate worms as holy creatures and recognized their pivotal role in agriculture. It is noteworthy that other regions of the Middle East were also utilizing manure and compost during this period. The early Arabic writer, Ibn al-Awwam, documented the utilization of animal manures and composting in his literary work, *Kitab al Falahah* (Renton 2020).

The dissemination of knowledge regarding composting occurred over centuries during the development of various civilizations. During the transition of events from the B.C. to the A.D. era, there is evidence of the documentation of practices such as composting. The earliest known evidence of composting for agricultural purposes goes back to the Akkadian Empire (2350 B.C.), as evidenced by clay tablets (Martin and Gershuny 1992). The utilization of organic resources in agriculture is documented in various religious texts, including the Bible, the Talmud, and the works of historical figures such as Cato the Elder (McGovern et al. 2003). However, during the Dark Ages, agricultural knowledge experienced a decline. Nevertheless, monasteries played a crucial role in preserving and documenting this knowledge and ensuring its transmission to future generations. The dissemination of composting knowledge to the Americas was facilitated by European immigrants. They contributed to the discovery of the old indigenous cultures of composting practices. The adoption of locally available materials, such as fish waste, by Native American traditions led to the modification of composting techniques among the settlers (Martich 2015).

### 2.1 *History of Composting in the United States*

The practice of composting was recognized as valuable by historical figures such as George Washington and Thomas Jefferson. These figures showed the articulation and the importance of composting in enhancing soil fertility on their respective farms. During the nineteenth century, there was a shift toward the utilization of chemical fertilizers. However, in the subsequent century, prominent authors such as Sir Albert Howard and Rudolph Steiner emphasized the importance of organic matter in soil, thereby influencing modern composting practices. As the twentieth century evolved, synthetic fertilizers gained prominence, giving rise to concerns regarding soil degradation. The increased usage of synthetic fertilizers was prompted by World War I and the heightened demand for food production. Despite this trend, some prominent individuals, such as Sir Albert Howard and Rudolph Steiner, continued in their support for composting and organic farming and encouraged modern composting practices (Howard 1935). Since the 1930s, composting has evolved and grown in relevance in environmentally friendly farming and environmental management. During this period, composting processes, soil research, and their applications experienced extensive advancements. The Dust Bowl, a period of significant environmental degradation caused by poor agricultural practices, severe drought, and substantial topsoil erosion, profoundly impacted the United States during the 1930s (Clive et al. 2020).

The University of California has been recognized for its contributions to the scientific understanding of composting, as evidenced by its research and publications in this field. Simultaneously, European research was mostly focused on the hygienic aspects of composting, particularly the extent to which it eradicates pathogenic organisms, and different innovative aeration techniques were introduced in the existing composting practices. The late 1960s and 1970s saw a rise in environmental awareness, which catalyzed heightened interest in composting. A novel composting technology known as the static pile composting method was developed in the United States in 1970. This approach modernized traditional composting practices by introducing a new technology for the mechanical turning of the feedstock materials. Through the amendment of this new system, the aeration process was achieved through a system of perforated pipes, which were embedded within each windrow to facilitate the continuous supply of fresh air and the extraction of exhaust gases from the composting process (Maynard 1994).

The materials that were frequently used in this method were biosolids and wood chips, with an innovative practice that facilitated the regulation of temperature and odors. Composting research significantly increased in the United States by the establishment of the Composting Council in 1989 and the substantial financial support provided by the Procter and Gamble Company. The Composting Council in Washington has placed a particular emphasis on research related to biosolid composting. Maynard (1994) conducted research on the use of waste for composting at the Connecticut Agricultural Experiment Station. Furthermore, Hyatt (1995) reported that from 1971 to 1993, the number of citations concerning compost rose to 11,353, both domestically and internationally, demonstrating a significant increase in research interest and activity. Figure 2 underscores the robust interest and research activity surrounding composting in the United States.

The report *Ancient Times: Early Civilizations of South America*, documented the utilization of animal and human waste as fertilizers by these civilizations such as (China, Japan, and India) Howard (1943–1880s). The Richmond Station of United

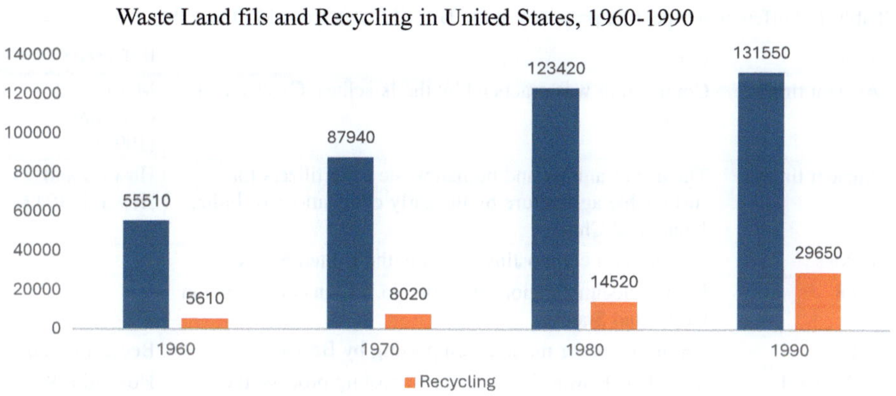


**Fig. 2** Municipal solid waste (MSW) management in the United States: 1960–1990. (Source: EPA 1996)

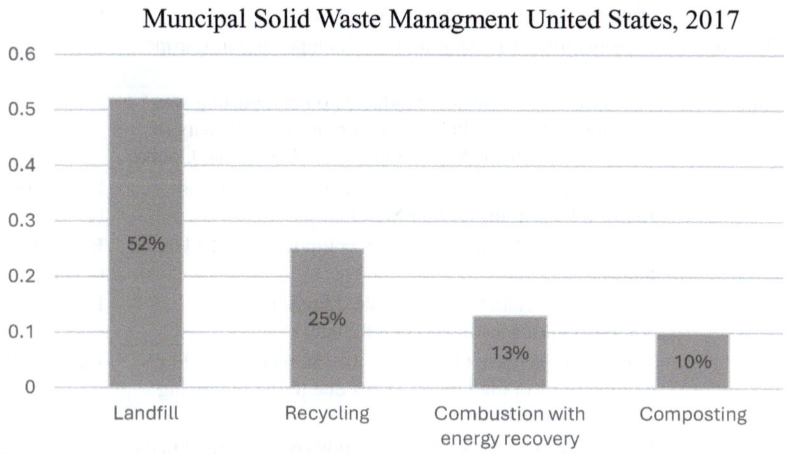


**Fig. 3** Advancing sustainable material management: 2017 fact sheet, EPA (2019)

States (1950–1960) primarily conducted limited research on composting at the University of California. Later on, due to the emerging trends in the late 1970s and early 1980s, Ohio State University initiated a series of studies on plant growth, highlighting the role of composting in relation to plant disease management. Hence, due to the increasing research trends in the 1980s (Fig. 3), the major composting firms of Europe entered the American composting market for further technological exchange, owing to commercial interest in the field of composting (Epstein 1997)

Composting practices had been initiated in ancient times in different geographical areas, where, mostly, traditional methods were used. Among the several methodologies, some are documented and presented in Table 1.

**Table 1** Different composting practices in the United States from 1880 to 1985

| Year | Event | Reference |
|---|---|---|
| Ancient times | Composting was practiced by the Israelites, Greeks, and Romans | Martin and Gershuny (1992) |
| Ancient times | The use of animal and human waste as fertilizers for sustainable agriculture by the early civilizations of India, Japan, and China | Howard and Howard (1943) |
| 1880s | Research on composting began in the United States | |
| 1888 | Earliest documentation (Bulletin No. 61) on composting in the United States | |
| 1922 | Groundwork for modern composting by Beccari | Beccari (1920) |
| 1924–1931 | Sir Albert Howard's modern composting process, the Institute of Plant Industry (Central India) | Howard (1935) |
| 1932 | The modern composting facility by the municipal solid waste management in Europe (Netherlands) | |
| 1939 | Van Vuren publications by using Sir Howard's principles of urban composting (South Africa) | Van Vuren (1949) |
| Post– World War II | Considerable increasing trends of composting in Europe | |
| 1950s and 1960s | The only major research conducted on composting was conducted at the Richmond Station of the University of California under the leadership of Drs. Gottas and Goulke | |
| 1960s | Two composting research stations at Gainesville and Johnson City in the United States | Breidenbach (1971) |
| 1969 | There were 18 facilities of MSW composting in the United States | Breidenbach (1971) |
| 1973 | The research centers at Beltsville, Maryland, on composting of biosolids (United States) | Hyatt (1995) |
| 1993 | Significant increase (11,353) in the international publication citations in the United States due to the increasing trend of composting | Hyatt (1995) |
| 1975 | Development of the aerated static pile (ASP) method in the United States | Epstein et al. (1976) |
| Late 1970s and early 1980s | Research by the University of Ohio on the interaction of compost in plant growth and diseases | Epstein et al. (1976) |
| 1985–1994 | Growth of biosolid-composting facilities in the United States | Epstein et al. (1976) |

# 3 History of Composting in Asia

## *3.1 Historical Development of Organic Farming in China*

Chinese agricultural practices have a long history in the traditional use of organic manure for improving soil fertility and enhancing crop productivity. Over centuries, Chinese farmers have developed extensive knowledge and practical experience in the collection, composting, storage, and application of organic manure. Organic fertilizers were used in many ancient agricultural regions to increase crop yield and mitigate soil degradation with the passage of time. Historical records indicate that the application of manure was not merely a routine agronomic practice but a well-understood and systematically implemented strategy for enhancing soil health. Notably, during the Chuen Qiu Zhan Guo period (also known as the Warring States period of 475–221 B.C.), Chinese scholars began to articulate the relationship between manure application and agricultural productivity. This early recognition of the nutrient value of organic waste highlights the advanced nature of traditional Chinese agricultural knowledge and its foundational role in sustainable land management practices.

The philosopher Meng Zi (Mencius) (circa 372–289 B.C.) mentioned the principle that the application of manure to one hundred mu (approximately 100 acres) of land could yield a harvest sufficient to feed nine individuals. The Han Dynasty (206 B.C.–25 A.D.) further elaborated on this, stating that the three most crucial elements for agricultural success are soil cultivation, manure application, and timely irrigation (Ju et al. 2005). During the Yuan Dynasty, Wang Zhen categorized six types of organic fertilizers, including human waste. This comprehensive classification not only described the diversity of organic inputs utilized by Chinese farmers but also highlighted their intentional management of nutrient sources to support crop production. The practices and principles stated by Wang Zhen informed about agricultural methods during both the Ming (1368–1644 A.D.) and Qing (1644–1911 A.D.) dynasties for sustainable soil fertility management (Cui et al. 2014).

In the early 1950s, the Chinese government implemented a policy of significant encouragement for farmers to collect and apply more animal manure, dunghill compost, and agronomic by-products on their farmlands. This initiative was undertaken to address the escalating demand for food in an expanding population. From ancient times through the Ming and Qing dynasties, wheat yields experienced a substantial increase, rising from approximately 800 to around 1400 kilograms per hectare (Table 2). This growth was predominantly attributable to the application of farmyard manure and the cultivation of green manure (Deng 2015).

During the period of mechanized agriculture in China, from the late 1950s to the early 1980s, the production and utilization of organic waste were largely reorganized at the household or village level. This localized approach to organic waste management was connected to sustainable agricultural production to discourage the use of synthetic fertilizers. In the mid- to late 1970s, there was a peak in the recycling of organic waste within rural agricultural systems. This era saw significant increases in crop yields due to the implementation of high-intensity multiple cropping systems. During this time, the significant relationship between organic waste recycling and intensified cropping patterns highlighted the vital role of organic inputs in soil fertility and food production. These practices were considered the foundation for sustainable nutrient management in Chinese agriculture (Zhu et al. 2019).

This happened when the collective farming system was replaced by the household responsibility system during the 1980s in China. After this transition, a decline

**Table 2** Ancient composting practices in China

| Period | Practice | Reference |
|---|---|---|
| Chuen Chiu Zhan Guo period (the Warring States period of 475–221 B.C.) | Realization of the effect of manure application on crop production | Ju et al. (2005) |
| Han Dynasty (206 B.C.–25 A.D.) | The book *Fan Sheng Zhi Shu*, precise time for manure application and irrigation practices | Ju et al. (2005) |
| Western Jin Dynasty (265–316 A. D.) | Green manuring in paddy rice fields | Ju et al. (2005) |
| Northern Wei Dynasty (386–534 A.D.) | Mung bean cultivation to increase soil fertility | Ju et al. (2005) |
| Tang Dynasty (681–907 A.D.) | The use of turnip *(Brassica rapa L.)* as a green manure | Ju et al. (2005) |
| Yuan Dynasty (1206–1368 A.D.) | The use of six major types of manure for sustainable soil fertility management | Ju et al. (2005) |
| Ming (1132–1227 A.D.) and Qing (1616–1911 A.D.) dynasties | The use of rapeseed (*Brassica napus L.*) as a manure crop | Ju et al. (2005) |
| Early 1950s | The Chinese government's initiatives for the use of manure and plant ash residues to increase the production of crops | Ju et al. (2005) |

in conventional compost preparation techniques, green manure practices were observed more in the developed coastal areas of China. Since the late 1980s, there has been a consistent increase in the scale of livestock production units (Yang 2006).

## 3.2 *The Status of Composting in China (1994–1997)*

The primary phase of composting, which occurred from the 1950s to the 1960s, was characterized by the utilization of open-air composting techniques. During this period, composting was a primary technology that was based on traditional composting practices in rural areas (Ju et al. 2005). The composting method involved creating an open stack, which was then covered with soil to preserve heat. The process was characterized by its reliance on natural ventilation, which was achieved by using open stacks that allowed for the circulation of air. The subsequent development and research phase (1970s–1980s) witnessed significant advancements in the field. The management of MSW emerged as a significant challenge, coinciding with the expansion of cities and the rise in population.

Consequently, this era is marked as a significant period in composting development and research. The subsequent phase, starting from the 1990s to the present, witnessed the dissemination of composting practices. This phase is characterized by a concerted effort to advance composting technology, driven by the recognition of its significance in MSW and sewage sludge management.

In 1990, the Chinese government formally recognized the importance of composting in sustainable agricultural development by identifying and promoting a range of composting technologies. Specifically, 11 types of composting technologies were designated trial technologies for further evaluation in diverse agricultural production sites (Wei et al. 2000). This initiative marked a significant step toward the industrialization of composting as a vital component of modern agricultural practices in China. The composting practices were primarily applied to the treatment of municipal solid waste (MSW) and sewage sludge. But later on, due to the advancement of composting technology and research, China was able to introduce advanced methods of composting, which are still used in the country.

### 3.3 Traditional Composting Applications in China

The conventional systems for the composting process were based on the production methodologies of windrow, in-vessel, and aerated static pipe systems. One-step and two-step fermentations are two of the important Chinese composting approaches and are broadly categorized into two primary systems, namely static and dynamic composting methods (Chen et al. 1987). These classifications reflect the technological diversity in the adaptation of composting practices in response to different waste compositions and treatment objectives. However, composting in China has been predominantly applied to the treatment of municipal solid waste (MSW) and sewage sludge to enhance the nutrient content and improve the stability of organic matter.

The Tianjin Hexi compost plant was one of the earliest facilities in China to implement a one-step fermentation process, signaling a shift toward more modernized composting operations. Among the two fermentation strategies, the two-step fermentation process is the most employed in MSW-composting facilities due to its enhanced microbial activity and improved decomposition ability (Chen 1994). In 1991, Beijing Dong Chun constructed a composter by using the windrow composting system, with a design capacity of 100 tons/day of MSW. However, the facility ceased operations due to its high operational costs. Constructed in 1991, the Beijing She Jinshan composting facility initially had a daily capacity of 100 tons of MSW (Shalini et al. 2021). Due to its success, it was expanded in 1994 to a capacity of 300 tons of MSW per day. Currently, the facility produces compound fertilizers by utilizing MSW compost and additional inorganic fertilizers. These developments demonstrate China's emerging efforts to modernize organic waste management through the adaptation of composting technologies.

The plant's compound fertilizer has proven to be a commercial success, as noted by Shi (1998). The Beijing Nangong composting facility was considered modern management practices of municipal solid waste (MSW) in China. The facility adopted a tunnel composting system, which was imported from Germany, characterized by controlled aeration and an enclosed environment to enhance the efficiency and stability of the composting process. Designed with a capacity to process 400 tons of MSW per day, the Beijing Nangong plant exemplifies the integration of international technological innovations into China's waste management

**Table 3** Establishment of different organizations in China for composting

| Year | Event | References |
|---|---|---|
| 1980s | The Chinese Ecological Agriculture Movement to enhance sustainability, which emerged in the 1980s | Xiao (2001) |
| 1990 | The organic certification of the tea factory in Zhejiang by a Netherlands organic certifying agency | Zong et al. (2003) |
| 1992 | The Green Food program in China | Zong et al. (2003) |
| 1995 | The State Environmental Protection Administration (SEPA) of China | Meng and Wu (2002) |
| 1999 | The Organic Tea Research and Development Center (OTRDC) for the processing and certification of organic tea | Yang and Tang (2002) |
| 2001 | Publication of guidelines for organic certification | Meng and Wu (2002) |
| 2004 | Implementation of the Technical Guidelines for Organic Foods | Meng and Wu (2002) |

infrastructure, marking a significant step toward improving the sustainability and environmental performance of urban organic waste treatment (Pandyaswargo and Premakumara 2014).

Despite the construction of numerous MSW-composting facilities in the 1980s in China, the majority of these facilities ceased operations after a few years (Table 3).

Chen and Chen (1992) conducted a study of the Dano composter, establishing its design and operation parameters.

## 4 History of Composting in Europe

The first extensive waste composting facility for waste management and recycling in Europe was established in the Netherlands in 1932. This facility, with indoor methods of composting, was operated by the Van Maanen process, which was originally designed in India. This plant was operated by Dutch enterprises, named as N.V. Vuilafvoer Maatschappij (VAM). The plant was implemented with a system in which municipal solid waste was arranged into large windrows and manually turned by overhead cranes to facilitate aerobic decomposition (Breidenbach 1971). European companies of composting, after World War II, experienced a notable expansion due to soil restoration, nutrient recovery, and waste reduction, contrasting US practices. After the war, Europeans paid attention to resource recovery and sustainable agriculture; hence, they emphasized the consideration of composting as an integral component of environmental and agricultural policy. Waste materials and their recycling in Europe was a significant step in this evolution process (Fig. 4).

During the 1970s and 1980s, a significant corpus of research on composting was produced, primarily from the United States and Europe. Building upon this substantial body of knowledge, Haug published an important manuscript on the subject,

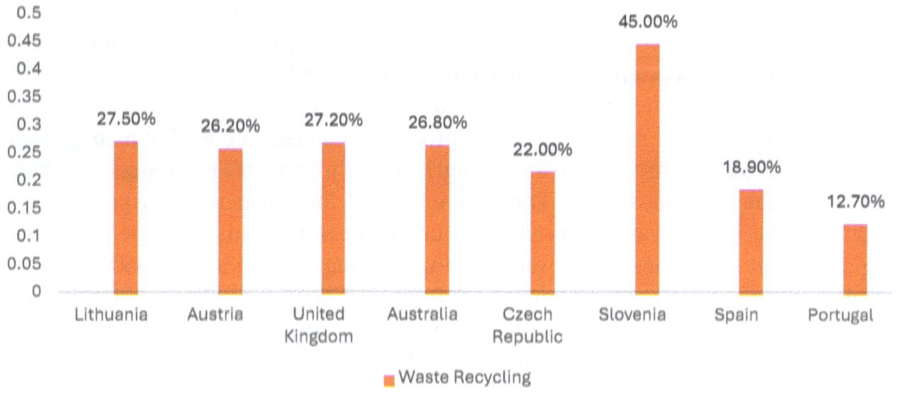


**Fig. 4** Waste recycling in Europe per inhabitant: Eurostat 2022. (Filipenco 2025)

entitled *The Practical Handbook of Compost Engineering* (Haug 2018). Initially published in the late 1980s, this text systematically synthesized a comprehensive array of data derived from composting practices and research endeavors, employing an engineering-oriented approach to facilitate a comprehensive synthesis.

In Europe, many of the early composting facilities, which incorporated a diverse array of mechanical systems, focused on the composting of mixed MSW to offer an alternative to landfill disposal. The proliferation of composting facilities worldwide over the past two decades is influenced by numerous factors. However, in Europe, the predominant factor has been the Landfill Directive (EC 1999), which imposes penalties and costs associated with landfill disposal, thereby incentivizing the adoption of composting as an alternative waste management strategy.

## 4.1 World War II, Recycling, and Composting

A significant proportion of kitchen waste was amassed in "pig bans," which were known for their smelly nature and were assumed to facilitate the transmission of certain diseases like foot and mouth. However, food waste collection continued in most municipalities until 1953. The war also significantly promoted composting, primarily in private gardens and agricultural allotments. During this period, the promotion of composting was supported through different channels, such as manuals, information leaflets, and media campaigns (videos and films), to educate gardeners on recycling (Billington and Easey 1943). However, World War II marked the emergence of an organic agricultural movement led by Lady Eve Balfour. Balfour strongly opposed the widespread use of synthetic chemical fertilizers and encouraged the advancement of composting in her work (*The Living Soil*) to maintain the fertility of British soils. However, after the war, there was more demand for food, which encouraged industrial-scale agricultural production and the use of synthetic fertilizers to increase the yield. Hence, the discussion between the organic school and the advocates of agricultural science intensified, and the assumptions of Balfour and her supporters were considered disapproved.

The years immediately following World War II and its agri-food crisis paved the way for an agricultural recycling. This approach demanded a collective management of waste through a process known as "agricultural treatment" in processing plants (or "accelerated composting"). During this event, long-distance transportation of solid organic waste was observed from urban areas to fields of wine-producing regions in Île-de-France. Later on, important research from June 1960, done by Louis Desvignes, Inspector General of Ponts-et-Chaussées, showed a historical perspective on the development of the composting process with innovative techniques (Desvignes 1960). In rural areas, the recommendation was to establish "controlled landfill sites," which encouraged the natural fermentation of waste and the production of compost. While the direct use of fresh waste was still occasionally observed, it was condemned later on in the name of hygiene (Dorfmann 1963). The French compost industry developed after World War II was better in knowledge about the application of organic materials (Bivar 2018), although there were a number of contradictions in the composting practices and urban development in France.

In the early 1970s, the French central government considered two methods, e.g., burning waste and then turning waste into compost, as effective waste treatment processes. But later on, due to the alteration in the industrial techniques of composting in 1972, there were difficulties in the marketing of composting in that era (Coudé du Foresto 1972). The consequences of these shifts in the composting dynamics of the Paris region were immediately seen. The number of fermentation plants declined from 1975 onward and then collapsed. Moreover, the new legislative frameworks regarding fertilizers almost negated the legal and economic importance of compost as a fertilizer. Surprisingly during the 1980s, researchers reported that "compost cannot be regarded as a fertilizer" (Segura 1984) due to the presence of packaging and plastic materials. Even in the late 1970s and early 1980s, due to the lower-quality materials, an engineer from the technical services of the Interprofessional Committee of Champagne Wines (CIVC) expressed similar concerns regarding the use of urban compost as an effective fertilizer for the Champagne vineyards (Perraud 1986).

However, during the 1990s, composting and the agricultural recycling of waste underwent a period of significant decline. In summary, the latter half of the twentieth century saw a single industrial approach to waste management, which was only the burning of waste. The demand for food production, which was significantly increased by World War II, led to the widespread use of chemical-based fertilizers, which were used for agricultural production. This transition, in conjunction with the mechanization of agriculture, led to a decline in the demand for conventional composting; hence, the need for animal manure, which was a key component in composting, was reduced.

## 5 Conventional Composting Process

The composting methods, such as windrow composting, aerated static piles, and in-vessel systems, have been widely adopted globally over the past decades for conventional composting (Wang et al. 2021). Several studies exist that discuss the importance of compost as a sustainable waste management strategy. In addition, the economic and environmental implications of the composting process are under ongoing investigation. This section will discuss the economic and environmental impact of compost, which is produced by various conventional composting processes. A notable method is windrow composting as an outdoor process with mechanical aeration. The organic waste materials are placed into elongated and slender piles, which are then subjected to physical agitation using a compost turner. Among the three systems, windrow composting is the least sophisticated. It is considered the most suitable option when managing large volumes of organic waste through composting (Vigneswaran et al. 2016).

This method is considered optimal for the large-scale composting of organic waste. However, this method requires more space and a protracted time frame till the maturation stage of compost. Aerated static pile (ASP) composting entails the forced introduction of ambient air into the compost pile, either through positive or negative pressure systems. The compost pile is constructed in a manner like the windrow system, yet it receives minimal or no mechanical turning. The application of ASP has been formally adopted for the composting of municipal sewage sludge (Sesay et al. 1998). To improve airflow and enhance pile porosity, the addition of bulking agents—such as wood chips—is commonly practiced. In contrast, in-vessel composting involves conducting the composting process within enclosed containers or vessels.

Each system employs distinct methodologies to ensure an efficient composting process (Anand et al. 2016). In comparison to windrow and ASP systems, in-vessel composting has been shown to be more efficient and require less land area (Michel et al. 2022). The speed with which compost is produced in an in-vessel composting system is an indication of the efficacy of the operational control mechanisms employed. However, it is considered that one of the most expensive of the systems examined before was in-vessel composting.

From an economic perspective, windrow systems can be a more viable option if more area is available, particularly for the composting of a large volume of waste. Conversely, windrow composting causes the release of more greenhouse gases (GHGs) into the environment (Bong et al. 2017). Cayuela et al. (2006) depicted that the process of turning can result in nitrogen loss of approximately 45% in windrows when compared to an aerated static pile (ASP) with forced aeration, which results in a loss of nitrogen of around 10%. It is hypothesized that the elevated temperatures and extended thermophilic phase result in a greater loss of nitrogen. Specifically, nitrogen loss occurs primarily through the volatilization of ammonium ($NH_4^+$) in the form of ammonia ($NH_3$), a process that can lead to the acidification of the substrate. Furthermore, to mitigate the emission of gases various amendments may be introduced into the composting system, which can be incorporated directly into the compost pile, blended with mature compost, or altered according to the temperature changes during the turning process (Ermolaev et al. 2012).

# 6 Different Composting Technologies

Aerobic composting is small-scale composting suitable for a small scale level. People use organic waste from the kitchen to make compost. Then, they use compost in their gardens and backyards. This method is an efficient and cost-effective approach to managing a limited quantity of kitchen waste (Liu et al. 2020). This approach typically results in a reduction in the overall volume of household waste requiring subsequent processing. But this system requires large-scale aerobic composting necessities, such as a large space, with a considerable investment in resources. This approach also requires additional financial investments and logistical efforts, including the collection, separation, and sorting of organic waste from their mixed sources. Hence, home-scale composting offers a viable solution for addressing domestic waste. It is an effective means of reducing the volume of biodegradable waste for landfill sites. Furthermore, it produces by-products that are of benefit to people and for agricultural purposes (Sayara et al. 2020).

# 7 Manual Composting Technology

Manual composting technology is a natural process that requires extended periods for the complete decomposition of organic materials to produce mature compost (Ayilara et al. 2020). It also requires outdoor, large, protected spaces to prevent the accumulation of highly odorous by-products. On the other hand, automatic composters partially mechanize key stages of the composting process. Automatic composting reduces labor costs and requires less space to make composting more suitable for urban environments. However, these systems require a continuous supply of electricity to accelerate decomposition. Manual composting, which operates through human labor and mechanical tools without automation, encompasses several widely used methods, including windrow composting, bin composting, in-vessel systems, and vermicomposting; a comparative analysis is presented in Table 4.

## *7.1 Windrow Composting*

Windrow composting is the process of creating long piles of raw organic materials for the purpose of composting. This method represents the fundamental technique in composting. This method is well-suited for the treatment of substantial volumes of

**Table 4** Comparison of manual composting systems. (Ayilara et al. 2020)

| Composting method | Description | Advantages | Drawbacks |
|---|---|---|---|
| Vessel composting | Composting in an enclosed area with forced aeration and mechanical turning | Enhances the composting process | Labor-intensive and expensive |
| Windrow composting | Raw materials in long, narrow piles, turned regularly to allow aeration | Rapid, retains heat | Difficult and costly to support |
| Vermicomposting | Earthworm-based compost | Earthworm excreta improve soil fertility | Degrades all kinds of organic matter |
| Static composting | Composting aerobically with passive aeration | Simple, low operational and capital costs | Time-consuming |
| Sheet composting | Organic matter spread thinly on the soil surface and tiled in | Cheap and straightforward | Decomposition of organic matter without building a composting pile |
| Indian Bangalore composting | Composting of night soil and refuse in pits, in alternate layers, and covered with a layer of refuse | Simple | Laborious, expensive, takes 6–8 months |
| Indian Indore composting | Agricultural by-products (farmyard manure and plant residues) | All organic waste available on a farm is composted | Labor-intensive, time-consuming, prone to flies and nutrient loss |

organic waste, resulting in the production of copious amounts of compost, as evidenced by numerous studies (Jalalipour et al. 2020). Notably, its simplicity contributes to its widespread adoption, particularly in domestic contexts where manual composting is prevalent.

Due to the substantial size of windrow compost heaps, aeration is required throughout the compost heaps (Arslan et al. 2011). Hence, facilitating effective aeration is a pivotal concern in the compost heaps for the better consumption of oxygen by microorganisms (Di et al. 2013). Moreover, the process of reducing the size of the compost pieces, whether manually or using machinery, has been demonstrated to accelerate the rate of decomposition. It is imperative to rotate the compost heaps regularly to restore the air and oxygen levels. Turning facilitates the rotation of compost materials, enabling the microbes located at the upper levels of the heap to migrate to the lower levels and vice versa, thereby facilitating the breakdown of the various components of the heap. The process of windrow composting is advantageous in terms of its cost-effectiveness and ease of maintenance. However, the process of composting requires sufficient spatial allocation for compost heaps. It is imperative that these heaps are distributed in a well-spaced configuration to ensure a better composting process. The process of composting with heaps requires a comparatively long time to obtain satisfactory compost, and it requires skilled manual labor for turning piles (Diaz et al. 2002).

## 7.2 *Passively Aerated Windrow Composting*

This system of composting has a significant improvement over the conventional windrow composting method. This is due to the incorporation of perforated pipes to facilitate the flow of air throughout the compost heap (Michel et al. 2022). The need for frequent turning can be avoided by adjusting the tube design to ensure adequate oxygenation of the microorganisms. However, to maintain sufficient air-filled porosity, it is essential to reduce the particle size before mixing the organic material in the form of a compost pile. Furthermore, insulating the heap with mature compost can help achieve and sustain thermophilic temperatures, even in the outer layers. A key advantage of this method is the ability to retain the internal heat of the pile while supplying oxygen through passive aeration, which may cause the reduction of overall composting duration as compared to traditional windrow systems.

## 7.3 *Bin Composting*

This technology is primarily employed within domestic households. This technology has the capacity to process a limited volume of waste, with the resulting compost intended for personal use (Bobeck 2010). The insertion of organic waste is typically initiated from the upper portion of a meticulously crafted container, characterized by a perforated wall that facilitates conventional air circulation. As the organic composting material is inserted into the container, it gradually degrades and becomes compacted, ultimately reaching the bottom of the container. The finished compost is then typically collected from the bottom of the container. Some bins may also be designed with an internal stirring operation to facilitate the mixing of the compost heap and increase its air-filled porosity. The perforated walls of a self-contained system allow heat to be retained and air to be circulated in the compost heap (Karnchanawong and Suriyanon 2011). Further advantage of the system is due to its requirement of less space than windrow composters (Arrigoni et al. 2018).

## 7.4 *In-Vessel Composting*

In-vessel composting is a controlled composting method in which organic materials are enclosed within a sealed container to enhance the regulation of environmental conditions (Manyapu et al. 2017). Regardless of the configuration, the technology facilitates enhanced control over airflow and temperature via the air portals strategically positioned on the vessel's exterior, facilitating the circulation of air and accelerating the composting process. Manual turning is required for 2 weeks during the initial process of composting to facilitate aeration and regulate temperature and moisture (Fraser 1997). This process has the advantage of being able to handle large batches of organic waste and requires less space.

In-vessel composting typically involves an initial reduction in the volume of organic waste, followed by a decomposition phase. This method reduces manual labor during the active composting phase, as mixing and aeration are conducted within the vessel itself. However, it requires frequent monitoring of internal conditions and periodic mechanical turning to maintain an optimal composting environment (El-Zein 2015).

### 7.5 Vermicomposting

Vermicomposting is a biological composting technique that utilizes specific microorganisms, in combination with selected earthworm species, to accelerate the decomposition of organic waste under ambient temperature conditions, thus enhancing the quality of the final compost product (Naik et al. 2024). Unlike conventional aerobic composting, vermicomposting involves the deliberate introduction of specialized epigeic earthworms—most commonly *Eisenia fetida*—into the organic substrate, where they contribute to both the fragmentation and biochemical transformation of the material. The worms have a voracious diet and a remarkable reproductive capacity, allowing them to efficiently digest the organic waste and excrete the produced vermicompost in the form of granulates (Kaur 2020). The casting of vermicomposting is highly enriched with nutrients. Their castings contain a high concentration of microbes, which facilitate the ongoing decomposition process and ultimately generate the final compost product. However, the worms require a comfortable space to live and work. To facilitate their habitation and activity within the worm bin, shredded paper or cardboard bedding materials must be prepared. Additionally, moisture and organic waste are essential for their sustenance.

The duration of this process is dependent on factors such as the number of worms, the temperature, and the amount of waste in the bin. Furthermore, the rapid rate of reproduction exhibited by these organisms can result in the worm bin quickly becoming full. Consequently, the necessity arises to transfer the worms to an alternative bin, thereby ensuring optimal conditions for their continued survival and well-being. While vermicomposting has been reported to reduce pathogens, it is not as effective as traditional composting because pathogens tend to be killed more quickly at higher temperatures. However, elevated temperatures can also lead to the demise of worms, a phenomenon that can complicate matters by facilitating the reproduction of pathogens. The method is highly economic, easy to maintenance, and requires less space (Lalander et al. 2015). The worms employed in this process have the capacity to consume organic matter at a rapid rate, thereby expediting the composting process, with the added benefit of microbes present in their castings. This method requires minimal labor input.

# 8 Automatic Composting Technology

The limited adoption of composting practices can be attributed to several reasons, such as insufficient public awareness, the extended duration required to complete the composting process, and limited knowledge of the biological mechanisms involved. Manual composting typically demands significant time to yield effective outcomes, along with consistent supervision and physical involvement, which may be difficult for individuals with demanding work schedules. Moreover, the requirement for substantial space renders composting methods unsuitable for urban dwellers. The development of automatic composters has been identified as a potential solution to address these challenges.

## *8.1 Forced Aerated Windrow Composting*

In a windrow composting system with air, blowers are installed at the bottom of perforated tubes. These blowers facilitate the movement of air across the compost heaps (Larney and Olson 2006). During the active stage of composting, these blowers are particularly important, as they inject air into the compost heaps, supplying the essential oxygen for the activity of microorganisms. This method plays a crucial role in breaking down of organic matter. It is also a common approach to insulate compost heaps to retain heat during the thermophilic stage in the entire mass of the composting heap (Michel et al. 2022). Heat retention is generally effective, and oxygen can be supplied without the need for turning, thereby reducing the overall composting duration. However, it is important to recognize that this system requires a high initial investment due to the need for blowers and air distribution channels to support ventilation and require sufficient space. The technology used in this method is similar to manual in-vessel composting, but it should be noted that these systems may differ in terms of structure, capacity, and mechanical setup.

## *8.2 Automatic-Turning In-Vessel Composting*

This process utilizes a motorized mechanism to rotate the vessels, thereby aiding in the aeration of the compost heaps. Moreover, the rotation of the vessels can be scheduled at predefined times and frequencies, thus enhancing the efficiency and reproducibility of the process (Mishra et al. 2021). In addition, certain in-vessel composting processes incorporate a motor to facilitate the rotation of the vessel, complemented by sensors that automatically monitor critical parameters such as temperature and humidity. This configuration significantly reduces the labor intensity of manual in-vessel composting processes. However, the utilization of a big motor to rotate the heavy waste-filled vessels necessitates a substantial investment and rigorous maintenance to ensure optimal performance and prevent malfunction. Notwithstanding the high investment and maintenance requirements, in-vessel composting is regarded as a space-efficient method that demands a limited workforce (Illahi et al. 2023).

### 8.3 Electrical Home Composter

An electrical composter is a machine that facilitates the decomposition of food waste into compost. This process involves the utilization of air, heat, and a specific method to reduce the size of the waste and mitigate the odor. The composter can be situated on a countertop in a domestic setting, though some models are designed for use in external locations. A domestic electrical composter, alternatively designated a food recycler, utilizes three-phase cycles to facilitate the breakdown of food waste (Fakharulrazi and Yakub 2020). Mostly, composters require between one day and several days to transform waste into simpler substances. Many electrical home composters employ the application of heat with the intention of reducing the duration of the mesophilic and thermophilic phases (Zaman et al. 2022). This process is designed to quickly reduce excess moisture in organic material. The materials are typically ground into a powder by the composter, which increases the surface area available for microbial action and thus accelerates the composting process. Following this, the materials are then mixed with soils or additives to encourage the growth of microorganisms. Electrical home composters may differ in terms of the processes they adopt, the quality of the end product they produce, and the duration of the process. However, many electrical home composters are based on three-phase cycles, which include the drying, grinding, and cooling phases (Martínez-Blanco et al. 2010). It is noteworthy that certain models may also yield compost that is not subjected to drying processes. This is achieved by incorporating an additional step, designated as the curing phase. It is assumed that incorporating this additional step amendment can prolong the composting process by up to 2 weeks (Kucbel et al. 2019).

Most food recyclers utilize ventilation and heat to accelerate the decomposition of food waste, resulting in a process analogous to that of conventional composting. However, the end product is typically characterized by its dryness, sterility, and immaturity, thus falling short of the criteria for adequate compost. The primary role of a domestic composter is to serve as an alternative to traditional composting methods, aiming to minimize the challenges in management, which are typically associated with compost heaps. It offers an odor-free composting solution suitable for confined or indoor environments, which are generally incompatible with conventional systems. These units are designed for the reduction of food and organic waste generated at the household level. Notwithstanding its expeditious operation, the initial financial outlay on the machine itself can be substantial, and its maintenance requirements are significant. Nevertheless, its low space and labor requirements make it an attractive alternative to conventional composting.

# 9 Comparative Summaries of the Manual and Automatic Composting Methods

Several research studies have shown that using an automated composting process could be good for a number of reasons. Technology could make composting faster, easier, and more convenient. Furthermore, it has the potential to motivate individuals to engage in domestic composting practices (Fig. 5) (Table 5).

This could potentially reduce the volume of waste disposed of in landfills, which would have environmental benefits. Despite the existence of numerous scientific studies addressing this subject, no studies have specifically employed patent analysis to examine composting as a new technology. However, research into patents is a useful source of information about new technologies for composting.

# 10 Recent Challenges to the Composting Process

Due to the progress in the technological aspects of the composting process, waste management has become easier nowadays. Different types of innovative processes are now being used to improve the quality of composting processes. However, composting process still has some limitations.

Agricultural by-products consist mostly of farm waste. Such waste may contain leaf litter, plant residues, and decayed plants. Some of these waste materials may contain recalcitrant (sclerophylly) compounds with a low nutrient content. The presence of these materials in composting can affect the rate of composting,

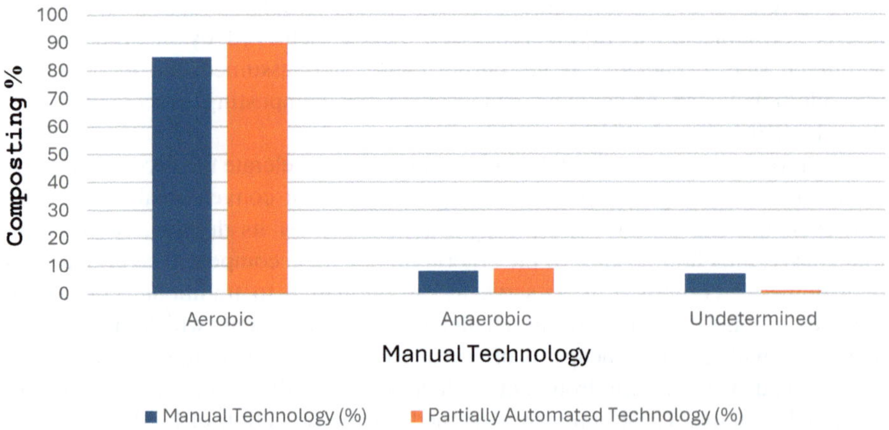


**Fig. 5** Comparison of composting processes by using manual technology and partially automated technology (Azis et al. 2022)

**Table 5** Comparison of different composting systems

| Feature | Windrow | ASP | In-Vessel |
|---|---|---|---|
| Preferable waste input | All types of waste but preferable for those with less emission of odor, such as plant-based wastes | Preferable for waste with more homogeneity and consistency, and bulking agent is required<br>Usually, it can accommodate around 1–5 t of waste for the whole composting process | All types of waste but preferable for easily degraded waste such as food waste |
| Space required | More space | Medium space | Low space |
| Area selection | Rural area | Rural area | Anywhere |
| Type of amendment can be considered | Adequate aeration with the amendment of chemical and microbial inoculants | Increase the airflow in the pile to enhance the microbial additive | Increase physical attributes to enhance the composting process |
| Composting period (with amendment) | Successfully applied | The effect of the other changes is going to be smaller than the windrow system's, even if they do allow more air flow | Can reduce the composting time by more than half, but the curing phase will still take around 4 to 8 weeks |
| Compost quality | Moderate | Moderate | Good |
| Composting period | Long | Long | Short |

particularly when they are added to a composting pile. The presence of excessive amounts of diverse chemical compounds such as lignin, cellulose, phenolics, tannins, and allergens in different ranges also affects the composting process.

Highly phenolic compounds in the substrate decompose more slowly during composting due to their structurally complex nature and high C:N ratio (Lewis and Brown 2010). To enhance the efficiency of composting and decomposition, co-composting is usually employed to achieve an optimum carbon-to-nitrogen (C:N) ratio and improve the nutrient quality of the final compost product (Arumugam et al. 2018). However, the effectiveness of co-composting can be dependent on the materials used for composting. While nutrient-rich substrates can stimulate microbial activity and thus speed up the composting process, the substrates with low-nutrient and cellulose-rich materials can delay the process because high-C/N-ratio materials are less readily utilized by microorganisms, resulting in slower decomposition. Therefore, careful management of the C:N ratio is essential to optimize microbial activity and ensure efficient composting (Masowa et al. 2018).

While different studies have already been done on composting, with a focus on optimizing the process and improving the compost quality, relatively few researchers have addressed the broader environmental implications of composting, especially related to gas emissions (Nasini et al. 2019). The composting of solid organic waste can release environmentally harmful gases such as methane ($CH_4$), nitrous oxide ($N_2O$), and ammonia ($NH_3$), which contribute to environmental pollution (Jiang et al. 2015). The amount and characteristics of odorous gases released during composting are influenced by the operational parameters and the nature of the input materials. However, there remains limited research detailing the correlation between composting conditions, feedstock properties, and the resulting odorous emissions (Delgado-Rodriguez et al. 2011).

## 11 Future Prospects and Conclusion

Composting represents an environmentally sustainable method of waste disposal, with a well-established historical foundation. Nevertheless, its large-scale implementation is hindered by several challenges. Key obstacles include the presence of pathogens, reduced nutrient density in the final product, prolonged processing times, and the generation of unpleasant gases such as methane ($CH_4$), nitrous oxide ($N_2O$), and ammonia ($NH_3$), which contribute to offensive odors. Due to these challenges, the use of chemical fertilizers has increased despite their consequences on the environment, contributing to greenhouse gas emissions and soil degradation, which ultimately harm the beneficial microflora of soil.

Current trends and future outlooks in composting endeavor to identify the existing difficulties and enhance the efficiency of the composting process. Further investigation is required to prioritize the rapid identification of pathogens and heavy metals. The incorporation of additional activators into the initial organic matter has also demonstrated potential for the improvement in the nutritional status of compost. Additionally, there is a growing interest in exploring the potential of slow-composting organic materials for annual and biennial cropping systems. Some insects (black soldier field larvae) are capable of degrading some types of organic wastes (e.g., catering wastes and manures). Hence, in the future, insects can be involved in a controlled manner in composting. Moreover, the energy demand for forced aeration in composting piles and reactors is an important element in composting, which can be addressed by using solar energy (Sundberg and Navia 2014).

The potential generation of greenhouse gases during the composting process highlights the importance of implementing effective management practices. To mitigate emissions of methane ($CH_4$) and nitrous oxide ($N_2O$), further research is required to better understand the composting process. The role of substrate composition, microbial communities, and physical conditions in composting still needs to be addressed to mitigate the drawbacks of composting. Addressing these challenges is essential not only for minimizing the environmental impact of composting but also for maximizing its benefits in enhancing soil health. Ultimately, improved composting practices can contribute to more sustainable agricultural production systems by reducing pollution and supporting long-term soil fertility.

**Acknowledgments** LIVING LAB 5.1 - Effluents and co-products of agricultural activities – Organic fertilizers (PRR-C05-i03-I-000218-LA 5.1), Portugal Recovery and Resilience Plan (PRR) under the grant Agreement No. BI/UTAD/52/2024, National Funds by FCT - Portuguese Foundation for Science and Technology, under the project UIDB/04033/ 2020 (https://doi.org/10.54499/UIDB/04033/2020), ECONUTRI project, the European Union Horizon Europe Innovation program under the Grant Agreement No. 101081858 are gratefully acknowledged.

## References

Ali AM (2025) Environmental pollution and climate change implications of agricultural fertilizer use. In: Agricultural nutrient pollution and climate change: challenges and opportunities. Springer Nature Switzerland, Cham, pp 1–28

Anand M, Nair MH, Singh IB (2016) In-vessel composting of food waste: a novel approach. Doctoral dissertation, Cochin University of Science and Technology

Arrigoni JP, Paladino G, Garibaldi LA, Laos F (2018) Inside the small-scale composting of kitchen and garden wastes: thermal performance and stratification effect in vertical compost bins. Waste Manag 76:284–293

Arslan EI, Ünlü A, Topal M (2011) Determination of the effect of aeration rate on composting of vegetable-fruit wastes. Clean—Soil Air Water 39:1014–1021

Arumugam K, Renganathan S, Babalola OO, Muthunarayanan V (2018) Investigation on paper cup waste degradation by bacterial consortium and Eudrillus eugeinea through vermicomposting. Waste Manag 74:185–193

Ayilara MS, Olanrewaju OS, Babalola OO, Odeyemi O (2020) Waste management through composting: challenges and potential. Sustainability 12(11):4456

Azis FA, Rijal M, Suhaimi H, Abas PE (2022) Patent landscape of composting technology: a review. Inventions 7(2):38

Beccari G (1920) Apparatus for fermenting garbage. US Pat, 1,329,105

Billington FH, Easey B (1943) Compost for garden plot or thousand-acre farm: a practical guide to modern methods. Faber & Faber

Bivar V (2018) Organic resistance: the struggle over industrial farming in postwar France. UNC Press Books

Bobeck M (2010) Organic household waste in developing countries: an overview of environmental and health consequences, and appropriate decentralised technologies and strategies for sustainable management, p. 30, URN: urn:nbn:se:miun:diva-11579

Bong CPC, Lim LY, Ho WS, Lim JS, Klemeš JJ, Towprayoon S, Ho CS, Lee CT (2017) A review on the global warming potential of cleaner composting and mitigation strategies. J Clean Prod 146:149–157

Breidenbach AW (1971) Composting of municipal solid wastes in the United States. Pub. SW-47r. U.S. Environmental Protection Agency, pp 1–112. ED092347

Cayuela ML, Sánchez-Monedero MA, Roig A (2006) Evaluation of two different aeration systems for composting two-phase olive mill wastes. Process Biochem 41(3):616–623

Chen ShH, Li GJ, Shao LM, Xue XK (1987) Rapid and high temperature method for composting municipal domesticsolid waste. Shanghai Environ Sci, 6(1):16–9

Chen SH (1994) Survey on municipal domestic wastes composting technology in mainland China. Chin J Environ Sci 15(1):53–56

Chen SHH, Chen JY (1992) Research into municipal solid waste dynamic composting (Dano technique). Shanghai Environ Sci 11(5):13–21

Clive AE, Lal R, Madden P, Miller RH, House G (2020) Sustainable agricultural systems. CRC Press. SBN 0-935734-21-X

Coudé du Foresto Y (1972) Rapport Général sur le projet de loi de finance pour 1973. Tome III. Les moyens des services et les dispositions spéciales. Annexe VI : Le traitement des ordures ménagères, Annexe au procès-verbal de la séance du 21 novembre 1972 du Sénat, novembre, pp 143–164

Cui Z, Dou Z, Chen X, Ju X, Zhang F (2014) Managing agricultural nutrients for food security in China: past, present, and future. Agron J 106(1):191–198

Delgado-Rodriguez M, Ruiz-Montoya M, Giraldez I, Lopez R, Madejon E, Diaz MJ (2011) Influence of control parameters in VOCs evolution during MSW trimming residues composting. J Agric Food Chem 59(24):13035–13042

Deng K (2015) Mapping China's growth and development in the long run, 221 BC to 2020. World Scientific

Desvignes L (1960) Destruction des ordures ménagères. AD75: 96WR1

Di Y, Liu J, Liu J, Liu S, Yan L (2013) Characteristic analysis for odor gas emitted from food waste anaerobic fermentation in the pretreatment workshop. J Air Waste Manag Assoc 63:1173–1181

Diaz LF, Savage GM, Golueke CG (2002) Composting of municipal solid wastes. In: Handbook of solid waste management, vol 2. McGraw-Hill, pp 12–11

Dorfmann R (1963) "Annexe J du PADOG [projet]", AD75: 132WR1. EC. (1999). Council directive 1999/31/EC of 26 April 1999 on the landfill of waste. Off J Eur Union L 182:1–19

El-Zein A, Seif MSH, Gooda E (2015) Effect of co-composting fish and banana wastes with organicmunicipal solid wastes on carbon/nitrogen ratio

EPA (2019) Advancing Sustainable Materials Management: 2017 fact sheet assessing trends in material generation, recycling, composting, combustion with energy recovery and landfilling in the United States

Epstein E (1997) The science of composting. Engineering & Technology. ISBN9780203736005. CRC press, Boca Raton. https://doi.org/10.1201/9780203736005

Epstein E, Willson GB, Burge WD, Mullen DC, Enkiri NK (1976) A forced aeration system for composting wastewater sludges. Journal (Water Pollution Control Federation) 48:688–694

Ermolaev E, Pell M, Smårs S, Sundberg C, Jönsson H (2012) Greenhouse gas emission from covered windrow composting with controlled ventilation. Waste Manag Res 30:155–160

European Communities (EC) (1999) 96/31/EC Official J Euro Parliment L182/01. Councildirective on the landfill of waste

Fakharulrazi AN, Yakub F (2020) Designing an automated composter for food waste Management with the implementation of internet of things. J Sustain Nat Resour 1:9–14

Filipenco D (2025) 'World waste: statistics by country and brief facts', Global Waste Index 2022Development Aid. https://www.developmentaid.org/news-stream/post/158158/worldwastestatistics-by-country

Flomo MR, Yana T, Zimba E (2023) Composting unearthed: a comprehensive review of challenges, historical evolution, prospects, applications, and processes. IJCRT 11

Fraser BS (1997) An investigation of in-vessel composting process control strategies. Doctoral dissertation, University of British Columbia

Haug R (2018) The practical handbook of compost engineering, 2nd edn. Routledge, CRC Press, Boca Raton, p 717

Howard A (1935) The manufacture of humus by the Indore process. J R Soc Arts 84:26–29

Howard SA, Howard A (1943) An agricultural testament, vol. 228. Chapter 9. http://krishikosh.egranth.ac.in/handle/1/5810086545

Hyatt GW (1995) Economic, scientific, and infrastructure basis for using municipal composts in agriculture. In: Agricultural utilization of urban and industrial by-products, vol 58. Wiley, pp 19–72

Illahi AAC, Co MA, Go JR, Magallanes FJ, Vong JM (2023) Development of an automated microcontroller-based in-vessel composting system for the household setting. In: 2023 IEEE 12th Global Conference on Consumer Electronics (GCCE). IEEE, pp 323–327

Jalalipour H, Jaafarzadeh N, Morscheck G, Narra S, Nelles M (2020) Potential of producing compostfrom source-separated municipal organic waste (A case study in Shiraz Iran). Sustain 12:1–17. 15,417.http://dx.doi.org/10.3390/su12229704
Jiang J, Liu X, Huang Y, Huang H (2015) Inoculation with nitrogen turnover bacterial agent appropriately increasing nitrogen and promoting maturity in pig manure omposting. Waste Manag 39:78–85
Ju X, Zhang F, Bao X, Römheld V, Roelcke M (2005) Utilization and management of organic wastes in Chinese agriculture: past, present and perspectives. Sci China Ser C Life Sci 48:965–979
Karnchanawong S, Suriyanon N (2011) Household organic waste composting using bins with different types of passive aeration. Resour Conserv Recycl 55:548–553
Kaur T (2020) Vermicomposting: an effective option for recycling organic wastes. In: Organic agriculture. IntechOpen, Hampshire, pp 1–17
Khan MN, Sial TA, Ali A, Wahid F (2024) Impact of agricultural wastes on the environment and possible management strategies. In: Frontier studies in soil science. Springer International Publishing, Cham, pp 79–108
Kucbel M, Raclavská H, Ruži˘cková J, Švédová B, Sassmanová V, Drozdová J, Raclavský K, Juchelková D (2019) Properties of composts from household food waste produced in automatic composters. J Environ Manage 236:657–666
Lalander CH, Komakech AJ, Vinnerås B (2015) Vermicomposting as manure management strategy for urban small-holder animal farms—Kampala case study. Waste Manag 39:96–103
Larney FJ, Olson AF (2006) Windrow temperatures and chemical properties during active and passive aeration composting of beef cattle feedlot manure. Can J Soil Sci 86:783–797
Lewis SE, Brown AV (2010) Comparative leaf decomposition rates including a non-native species in an urban Ozark stream. J Arkansas Acad Sci 64:92–96
Liaqat I, Bashir F, Zafar U, Hanif U, Andleeb S, Saleem S, Arshad M (2023) Role of microbes in sustainable utilization of animal wastes. In: Climate changes mitigation and sustainable bioenergy harvest through animal waste: sustainable environmental implications of animal waste. Springer Nature Switzerland, Cham, pp 385–411
Liu T, Ren X, Zhao J, Chen H, Wang Q, Awasthi SK, Duan Y, Pandey A, Taherzadeh MJ, Awasthi MK et al (2020) Sustainability analysis of large-scale food waste composting. In: Current developments in biotechnology and bioengineering. Elsevier, Amsterdam, pp 301–322
Manyapu V, Shukla S, Kumar S, Rajendra K (2017) In-vessel composting: a rapid technology for conversion of biowaste into compost. Open Access Int J Sci Eng 2:58–63
Martich T (2015) Impacts of fish waste piles in Alaska. University of Alaska Anchorage. http://hdl.handle.net/11122/6543
Martin DL, Gershuny G (1992) The Rodale book of composting: easy methods for every gardener. Rodale
Martínez-Blanco J, Colón J, Gabarrell X, Font X, Sánchez A, Artola A, Rieradevall J (2010) The use of life cycle assessment for the comparison of biowaste composting at home and full scale. Waste Manag 30(6):983–994
Masowa MM, Kutu FR, Babalola OO, Mulidzi AR (2018) Physico-chemical properties and phytotoxicity assessment of cocomposted winery solid wastes with and without effective microorganism inoculation. ResCrops 19:549–559
Maynard AA (1994) Seventy years of research on waste composting and utilization at the Connecticut agricultural experiment station. Compost Sci Util 2(2):13–21
McGovern PE, Fleming SJ, Katz SH (2003) The origins and ancient history of wine: food and nutrition in history and antropology. Routledge
Meng FQ, Wu WL (2002) Production, trade and law management system of organic food home and abroad. Chin J Eco-Agric 10:4–6
Michel F, O'Neill T, Rynk R, Bryant-Brown M, Calvez V, Li J, Paul J (2022) Contained and in-vessel composting methods and methods summary. In: The composting handbook. Academic Press, pp 271–305

Mishra S, Mishra SK, Mishra SK, Yadav KD (2021) Energy assessment of rotary drum and aerated in-vessel for composting of garden waste. Res Sq 1–16. https://doi.org/10.21203/rs.3.rs-302722/v1

Mondal S, Rao Y (2025) Biochemistry of composting suchhanda Mondal, yv Rao and Umalaxmi thingujam. In: Perspectives on waste recycling in Indian agriculture. Cambridge Scholar Publishing, p 326

Naik J, Mahanaik T, Pradhan M, Das AK, Symon K, Rout M, Das S (2024) Vermicasting: the processing of organic wastes through earthworms. Int J Biol Pharma Sci 10(1):1–12

Nasini L, De Luca G, Ricci A, Ortolani F, Caselli A, Massaccesi L, Regni L, Page A (2019) Fertility from urban wastes? The case for composting in Great Britain, 1920s-1960s. Environ History 25(1):3–22

Pandyaswargo AH, Premakumara DGJ (2014) Financial sustainability of modern composting: the economically optimal scale for municipal waste composting plant in developing Asia. Int J Recycl Org Waste Agric 3:1–14

Pardo RNC, Rojas GMA, Florez LM (2021) Thermal analysis of the physicochemical properties of organic waste to application in the compost process. Biomass Convers Biorefin 13:1–13

Perraud A (1986) Le compost urbain dans le vignoble cham-penois. Compost Information, N°25:6–8

Renton K (2020) Mediterranean horse cultures: Greek, Roman and Arabic equine texts in late medieval and early moder Andalusia. Pedralbes: Revista d'historia moderna 40:97–122

Sayara T, Basheer-Salimia R, Hawamde F, Sánchez A (2020) Recycling organic waste throughcomposting: Process performance and compost application in agriculture. Agronomy, 10(11), p.1838

Segura J (1984) Le compostage des ordures ménagères et ses débouchés agronomiques. Biofutur. Le mensuel européen de biotechnologie, N°21:19–29

Sesay AA, Lasaridi KE, Stentiford EI (1998) Aerated static pile composting of municipal solid waste (MSW): a comparsion of positive pressure aeration with hybrid positive and negative aeration. Waste Manag Res 16(3):264–272

Shalini S, Joseph K, Yan B, Karthikeyan OP, Palanivelu K, Ramachandran A (2021) Solid waste Management practices in India and China: sustainability issues and opportunities. In: Waste management policies and practices in BRICS Nations. CRC Press, pp 73–114

Shi Y (1998) Current status and counter measures of refuse in Beijing. Res Environ Sci 11(3):40–56

Smout TC (2019) People and woods in Scotland: a history. Edinburgh University Press

Sundberg C, Navia R (2014) Is there still a role for composting? Waste Manag Res 32(6):459–460

US Environmental Protection Agency (EPA) (1996) ECO Update: Ecotox Thresholds. Office of Emergencyand Remedial Response, EPA 540/F-95/038, 1996002E

Van Vuren JPJ (1949) Soil fertility and sewage. Faber and Faber Ltd., London

Vigneswaran S, Kandasamy J, Johir MAH (2016) Sustainable operation of composting in solid waste management. Proc Environ Sci 35:408–415

Wang LK, Wang MHS, Cardenas RR, Sabiani NHM, Yusoff MS, Hassan SH, Kamaruddin MA, George FO, Hung YT (2021) Composting processes for disposal of municipal and agricultural solid wastes. In: Solid waste engineering and management, volume 1. Springer, pp 399–523

Wei YS, Fan YB, Wang MJ, Wang JS (2000) Composting and compost application in China. Resour Conserv Recycl 30(4):277–300

Xiao XJ (2001) A brief introduction to development of organic agriculture and organic food in China and foreign countries. Time Organic Food 4:20–31

Yang HS (2006) Resource management, soil fertility and sustainable crop production: experiences of China. Agric Ecosyst Environ 116(1–2):27–33

Yang LW, Tang XM (2002) The status and prospect of organic agriculture. Chinese Seeds Industry 8:8–9

Zaman B, Hardyanti N, Ramadan BS (2022) An innovative thermal composter to accelerate food waste decomposition at the household level. Bioresour Technol Rep 19:101203

Zhang Z, Yang H, Linghu M, Li J, Chen C, Wang B (2024) Cattle manure composting driven by a microbial agent: a coupled mechanism involving microbial community succession and organic matter conversion. Sci Total Environ 952:175953
Zhu J, Fan C, Shi H, Shi L (2019) Efforts for a circular economy in China: a comprehensive review of policies. J Ind Ecol 23(1):110–118
Zong LG, Lu D, Yang YG, Xiao XJ, Zhou ZJ (2003) The prospect of organic agriculture in China. Chin J Eco-Agric 11:152–154

# Overview of Different Nature-Based Ingredients in the Composting Process and Their Potential Value Addition

**Miguel Baltazar, Sandra Pereira, Lia Dinis, Sandra Martins, and Cátia Brito**

## 1 Introduction

As the global population is projected to rise from 7.2 billion to approximately 9.6 billion by 2050 (Gerland et al. 2014), agricultural output must increase by 70% compared to the past two decades. Agriculture today faces two major challenges that threaten global food security: climate change, which disrupts weather patterns and growing conditions, and soil degradation, which reduces land productivity and sustainability (Shukla et al. 2022; Froehlich et al. 2022). Shifts in global temperatures and precipitation patterns and the increased frequency of extreme weather events, such as droughts, floods, and heat waves, are already affecting crop development and productivity worldwide (Ghosh et al. 2021; Shukla et al. 2022). These worsening conditions directly impact soil functions by accelerating erosion, depleting organic carbon (C) reserves, and disrupting nutrient cycling, further compromising soil fertility (Brevik 2012). Soil is a crucial yet finite natural resource that supports plant growth, facilitates nutrient cycling, degrades pollutants, and serves as a C reservoir (El Mujtar et al. 2019).

Agriculture significantly exacerbates these problems, mainly due to the reliance on traditional agronomic practices, such as conventional soil management, the excessive use of fertilizers and pesticides, and the improper management of agricultural residues (Banias et al. 2017; Sanaullah et al. 2020). Globally, agrifood systems account for about one-third of the total anthropogenic greenhouse gas (GHG) emissions. Specifically, in 2022, global agrifood systems' emissions were 16.2

M. Baltazar · S. Pereira · L. Dinis · S. Martins · C. Brito (✉)
CITAB–Centre for the Research and Technology of Agro-Environmental and Biological Sciences, University of Trás-os-Montes and Alto Douro (UTAD), Vila Real, Portugal

Inov4Agro–Institute for Innovation, Capacity Building and Sustainability of Agri-Food Production, University of Trás-os-Montes and Alto Douro (UTAD), Vila Real, Portugal
e-mail: cvqbrito@utad.pt



R. K. Singh et al. (eds.), *Composting of Farm Residues by Nature-Based Solutions*,
https://doi.org/10.1007/978-3-032-12255-1_2

billion tonnes of carbon dioxide ($CO_2$) equivalent (Gt $CO_2$ eq.), representing an increase of 10% since 2000 (FAO 2025). Despite the agricultural sector's substantial contribution to climate change, it also holds significant potential for mitigation through the adoption of sustainable practices, which are crucial for reducing GHG emissions, optimizing the use of natural resources, and safeguarding ecosystem services (Smith and Olesen 2010).

Composting is an eco-friendly approach that not only reduces the amount of organic waste sent to landfills but also generates a valuable resource that enhances soil health (Amuah et al. 2022). When incorporated into the soil, compost can enhance soil health and reduce the dependence on synthetic inputs by facilitating organic residue recycling, improving nutrient cycling, and promoting C sequestration (Aguilar-Paredes et al. 2023). Thus, composting aligns with the principles of the circular economy (Bhorkar et al. 2023). This natural process relies on the decomposition of organic materials by specialized aerobic microorganisms to create nutrient-rich organic matter (OM) (Raklami et al. 2021; Huang et al. 2022; Aguilar-Paredes et al. 2023). The composting process consists of four distinct phases: mesophilic, thermophilic, cooling, and maturation, each driven by specific microbial communities (Waqas et al. 2023). During the mesophilic phase, microbial activity initiates decomposition, gradually raising the temperature, while the breakdown of carbohydrates and lipids produces organic acids, leading to pH reduction (Tuomela 2000). When temperatures exceed 40 °C, mesophilic microorganisms are progressively replaced by thermophiles, which dominate the thermophilic phase (50–60 °C). In this stage, microbial degradation of proteins releases ammonia ($NH_3$), leading to a pH increase. As high-energy compounds are depleted, the cooling phase begins, during which temperatures decline, and mesophilic microorganisms reestablish, contributing to compost stabilization. Finally, during the maturation phase, the compost reaches the ambient temperature, allowing mesophilic bacteria and fungi to complete humification, enhancing the compost's stability and agronomic value (Makan et al. 2013; Azim et al. 2018). A successful composting process relies on the careful selection of feedstock to ensure an optimal carbon/nitrogen (C/N) ratio, which is essential for sustaining microbial activity (Agnew and Leonard 2003; Azim et al. 2018). These materials, commonly classified as C-rich and nitrogen (N)-rich, play distinct yet complementary roles in maintaining decomposition efficiency and promoting a stable, high-quality compost (Schwarz and Bonhotal 2011; Hamid et al. 2019). The C source provides energy and serves as a key structural component for microorganisms, while the N source supplies protein synthesis and enzymatic functions, essential for organic matter synthesis (Schwarz and Bonhotal 2011).

The present chapter provides a comprehensive overview of several nature-based ingredients used in composting and their potential to improve the composting process and the sustainability of agricultural systems. It explores the roles of C-rich materials (woody materials and crop residues) and N-rich materials (green manures, food waste, and animal manures) in optimizing microbial activity and decomposition efficiency and discusses the importance of bulking agents and additives in improving the composting process and compost quality. In addition, this chapter highlights the added value of nature-based ingredients and their benefits in soil health improvement, nutrient cycling and efficiency, waste reduction and environmental impact, and biodiversity support.

# 2 Nature-Based Ingredients Used in Composting

## *2.1 Carbon-Rich Materials*

C-rich materials play a crucial role in composting, serving as both an energy source and a structural component for microorganisms (Schwarz and Bonhotal 2011). Compost rich in organic matter (OM) provides nutrients and enhances soil structure, improves water retention, supports microbial activity, and contributes to C sequestration (Sharma et al. 2024). The nutrient content and availability in compost depend on factors such as feedstock composition, composting conditions, and soil properties. In agriculture, C-rich amendments sustainably enhance crop yields and vegetable quality by influencing plant–microbe interactions and nutrient acquisition pathways (González-Coria et al. 2022). These materials significantly contribute to OM accumulation and biological activity in soils, though their initial application may temporarily reduce nutrient availability (Barthès et al. 2010). Nevertheless, composts derived from C-rich materials improve soil fertility and structure, increase water retention, and stimulate microbial activity, thereby reducing reliance on chemical fertilizers and mitigating environmental impacts (González-Coria et al. 2022).

Additionally, incorporating crop residues in (co)composting facilitates C recycling, playing a key role in GHG emissions and promoting soil C sequestration. As a result, C-rich soil amendments represent a sustainable farming strategy that balances productivity with environmental benefits (González-Coria et al. 2022).

### 2.1.1 Crop Residues and Other Woody Materials

Crop residues and woody materials are valuable C-rich inputs in composting due to their high C content. Their decomposition reduces agricultural and forestry waste while transforming it into nutrient-rich compost. Crop residues, such as straw, husks, pruning residues, and leaves, are abundant agricultural by-products that serve as significant sources of OM for composting. Additionally, materials like sawdust, wood chips, and shredded paper contribute to the C pool, further supporting microbial activity and soil enrichment.

The C/N ratio is a critical factor influencing composting efficiency. Optimal aerobic composting requires an initial C/N ratio between 25 and 35 (Doublet et al. 2011; Ren et al. 2023). A low C/N ratio leads to excessive nitrogen release as $NH_3$, causing N depletion and the production of strong odors (Azim et al. 2018). On the other hand, a high C/N ratio limits microbial activity, slowing humus formation and

**Fig. 1** Compost made with vine pruning wood

compost maturation (Liu et al. 2017a). Additionally, the C/N ratio influences the compost pile temperature, which is critical for pathogen elimination (Chen et al. 2018). Thus, selecting an appropriate initial C/N ratio is crucial for optimizing the aerobic composting of agricultural residues.

Incorporating composted straw into soil improves productivity by enhancing soil physicochemical properties, such as water retention, porosity, and nutrient content. It also supports soil health by fostering beneficial microbial communities while suppressing stress-tolerant microbes and plant pathogens (Song et al. 2023). *Acacia* compost is another viable soil amendment (Brito et al. 2013), but its high pH and electrical conductivity (EC) exceed the recommended limits for horticultural substrates (Abad et al. 2001), and its OM content is lower than that of peat or pine bark compost. Co-composting *Acacia* with materials of lower EC and pH and a higher OM content, such as pine bark, creates a well-structured, biodegradable feedstock, making it suitable for large-scale composting. Additionally, OM mineralization enriches compost with essential nutrients like phosphorus (P), potassium (K), calcium (Ca), magnesium (Mg), and iron (Fe) (Brito et al. 2015).

Several studies highlight the agronomic benefits of compost applications. Jaime-Rodríguez et al. (2025) evaluated compost from woody plant residues (municipal pruning residues) and its effects on orange-fleshed sweet potato growth, yield, and bioactive compounds. Compost application significantly reduced phenolic compounds and carotenoids in storage roots, potentially due to increased soil microbial activity affecting secondary metabolism. The compost dose was calculated based on the maximum amount of N allowed in composts according to the EU fertilizer regulation 2019/1009, which is 170 N kg/hm$^2$/yr$^1$. Vine pruning waste is another viable composting material (Fig. 1). Gaiotti et al. (2017) found that this kind of compost, with a C/N ratio of approximately 20, applied at 4 t/ha (fresh weight) annually, enhanced soil fertility, root growth, vegetative development, yield, and grape quality. Beyond crop improvements, compost application positively impacts soil health, underscoring its role in sustainable vineyard management.

A good example of the sustainable management of agroforestry residues includes the establishment of Residual Biomass Collection Centers, where organic waste from agriculture (e.g., cereals, rice, orchards, olive groves, and vineyards), forestry (e.g., stumps, roots, leaves, and branches), and gardens is collected for composting (Silva et al. 2024). Organic conditioners derived from these sources, particularly in stone pine plantations, have been shown to enhance soil quality by increasing OM, C, N, and water retention while minimizing environmental impact. Cornstalks have also been extensively used for co-composting with livestock manure and municipal organic solid waste, improving the compost quality and efficiency (Guo et al. 2012). Additionally, they function as filtering materials to remove suspended solids and OM from anaerobically digested centrate. Through filtration, the C/N ratio of cornstalks decreases from 41.50 to 16.81. However, these filtered cornstalks can be directly composted to produce organic fertilizers, during which the C/N ratio further declines to 13.62, and approximately 35.4% of cellulose is biodegraded (Guo et al. 2012).

Sugarcane bagasse, co-composted with exhausted grape marc, enhances the compost quality by prolonging thermophilic phase duration, regulating pH and EC, boosting microbial populations and enzyme activity, and facilitating lignocellulose breakdown. This process improves the final compost characteristics, including refined particle size distribution, enriched macro- and micronutrient levels, and increased seed germination rates (Zhang and Sun 2016). Similarly, sugarcane filter mud, when combined with *Brassica* residue compost and household waste compost, significantly enhances sweet corn growth and yield (Salman et al. 2023). Applying sugarcane filter mud at a rate of 2 t/ha, alongside compost, further boosts crop performance, demonstrating the potential of integrating locally sourced compost with sugarcane by-products to improve soil fertility and recycle organic waste.

Despite their agricultural benefits, some C-rich materials decompose slowly. Wheat straw, for example, has a high lignocellulose content, limiting microbial degradation (Zhang and Sun 2016). Wheat straw, for example, has a high lignocellulose content, limiting microbial degradation (Zhang and Sun 2016). In that study, after 25 days of composting, hemicellulose and cellulose content decreased by only 14% and 33%, respectively, indicating incomplete breakdown. Microbial analysis revealed minimal structural shifts, with Proteobacteria and Bacteroidetes remaining dominant, and pathogenic bacteria such as Pseudomonas and Flexibacter detected, suggesting incomplete sanitization. Moreover, the presence of Thermomyces, a key lignocellulose degrader, accounted for only 19% of fungal communities, further highlighting the slow degradation of wheat straw (Zhang and Sun 2016, Sadeghi et al. 2024).

Co-composting olive industry waste, including olive mill pomace and solid husk as C sources, with poultry manure (N source), green waste (bulking agent), and olive mill wastewater (humidifier) produces high-quality compost. This mixture enhances microbial activity, accelerates temperature rise, and supports aerobic microorganism proliferation, leading to a steady C/N ratio decline over time (Bargougui et al. 2020).

The final compost had a fine texture (70.41% <2 mm), a neutral pH (6.69), and a C/N ratio near 10. It was also rich in essential mineral nutrients (P, K, and Ca). Germination tests on tomato (*Solanum lycopersicum*), cress (*Lepidium sativum*), sorghum (*Sorghum bicolor*), and alfalfa (*Medicago sativa*) confirmed its non-phytotoxicity, with germination index values exceeding 85%. These findings support olive industry waste and poultry manure co-composting as an effective strategy for high-quality organic fertilizer production (Bargougui et al. 2020).

The wine industry also generates substantial waste, and composting offers a sustainable method for recycling these by-products. Pinto et al. (2023) studied grape marc and stalk composting under static and turned pile conditions. Thermophilic temperatures were reached quickly, with maximum temperatures of 70.5–71.8 °C in turned piles. However, after 42 days, the moisture content in these piles dropped below the recommended levels. This, along with excessive heat, hindered OM mineralization. In contrast, static piles maintained adequate porosity, yielding a higher proportion of mineralizable OM (568 g/kg). After 140 days, both methods produced stable and mature compost ($NH_4^+$-N/$NO_3^-$-N < 0.5), suitable as a soil amendment. Notably, the static pile required no rewetting or turning, reducing both agronomic and environmental costs. Alises et al. (2025) examined compost primarily composed of grape stalks and spent grape marc (C/N = 11.3; OM = 592 g/kg; EC = 1.20 mS/cm; pH = 9.71) and its impact on vineyard soils and wine quality. Incorporating different organic amendments into vineyard soil influenced the wine aroma composition. Applying 1–2 kg/linear meter of grape marc compost increased esters associated with fruity notes, while 3 kg/linear meter slightly elevated C6 compounds, contributing to green aromas. Additionally, applying 2 kg/linear meter of wine-distillery compost significantly enhanced terpenes and C13-norisoprenoids, compounds linked to greater aromatic typicity. These findings suggest that wine-distillery compost, particularly at 2 kg/linear meter, is a viable alternative to conventional fertilization, improving both soil quality and wine sensory attributes.

Rice husk also plays a role in composting. In sewage sludge composting, incorporating 25% rice husk minimized total organic C and N losses to 52.49% and 23.24%, respectively, and resulted in a mature compost with a C/N ratio of 18.82 (Temel 2023). Additionally, rice husk reduced the moisture content, optimizing the composting process in sewage sludge treatment. While the final compost in the control group had a moisture content of 50.72%, it decreased to 31.73% and 28.18% in reactors containing 10% and 25% rice husk, respectively.

### 2.1.2 Green Waste

Green waste comprises plant-based organic material discarded as waste, including branch cuttings, grass clippings, pruning debris, small amounts of soil, prunings from young trees and shrubs, and dead and green leaves, originating from municipal parks, gardens, reserves, and domestic sources (Reyes-Torres et al. 2018; Parab et al. 2023). Unlike other waste types, untreated green waste generally presents minimal environmental risks (Parab et al. 2023). Its composition varies depending on the vegetation type, seasonal changes, and collection methods, but it remains a valuable feedstock for composting. Rich in N, P, and Ca, green waste composting offers a cost-effective method for converting organic residues into nutrient-rich compost (Zhang and Sun 2016; Wei et al. 2017). Properly composted green waste enhances soil fertility, promotes microbial activity, and reduces dependence on synthetic fertilizers (Yang and Zhang 2022; Parab et al. 2023). Additionally, it reintegrates nutrients into the ecosystem while mitigating waste disposal issues and GHG emissions (Morales et al. 2016; Zhang and Sun 2016; Wei et al. 2017).

Green waste composting is advantageous due to its low micropollutant content, allowing compliance with organic farming standards (Bustamante et al. 2016). The thermophilic phase of composting effectively eliminates pathogens and stabilizes OM, resulting in a biologically stable soil amendment (Zhang et al. 2020b). Given its minimal micropollutant content, compost from green waste is particularly well-suited for organic farming at both household and community levels, offering an environmentally friendly alternative for soil management (Bai et al. 2010).

Several studies highlight the benefits of green waste compost. For instance, Yousaf et al. (2022) evaluated its role in remediating petroleum hydrocarbon–contaminated soil. Compost from *Poaceae* and *Fabaceae* plants improved plant growth and significantly degraded hydrocarbons, with ryegrass removing 68.5%. The compost contained 1.19% N, 0.309% P, 3.3% K, 18% organic C, and trace elements like Zn, Cu, Fe, and Mn. It also boosted microbial activity, moisture retention, and nutrient availability, facilitating contaminant breakdown. In vineyards, long-term green waste compost applications improved soil organic C from 15 to 52.6 g/kg and increased the C/N ratio from 5.5 to 13, enhancing soil properties while sustaining plant growth and grape quality (Mugnai et al. 2012). Dalal et al. (2010) investigated green waste compost derived from urban garden waste in sorghum cultivation. With a higher C/N ratio and lignin content than feedlot manure, it released N more gradually, reducing nitrous oxide ($N_2O$) emissions below untreated soil levels. When combined with manure, emissions dropped by 64%, while nitrate ($NO_3^-$) leaching decreased, improving N retention.

Despite its benefits, green waste composting presents challenges. Its high ligno-cellulose content, including wood, bark, and stems, slows microbial decomposition, leading to prolonged compost maturation and lower humic substance quality (Ma et al. 2020; Hoang et al. 2021). Additionally, green waste composition varies with the source vegetation, seasonality, and collection methods. Typically, it has a C/N ratio above 25 and is deficient in key nutrients such as N (0.5–1.5% dry basis), P (0.1–0.2%), and K (0.4–0.8%) while being rich in recalcitrant compounds like lignin. To address these challenges, research has focused on optimizing green waste composting by reducing the processing time, minimizing gas emissions, and improving the compost quality (Zhang et al. 2013; Bustamante et al. 2016; Zhang and Sun 2016). Effective strategies include shredding to reduce particle size, adding energy amendments (e.g., non-refined sugar and phosphate rock), incorporating bulking agents (e.g., biochar and wood chips), and adjusting aeration and temperature control. These techniques accelerate decomposition, shorten the processing time, and enhance the compost quality (Reyes-Torres et al. 2018).

## 2.2 Nitrogen-Rich Materials

N-rich materials are essential for composting because they provide the N required for protein synthesis and enzymatic functions, which accelerates OM decomposition, enhances nutrient cycling, and leads to the formation of humus. This process ultimately improves soil fertility and the overall quality of the final compost (Zakarya et al. 2023; Xu et al. 2024).

Green manures are important contributors as N-rich materials. The growing global population and evolving lifestyles have led to a significant increase in waste production in recent years (Ebrahimi et al. 2024). Composting N-rich materials, such as animal manure and food waste, offers a nature-based solution for the sustainable management of these organic wastes.

### 2.2.1 Green Manure

These crops, often grown for incorporation into the soil, either directly or after composting, are a valuable tool in sustainable agriculture. They enhance soil organic matter, contribute to nutrient cycling, and promote soil health.

Green manure can be categorized into two types: in situ, where crops are grown and integrated into the soil on-site, and ex situ, where plant residues are harvested and added to the soil before planting the main crop (Meena et al. 2018). Common green manure plants include legumes such as peas, beans, lupines, soybeans, clover, alfalfa, and vetch, which are well-known for their ability to fix N in the soil. Grasses like rye and oats are also frequently used in combination with legumes because they improve soil properties and reduce soil erosion (Fageria 2007). Green manure offers several ecological and agricultural benefits. It enhances N uptake efficiency, reduces nutrient leaching (Thorup-Kristensen et al. 2012), and increases microbial biomass (Elfstrand et al. 2007). Additionally, it contributes to mitigating GHG emissions, controlling pests, suppressing weeds, and promoting biodiversity (Melander et al. 2020).

Historically, before the widespread use of synthetic fertilizers, green manure was vital for maintaining soil fertility and ensuring food security (Meena et al. 2018). Today, it remains a key component of sustainable farming practices, reducing the dependence on chemical fertilizers and supporting ecological sustainability (Spiertz 2010). Green manure plants decompose quickly due to their high N content and low C/N ratios. This accelerates the composting process, stimulates microbial activity, and improves the compost quality. Additionally, these plants enhance the structure of the compost, improving aeration, drainage, and nutrient availability, which benefits both plant growth and soil health. For example, a study by Tejada et al. (2008) evaluated the effects of green manure composted with beet vinasse (*Trifolium pratense* L.) on soil restoration. When applied for over four years in a 2:1 compost ratio, this compost improved soil structural stability by 10.5%, reduced bulk density by 13.5%, and significantly increased microbial biomass, soil respiration, and enzymatic activities. Notable increases in dehydrogenase (+97.5%), urease (+98.4%), and β-glucosidase (+99.1%) highlighted its effectiveness in enhancing soil health.

However, green manure presents certain challenges. If allowed to mature before composting, these plants may introduce weed seeds. To mitigate this, crops should be harvested before seed formation, and thermophilic composting methods can be employed to destroy weed seeds. Excessive N in green manure can also lead to $NH_3$ odors, which can be mitigated by balancing with C-rich materials to maintain an optimal C/N ratio of 25/1 to 35/1. Additionally, the nutrient content of green manure can vary based on species, growth stages, and environmental conditions, affecting the final compost quality. Using a diverse mix of green manure species and conducting regular testing can help ensure a balanced nutrient profile.

### 2.2.2 Animal Manures

The livestock industry is responsible for around 14.5% of the global anthropogenic $CO_2$ emissions, a percentage expected to increase until 2030 (Alex Thumba et al. 2022; Yu et al. 2024). Along with $CO_2$, this industry significantly contributes to $CH_4$ and $N_2O$ emissions, both of which are potent GHGs linked to global warming and climate change (Grossi et al. 2019; He et al. 2023). One of the most abundant by-products of this industry is animal manure, which is usually stored for prolonged periods (Loyon 2018). However, its improper management can lead to soil and water contamination (Li et al. 2022c; Zhou et al. 2022; Qi et al. 2023). Thus, implementing effective management strategies and alternative utilization approaches, such as composting, is crucial to mitigating the environmental impact of livestock production (Symeon et al. 2025).

Animal manure is rich in OM and essential nutrients such as N, P, and K, among others. Thus, it has been historically used in soil fertilization to improve crop productivity. Besides its direct use, the composting of animal manure could also become one of the most widely adopted agricultural practices, serving as a natural source of N and producing nutrient-rich compost, leading to the reduction of the use of synthetic fertilizers. Moreover, when compared to conventional manure storage, composting has been shown to reduce overall GHG emissions despite still leading to the production of $N_2O$ (Brown et al. 2008). Furthermore, through the composting process, many of the problems associated with the storage and direct soil application of animal manure can be mitigated, namely, the decrease in moisture content; the lack of conversion of C, N, and P into more stable forms; and the management of the pathogenic content (Ucaroglu and Ozbek 2024). The composting process offers both economic and environmental advantages by reducing the challenges associated with animal manure storage and disposal while also facilitating the reintegration of essential compounds back into the agricultural chain (Urra et al. 2019). As manure possesses high OM and nutrient content, its composting could reduce the nutrient runoff and pollution associated with animal husbandry (Szogi et al. 2015).

Animal manure often contains pathogenic bacteria, antibiotics, heavy metals, and other compounds associated with animal husbandry. Different livestock species produce manure with varying physicochemical properties, necessitating tailored composting approaches for optimal results. Swine, cattle, and poultry manure, for example, each requires distinct management strategies (Millner et al. 2014; Deng et al. 2020). Huang et al. (2017) observed that the composting process of swine, cattle, and poultry was inherently different, requiring tailored approaches and leading to composts with differing qualities.

Swine manure is known to contain elevated levels of Cu, Mn, and Zn due to feed additives (Bolan et al. 2004; Shurson et al. 2022), which can negatively impact soil health if not properly managed. However, composting swine manure has been shown to improve the final compost quality (Matiz-Villamil et al. 2023). For instance, co-composting swine manure with green waste has resulted in stable composts, with N concentrations ranging from 1.7% to 2.4%, while reducing heavy metals and pathogens (Arias et al. 2021). Meng et al. (2018b) also observed that the addition of swine manure to mushroom substrate and rice husks improved the thermophilic phase of the composting process, provided essential microorganisms, aided in the C decomposition, and increased the nutritional content of the compost. Composting with swine manure has also been linked to improved plant growth and seedling quality, with effects comparable to synthetic fertilizers (Meng et al. 2018a). This has also been observed by Das et al. (2017) with both swine and cattle manure composts. These authors noted that these animal manure composts increased soil pH, nutrient availability, and microbial activity in a rice cropping system, leading to improved soil fertility and crop productivity.

Unlike swine manure, cattle manure is generally lower in nutrient concentration while also having a particularly high fiber content (Li et al. 2021). Adding to this, it is usually riddled with antibiotics due to the inherent problems of cattle husbandry (Ferroni et al. 2020; Zalewska et al. 2021). However, composting cattle manure can reduce antibiotic concentrations (Ho et al. 2013; Tasho and Cho 2016) and the dissemination of antibiotic resistance genes and their potential hosts (Gou et al. 2018; Zhang et al. 2020a). Furthermore, the addition of cattle manure compost back into the agricultural chain provides several benefits, including the amelioration of soil quality and the improvement of plant growth, without any of the negative side effects of direct soil application. For instance, increasing the depth or application rate of cattle manure compost alleviates soil salinity and alkalinity, increases organic matter, and boosts corn root growth and yield (Li et al. 2022b). Similar increases in corn yield were observed by Wolka and Melaku (2015), with their results revealing that compost resulting from the co-composting of food waste and cattle manure led to significant increases in crop yield, even when compared to synthetic fertilizers.

Poultry manure can also be composted to be added back into the agricultural chain, with benefits such as increasing nutrient availability and reducing pathogenic bacteria, antimicrobial residues, and antimicrobial resistance genes (Rizzo et al. 2015; Esperón et al. 2020). However, the composting of this type of animal manure is relatively challenging due to its high moisture content, low C/N and N/P ratios, and high pH, all of which can reduce the process efficiency (Ogunwande et al. 2008;

Chang et al. 2020). This can be amended by improving the choice of feedstock or by co-composting with other materials. For example, the co-composting of poultry manure with olive industry wastes led to non-phytotoxic compost rich in P, K, and Ca, with a neutral pH and a C/N ratio close to 10 (Bargougui et al. 2020). Vandecasteele et al. (2014) also demonstrated that the choice of the poultry feedstock significantly altered the compost properties and that around 10% of poultry manure was the composting threshold for avoiding nutrient losses and decreases in the N/P ratio. Nonetheless, similarly to the aforementioned animal manure composts, poultry manure compost can also increase the available OM and nutrients, leading to increases in plant biomass and overall production (Ch'ng et al. 2013; Rizzo et al. 2022).

Effective manure composting presents a vital strategy for mitigating the environmental impacts of the livestock industry. Transforming animal waste into a nutrient-rich resource not only reduces GHG and limits soil and water contamination but also enhances soil fertility and crop productivity. Tailoring composting methods to the distinct properties of swine, cattle, and poultry manure further optimizes these benefits.

### 2.2.3 Food Waste

Food waste is a major global concern, as nearly one-third of all food produced for human consumption is lost or discarded annually (Urugo et al. 2024). It encompasses a variety of organic kitchen leftovers, such as vegetable scraps, fruit peels, coffee grounds, tea bags, eggshells, stale bread, and other food remnants. Traditionally, food waste has been disposed of in landfills or incinerated, contributing to environmental problems, such as $CH_4$ emissions, resource wastage, and toxic leachate production (Chia et al. 2020; Ebrahimi et al. 2024). However, composting food waste offers a sustainable solution, with ecological and agricultural benefits (Cerda et al. 2018).

Composed primarily of organic acids, carbohydrates, proteins, and lipids, food waste is rich in biodegradable nutrients and stimulates microbial activity, making it an excellent candidate for aerobic composting (Chew et al. 2018; Wang et al. 2020). It also contains trace inorganic compounds that further support composting (Palaniveloo et al. 2020). The high moisture content (over 70%) of food waste fosters microbial growth, especially when combined with C-rich materials like straw, leaves, or shredded paper. This combination promotes the decomposition of organic matter, minimizes N loss, and supports N-fixing microbes, enhancing the overall composting process (Siles-Castellano et al. 2021). With a moisture content exceeding 70%, food waste creates an ideal environment for microbial growth when combined with C-rich materials like straw, leaves, or shredded paper. Additionally, its highly bioavailable OM aids in humus formation and lignocellulose recycling (Nguyen et al. 2020; Qin et al. 2021). Food waste composting not only reduces GHG emissions from landfills but also improves soil structure and fertility by enriching the soil with essential nutrients, such as C, N, and phosphorus (P); soil water retention;

and microbial activity, key factors for nutrient cycling and soil health (Bernstad Saraiva Schott et al. 2016; Cao et al. 2023). Additionally, food waste compost helps remediate pollutants, reduces soil toxicity, and supports plant growth by offering protection against environmental stress and diseases (Cao et al. 2023).

Several studies have explored the effectiveness of food waste composting. Vázquez and Soto (2017) evaluated home composting in eight rural areas, reporting an average efficiency of 77%, with each person composting 126 kg of biowaste annually, including meat and fish leftovers. The compost produced exhibited a low C/N ratio, minimal contaminants, and a high nutrient content, meeting agricultural standards. Similarly, Manu et al. (2019) studied composting household biodegradable waste with garden waste, enhanced with a microbial inoculum to accelerate composting. After 2 weeks of drying, the compost was mixed with soil in concentrations ranging from 0% to 30%. Plant growth increased by up to 20% but declined slightly at the 30% concentration due to excess compounds. Song et al. (2021) found that co-composting food waste digestate with sawdust and mature compost produced high-quality compost, reducing ammonium nitrogen ($NH_4^+$-N) levels and achieving a seed germination index above 80%.

Food waste compost also offers significant environmental benefits. Beiyuan et al. (2018) demonstrated that it could mitigate soil pollution by adsorbing heavy metals like lead (Pb), Cu, and Zn, thanks to its high surface charge and functional groups. Moreover, it was rich in OM and functional groups, such as carboxylic, phenolic, and lactonic groups, which interact with metals and release dissolved OM. This ability to reduce metal mobility and enhance nutrient availability aids in soil remediation. Moreover, composting food waste from diverse sources, such as dairy, grains, fruits, vegetables, meat scraps, and seafood, has been shown to lower contaminants like fuel oil in soil, making it an effective tool for soil remediation (Tran et al. 2018).

In addition to its environmental benefits, food waste composting has gained attention for its cost-effectiveness and practicality. For instance, several countries have implemented successful initiatives to promote food waste composting. Vermont's Universal Recycling Law requires the diversion of food waste from landfills. Since 2020, disposing of food scraps in regular trash has been prohibited, promoting composting and recycling. This regulation compels businesses, households, and institutions to separate food waste, reducing $CH_4$ emissions and supporting composting programs. By increasing recycling rates, the law fosters more sustainable waste management practices across the state (Vermont ANR 2018). South Korea, China, Germany, and the Netherlands also have policies in place to encourage organic waste recycling, further advancing circular economies (Nguyen et al. 2017). In South Korea, a mandatory composting program requires residents to separate food waste, achieving a 95% recycling rate. Over the past decade, the South Korean government has implemented stringent policies to enhance sustainable waste management, including legislation, regulations, and disposal fees based on volume, to reduce food waste generation and increase recycling (Nguyen et al. 2017).

Despite these positive outcomes, challenges remain in food waste composting. The high moisture content and imbalanced nutrient composition of food waste can lead to unpleasant odors and leachate (Calabi-Floody et al. 2019). Additionally, food waste is often composted with other materials, such as green waste or bulking agents, to improve composting efficiency and avoid nutrient imbalances (Chen et al. 2021). Co-composting food waste with green waste helps overcome the variable characteristics of green waste, such as material proportions and degradation level, and accelerates the thermophilic phase, resulting in better compost quality (Oviedo-Ocaña et al. 2019). For instance, food–green waste mixtures show improved nutrient profiles, with a more balanced C/N ratio and higher P levels, lower ash content, and better fertility, making them more suitable for agricultural use (Oviedo-Ocaña et al. 2019).

Despite these challenges, food waste composting remains a cost-effective, locally available solution with minimal heavy metal and antibiotic residues (Manu et al. 2019; Song et al. 2021). By integrating food waste into composting systems, we can support circular agriculture, reduce environmental impacts, and enhance soil health, ultimately benefiting both the economy and the environment.

## 2.3 *Bulking Agents and Additives*

The selection of bulking agents and additives is a crucial factor in composting efficiency, nutrient stabilization, and environmental impact. These components influence microbial activity, organic matter degradation, and the final compost quality, thereby optimizing the composting process.

### 2.3.1 Role of Bulking Agents and Additives in the Composting Processing

A well-balanced composting process requires an optimal C/N ratio, pH levels, moisture content, aeration, and temperature to support microbial degradation. Adding specific materials to composting feedstocks can significantly enhance the composting process and enhance the final compost quality (Ren et al. 2023; Noor et al. 2024). These materials, commonly known as compost conditioners, include bulking agents (C-rich materials such as sawdust, straw, and garden waste) and additives (microbial and mineral-based enhancers) (Ren et al. 2023).

Bulking agents and additives function synergistically in composting, working to create optimal conditions for microbial activity and organic matter decomposition. Their roles include optimizing key composting parameters, modulating microbial activity, reducing GHG emissions, and enhancing nutrient retention (Batham et al. 2013; Ren et al. 2023; Noor et al. 2024). The selection of these materials should consider factors such as availability, cost-effectiveness, and sustainability (Batham et al. 2013; Ren et al. 2023; Noor et al. 2024). However, economic constraints and resource availability may present challenges, which can be mitigated by using locally available materials or integrating multiple bulking agents to achieve optimal composting performance.

Bulking agents play a crucial role in adjusting the physical structure of the compost pile. Their primary functions include enhancing aeration by maintaining air spaces to support microbial activity, regulating moisture by absorbing excess water or retaining sufficient hydration, and providing a carbon source for microbial metabolism (Batham et al. 2013; Ren et al. 2023; Noor et al. 2024). A variety of organic materials have been evaluated for their efficacy as bulking agents, including leaves, tree prunings, corn stalks, garden waste, sawdust, wood chips, cornstalks, wheat straw, and rice husks (Batham et al. 2013; Ren et al. 2023).

Given that livestock and poultry manures often have a low C/N ratio due to their high N content, C-rich bulking agents are required to adjust this ratio to the optimal range of 25–35 (Doublet et al. 2011; Ren et al. 2023). Studies have demonstrated that incorporating materials such as cornstalks effectively adjusts the C/N ratio of cow dung compost to 24.84, accelerating decomposition (Wei et al. 2020). Similarly, rice husk additions have been shown to increase the C/N ratio from 7.1–7.3 to 17.0–24.0, thereby stimulating microbial activity (Iranzo et al. 2004). Straw amendments have also been beneficial in C/N ratio adjustment while simultaneously improving ventilation, decomposition rates, and N retention (Li et al. 2022a).

Beyond regulating the nutrient balance, bulking agents contribute to compost pile structural integrity by preventing excessive compaction and ensuring sufficient oxygen diffusion. By creating free air spaces, they facilitate aerobic microbial activity (Batham et al. 2013; Ren et al. 2023). The effectiveness of different bulking agents in promoting aeration varies. For instance, under identical pressure conditions, wood chips generate more free air space than leaves, thus improving oxygen diffusion within the compost pile (McCartney and Chen 2001). Co-composting green waste with coconut husk fiber biochar and spent mushroom substrate has been shown to improve particle size distribution, free air space, nutrient content, and cation exchange capacity while reducing phytotoxicity (Zhang and Sun 2014). Similarly, bamboo biochar enhances porosity, oxygen supply, and microbial activity (Liu et al. 2017a).

Moisture content is another critical factor in composting, as biochemical reactions primarily occur in the liquid phase. Adequate moisture is essential for microbial activity; however, excess moisture reduces air permeability, leading to anaerobic conditions that hinder decomposition (Ren et al. 2023). Bulking agents such as wood chips and straw help regulate moisture levels by absorbing excess water, thereby maintaining an optimal moisture content range of 55–65% for efficient composting (Batham et al. 2013).

While bulking agents primarily influence the physical structure of compost, additives regulate its chemical and biological composition. Additives serve to adjust pH, conserve nutrients, and introduce beneficial microorganisms, thereby optimizing decomposition conditions (Ren et al. 2023). These can be classified into chemical, microbial, and mineral-based additives (Barthod et al. 2018; Ren et al. 2023).

The pH of compost fluctuates during the decomposition process, typically decreasing in the early stages and increasing in later ones (Onwosi et al. 2017). Microbial activity is highly pH-sensitive, with an optimal range of 6–8 ensuring effective organic matter decomposition (Chang and Chen 2010; Batham et al. 2013). The addition of pH-regulating conditioners can mitigate pH imbalances. For instance, Ca-bentonite has been shown to increase pH levels during pig manure composting (Wang et al. 2016), while bamboo charcoal and zeolite counteract excessive pH rises during the thermophilic phase by adsorbing $NH_3$ (Venglovsky et al. 2005; Chen et al. 2010). In poultry waste composting, increasing the proportion of pine chip biochar raised pH levels and enhanced peak pile temperatures (Steiner et al. 2010). The addition of an inoculum consortium resulted in a pH increase from 4.3 to 6.3 during food waste composting, which may be explained by the degradation of acids, along with OM, through enhanced biological activity (Manu et al. 2017).

Microbial communities play a dominant role in OM degradation and nutrient retention during composting. Several additives enhance microbial populations by modulating temperature, moisture, and aeration while providing readily available nutrients and carbon sources (Barthod et al. 2018). For example, Li et al. (2019) demonstrated that pine leaf biochar significantly increased microbial diversity, particularly *Firmicutes*, *Actinobacteria*, and *Proteobacteria*, which correlated with improved composting performance (temperature, pH, total organic C, and $CO_2$ and $NH_3$ emissions). Similarly, the application of medical stone during composting decreased ammonia-oxidizing bacteria and archaea during the initial 28 days but increased their abundance during the maturation phase, indicating their role in nitrification (Awasthi et al. 2018). The inoculation of beneficial microbial agents can further enhance microbial dynamics in the early stages of composting, improving OM degradation, enzymatic activity, and compost maturation (Awasthi et al. 2018). The application of *Bacillus stearothermophilus* has been shown to modify the microbial community structure, increasing lactic acid *Bacillus* and nitrifying bacteria, thereby reducing pH fluctuations during nitrification and reducing $NH_3$ emissions (Wang et al. 2019). Zhao et al. (2022) observed that a commercial microbial inoculum increased the strain abundance by 3.4 times, prolonged the thermophilic phase, and enhanced lignocellulose degradation by 1.2 times. Also, the inoculation with enriched thermotolerant nitrifying bacteria improved the sewage sludge compost quality (temperature, pH, organic matter degradation, C/N ratio, and germination index) (Zhao et al. 2020).

Beyond microbial activity, additives also influence nutrient conservation and heavy metal stabilization. Conditions that promote humification and microbial diversity reduce heavy metal bioavailability (Ejileugha et al. 2024). Biochar-based products, for example, can reduce N loss and stabilize heavy metals. For instance, bamboo biochar has been reported to reduce nitrogen loss by 64.1% while stabilizing $Cu^{2+}$ and $Zn^{2+}$ (Hua et al. 2009). Similarly, rice husk biochar slows poultry manure decomposition, improving nutrient retention (Theeba et al. 2016). In poultry waste composting, increasing the proportion of pine chip biochar reduced the total N loss by 52% (Steiner et al. 2010). Adding lignite at 15% enhanced N retention during poultry litter composting due to high moisture content and low $NH_3$ availability that could suppress ammonia oxidizers, therefore improving N retention (Cao et al. 2020). Rice straw has also enhanced microbial diversity and suppressed heavy metal availability in composting river sediments mixed with agricultural waste (Chen et al. 2017). The aerobic composting of pig manure with maggot and chicken manure further inactivated Pb, Zn, Cu, Chromium (Cr), and cadmium (Cd) (Xu et al. 2021b). Moreover, the addition of shell powder and biochar increases humic acid synthesis, influencing microbial abundance and further reducing heavy metal bioavailability (Li et al. 2024). The use of Ca-bentonite also reduces metal availability through adsorption and complexation of the Zn and Cu metal ions on inorganic components (Wang et al. 2016). Composting of cow manure and tree litter with or without the inoculation of beneficial microbes reduced the solubility of Pb, Cu, and Zn (Reyes Pinto et al. 2020). The use of microbial additives, such as *Thiobacillus thioparus* 1904, has been shown to enhance N retention in chicken manure composting (Lu et al. 2018), while the introduction of thermophilic nitrifying bacteria improves sewage sludge compost quality (Zhao et al. 2020). Additionally, earthworms contribute to heavy metal stabilization through enzymatic processes and accumulation in their gut and intestines, reducing their bioavailability (Swati and Hait 2017).

### 2.3.2 Role of Bulking Agents and Additives in Agro-Environmental Issues

Compost maturity refers to the extent to which organic materials fully decompose and stabilize during the composting process. Finished compost should be both stable (resistant to further decomposition) and mature (suitable for its intended application) to ensure that it does not cause adverse effects upon use. Indeed, compost maturity is a crucial factor in determining its quality and safety for agricultural and horticultural purposes. Parameters such as temperature, C/N ratio, and dissolved organic carbon are commonly used to assess compost stability, while seed germination and gaseous emission tests evaluate maturity. Immature compost poses risks, including phytotoxicity, N immobilization, and odor generation. The continued decomposition of unstable compost can lead to N deficiency, as microbial activity consumes the available N, along with the emission of unpleasant odors (Mahapatra et al. 2022).

The germination index is a biological indicator used to assess compost toxicity and maturity. A germination index exceeding 50% indicates a compost free from phytotoxins (Barthod et al. 2018). The presence of additives significantly influences this index by affecting the nutrient availability. For example, bamboo charcoal and bamboo vinegar added during pig manure composting increased the germination index to 95%, suggesting that charcoal may act as a plant growth stimulant (Chen et al. 2010).

The effectiveness of bulking agents in promoting compost maturity varies with the material used. Studies show that cornstalks, sawdust, and spent mushroom substrate influence compost maturity, with spent mushroom substrate proving particularly effective by reducing the C/N ratio from 23 to 16 and increasing the germination index from 53% to 111% (Yang et al. 2013). Similarly, in kitchen waste composting, additives such as tea waste and biochar have been found to accelerate maturation, enhance lignocellulose degradation, and improve water-holding capacity, porosity, nutrient content, and seed germination indices (Zahra et al. 2023). The inoculation of compost with thermotolerant nitrifying bacteria has been shown to enhance the compost quality and germination index of sewage sludge compost (Zhao et al. 2020). The selection of bulking agents depends on regional availability and economic feasibility. In Korea, for instance, sawdust is the predominant bulking agent for food waste composting, but its reliance on imports raises concerns regarding economic sustainability. Studies suggest that mushroom waste and ginkgo leaves are viable alternatives, as they accelerate compost maturation more efficiently than sawdust (Lee et al. 2020). Additionally, the use of Ca-bentonite during pig manure composting has been found to enhance the process, leading to increased Chinese cabbage yield and significantly reduced Cu and Zn accumulation in plants compared to control samples (Wang et al. 2016).

Composting generates GHGs such as $CH_4$, $N_2O$, and $CO_2$, as well as odorous emissions, including $NH_3$ and $H_2S$ (Barthod et al. 2018; Ren et al. 2023). Variations in feedstock composition and aeration conditions affect the extent of these emissions.

The heterogeneity of organic waste and ventilation limitations within compost piles can cause alternating aerobic and anaerobic conditions, leading to fluctuations in $CH_4$ and $N_2O$ emissions (Ren et al. 2023). Proper aeration, achieved through mechanical turning or the addition of bulking agents, can mitigate emissions. N losses in the form of $N_2O$ occur under aerobic conditions due to incomplete nitrification/denitrification or under anaerobic conditions due to oxygen limitations, leading to $NO_3^-$ accumulation (Barthod et al. 2018). The type of additive incorporated into compost can significantly impact $CO_2$ emissions. Readily decomposable organic materials increase $CO_2$ emissions, while those rich in lignin or other slowly degrading compounds can suppress them (Chowdhury et al. 2014; Barthod et al. 2018).

Various materials have been found to mitigate GHG emissions, reducing the compost's environmental footprint. Sawdust has been reported to reduce $CH_4$ and $N_2O$ emissions in kitchen waste composting more effectively than cornstalks and spent mushroom substrate (Yang et al. 2013). The incorporation of garden waste into composting piles can lower gaseous emissions by up to 80% while simultaneously promoting microbial activity beneficial for organic degradation (Xu et al. 2021a). Bamboo biochar has been shown to further reduce $NH_3$, $CO_2$, $CH_4$, and $N_2O$ emissions (Liu et al. 2017b). In poultry waste composting, pine chip biochar has been found to reduce $NH_3$ emissions and decrease the total N loss by 52% (Steiner et al. 2010). Zeolite mitigates N loss during composting by adsorbing $NH_3/NH_4+$, lowering the compost pH, and enhancing microbial nitrification (Shan et al. 2021). *Bacillus stearothermophilus* has been found to reduce $NH_3$ emissions in layer manure composting (Wang et al. 2019). Similarly, nitrogen-retaining microbial agents (including *Paucisalibacillus, Sporosarcina, Sphingobacterium,* and

*Oceanobacillus*) reduced $NH_3$ emissions by 58.8% in chicken manure composting while also shortening the composting cycle by 15 days (Qiu et al. 2021). The addition of 10% medical stone was found to enhance organic degradation and reduce greenhouse gas emissions, lowering $CH_4$ emissions by 48.76% and $N_2O$ emissions by 85.26% (Awasthi et al. 2018). Vermicomposting generally emits lower levels of GHGs than conventional composting, particularly in terms of $CH_4$ reduction. However, earthworms contribute significantly to $N_2O$ emissions, necessitating a balanced approach to composting to minimize overall greenhouse gas emissions (Yasmin et al. 2022).

Odor generation is another critical environmental concern, as it affects air quality and public health. The addition of specific conditioners can enhance composting efficiency and limit anaerobic fermentation, thereby reducing odor production (Ren et al. 2023). For instance, incorporating cornstalks into kitchen waste composting was found to lower $H_2S$ emissions, though it had minimal effect on $NH_3$ emissions (Yuan et al. 2015). Biochar and spent mushroom substrate have been shown to reduce $NH_3$, $H_2S$, and dimethyl disulfide emissions during pig manure composting (Liu et al. 2020). Sawdust reduced $CH_4$ emissions more effectively than cornstalks and spent mushroom in pig manure composting, although $NH_3$ and $N_2O$ emissions were higher when mixed with cornstalks or sawdust compared to spent mushroom compost (Sun et al. 2014). Another way to reduce the odor emissions is to trap the N excess and thus reduce the $NH_3$ contents in the composting pile, for example, by using zeolite (Awasthi et al. 2016). Awasthi et al. (2016) also reported that the addition of higher doses of zeolite + lime drastically reduces $CH_4$ and $N_2O$ emissions and $NH_3$ loss.

In summary, Fig. 2 provides an overview of the potential value addition of different nature-based ingredients in the composting process.

## 3 Conclusions and Future Direction

This chapter has provided a comprehensive overview of the diverse nature-based ingredients utilized in composting, emphasizing their crucial role in enhancing compost quality and promoting sustainable agricultural practices. By exploring the characteristics and contributions of C-rich materials, such as crop residues, woody materials, and green waste, and N-rich materials, including green manures, animal manures, and food waste, this discussion has underscored the importance of carefully selecting and combining these ingredients to optimize the composting process. Building upon this understanding, it has also highlighted that compost derived from these components contributes to improved soil health, reduced reliance on synthetic fertilizers, a lower environmental footprint associated with organic waste, and enhanced crop growth and productivity. Furthermore, the strategic use of bulking agents and additives plays a fundamental role in composting efficiency. Bulking agents maintain optimal physical conditions within the compost pile by ensuring adequate aeration and moisture content, while additives regulate pH, enhance

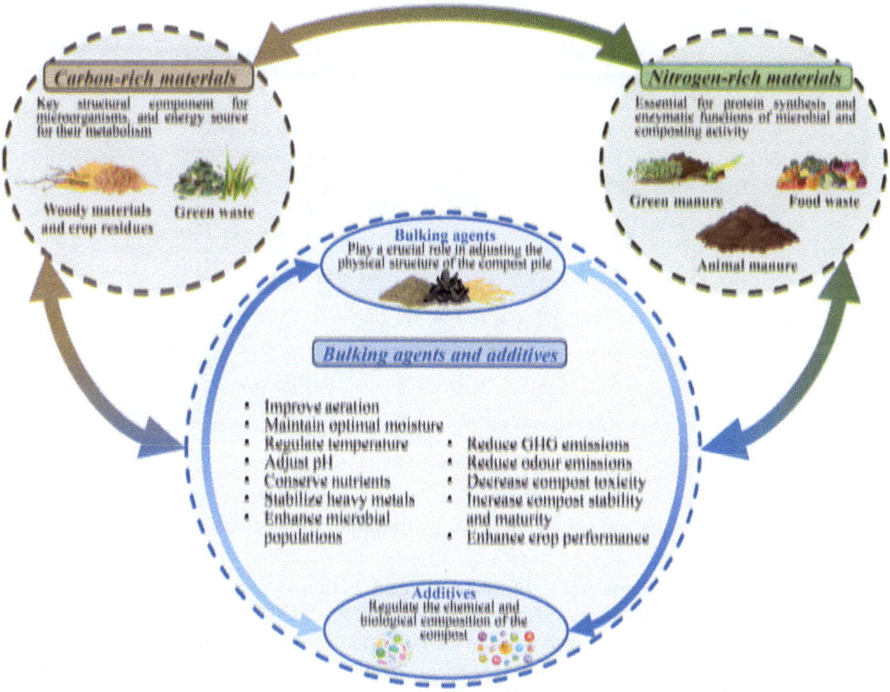


**Fig. 2** Overview of how different nature-based ingredients contribute to composting efficiency and stability

microbial activity, conserve nutrients, and stabilize heavy metals. Additionally, these components contribute to compost stability and maturity and help mitigate environmental concerns such as GHG and odor emissions, further reinforcing the sustainability of composting as a waste management strategy. Ultimately, the thoughtful application of these ingredients in composting represents a critical step toward a more circular and environmentally responsible agricultural system (summarized in Fig. 3).

Future research should focus on identifying cost-effective and locally available C- and N-rich materials, as well as bulking agents and additives that enhance composting efficiency while minimizing environmental impact. The selection of these materials should prioritize economic viability, sustainability, and resource availability to improve composting outcomes and reduce waste management costs. Addressing these aspects will help develop more sustainable composting practices, ultimately benefiting soil health, agricultural productivity, and environmental conservation.

**Acknowledgments** This work is supported by National Funds by the Portuguese Foundation for Science and Technology (FCT), under the projects UID/04033/2025: Centre for the Research and

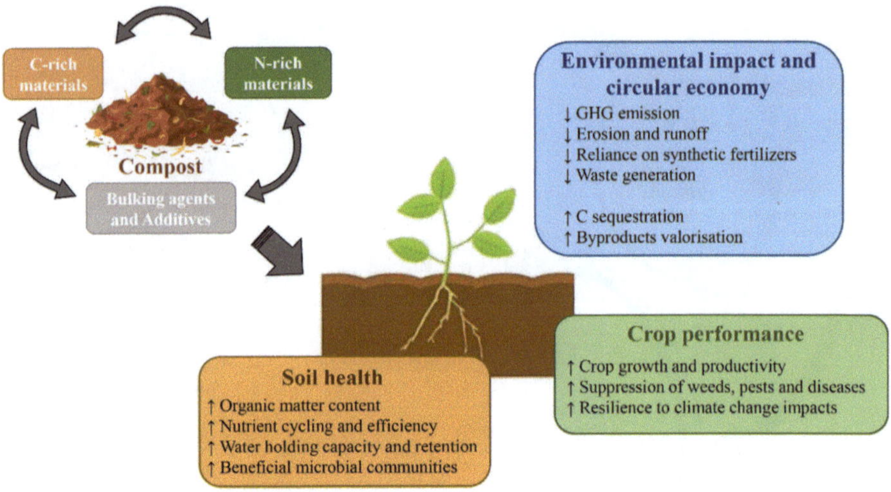


**Fig. 3** Ecosystem services and goods improved by the application of compost use

Technology of Agro-Environmental and Biological Sciences (https://doi.org/10.54499/UID/04033/2025) and LA/P/0126/2020 (https://doi.org/10.54499/LA/P/0126/2020).

The authors also thank the Project Vine&Wine Portugal - Driving Sustainable Growth Through Smart Innovation, Ref. no. C644866286-00000011, cofinanced within the scope of Mobilization Agendas for Business Innovation, under Reg. (EU) 2021/241, in the Recovery and Resilience Plan (PRR) of Portugal, namely, the subproject BFF_Bioprocessed Fertilizers: an ecoFriendly technology for nutrient recycling and environmental sustainability, for the knowledge and support in the thematic area and for Fig. 1.

## References

Abad M, Noguera P, Burés S (2001) National inventory of organic wastes for use as growing media for ornamental potted plant production: case study in Spain. Bioresour Technol 77(2):197–200. https://doi.org/10.1016/S0960-8524(00)00152-8

Agnew JM, Leonard JJ (2003) The physical properties of compost. Compost Sci Util 11(3): 238–264. https://doi.org/10.1080/1065657X.2003.10702132

Aguilar-Paredes A, Valdés G, Araneda N, Valdebenito E, Hansen F, Nuti M (2023) Microbial community in the composting process and its positive impact on the soil biota in sustainable agriculture. Agronomy 13(2):542. https://doi.org/10.3390/agronomy13020542

Alex Thumba D, Lazarova-Molnar S, Niloofar P (2022) Comparative evaluation of data requirements and level of decision support provided by decision support tools for reducing livestock-related greenhouse gas emissions. J Clean Prod 373:133886. https://doi.org/10.1016/j.jclepro.2022.133886

Alises MO, Palomo ES, González-Viñas MÁ (2025) Wine waste compost addition to a vineyard of Chelva grape of Central Spain: effect on wine volatile compounds. Horticulturae 11(2):219. https://doi.org/10.3390/horticulturae11020219

Amuah EEY, Fei-Baffoe B, Sackey LNA, Douti NB, Kazapoe RW (2022) A review of the principles of composting: understanding the processes, methods, merits, and demerits. Org Agric 12(4):547–562. https://doi.org/10.1007/s13165-022-00408-z

Arias O, Viña S, Soto M (2021) Co-composting of forest and industrial wastes watered with pig manure. Environ Technol 42(5):705–716. https://doi.org/10.1080/09593330.2019.1643411

Awasthi MK, Wang Q, Huang H, Ren X, Lahori AH, Mahar A, Ali A, Shen F, Li R, Zhang Z (2016) Influence of zeolite and lime as additives on greenhouse gas emissions and maturity evolution during sewage sludge composting. Bioresour Technol 216:172–181. https://doi.org/10.1016/j.biortech.2016.05.065

Awasthi MK, Wang Q, Awasthi SK, Wang M, Chen H, Ren X, Zhao J, Zhang Z (2018) Influence of medical stone amendment on gaseous emissions, microbial biomass and abundance of ammonia oxidizing bacteria genes during biosolids composting. Bioresour Technol 247:970–979. https://doi.org/10.1016/j.biortech.2017.09.201

Azim K, Soudi B, Boukhari S, Perissol C, Roussos S, Thami Alami I (2018) Composting parameters and compost quality: a literature review. Org Agric 8(2):141–158. https://doi.org/10.1007/s13165-017-0180-z

Bai J, Shen H, Dong S (2010) Study on eco-utilization and treatments of highway greening waste. Procedia Environ Sci 2:25–31. https://doi.org/10.1016/j.proenv.2010.10.005

Banias G, Achillas C, Vlachokostas C, Moussiopoulos N, Stefanou M (2017) Environmental impacts in the life cycle of olive oil: a literature review: environmental impacts in the life cycle of olive oil. J Sci Food Agric 97(6):1686–1697. https://doi.org/10.1002/jsfa.8143

Bargougui L, Guergueb Z, Chaieb M, Mekki A (2020) Co-composting of olive industry wastes with poultry manure and evaluation of the obtained compost maturity. Waste Biomass Valoriz 11(11):6235–6247. https://doi.org/10.1007/s12649-019-00901-9

Barthès BG, Manlay RJ, Porte O (2010) Effects of ramial wood amendments on crops and soil: a synthesis of experimental results. Cah Agric 19(4):280–287. https://doi.org/10.1684/agr.2010.0412

Barthod J, Rumpel C, Dignac M-F (2018) Composting with additives to improve organic amendments. A review. Agron Sustain Dev 38(2):17. https://doi.org/10.1007/s13593-018-0491-9

Batham M, Gupta R, Tiwari A (2013) Implementation of bulking agents in composting: a review. J Bioremed Biodegr 04(07):205. https://doi.org/10.4172/2155-6199.1000205

Beiyuan J, Tsang DCW, Bolan NS, Baek K, Ok YS, Li X-D (2018) Interactions of food waste compost with metals and metal-chelant complexes during soil remediation. J Clean Prod 192:199–206. https://doi.org/10.1016/j.jclepro.2018.04.239

Bernstad Saraiva Schott A, Wenzel H, La Cour Jansen J (2016) Identification of decisive factors for greenhouse gas emissions in comparative life cycle assessments of food waste management – an analytical review. J Clean Prod 119:13–24. https://doi.org/10.1016/j.jclepro.2016.01.079

Bhorkar M, Samarth A, Lanjewar A, Narkhede H, Agashe J, Nirwan N (2023) Development of modified compost unit for domestic biodegradable waste. Mater Today Proc. https://doi.org/10.1016/j.matpr.2023.11.024

Bolan N, Adriano D, Mahimairaja S (2004) Distribution and bioavailability of trace elements in livestock and poultry manure by-products. Crit Rev Environ Sci Technol 34(3):291–338. https://doi.org/10.1080/10643380490434128

Brevik EC (2012) Soils and climate change: gas fluxes and soil processes. Soil Horiz 53(4):12. https://doi.org/10.2136/sh12-04-0012

Brito LM, Mourão I, Coutinho J, Smith S (2013) Composting for management and resource recovery of invasive *Acacia* species. Waste Manag Res J Sustain Circ Econ 31(11):1125–1132. https://doi.org/10.1177/0734242X13502384

Brito LM, Mourão I, Coutinho J, Smith SR (2015) Co-composting of invasive *Acacia longifolia* with pine bark for horticultural use. Environ Technol 36(13):1632–1642. https://doi.org/10.1080/09593330.2014.1002863

Brown S, Kruger C, Subler S (2008) Greenhouse gas balance for composting operations. J of Env Quality 37(4):1396–1410. https://doi.org/10.2134/jeq2007.0453

Bustamante MA, Ceglie FG, Aly A, Mihreteab HT, Ciaccia C, Tittarelli F (2016) Phosphorus availability from rock phosphate: combined effect of green waste composting and sulfur addition. J Environ Manag 182:557–563. https://doi.org/10.1016/j.jenvman.2016.08.016

Calabi-Floody M, Medina J, Suazo J, Ordiqueo M, Aponte H, Mora MDLL, Rumpel C (2019) Optimization of wheat straw co-composting for carrier material development. Waste Manag 98: 37–49. https://doi.org/10.1016/j.wasman.2019.07.041

Cao Y, Bai M, Han B, Impraim R, Butterly C, Hu H, He J, Chen D (2020) Enhanced nitrogen retention by lignite during poultry litter composting. J Clean Prod 277:122422. https://doi.org/10.1016/j.jclepro.2020.122422

Cao X, Williams PN, Zhan Y, Coughlin SA, McGrath JW, Chin JP, Xu Y (2023) Municipal solid waste compost: global trends and biogeochemical cycling. Soil Environ Health 1(4):100038. https://doi.org/10.1016/j.seh.2023.100038

Cerda A, Artola A, Font X, Barrena R, Gea T, Sánchez A (2018) Composting of food wastes: status and challenges. Bioresour Technol 248:57–67. https://doi.org/10.1016/j.biortech.2017.06.133

Ch'ng HY, Ahmed OH, Kassim S, Majid NMA (2013) Co-composting of pineapple leaves and chicken manure slurry. Int J Recycl Org Waste Agric 2(1):23. https://doi.org/10.1186/2251-7715-2-23

Chang JI, Chen YJ (2010) Effects of bulking agents on food waste composting. Bioresour Technol 101(15):5917–5924. https://doi.org/10.1016/j.biortech.2010.02.042

Chang R, Li Y, Chen Q, Gong X, Qi Z (2020) Effects of carbon-based additives and ventilation rate on nitrogen loss and microbial community during chicken manure composting. PLoS One 15(9):e0229880. https://doi.org/10.1371/journal.pone.0229880

Chen Y-X, Huang X-D, Han Z-Y, Huang X, Hu B, Shi D-Z, Wu W-X (2010) Effects of bamboo charcoal and bamboo vinegar on nitrogen conservation and heavy metals immobility during pig manure composting. Chemosphere 78(9):1177–1181. https://doi.org/10.1016/j.chemosphere.2009.12.029

Chen Y, Liu Y, Li Y, Wu Y, Chen Y, Zeng G, Zhang J, Li H (2017) Influence of biochar on heavy metals and microbial community during composting of river sediment with agricultural wastes. Bioresour Technol 243:347–355. https://doi.org/10.1016/j.biortech.2017.06.100

Chen Z, Kim J, Jiang X (2018) Survival of *Escherichia coli* O157:H7 and *Salmonella enterica* in animal waste-based composts as influenced by compost type, storage condition and inoculum level. J Appl Microbiol 124(5):1311–1323. https://doi.org/10.1111/jam.13719

Chen Y, Chen Y, Li Y, Liu Y, Li H, Jiang H, Luo X, Tang P, Chen L, Yan H (2021) Evolution of humic substances and the forms of heavy metals during co-composting of rice straw and sediment with the aid of Fenton-like process. Bioresour Technol 333:125170. https://doi.org/10.1016/j.biortech.2021.125170

Chew KW, Chia SR, Show PL, Ling TC, Arya SS, Chang J-S (2018) Food waste compost as an organic nutrient source for the cultivation of *Chlorella vulgaris*. Bioresour Technol 267:356–362. https://doi.org/10.1016/j.biortech.2018.07.069

Chia WY, Chew KW, Le CF, Lam SS, Chee CSC, Ooi MSL, Show PL (2020) Sustainable utilization of biowaste compost for renewable energy and soil amendments. Environ Pollut 267:115662. https://doi.org/10.1016/j.envpol.2020.115662

Chowdhury MA, De Neergaard A, Jensen LS (2014) Potential of aeration flow rate and bio-char addition to reduce greenhouse gas and ammonia emissions during manure composting. Chemosphere 97:16–25. https://doi.org/10.1016/j.chemosphere.2013.10.030

Dalal RC, Gibson I, Allen DE, Menzies NW (2010) Green waste compost reduces nitrous oxide emissions from feedlot manure applied to soil. Agric Ecosyst Environ 136(3–4):273–281. https://doi.org/10.1016/j.agee.2009.06.010

Das S, Jeong ST, Das S, Kim PJ (2017) Composted cattle manure increases microbial activity and soil fertility more than composted swine manure in a submerged Rice Paddy. Front Microbiol 8: 1702. https://doi.org/10.3389/fmicb.2017.01702

Deng W, Zhang A, Chen S, He X, Jin L, Yu X, Yang S, Li B, Fan L, Ji L, Pan X, Zou L (2020) Heavy metals, antibiotics and nutrients affect the bacterial community and resistance genes in chicken manure composting and fertilized soil. J Environ Manag 257:109980. https://doi.org/10.1016/j.jenvman.2019.109980

Doublet J, Francou C, Poitrenaud M, Houot S (2011) Influence of bulking agents on organic matter evolution during sewage sludge composting; consequences on compost organic matter stability and N availability. Bioresour Technol 102(2):1298–1307. https://doi.org/10.1016/j.biortech.2010.08.065

Ebrahimi M, Gholipour S, Mostafaii G, Yousefian F (2024) Biochar-amended food waste compost: a review of properties. Results Eng 24:103118. https://doi.org/10.1016/j.rineng.2024.103118

Ejileugha C, Onyegbule UO, Osuoha JO (2024) Use of additives in composting promotes passivation and reduction in bioavailability of heavy metals (HMs) in compost. Rev Environ Contam Toxicol 262(1):2. https://doi.org/10.1007/s44169-023-00055-9

El Mujtar V, Muñoz N, Prack Mc Cormick B, Pulleman M, Tittonell P (2019) Role and management of soil biodiversity for food security and nutrition; where do we stand? Glob Food Secur 20:132–144. https://doi.org/10.1016/j.gfs.2019.01.007

Elfstrand S, Hedlund K, Mårtensson A (2007) Soil enzyme activities, microbial community composition and function after 47 years of continuous green manuring. Appl Soil Ecol 35(3): 610–621. https://doi.org/10.1016/j.apsoil.2006.09.011

Esperón F, Albero B, Ugarte-Ruíz M, Domínguez L, Carballo M, Tadeo JL, Del Mar DM, Moreno MÁ, De La Torre A (2020) Assessing the benefits of composting poultry manure in reducing antimicrobial residues, pathogenic bacteria, and antimicrobial resistance genes: a field-scale study. Environ Sci Pollut Res 27(22):27738–27749. https://doi.org/10.1007/s11356-020-09097-1

Fageria NK (2007) Green manuring in crop production. J Plant Nutr 30(5):691–719. https://doi.org/10.1080/01904160701289529

FAO (2025) Greenhouse gas emissions from agrifood systems. Global, regional and country trends, 2000–2022. In: Statistics. https://www.fao.org/statistics/highlights-archive/highlights-detail/greenhouse-gas-emissions-from-agrifood-systems.-global%2D%2Dregional-and-country-trends%2D%2D2000-2022/en. Accessed 24 Mar 2025

Ferroni L, Lovito C, Scoccia E, Dalmonte G, Sargenti M, Pezzotti G, Maresca C, Forte C, Magistrali CF (2020) Antibiotic consumption on dairy and beef cattle farms of Central Italy based on paper registers. Antibiotics 9(5):273. https://doi.org/10.3390/antibiotics9050273

Froehlich HE, Koehn JZ, Holsman KK, Halpern BS (2022) Emerging trends in science and news of climate change threats to and adaptation of aquaculture. Aquaculture 549:737812. https://doi.org/10.1016/j.aquaculture.2021.737812

Gaiotti F, Marcuzzo P, Belfiore N, Lovat L, Fornasier F, Tomasi D (2017) Influence of compost addition on soil properties, root growth and vine performances of *Vitis vinifera* cv Cabernet sauvignon. Sci Hortic 225:88–95. https://doi.org/10.1016/j.scienta.2017.06.052

Gerland P, Raftery AE, Ševčíková H, Li N, Gu D, Spoorenberg T, Alkema L, Fosdick BK, Chunn J, Lalic N, Bay G, Buettner T, Heilig GK, Wilmoth J (2014) World population stabilization unlikely this century. Science 346(6206):234–237. https://doi.org/10.1126/science.1257469

Ghosh UK, Islam MN, Siddiqui MN, Khan MAR (2021) Understanding the roles of osmolytes for acclimatizing plants to changing environment: a review of potential mechanism. Plant Signal Behav 16(8):1913306. https://doi.org/10.1080/15592324.2021.1913306

González-Coria J, Lozano-Castellón J, Jaime-Rodríguez C, Olmo-Cunillera A, Laveriano-Santos EP, Pérez M, Lamuela-Raventós RM, Puig J, Vallverdú-Queralt A, Romanyà J (2022) The effects of differentiated organic fertilization on tomato production and phenolic content in traditional and high-yielding varieties. Antioxidants 11(11):2127. https://doi.org/10.3390/antiox11112127

Gou M, Hu H-W, Zhang Y-J, Wang J-T, Hayden H, Tang Y-Q, He J-Z (2018) Aerobic composting reduces antibiotic resistance genes in cattle manure and the resistome dissemination in agricultural soils. Sci Total Environ 612:1300–1310. https://doi.org/10.1016/j.scitotenv.2017.09.028

Grossi G, Goglio P, Vitali A, Williams AG (2019) Livestock and climate change: impact of livestock on climate and mitigation strategies. Anim Front 9(1):69–76. https://doi.org/10.1093/af/vfy034

Guo R, Li G, Jiang T, Schuchardt F, Chen T, Zhao Y, Shen Y (2012) Effect of aeration rate, C/N ratio and moisture content on the stability and maturity of compost. Bioresour Technol 112:171–178. https://doi.org/10.1016/j.biortech.2012.02.099

Hamid H, Harun H, Sunar N, Muhammad MS, Hamidon N, Ali R, Ahmad FH, Lim PQ (2019) Development of organic fertilizer from food waste by composting in UTHM Campus Pagoh. FAZ Publishing/Advanced Technology Centre (ATC), Faculty of Engineering Faculty of Technology, Universiti Tun Hussein Onn, Malaysia

He D, Deng X, Wang X, Zhang F (2023) Livestock greenhouse gas emission and mitigation potential in China. J Environ Manag 348:119494. https://doi.org/10.1016/j.jenvman.2023.119494

Ho YB, Zakaria MP, Latif PA, Saari N (2013) Degradation of veterinary antibiotics and hormone during broiler manure composting. Bioresour Technol 131:476–484. https://doi.org/10.1016/j.biortech.2012.12.194

Hoang AT, Nižetić S, Ong HC, Mofijur M, Ahmed SF, Ashok B, Bui VTV, Chau MQ (2021) Insight into the recent advances of microwave pretreatment technologies for the conversion of lignocellulosic biomass into sustainable biofuel. Chemosphere 281:130878. https://doi.org/10.1016/j.chemosphere.2021.130878

Hua L, Wu W, Liu Y, McBride MB, Chen Y (2009) Reduction of nitrogen loss and Cu and Zn mobility during sludge composting with bamboo charcoal amendment. Environ Sci Pollut Res 16(1):1–9. https://doi.org/10.1007/s11356-008-0041-0

Huang J, Yu Z, Gao H, Yan X, Chang J, Wang C, Hu J, Zhang L (2017) Chemical structures and characteristics of animal manures and composts during composting and assessment of maturity indices. PLoS One 12(6):e0178110. https://doi.org/10.1371/journal.pone.0178110

Huang D, Gao L, Cheng M, Yan M, Zhang G, Chen S, Du L, Wang G, Li R, Tao J, Zhou W, Yin L (2022) Carbon and N conservation during composting: a review. Sci Total Environ 840:156355. https://doi.org/10.1016/j.scitotenv.2022.156355

Iranzo M, Cañizares JV, Roca-Perez L, Sainz-Pardo I, Mormeneo S, Boluda R (2004) Characteristics of rice straw and sewage sludge as composting materials in Valencia (Spain). Bioresour Technol 95(1):107–112. https://doi.org/10.1016/j.biortech.2004.01.013

Jaime-Rodríguez C, Pérez M, Pérez-Llorca M, Lozano-Castellón J, Lamuela-Raventós RM, Vallverdú-Queralt A, González-Coria J, Chantry O, Hernandez R, Romanyà J (2025) Impacts of woody residue amendments and compost on 'Beauregard' Orange fleshed sweet potato. Hortic Plant J. https://doi.org/10.1016/j.hpj.2024.11.003

Lee J-H, Luyima D, Lee C-H, Park S-J, Oh T-K (2020) Efficiencies of unconventional bulking agents in composting food waste in Korea. Appl Biol Chem 63(1):68. https://doi.org/10.1186/s13765-020-00554-6

Li J, Xing W, Bao H, Wang J, Tong X, Zhang H, Luo W, Wu F (2019) Impact of pine leaf biochar amendment on bacterial dynamics and correlation of environmental factors during pig manure composting. Bioresour Technol 293:122031. https://doi.org/10.1016/j.biortech.2019.122031

Li Y, Zhao J, Krooneman J, Euverink GJW (2021) Strategies to boost anaerobic digestion performance of cow manure: laboratory achievements and their full-scale application potential. Sci Total Environ 755:142940. https://doi.org/10.1016/j.scitotenv.2020.142940

Li D, Yuan J, Ding J, Wang H, Shen Y, Li G (2022a) Effects of carbon/nitrogen ratio and aeration rate on the sheep manure composting process and associated gaseous emissions. J Environ Manag 323:116093. https://doi.org/10.1016/j.jenvman.2022.116093

Li S, Liu Z, Li J, Liu Z, Gu X, Shi L (2022b) Cow manure compost promotes maize growth and ameliorates soil quality in saline-alkali soil: role of fertilizer addition rate and application depth. Sustainability 14(16):10088. https://doi.org/10.3390/su141610088

Li Y, Wang M, Chen X, Cui S, Hofstra N, Kroeze C, Ma L, Xu W, Zhang Q, Zhang F, Strokal M (2022c) Multi-pollutant assessment of river pollution from livestock production worldwide. Water Res 209:117906. https://doi.org/10.1016/j.watres.2021.117906

Li T, Yan Z, Sun Y, Hu X, Peng C, Zhao S, Xu D, Liu D, Shen Q (2024) Inorganic and organic additives differently regulate compost microbiomes in response to heavy metals immobilization. Chem Eng J 502:158087. https://doi.org/10.1016/j.cej.2024.158087

Liu N, Zhou J, Han L, Huang G (2017a) Characterization of lignocellulosic compositions' degradation during chicken manure composting with added biochar by phospholipid fatty acid (PLFA) and correlation analysis. Sci Total Environ 586:1003–1011. https://doi.org/10.1016/j.scitotenv.2017.02.081

Liu N, Zhou J, Han L, Ma S, Sun X, Huang G (2017b) Role and multi-scale characterization of bamboo biochar during poultry manure aerobic composting. Bioresour Technol 241:190–199. https://doi.org/10.1016/j.biortech.2017.03.144

Liu Y, Ma R, Li D, Qi C, Han L, Chen M, Fu F, Yuan J, Li G (2020) Effects of calcium magnesium phosphate fertilizer, biochar and spent mushroom substrate on compost maturity and gaseous emissions during pig manure composting. J Environ Manag 267:110649. https://doi.org/10.1016/j.jenvman.2020.110649

Loyon L (2018) Overview of animal manure management for Beef, Pig, and poultry farms in France. Front Sustain Food Syst 2:36. https://doi.org/10.3389/fsufs.2018.00036

Lu Y, Gu W, Xu P, Xie K, Li X, Sun L, Wu H, Shi C, Wang D (2018) Effects of Sulphur and Thiobacillus thioparus 1904 on nitrogen cycle genes during chicken manure aerobic composting. Waste Manag 80:10–16. https://doi.org/10.1016/j.wasman.2018.08.050

Ma C, Lo PK, Xu J, Li M, Jiang Z, Li G, Zhu Q, Li X, Leong SY, Li Q (2020) Molecular mechanisms underlying lignocellulose degradation and antibiotic resistance genes removal revealed via metagenomics analysis during different agricultural wastes composting. Bioresour Technol 314:123731. https://doi.org/10.1016/j.biortech.2020.123731

Mahapatra S, Ali MH, Samal K (2022) Assessment of compost maturity-stability indices and recent development of composting bin. Energy Nexus 6:100062. https://doi.org/10.1016/j.nexus.2022.100062

Makan A, Assobhei O, Mountadar M (2013) Effect of initial moisture content on the in-vessel composting under air pressure of organic fraction of municipal solid waste in Morocco. Iran J Environ Health Sci Eng 10(1):3. https://doi.org/10.1186/1735-2746-10-3

Manu MK, Kumar R, Garg A (2017) Performance assessment of improved composting system for food waste with varying aeration and use of microbial inoculum. Bioresour Technol 234:167–177. https://doi.org/10.1016/j.biortech.2017.03.023

Manu MK, Kumar R, Garg A (2019) Decentralized composting of household wet biodegradable waste in plastic drums: effect of waste turning, microbial inoculum and bulking agent on product quality. J Clean Prod 226:233–241. https://doi.org/10.1016/j.jclepro.2019.03.350

Matiz-Villamil A, Méndez-Carranza KJ, Pascagaza-Pulido AF, Rendón-Rendón T, Noriega-Noriega J, Pulido-Villamarín A (2023) Trends in the management of organic swine farm waste by composting: a systematic review. Heliyon 9(8):e18208. https://doi.org/10.1016/j.heliyon.2023.e18208

McCartney D, Chen H (2001) Using a biocell to measure effect of compressive settlement on free air space and microbial activity in windrow composting. Compost Sci Util 9(4):285–302. https://doi.org/10.1080/1065657X.2001.10702048

Meena BL, Fagodiya RK, Prajapat K, Dotaniya ML, Kaledhonkar MJ, Sharma PC, Meena RS, Mitran T, Kumar S (2018) Legume green manuring: an option for soil sustainability. In: Meena RS, Das A, Yadav GS, Lal R (eds) Legumes for soil health and sustainable management. Springer Singapore, Singapore, pp 387–408

Melander B, Rasmussen IA, Olesen JE (2020) Legacy effects of leguminous green manure crops on the weed seed bank in organic crop rotations. Agric Ecosyst Environ 302:107078. https://doi.org/10.1016/j.agee.2020.107078

Meng X, Dai J, Zhang Y, Wang X, Zhu W, Yuan X, Yuan H, Cui Z (2018a) Composted biogas residue and spent mushroom substrate as a growth medium for tomato and pepper seedlings. J Environ Manag 216:62–69. https://doi.org/10.1016/j.jenvman.2017.09.056

Meng X, Liu B, Xi C, Luo X, Yuan X, Wang X, Zhu W, Wang H, Cui Z (2018b) Effect of pig manure on the chemical composition and microbial diversity during co-composting with spent mushroom substrate and rice husks. Bioresour Technol 251:22–30. https://doi.org/10.1016/j.biortech.2017.09.077

Millner P, Ingram D, Mulbry W, Arikan OA (2014) Pathogen reduction in minimally managed composting of bovine manure. Waste Manag 34(11):1992–1999. https://doi.org/10.1016/j.wasman.2014.07.021

Morales AB, Bustamante MA, Marhuenda-Egea FC, Moral R, Ros M, Pascual JA (2016) Agri-food sludge management using different co-composting strategies: study of the added value of the composts obtained. J Clean Prod 121:186–197. https://doi.org/10.1016/j.jclepro.2016.02.012

Mugnai S, Masi E, Azzarello E, Mancuso S (2012) Influence of long-term application of green waste compost on soil characteristics and growth, yield and quality of grape (*Vitis vinifera* L.). Compost Sci Util 20(1):29–33. https://doi.org/10.1080/1065657X.2012.10737019

Nguyen DD, Yeop JS, Choi J, Kim S, Chang SW, Jeon B-H, Guo W, Ngo HH (2017) A new approach for concurrently improving performance of south Korean food waste valorization and renewable energy recovery via dry anaerobic digestion under mesophilic and thermophilic conditions. Waste Manag 66:161–168. https://doi.org/10.1016/j.wasman.2017.03.049

Nguyen V-T, Le T-H, Bui X-T, Nguyen T-N, Vo T-D-H, Lin C, Vu T-M-H, Nguyen H-H, Nguyen D-D, Senoro DB, Dang B-T (2020) Effects of C/N ratios and turning frequencies on the composting process of food waste and dry leaves. Bioresour Technol Rep 11:100527. https://doi.org/10.1016/j.biteb.2020.100527

Noor RS, Shah AN, Tahir MB, Umair M, Nawaz M, Ali A, Ercisli S, Abdelsalam NR, Ali HM, Yang SH, Ullah S, Assiri MA (2024) Recent trends and advances in additive-mediated composting technology for agricultural waste resources: a comprehensive review. ACS Omega 9(8):8632–8653. https://doi.org/10.1021/acsomega.3c06516

Ogunwande GA, Osunade JA, Adekalu KO, Ogunjimi LAO (2008) Nitrogen loss in chicken litter compost as affected by carbon to nitrogen ratio and turning frequency. Bioresour Technol 99(16):7495–7503. https://doi.org/10.1016/j.biortech.2008.02.020

Onwosi CO, Igbokwe VC, Odimba JN, Eke IE, Nwankwoala MO, Iroh IN, Ezeogu LI (2017) Composting technology in waste stabilization: on the methods, challenges and future prospects. J Environ Manag 190:140–157. https://doi.org/10.1016/j.jenvman.2016.12.051

Oviedo-Ocaña ER, Dominguez I, Komilis D, Sánchez A (2019) Co-composting of green waste mixed with unprocessed and processed food waste: influence on the composting process and product quality. Waste Biomass Valoriz 10(1):63–74. https://doi.org/10.1007/s12649-017-0047-2

Palaniveloo K, Amran MA, Norhashim NA, Mohamad-Fauzi N, Peng-Hui F, Hui-Wen L, Kai-Lin Y, Jiale L, Chian-Yee MG, Jing-Yi L, Gunasekaran B, Razak SA (2020) Food waste composting and microbial community structure profiling. Processes 8(6):723. https://doi.org/10.3390/pr8060723

Parab C, Yadav KD, Prajapati V (2023) Genomics and microbial dynamics in green waste composting: a mini review. Ecol Genet Genomics 29:100206. https://doi.org/10.1016/j.egg.2023.100206

Pinto R, Correia C, Mourão I, Moura L, Brito LM (2023) Composting waste from the white wine industry. Sustainability 15(4):3454. https://doi.org/10.3390/su15043454

Qi J, Yang H, Wang X, Zhu H, Wang Z, Zhao C, Li B, Liu Z (2023) State-of-the-art on animal manure pollution control and resource utilization. J Environ Chem Eng 11(5):110462. https://doi.org/10.1016/j.jece.2023.110462

Qin R, Su C, Mo T, Liao L, Zhu F, Chen Y, Chen M (2021) Effect of excess sludge and food waste feeding ratio on the nutrient fractions, and bacterial and fungal community during aerobic co-composting. Bioresour Technol 320:124339. https://doi.org/10.1016/j.biortech.2020.124339

Qiu Z, Li M, Song L, Wang C, Yang S, Yan Z, Wang Y (2021) Study on nitrogen-retaining microbial agent to reduce nitrogen loss during chicken manure composting and nitrogen transformation mechanism. J Clean Prod 285:124813. https://doi.org/10.1016/j.jclepro.2020.124813

Raklami A, Tahiri A, Bechtaoui N, Abdelhay EG, Pajuelo E, Baslam M, Meddich A, Oufdou K (2021) Restoring the plant productivity of heavy metal-contaminated soil using phosphate sludge, marble waste, and beneficial microorganisms. J Environ Sci 99:210–221. https://doi.org/10.1016/j.jes.2020.06.032

Ren X, Jiao M, Chen X, Liu T, Zhang Y, Zhang Z (2023) Role of bulking agents and additive on composting. In: Pandey A, Awasthi M, Zhang Z (eds) Current developments in biotechnology and bioengineering. Elsevier, pp 127–142

Reyes Pinto K, Meza-Contreras V, Alegre-Orihuela JC, Réategui-Romero W (2020) Bioavailability and solubility of heavy metals and trace elements during composting of cow manure and tree litter. Appl Environ Soil Sci 2020:1–20. https://doi.org/10.1155/2020/5680169

Reyes-Torres M, Oviedo-Ocaña ER, Dominguez I, Komilis D, Sánchez A (2018) A systematic review on the composting of green waste: feedstock quality and optimization strategies. Waste Manag 77:486–499. https://doi.org/10.1016/j.wasman.2018.04.037

Rizzo PF, Torre VD, Riera NI, Crespo D, Barrena R, Sánchez A (2015) Co-composting of poultry manure with other agricultural wastes: process performance and compost horticultural use. J Mater Cycles Waste Manag 17(1):42–50. https://doi.org/10.1007/s10163-013-0221-y

Rizzo PF, Young BJ, Pin Viso N, Carbajal J, Martínez LE, Riera NI, Bres PA, Beily ME, Barbaro L, Farber M, Zubillaga MS, Crespo DC (2022) Integral approach for the evaluation of poultry manure, compost, and digestate: amendment characterization, mineralization, and effects on soil and intensive crops. Waste Manag 139:124–135. https://doi.org/10.1016/j.wasman.2021.12.017

Sadeghi E, Ghorbani Nasrabadi R, Movahedi Naini SA, Barani Motlagh M, Khoshhal Sarmast M (2024) Assessing the co-composting of wheat straw: a study on compost quality and phosphorus levels for improved soil biochemical responses and maize growth. J Soil Sci Plant Nutr 24(4): 6248–6267. https://doi.org/10.1007/s42729-024-01940-5

Salman M, Inamullah JA, Mihoub A, Saeed MF, Radicetti E, Ahmad I, Naeem A, Ullah J, Pampana S (2023) Composting sugarcane filter mud with different sources differently benefits sweet maize. Agronomy 13(3):748. https://doi.org/10.3390/agronomy13030748

Sanaullah M, Usman M, Wakeel A, Cheema SA, Ashraf I, Farooq M (2020) Terrestrial ecosystem functioning affected by agricultural management systems: a review. Soil Tillage Res 196: 104464. https://doi.org/10.1016/j.still.2019.104464

Schwarz M, Bonhotal J (2011) Composting at home-the green and Brown alternative. Cornell Waste Management Institute. Department of Crop and Soil Sciences, Ithaca

Shan G, Li W, Gao Y, Tan W, Xi B (2021) Additives for reducing nitrogen loss during composting: a review. J Clean Prod 307:127308. https://doi.org/10.1016/j.jclepro.2021.127308

Sharma A, Soni R, Soni SK (2024) From waste to wealth: exploring modern composting innovations and compost valorization. J Mater Cycles Waste Manag 26(1):20–48. https://doi.org/10.1007/s10163-023-01839-w

Shukla PR, Skea J, van Diemen R, Haughey E, Malley J, Pathak M, Portugal Pereira J (2022) Technical summary. In: Climate Change and Land: IPCC special report on climate change, desertification, land degradation, sustainable land management, food security, and greenhouse gas fluxes in terrestrial ecosystems. Intergovernmental Panel on Climates Change, Cambridge University, Cambridge, pp 37–74

Shurson GC, Urriola PE, Hung Y-T (2022) Too much of a good thing: rethinking feed formulation and feeding practices for zinc in swine diets to achieve one health and environmental sustainability. Animals 12(23):3374. https://doi.org/10.3390/ani12233374

Siles-Castellano AB, López-González JA, Jurado MM, Estrella-González MJ, Suárez-Estrella F, López MJ (2021) Compost quality and sanitation on industrial scale composting of municipal solid waste and sewage sludge. Appl Sci 11(16):7525. https://doi.org/10.3390/app11167525

Silva MEF, Saetta R, Raimondo R, Costa JM, Ferreira JV, Brás I (2024) Forest waste composting—operational management, environmental impacts, and application. Environ Sci Pollut Res. https://doi.org/10.1007/s11356-024-32279-0

Smith P, Olesen JE (2010) Synergies between the mitigation of, and adaptation to, climate change in agriculture. J Agric Sci 148(5):543–552. https://doi.org/10.1017/S0021859610000341

Song B, Manu MK, Li D, Wang C, Varjani S, Ladumor N, Michael L, Xu Y, Wong JWC (2021) Food waste digestate composting: feedstock optimization with sawdust and mature compost. Bioresour Technol 341:125759. https://doi.org/10.1016/j.biortech.2021.125759

Song M, Li J, Gao L, Tian Y (2023) Comprehensive evaluation of effects of various carbon-rich amendments on overall soil quality and crop productivity in degraded soils. Geoderma 436: 116529. https://doi.org/10.1016/j.geoderma.2023.116529

Spiertz JHJ (2010) Nitrogen, sustainable agriculture and food security. A review. Agron Sustain Dev 30(1):43–55. https://doi.org/10.1051/agro:2008064

Steiner C, Das KC, Melear N, Lakly D (2010) Reducing nitrogen loss during poultry litter composting using biochar. J Environ Qual 39(4):1236–1242. https://doi.org/10.2134/jeq2009.0337

Sun X, Lu P, Jiang T, Schuchardt F, Li G (2014) Influence of bulking agents on CH4, N2O, and NH3 emissions during rapid composting of pig manure from the Chinese Ganqinfen system. J Zhejiang Univ Sci B 15(4):353–364. https://doi.org/10.1631/jzus.B13a0271

Swati A, Hait S (2017) Fate and bioavailability of heavy metals during vermicomposting of various organic wastes—a review. Process Saf Environ Prot 109:30–45. https://doi.org/10.1016/j.psep.2017.03.031

Symeon GK, Akamati K, Dotas V, Karatosidi D, Bizelis I, Laliotis GP (2025) Manure management as a potential mitigation tool to eliminate greenhouse gas emissions in livestock systems. Sustainability 17(2):586. https://doi.org/10.3390/su17020586

Szogi AA, Vanotti MB, Ro KS (2015) Methods for treatment of animal manures to reduce nutrient pollution prior to soil application. Curr Pollut Rep 1(1):47–56. https://doi.org/10.1007/s40726-015-0005-1

Tasho RP, Cho JY (2016) Veterinary antibiotics in animal waste, its distribution in soil and uptake by plants: a review. Sci Total Environ 563–564:366–376. https://doi.org/10.1016/j.scitotenv.2016.04.140

Tejada M, Gonzalez JL, García-Martínez AM, Parrado J (2008) Application of a green manure and green manure composted with beet vinasse on soil restoration: effects on soil properties. Bioresour Technol 99(11):4949–4957. https://doi.org/10.1016/j.biortech.2007.09.026

Temel FA (2023) Evaluation of the influence of rice husk amendment on compost quality in the composting of sewage sludge. Bioresour Technol 373:128748. https://doi.org/10.1016/j.biortech.2023.128748

Theeba M, Husni MA, Samsuri AW, Robert TB, Illani ZH (2016) Nutrient retention capacity of rice husk biocharcoal in co-composted poultry manure. J Trop Agric and Fd Sc 44(2):197–209

Thorup-Kristensen K, Dresbøll DB, Kristensen HL (2012) Crop yield, root growth, and nutrient dynamics in a conventional and three organic cropping systems with different levels of external inputs and N re-cycling through fertility building crops. Eur J Agron 37(1):66–82. https://doi.org/10.1016/j.eja.2011.11.004

Tran H-T, Vu C-T, Lin C, Bui X-T, Huang W-Y, Vo T-D-H, Hoang H-G, Liu W-Y (2018) Remediation of highly fuel oil-contaminated soil by food waste composting and its volatile organic compound (VOC) emission. Bioresour Technol Rep 4:145–152. https://doi.org/10.1016/j.biteb.2018.10.010

Tuomela M (2000) Biodegradation of lignin in a compost environment: a review. Bioresour Technol 72(2):169–183. https://doi.org/10.1016/S0960-8524(99)00104-2

Ucaroglu S, Ozbek B (2024) Bioconversion of organic wastes: treatment sludges, animal manures, and agricultural wastes. Biomass Convers Biorefinery. https://doi.org/10.1007/s13399-024-05883-w

Urra J, Alkorta I, Lanzén A, Mijangos I, Garbisu C (2019) The application of fresh and composted horse and chicken manure affects soil quality, microbial composition and antibiotic resistance. Appl Soil Ecol 135:73–84. https://doi.org/10.1016/j.apsoil.2018.11.005

Urugo MM, Teka TA, Gemede HF, Mersha S, Tessema A, Woldemariam HW, Admassu H (2024) A comprehensive review of current approaches on food waste reduction strategies. Compr Rev Food Sci Food Saf 23(5):e70011. https://doi.org/10.1111/1541-4337.70011

Vandecasteele B, Reubens B, Willekens K, De Neve S (2014) Composting for increasing the fertilizer value of chicken manure: effects of feedstock on P availability. Waste Biomass Valoriz 5(3):491–503. https://doi.org/10.1007/s12649-013-9264-5

Vázquez MA, Soto M (2017) The efficiency of home composting programmes and compost quality. Waste Manag 64:39–50. https://doi.org/10.1016/j.wasman.2017.03.022

Venglovsky J, Sasakova N, Vargova M, Pacajova Z, Placha I, Petrovsky M, Harichova D (2005) Evolution of temperature and chemical parameters during composting of the pig slurry solid fraction amended with natural zeolite. Bioresour Technol 96(2):181–189. https://doi.org/10.1016/j.biortech.2004.05.006

Vermont ANR (2018) Universal Recycling Law (Act 148) | summary sheet

Wang Q, Li R, Cai H, Awasthi MK, Zhang Z, Wang JJ, Ali A, Amanullah M (2016) Improving pig manure composting efficiency employing Ca-bentonite. Ecol Eng 87:157–161. https://doi.org/10.1016/j.ecoleng.2015.11.032

Wang Y, Bi L, Liao Y, Lu D, Zhang H, Liao X, Liang JB, Wu Y (2019) Influence and characteristics of Bacillus stearothermophilus in ammonia reduction during layer manure composting. Ecotoxicol Environ Saf 180:80–87. https://doi.org/10.1016/j.ecoenv.2019.04.066

Wang X, Kong Z, Wang Y, Wang M, Liu D, Shen Q (2020) Insights into the functionality of fungal community during the large scale aerobic co-composting process of swine manure and rice straw. J Environ Manag 270:110958. https://doi.org/10.1016/j.jenvman.2020.110958

Waqas M, Hashim S, Humphries UW, Ahmad S, Noor R, Shoaib M, Naseem A, Hlaing PT, Lin HA (2023) Composting processes for agricultural waste management: a comprehensive review. Processes 11(3):731. https://doi.org/10.3390/pr11030731

Wei Y, Li J, Shi D, Liu G, Zhao Y, Shimaoka T (2017) Environmental challenges impeding the composting of biodegradable municipal solid waste: a critical review. Resour Conserv Recycl 122:51–65. https://doi.org/10.1016/j.resconrec.2017.01.024

Wei Y, Yuan H, Wachemo AC, Li X (2020) Anaerobic co-digestion of cattle manure and liquid fraction of digestate (LFD) pretreated corn stover: pretreatment process optimization and evolution of microbial community structure. Bioresour Technol 296:122282. https://doi.org/10.1016/j.biortech.2019.122282

Wolka K, Melaku B (2015) Exploring selected plant nutrient in compost prepared from food waste and cattle manure and its effect on soil properties and maize yield at Wondo Genet, Ethiopia. Environ Syst Res 4(1):9. https://doi.org/10.1186/s40068-015-0035-0

Xu H, Hong C, Yao Y, Liu L, Wang W, Zhu W, Hong L, Weng J, Zhou Y, Zhu F (2021a) The process of biotransformation can produce insect protein and promote the effective inactivation of heavy metals. Sci Total Environ 776:145864. https://doi.org/10.1016/j.scitotenv.2021.145864

Xu Z, Ma Y, Zhang L, Han Y, Yuan J, Li G, Luo W (2021b) Relating bacterial dynamics and functions to gaseous emissions during composting of kitchen and garden wastes. Sci Total Environ 767:144210. https://doi.org/10.1016/j.scitotenv.2020.144210

Xu Z, Gao X, Li G, Nghiem LD, Luo W, Zhang F (2024) Microbial sources and sinks of nitrous oxide during organic waste composting. Environ Sci Technol 58(17):7367–7379. https://doi.org/10.1021/acs.est.3c10341

Yang W, Zhang L (2022) Addition of mature compost improves the composting of green waste. Bioresour Technol 350:126927. https://doi.org/10.1016/j.biortech.2022.126927

Yang F, Li GX, Yang QY, Luo WH (2013) Effect of bulking agents on maturity and gaseous emissions during kitchen waste composting. Chemosphere 93(7):1393–1399. https://doi.org/10.1016/j.chemosphere.2013.07.002

Yasmin N, Jamuda M, Panda AK, Samal K, Nayak JK (2022) Emission of greenhouse gases (GHGs) during composting and vermicomposting: measurement, mitigation, and perspectives. Energy Nexus 7:100092. https://doi.org/10.1016/j.nexus.2022.100092

Yousaf U, Ali Khan AH, Farooqi A, Muhammad YS, Barros R, Tamayo-Ramos JA, Iqbal M, Yousaf S (2022) Interactive effect of biochar and compost with Poaceae and Fabaceae plants on remediation of total petroleum hydrocarbons in crude oil contaminated soil. Chemosphere 286:131782. https://doi.org/10.1016/j.chemosphere.2021.131782

Yu X, Zhao L, Yao Z, Zhao Y, Yu J, Feng J, Luo J, Luo L, Huo L (2024) Methodological study on carbon sequestration accounting for emission reductions from the whole-chain utilization of livestock and poultry manure. Environ Res 263:120269. https://doi.org/10.1016/j.envres.2024.120269

Yuan J, Yang Q, Zhang Z, Li G, Luo W, Zhang D (2015) Use of additive and pretreatment to control odors in municipal kitchen waste during aerobic composting. J Environ Sci 37:83–90. https://doi.org/10.1016/j.jes.2015.03.028

Zahra K, Farhan M, Kanwal A, Sharif F, Hayyat MU, Shahzad L, Ghafoor GZ (2023) Investigating the role of bulking agents in compost maturity. Sci Rep 13(1):16003. https://doi.org/10.1038/s41598-023-41891-y

Zakarya IA, Muniandy S, Tengku Izhar TN, Md Nasir NAN, Mohamad M, Tudor G (2023) Performance on nitrogen rich component for composting of food waste. E3S Web Conf 437:04003. https://doi.org/10.1051/e3sconf/202343704003

Zalewska M, Błażejewska A, Czapko A, Popowska M (2021) Antibiotics and antibiotic resistance genes in animal manure – consequences of its application in agriculture. Front Microbiol 12:610656. https://doi.org/10.3389/fmicb.2021.610656

Zhang L, Sun X (2014) Changes in physical, chemical, and microbiological properties during the two-stage co-composting of green waste with spent mushroom compost and biochar. Bioresour Technol 171:274–284. https://doi.org/10.1016/j.biortech.2014.08.079

Zhang L, Sun X (2016) Improving green waste composting by addition of sugarcane bagasse and exhausted grape marc. Bioresour Technol 218:335–343. https://doi.org/10.1016/j.biortech.2016.06.097

Zhang L, Sun X, Tian Y, Gong X (2013) Effects of brown sugar and calcium superphosphate on the secondary fermentation of green waste. Bioresour Technol 131:68–75. https://doi.org/10.1016/j.biortech.2012.10.059

Zhang L, Li L, Sha G, Liu C, Wang Z, Wang L (2020a) Aerobic composting as an effective cow manure management strategy for reducing the dissemination of antibiotic resistance genes: an integrated meta-omics study. J Hazard Mater 386:121895. https://doi.org/10.1016/j.jhazmat.2019.121895

Zhang W, Yu C, Wang X, Hai L (2020b) Increased abundance of nitrogen transforming bacteria by higher C/N ratio reduces the total losses of N and C in chicken manure and corn stover mix composting. Bioresour Technol 297:122410. https://doi.org/10.1016/j.biortech.2019.122410

Zhao Y, Li W, Chen L, Meng L, Zheng Z (2020) Effect of enriched thermotolerant nitrifying bacteria inoculation on reducing nitrogen loss during sewage sludge composting. Bioresour Technol 311:123461. https://doi.org/10.1016/j.biortech.2020.123461

Zhao Y, Zhuge C, Weng Q, Hu B (2022) Additional strains acting as key microbes promoted composting process. Chemosphere 287:132304. https://doi.org/10.1016/j.chemosphere.2021.132304

Zhou L, Li S, Li F (2022) Damage and elimination of soil and water antibiotic and heavy metal pollution caused by livestock husbandry. Environ Res 215:114188. https://doi.org/10.1016/j.envres.2022.114188

# Earthworm-Mediated Bio-composting and Basic Transformation Mechanisms in Organic Matter

**Elisabete Nascimento-Gonçalves, Tiago Azevedo, Catarina Medeiros, Maria Cristina Morais, Anabela R. Reis, João Ricardo Sousa, Paula A. Oliveira, Ana M. Coimbra, and Marta Roboredo**

## 1 Introduction

Sustainable agriculture and efficient waste management are crucial for addressing environmental challenges and ensuring food security (Wang et al. 2021). As the global population increases, residue generation across various human activities also rises (Singh et al. 2014; Soukiazis and Proença 2020; Shah et al. 2023) and, combined with improper management and disposal practices, leads to environmental pollution and depletion of natural resources (Hajam et al. 2023). These circumstances require adaptation and mitigation measures, and consequently, there is a growing need to transform residues into value-added products through innovative techniques. Vermicomposting is a highly efficient process of bio-oxidation and stabilization of organic residues through a symbiotic relationship between earthworms and mesophilic microorganisms (Abad and Shafiqi 2024). Earthworms act as

Authors Elisabete Nascimento-Gonçalves, Tiago Azevedo, and Catarina Medeiros have equally contributed to this chapter.

E. Nascimento-Gonçalves · M. C. Morais · J. R. Sousa · P. A. Oliveira (✉) · A. M. Coimbra · M. Roboredo
Centre for the Research and Technology of Agro-Environmental and Biological Sciences (CITAB), University of Trás-os-Montes and Alto Douro (UTAD), Vila Real, Portugal

Institute for Innovation, Capacity Building and Sustainability of Agri-Food Production (Inov4Agro), UTAD, Vila Real, Portugal
e-mail: pamo@utad.pt

T. Azevedo · C. Medeiros
Centre for the Research and Technology of Agro-Environmental and Biological Sciences (CITAB), University of Trás-os-Montes and Alto Douro (UTAD), Vila Real, Portugal

A. R. Reis
CGeo – Geosciences Centre, University of Coimbra (Polo II), Coimbra, Portugal



R. K. Singh et al. (eds.), *Composting of Farm Residues by Nature-Based Solutions*,
https://doi.org/10.1007/978-3-032-12255-1_3

the primary drivers of the vermicomposting process (Abad and Shafiqi 2024; Santana et al. 2020; Domínguez et al. 2019), as they promote aeration, fragmentation, and digestion of the residues while creating optimal conditions for the microbial biochemical degradation of organic matter (Hanc et al. 2019). The joint activity of earthworms and microbes transforms organic residues, such as food remnants, agricultural waste, and animal manure, into stabilized humus, a microbiologically active organic amendment enriched in macronutrients, micronutrients, vitamins, growth hormones, and enzymes (e.g., proteases, amylases, lipases, cellulases, and chitinases) (Abad and Shafiqi 2024; Ma et al. 2024; Gupta et al. 2022).

The vermicomposting process is typically divided into two stages related to earthworm activity (Gómez-Roel et al. 2024). During the first stage, known as the active stage (Domínguez et al. 2021), earthworms ingest and fragment the organic waste (Santana et al. 2020). After being broken down, organic matter moves through the earthworm's gut, where it mixes with gut microbiome and digestive enzymes to form a partially digested product called vermicast or worm cast (Lazcano et al. 2008). This process results in modifications of the physical, chemical, and microbial composition of the organic matter (Hanc et al. 2019). The duration of this stage is variable and depends on several factors, including the earthworm species, their density, and the rate of organic matter consumption and processing (Lazcano et al. 2008), as well as the type of organic residue introduced (Baral et al. 2015). Throughout the second stage, known as the maturation or aging stage, earthworm activity is reduced, and microbial activity prevails (Aira et al. 2007; Aira et al. 2011). In this stage, occurring after the excretion of the vermicasts, the transformation of the organic compounds digested by earthworms continues (Gómez-Brandón et al. 2020), resulting in a final product with a high nutrient content, along with decreased microbial populations and lower enzyme activities (Hanc et al. 2020). The active stage is linked to the movement of residues through the earthworms' intestine (i.e., gut-associated processes (GAPs)), while the maturation stage corresponds to the aging of the worm casting (i.e., cast-associated processes (CAPs)) (Gómez Brandón et al. 2019).

Vermicomposting is a self-regulating, self-sustaining, and self-improving (Singh and Sinha 2022), ancient yet innovative, biotechnological process that not only provides an efficient solution for organic residue management but also yields a nutrient-rich soil organic amendment (Abad and Shafiqi 2024; Ma et al. 2024). The joint activity of earthworms and microorganisms is crucial for maintaining optimal conditions for the decomposition of organic matter over time (Suthar 2009). Furthermore, the continuous addition of organic residues sustains the earthworm population, allowing it to maintain a high level of efficiency (Dominguez and Edwards 2010), thus contributing to the improvement of the process.

The addition of vermicompost to soil enhances the soil structure, improves soil fertility, and promotes overall soil health and quality (Abad and Shafiqi 2024; Gazi et al. 2024; Danish Toor et al. 2024), contributing to faster seed germination, accelerated seedling growth and development, and increased yield of various agricultural crops, highlighting its multifaceted benefits for agricultural productivity (Sundararasu 2019; Thirunavukkarasu et al. 2023; Azarmi et al. 2008; Zuo et al.

2018). Earthworm-mediated bio-composting facilitates residue valorization, enhances soil restoration, and promotes climate-resilient crops (Toor et al. 2024).

This chapter aims to discuss the importance of managing organic residues within the context of sustainable agriculture, the potential of vermicomposting, and the contribution of earthworms to producing high-quality organic amendments for enhancing soil health and quality and crop productivity.

## 2 Earthworm Species

Earthworms are eucoelomate, vermiform, metamerically segmented, and hermaphroditic invertebrates, scientifically classified under the phylum Annelida and class Oligochaeta (Jesna and Livingstone 2023; Peramban et al. 2023). These organisms are widespread across the globe, except in regions with extremely harsh conditions, such as severe drought, flooding, and extreme cold (e.g., Antarctica) (Jesna and Livingstone 2023; Peramban et al. 2023; Domínguez 2018; Misirlioğlu et al. 2023; Singh et al. 2019). In many temperate ecosystems, earthworms often represent the largest soil macroinvertebrate biomass and play a key role in shaping the physical, chemical, and biological properties of the soil (Jesna and Livingstone 2023; Peramban et al. 2023; Domínguez 2018; Misirlioğlu et al. 2023; Singh et al. 2019). They have been successfully used in vermicomposting processes to recycle urban, industrial, and agricultural organic residues for the production of organic fertilizers and as a protein source for animal feed (Ma et al. 2024; Lirikum et al. 2022).

Comprising over 6000 existing species, earthworms can be categorized into three ecological groups: epigeic, anecic, and endogeic (Vuković et al. 2021), based on their unique physical and behavioral traits (Peramban et al. 2023; Domínguez 2018). Epigeic earthworms are pigmented, inhabit the surface litter layer, lack permanent burrows, and feed on litter and decaying organic materials (Vuković et al. 2021). Endogeic earthworms, on the other hand, have little pigmentation, and they are found deeper in the soil profile, where they feed on soil and the associated organic matter and build complex horizontal burrows (Domínguez 2018). Anecic species are relatively large and darker brown and construct vertical burrows extending several meters into the soil profile. They surface at night to feed on litter, animal feces, and decomposed organic matter, which they transport into their burrows (Domínguez 2018; Vuković et al. 2021).

Several earthworm species have been evaluated for their potential use in vermicomposting through their capacity to decompose organic matter, including *Eisenia fetida*, *Eisenia andrei*, *Dendrobaena veneta*, *Dendrobaena hortensis*, *Eudrilus eugeniae*, and *Perionyx excavatus* (Domínguez 2018; Lirikum et al. 2022; Khwairakpam and Bhargava 2009). Among these, the epigeic species *E. fetida* and *E. andrei* are the most commonly used in vermicomposting systems around the world (Domínguez 2018; Vuković et al. 2021; Khwairakpam and Bhargava 2009) due to their exceptional tolerance to varying environmental conditions like temperature and moisture, high feeding and reproductive rates, short life cycles, and ease of management (Nascimento-Gonçalves et al. 2024; Bhat et al. 2018). *E. andrei* and *E. fetida* are able to consume and digest large amounts of organic residues quickly and efficiently, thus producing holorganic casts. They are the preferred species and are invaluable to the vermicomposting systems (Domínguez 2018; Lirikum et al. 2022; Mupambwa and Mnkeni 2018).

## 3 Organic Matter Transformation Mechanisms in Vermicomposting

Organic matter transformation in vermicomposting is a dynamic biological process, involving the physical breakdown of residues by earthworms and the biochemical degradation by microorganisms and earthworm-derived enzymes (Fig. 1).

### *3.1 Physical Processes*

Physical degradation is a fundamental step in the vermicomposting process. The movement of earthworms through the substrate plays an essential role in mixing and aeration. Their burrowing, tunneling, and feeding activities continuously mix fresh organic residues with the partially decomposed material, redistributing nutrients and microbial populations, ensuring a homogeneous decomposition environment while maintaining aerobic conditions (Vuković et al. 2021; Edwards and Arancon 2022a). This turnover improves substrate porosity, stability, and aggregation, facilitating the penetration of water, oxygen, and microbes (Vuković et al. 2021; Domínguez et al. 2003; Domínguez 2004). Aeration supports the metabolic activity of aerobic microorganisms and prevents undesirable odors and the production of greenhouse gases (such as methane or nitrogen gases) (Chan et al. 2011). Further physical degradation

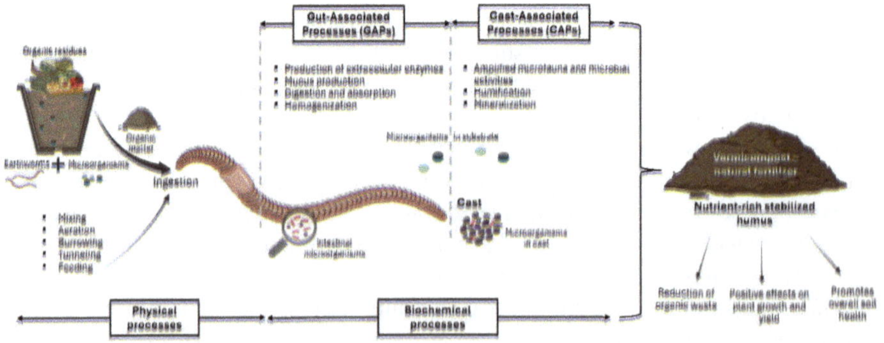


**Fig. 1** Overview of the vermicomposting process: physical and biochemical processes

occurs as earthworms consume organic matter, which is ground within their muscular gizzard, where ingested particles are mixed with ingested soil, sand, and gut secretions, resulting in smaller particles with greater accessibility to microbial communities (Shakeel et al. 2024; Kiyasudeen et al. 2020). These mechanical actions enlarge the surface area available for microbial activity, facilitate the organic matter turnover, and enhance aeration (Lirikum et al. 2022; Vuković et al. 2021).

## 3.2 *Biochemical Processes*

Beyond ingestion, biochemical processes occurring in the organic matter involve enzymes produced by earthworms and their gut microbiome (Anand and Sinha 2020). These processes take place during two main stages: gut-associated processes (GAPs) and cast-associated processes (CAPs).

### 3.2.1 Gut-Associated Processes (GAPs)

Initially, biochemical transformations occur within the earthworm's gut, where decaying organic matter and associated microorganisms undergo various modifications (Domínguez et al. 2010). These include nutrient transformation, changes in microbial diversity and activity, homogenization of the organic material, and mucus production, digestion, and assimilation (Domínguez 2004). The selective effects of gut passage on microbes may result from the competition between microorganisms that are ingested and the endosymbiotic microbes that live in the gut (Gómez-Brandón and Domínguez 2014). Endosymbiotic microbes contribute to decomposition by producing extracellular enzymes, which degrade complex organic molecules, such as cellulose and phenolics (Domínguez et al. 2010). For instance, cellulose and hemicellulose are hydrolyzed into simpler carbohydrates (monosaccharides), proteins are metabolized into amino acids, and lipids are broken down into simpler fatty acids and glycerol (Wang et al. 2024a). Subsequently, the partially digested organic matter is expelled from the earthworm's body in the form of casts, enriched with microbial populations (Domínguez et al. 2010).

### 3.2.2 Cast-Associated Processes (CAPs)

Once expelled, earthworm casts offer an optimal environment for microbial colonization (Gómez-Brandón and Domínguez 2014). This transition marks the beginning of CAPs, in which the surrounding microbial communities are further stimulated by favorable moisture levels and the intense mixing that occurred within the gut (Domínguez et al. 2010). Casts show distinct microbial communities and nutrient compositions compared to the original organic material prior to ingestion. The production of organic acids, which lowers the pH around the decomposing material, enhances the dissolution of P, K, and Ca compounds (Wang et al. 2024a), and organic matter decomposition is further stimulated, with casts exhibiting high mineralization rates (Domínguez 2004; Domínguez et al. 2010). Besides, throughout these processes, the organic matter is continuously stabilized and humified (Wang et al. 2022). During CAPs, humic substances—humic acids, fulvic acids, and humins—are formed through the biochemical degradation and stabilization of organic matter by microbial metabolism and chemical condensation reactions (Wang et al. 2022; Dores-Silva et al. 2018). This process entails the progressive decomposition of more labile organic compounds, accompanied by an increase in aromatic carbon, relative to aliphatic carbon, indicating a higher degree of humification (Aguiar et al. 2013; Jindo et al. 2012). Over time, the combined action of microbial metabolism and biochemical interactions results in a structurally and chemically altered organic matrix, with improved stability and soil-conditioning properties (Anand and Sinha 2020; Dores-Silva et al. 2018).

In summary, the interplay of mechanical, enzymatic, and microbial actions leads to significant changes in the organic matter's composition and quality. In terms of physical properties, the substrate becomes finer and more homogeneous due to fragmentation. Biochemically, the transformation of organic matter is driven by the sequential action of GAPs and CAPs, where earthworms and their gut microbiota initiate decomposition and casts provide a nutrient-rich, microbially active environment that sustains further degradation. Additionally, microbial diversity and activity are enhanced, contributing to the suppression of pathogens and the promotion of beneficial soil microorganisms.

As a result, the mature vermicompost presents increased nutrient levels and a reduced C:N ratio, indicative of high stability and maturity.

## 4 Advantages and Limitations of Vermicomposting Compared to Composting

When considering organic residue management strategies, traditional composting is still more widely implemented than vermicomposting, even though both approaches are effective in recycling organic materials into nutrient-rich soil amendments, with the potential to address most of the major constraints to agricultural production (Abbott et al. 2018). Vermicomposting has both advantages and limitations over traditional composting (Table 1), though results depend on the input materials and technology used. Vermicompost processing is faster, having a higher nutrient content and improved pathogen reduction, while being more sensitive to temperature and requiring higher initial costs. The choice between methods depends on resources, climate, and residue management goals, with a combined approach (pre-composting, followed by vermicomposting) often yielding the best results in terms of processing time, nutrient quality, and humus content (Dume et al. 2023; Fornes et al. 2012; Yu et al. 2022). These topics will be examined in detail in the

**Table 1** Main differences between traditional composting and vermicomposting

| | Vermicomposting | Traditional composting |
|---|---|---|
| Biological agents | Earthworms and microorganisms | Microorganisms |
| Processing time | Faster | Slower |
| Nutrient | Higher | Lower |
| Pathogen | More effective | Less effective |
| Temperature | Narrow range (<25 °C) | Wider range (45–65 °C) |
| Moisture | 70–90% | 40–60% |
| Initial investment | Higher due to earthworm acquisition and environmental control | Lower, with minimal infrastructure needs |
| Maintenance requirements | Higher due to environmental control and monitoring | Lower, with less intervention required |
| Final product | Fine and homogeneous, with higher humus content | Coarse and agglomerated |

following sections of this subchapter, providing a comprehensive analysis of the factors that influence the choice between traditional composting and vermicomposting.

## 4.1 *Process Dynamics and Duration*

Traditional composting relies solely on indigenous thermophilic and mesophilic microorganisms to transform the organic matter into stabilized and sanitized compost (Bohacz 2018). An early thermophilic phase ensures sanitization, and the later mesophilic and cure phases, also known as maturation, slowly promote the degradation of recalcitrant organic compounds, generating humic acids and eliminating phytotoxicity (Kuhad et al. 2011). Composting usually requires moisture and temperature control and physical aeration to ensure aerobic conditions (Fornes et al. 2012). In contrast, vermicomposting involves the synergistic action of earthworms and microorganisms and is usually divided into two distinct stages. During the first stage, also known as the active phase, earthworms process the organic residues, which changes their physical structure and microbial composition. Then, during the second stage, called the maturation phase, microorganisms complete the decomposition process as the earthworms move on to fresher layers (Gómez-Brandón et al. 2011a), as previously described. This process only requires moisture control, as the earthworms' activity promotes substrate aeration.

Regarding the duration of the processes, the literature does not present a clear consensus, as vermicomposting is often described as a faster process without robust scientific validation. The duration of each process depends highly on factors such as the composition and quantity of the initial residues, among others (Campitelli and Ceppi 2008). For instance, Fornes et al. (2012) described substantial differences in the duration of composting and vermicomposting of tomato crop residues, with the former taking 180 days and the latter extending up to 261 days to be considered mature (Fornes et al. 2012). Additionally, several authors argue that a pre-composting phase can help mitigate the unpredictability of the vermicomposting duration by improving earthworm activity conditions (e.g., reducing potential toxicity and pathogen load) and, in turn, enhancing compost quality (Nair et al. 2006; Sharma and Garg 2020; Kauser and Khwairakpam 2022).

## *4.2 Physicochemical Properties and Nutrient Transformation*

Both processes are significantly influenced by the initial C:N ratio, which reflects the proportions of elements required for microbial growth and activity, directly impacting the organic matter mineralization and other processes such as nitrification (Nigussie et al. 2016; Hasnat et al. 2022). Generally, an initial C:N ratio between 25 and 30 is regarded as optimal, as it maintains a balance between N mineralization and immobilization, stimulating microbial activity while minimizing N losses (Getino-Álvarez et al. 2023). In vermicomposting, maintaining this ratio is important for protecting earthworms from elevated $NH_4^+$ levels, particularly when pre-composting is not done (Hughes et al. 2008). In traditional composting, due to the rapid C oxidation during the thermophilic phase, mature compost typically exhibits a lower C:N ratio, approximately 10 (Wang et al. 2024b; Zhao et al. 2024; EL-Mogy et al. 2024), while the mature vermicompost usually has a final C:N ratio ranging from 10 to 20 (EL-Mogy et al. 2024; Majlessi et al. 2012), depending on the initial materials and process conditions. Vermicompost, offering a slightly higher and more balanced C:N ratio, acts as a slow-release nutrient fertilizer, thereby supporting sustained plant growth and long-term soil health (Ozer et al. 2024).

Earthworms are also more sensitive to pH changes since extreme values can cause death and substrate avoidance. Thus, vermicomposting may require a thorough selection of initial materials (Kaur 2020; Ali et al. 2015; Garg and Gupta 2011). Usually, vermicompost presents a slightly lower pH compared to traditional compost, which may result from N and P mineralization and microbial decomposition that produces intermediate organic, fulvic, and humic acids (Lazcano et al. 2008). A fertilizer with mild acidity might be considered advantageous in calcareous soils or as a plant growth medium (Cáceres et al. 2016; Afshari et al. 2023). Vermicompost also has lower electrical conductivity values than compost, reducing phytotoxicity risks, likely due to more efficient nutrient cycling, lower $NH_4^+$ concentrations resulting from immobilization into microbial biomass, and the precipitation of dissolved salts (Lazcano et al. 2008; Frederickson et al. 2007).

These factors, C:N ratio, initial pH, and electrical conductivity, require careful consideration as they impact the earthworm activity and shape the resident microbial communities, thereby influencing the efficiency of the degradation processes, and consequently determining the quality for soil application.

The nutrient composition of the mature vermicompost typically presents higher levels of N, P, K, Ca, and Mg, as well as micronutrients such as Fe, Zn, Mn, Cu, and B, in comparison with traditional compost (Dume et al. 2023; Pottipati et al.

2022; Patra et al. 2022). Regarding N, vermicomposting generally leads to faster and more efficient mineralization (Dume et al. 2023; Tognetti et al. 2005) due to the combined action of earthworms and microorganisms, promoting an environment rich in ammonia-oxidizing microorganisms and depleted in denitrifying bacteria (Lv et al. 2019). Phosphorus transformations in soil are predominantly controlled by the activity of phosphatase-producing microorganisms. Vermicompost is especially rich in these microorganisms, thereby enhancing P bioavailability through phosphatase-mediated mineralization and microbial solubilization processes (Chiba et al. 2024). However, the inoculation of compost and vermicompost with phosphate-solubilizing bacteria has been identified as a strategy even more effective for increasing the bioavailable P pool, facilitating P assimilation by plants, and improving overall soil fertility (Parastesh et al. 2019; Wu and Wan 2023).

Humification, which occurs mainly during the maturation phase, is considered faster and more efficient in vermicomposting due to earthworm digestion and enhanced microbial activity, resulting in fine, peat-like particles rich in humic acids (Aguiar et al. 2013; Gong et al. 2018). Vermicompost, through the interactions between earthworms, microorganisms, and organic matter, is also richer in plant growth regulators, such as auxins, gibberellins, and cytokinins (Oyege and Balaji Bhaskar 2023; Rehman et al. 2023; Zhang et al. 2000).

## *4.3 Microbial Communities*

Microbial communities are affected by the composting and vermicomposting processes. Vermicomposting increases the abundance of beneficial microorganisms, thereby enhancing soil health, supporting nutrient cycling, and promoting disease suppression (Pathma and Sakthivel 2012). Microbial composition is dependent on the initial materials. While in composting, the main identified phyla are Actinobacteria and Firmicutes (Fracchia et al. 2006), in vermicomposting, occurring under mesophilic conditions, Actinobacteria (e.g., *Streptomyces* spp.), Firmicutes (e.g., *Bacillus* spp.), Proteobacteria, Bacteroidetes, and fungi such as *Trichoderma* spp. are the most abundant (Domínguez et al. 2019; Koskey et al. 2023; Usmani et al. 2019). The gut microbiota of earthworms is rich in antagonistic microorganisms that suppress soilborne pathogens' activity by producing secondary metabolites, such as siderophores, and lytic enzymes, which induce plant host resistance (Grantina-Ievina et al. 2013; Pathma and Sakthivel 2013; Javoreková et al. 2019; Ascencion et al. 2015). Additionally, the diverse microbial community associated with earthworms plays a crucial role in promoting plant growth by facilitating the solubilization of nutrients and increasing their bioavailability for plant development, (Dume et al. 2023; Tognetti et al. 2005; Parastesh et al. 2019; Wu and Wan 2023), as well as through the biosynthesis of phytohormones, such as indole 3-acetic acid, gibberellic acid, and kinetin (Ravindran et al. 2016; Zhang et al. 2015).

## 4.4 Pathogen Reduction and Plant Disease Suppression

For compost and vermicompost to be considered safe for soil application, they must undergo sanitization to ensure the absence of pathogenic microorganisms, including *Salmonella* spp., enteric viruses, viable ova of parasitic helminths, and indicator organisms like fecal coliform (*Escherichia coli*) and streptococci (*Enterococcus* spp.). This reduction in microbial load is critical to prevent the contamination of food crops via soil–plant pathways, mitigating potential zoonotic and anthropozoonotic transmission risks to human and animal populations. Compliance with established microbiological safety thresholds is essential to meet regulatory standards for organic soil amendments.

The presence of pathogens is largely dependent on the composition and contamination level of the initial materials used. Both composting and vermicomposting effectively reduce the load of pathogenic microorganisms but through different mechanisms. In traditional composting, the high temperatures (45–70 °C) eliminate pathogens during the thermophilic phase (Fornes et al. 2012). Proper management, aeration, and frequent turning prevent the formation of cold spots where pathogenic microorganisms may persist, thus ensuring sanitization (Avery et al. 2012; Ahn et al. 2007). Composting is highly effective against heat-sensitive pathogens, yet it may favor the selection of thermotolerant strains or fail to eliminate them due to poor homogenization of the compost pile (Peng et al. 2024; Inglis et al. 2010; Singh et al. 2010).

Vermicomposting, on the other hand, relies on microbial antagonism and the enzymatic digestion in the earthworm gut, along with the action of earthworms' immune system (coelomic cells and secreted fluids), to eliminate harmful pathogens and inactivate viruses (Hua et al. 2011; Monroy et al. 2009; Roubalová et al. 2020; Ghandehari Yazdi et al. 2024). For example, in comparison to composting, vermicomposting is more effective in the overall reduction of pathogens in municipal solid waste and sewage sludge (Domínguez et al. 2021; Soobhany 2018).

Both vermicompost and compost can also suppress plant pathogens, primarily through microbial antagonism between their microbiome and pathogenic organisms, namely *Phytophthora nicotianae* (Szczech and Smolińska 2001; Blaya et al. 2016), *Meloidogyne incognita* (Ahamad et al. 2023; Zuhair et al. 2022; Siddiqui 2004), *Rhizoctonia solani* (González-Hernández et al. 2022; Simsek Ersahin et al. 2009), and *Fusarium oxysporum* (Zhao et al. 2019; Suárez-Estrella et al. 2007), among many others. Additionally, earthworm coelomic fluid exhibits antifungal and antibacterial properties (Sethulakshmi et al. 2018; Ečimović et al. 2021; Plavšin et al. 2017), but its efficacy varies between earthworm species, adding an additional layer of variability to vermicomposting outcomes that needs to be considered.

Given these factors, if (vermi)composts were to be used as alternatives to pesticides, they could significantly reduce agricultural workers' exposure to chemicals (Singh et al. 2024). However, their effectiveness appears to depend more on the process management, raw materials, the earthworm species (in the case of vermicomposting), and the specific microbial communities, with vermicomposting potentially offering greater pathogen reduction through its unique biological mechanisms.

## 4.5 Metal Bioavailability

Metals may be present in different phases associated with the mineral and organic fractions in municipal and industrial residues, as well as in vermicompost (Swati and Hait 2017; Förstner 2004): The available forms include, in decreasing order of mobility: metals weakly adsorbed by electrostatic forces; metals bound to carbonates or acid extractable (susceptible to changes in pH); in reducible fraction, unstable under anoxic conditions; metals bound to organic matter (which can be an be oxidized); and the residual fraction, which is permanently included in the crystal lattice of the mineral components, either original or formed during vermicomposting.

The activity of earthworms goes beyond organic matter decomposition. During the process of vermicomposting, hydrolytic enzymes secreted by the earthworms' digestive system and associated microbial communities release metals bound to ions and carbonates (in the mineral fractions or bound to organic matter in residues). This leads to an increase in mobile/available metal fractions (Morgan and Morgan 1999), which are easily absorbed by the cutaneous tissues of earthworms and accumulate there. Also, to reduce metal toxicity, earthworms produce phytochelatin and metallothionein, molecules that are involved in homeostasis and the detoxification of inorganic materials (Hussain et al. 2021). Phytochelatin, produced in response to metals such as cadmium or metalloids such as arsenic, chelates these ions using glutathione (Liebeke et al. 2013). Additionally, metallothionein neutralizes metals like copper, nickel, and zinc by forming chelation products (Hussain et al. 2021).

As such, earthworms reduce metal bioavailability, either by bioaccumulation or by excretion in less mobile or bioavailable forms in the casts (Li et al. 2009; Suthar and Singh 2008). However, in the presence of high metal contents, to overcome the stress condition, earthworms increase their metal excretion rate (Li et al. 2009), which increases the metal content in the vermicasts (and vermicompost).

During the mineralization of organic matter, metals bind to the humic acids produced, forming stable organo-metal complexes (chelates), or remain as silicate fractions (Wang et al. 2013; Hait and Tare 2012), leading to metal immobilization. Increased metal levels in vermicompost relative to the initial values may indicate that the effect of concentration caused by organic matter stabilization is greater than the bioaccumulated metals in earthworms during vermicomposting (Rékási et al. 2023; Lv et al. 2016; He et al. 2016).

Metals in vermicompost may be present in different chemical forms with different mobility and bioavailability potential. Earthworms' activity decreases metal mobility by changing their chemical form during vermicomposting. Microbial communities in vermicasts, by enhancing enzymatic activity, also contribute to the decrease of metal mobility. This may allow vermicomposts to properly match the agricultural threshold limits (Singh and Kalamdhad 2016). For example, *E. fetida* and *E. eugeniae* have been shown to remediate Zn, Cu, and Pb in municipal solid waste and textile sludge. The use of these earthworm species in vermicomposting resulted in a significantly higher remediation rate of these elements in comparison with aerobic composting (Paul et al. 2020) due to the presence of earthworm-specific biochemical detoxification mechanisms.

## 4.6 Operational Requirements and Scalability

Despite the previously mentioned advantages, vermicomposting has some limitations compared with traditional composting, primarily due to the scalability and requirements of the process. Earthworms depend upon specific environmental conditions, including a dark environment with optimal temperature (15–25 °C), moisture (75–90%), and pH (close to neutral) and low $NH_4^+$ and salt levels (Kaur 2020). They are sensitive to temperature fluctuations, with their activity declining above 30 °C due to increased microbial activity that results in a depleted oxygen environment (Lirikum et al. 2022). Additionally, moisture content must be controlled and maintained between 75% and 90% for optimal growth and reproductive response, as the surrounding soil water potential influences the water content of earthworms (Lirikum et al. 2022). Environmental conditions limit the type and quantity of organic residues that can be processed and may require an adaptation period for the earthworms to certain residues. In contrast, traditional composting is less constrained because it does not need to meet the specific biological needs of earthworms (Liang et al. 2003).

Vermicomposting may need a larger surface area than traditional composting, as most earthworm species primarily operate at the substrate's surface (Edwards and Arancon 2022b). Initial setup costs may be higher due to the expense of acquiring and maintaining a healthy earthworm population (Lim et al. 2016). However, traditional composting facilities may also require infrastructure investment to manage aeration, turning, and temperature (Manea et al. 2024). Additionally, manual extraction of earthworms is labor-intensive, particularly in large-scale operations, posing a logistical challenge for automation (Gutiérrez-López et al. 2016).

To address these challenges, smart vermicompost reactors are being developed to automate earthworm extraction and regulate key parameters like moisture and temperature (Ghorbani et al. 2021; Hu et al. 2021a). While traditional composting remains the preferred choice for large-scale operations due to its ability to handle high volumes efficiently in less space (Stipniece et al. 2022), vermicomposting offers a more sustainable approach for smaller-scale applications, with the added benefit of producing a higher-quality amendment. Its scalability remains a key limitation, and widespread adoption will depend on continued technological advancements that improve efficiency and reduce labor.

# 5 Examples of Residues Used in Vermicomposting

Vermicomposting can use a wide range of organic residues. These include residues derived from agricultural practice, garden residues, kitchen waste, animal manure, industrial and commercial residues, and forestry and agro-industrial by-products to produce nutrient-rich vermicompost (Mohite et al. 2024; Ma et al. 2022). Pretreatments, such as chopping or partial decomposition, often help accelerate the vermicomposting process, while the combination of different residues, based on their chemical compositions, can optimize nutrient balance, decomposition rates, and vermicompost quality. The choice of residues should consider the availability, composition, pretreatment needs, and environmental risks. However, it is essential to ensure that materials are free from nonbiodegradable items, such as plastics, glass, and metals, that are not suitable for vermicomposting (Adam et al. 2024; Peneva et al. 2025). It is also necessary to be careful with certain organic residues, including acidic fruits, garlic, meat, bones, processed foods, and dairy products (Stanford et al. 2009).

## *5.1 Agricultural Residues and Agro-Industrial by-Products*

Sustainability in agriculture stresses the effective use of resources, reduction of waste, and improvement of soil health and quality (Sharma et al. 2024). Agricultural residues, including crop residues, prunings, straw, husks, and chaff, are among the most common substrates for vermicomposting (Nascimento-Gonçalves et al. 2024; Mohite et al. 2024). These materials are typically rich in lignocellulose, providing a steady source of carbon for microbial activity. However, their high C:N ratio often slows decomposition and requires them to be mixed with nitrogen-rich residues, such as animal manure. For example, rice husks and wheat straw, when combined with livestock manure, have shown enhanced decomposition rates and nutrient profiles in vermicompost (Srivastava et al. 2024; Cao et al. 2021). Similarly, agro-industrial by-products, such as sugarcane bagasse, grape marc, wine cellar by-products, coffee pulp, and olive pomace, are increasingly recognized as valuable inputs for vermicomposting (Nascimento-Gonçalves et al. 2024; Garg and Gupta 2009). These materials often contain high levels of lignin and polyphenols, which present specific challenges for the vermicomposting process (Ganesh et al. 2009). Lignin is a complex polymer that is difficult for microorganisms and earthworms to break down, slowing down the process (Ganesh et al. 2009). Some phenolic compounds can be toxic to earthworms, affecting their health and activity. In addition, polyphenols can also inhibit the activity of microorganisms essential for organic matter breakdown, reducing vermicomposting efficiency (Pundee et al. 2023). Partial composting or combining different substrates can overcome these limitations, producing high-quality vermicompost while minimizing negative effects on earthworms (Vyas et al. 2022; Xu et al. 2023). For example, vermicomposting effectively enhanced the humification of organic matter in cattle manure, especially when combined with olive pomace. This improves the quality of these materials as organic soil amendments (Plaza et al. 2008).

## 5.2 *Livestock Manure*

Animal manures, such as those from cows, pigs, poultry, and goats, are rich in nutrients like N that support plant growth, making them valuable resources for agricultural purposes (Shober and Maguire 2018). In the context of vermicomposting, these manures are excellent substrates for earthworms, as the N content supports microbial activity and accelerates the decomposition of organic matter (Nasiru et al. 2014).

However, there are a few challenges associated with using fresh manure directly in vermicomposting that must be considered (Moinard et al. 2021). Poultry manure may be less appropriate due to its high $NH_4^+$ levels and elevated salinity, which can negatively affect earthworm activity (Ravindran et al. 2017). Pig manure may contain high concentrations of metals and antibiotic residues, which can disrupt the microbiota of the vermicompost (Rasschaert et al. 2020). Cow manure is relatively suitable but may have moderate $NH_4^+$ levels and drug residues, which could impact decomposition (Nleya et al. 2023). Goat manure stands out as one of the most suitable options for vermicomposting, as it generally contains lower $NH_4^+$ levels and lower salinity, making it the safest substrate for earthworms (Loh 2005).

In manure, the presence of pathogens, such as bacteria, or pharmaceutical contaminants also carries risks for human handling and may affect the earthworms (Wang et al. 2024c).

As previously mentioned, to adequately vermicompost livestock manure, it is crucial to balance the C:N ratio through the addition of carbon-rich materials (e.g., straw, sawdust, or biochar) (Cao et al. 2021). The strategy of combining different materials is a valuable tool for reducing the environmental footprint of livestock farming, converting animal residues into a resource that supports soil health and reduces the need for synthetic fertilizers.

## 5.3 *Industrial and Wastewater Treatment Plant Residues*

Vermicomposting has also gained attention as a sustainable management approach for industrial and wastewater treatment plant residues. These often contain a mix of organic and inorganic materials, making them suitable candidates for vermicomposting, provided that certain precautions are taken to avoid environmental and health risks.

Industrial residues, such as paper mill sludge, brewery residues, and textile effluents, are particularly rich in organic matter and thus valuable substrates that can be converted into nutrient-rich humus through vermicomposting (Badhwar et al. 2021). For example, vermicompost derived from paper sludge has been shown to significantly enhance ginseng root yield when compared to untreated soil (Eo and Park 2019). Vermicomposting not only reduces the volume of industrial residues but also contributes to environmental sustainability by producing high-quality organic fertilizer. However, industrial residues may contain harmful chemicals, metals, dyes, and other contaminants (Yadav and Garg 2011), and monitoring vermicompost composition is often necessary to ensure that no hazardous substances, which could affect soil health or enter food chains, remain.

In the case of wastewater treatment plant residues, sewage sludge (biosolids) can be an abundant substrate for vermicomposting. Sewage sludge contains significant amounts of organic matter and nutrients, such as N, P, and K, beneficial for soil fertility (Nunes et al. 2021). Due to the high N content of sewage sludge, the addition of bulking materials is often necessary to balance the C:N ratio. This practice accelerates the decomposition and stabilization of the sludge while reducing potential phytotoxicity (Hu et al. 2021b).

Vermicomposting sewage sludge also presents challenges because of the presence of pathogens, metals (Cd, Pb, and Hg), and organic pollutants, including pharmaceutical and personal care products (Fijalkowski et al. 2017). These pose risks to both the environment and human health if not adequately removed or detoxified, and the direct application to soil is subject to regulation in several countries (Iticescu et al. 2021). One significant advantage of vermicomposting sewage sludge is the earthworms' capacity to bioaccumulate metals in their tissues, which may aid in the detoxification of the final product (Dume et al. 2022). Nevertheless, it is important to continuously monitor metal levels, both in earthworm biomass and the final vermicompost, ensuring its safety for agricultural use. Sewage sludge, when improperly managed, may present health risks due to the presence of pathogens. As previously emphasized, vermicomposting is effective in sanitizing sewage sludge, notably reducing pathogenic bacteria, including coliforms (Ghahdarijani et al. 2022).

Continuous monitoring and careful management of the process are essential for ensuring the safe and effective use of vermicomposted sewage sludge in agriculture.

## 5.4 *Vermicompost in Agriculture*

In agriculture, vermicompost is primarily used as a soil fertilizer (Oyege and Balaji Bhaskar 2023). Vermicompost application to land enhances soil organic matter, leading to improved soil aggregation, porosity, and structural stability (Zhao et al. 2017; Lim et al. 2015), thus optimizing water retention and infiltration while reducing compaction and erosion (Baghbani-Arani et al. 2021; Rivier et al. 2022; Liu et al. 2022).

Vermicompost is a valuable source of plant nutrients, and its use reduces the need for chemical fertilizers, fostering more sustainable and ecologically balanced farming practices (Hajam et al. 2023). This is an important step toward sustainable agriculture since long-term dependence on chemical fertilizers, which provides quick and concentrated nutrient delivery, promotes soil degradation, nutrient imbalances, reduced microbial diversity, and environmental concerns (Tyagi et al. 2022). Vermicompost application introduces beneficial microorganisms, such as bacteria, fungi, and actinomycetes, which enhance nutrient cycling and increase nutrient bioavailability (Wei et al. 2024). Moreover, the application of biofertilizers derived from vermicompost increases the variety of fungal and bacterial communities in the soil and boosts the presence of beneficial microorganisms, supporting soil fertility and ecosystem balance (Pathma and Sakthivel 2012; Tian et al. 2024; Gómez-Brandón et al. 2011b).

Vermicompost plays a crucial role in stimulating the production of plant hormones, thereby helping plants cope with both abiotic and biotic stresses (Rehman et al. 2023). Acting as a natural plant growth regulator, it enhances resilience against environmental challenges, such as salinity and drought, while contributing to the suppression of diseases and pests (Ahmad et al. 2022; Sahab et al. 2021; Yatoo et al. 2021). Rich in essential nutrients, humic acids, plant growth–regulating hormones, beneficial microorganisms, and enzymes, vermicompost significantly improves plant nutrition, enhances photosynthesis, and supports overall plant health (Rehman et al. 2023).

Vermicompost tea, the fermented aqueous extract of mature vermicompost, is an organic biofertilizer. Vermicompost tea can be produced either by aerated or non-aerated processes (Yatoo et al. 2021). Producing vermicompost tea requires brewing mature compost in water for a specific time and then filtering it before application to plants (Yatoo et al. 2021; St. Martin et al. 2012; Ingham 2005). It delivers soluble nutrients, microbial metabolites, and plant hormones that support plant health through foliar and soil applications (Yatoo et al. 2024; Pant et al. 2011). Vermicompost tea is particularly effective in disease suppression through the increased production of secondary metabolites (terpenoids, phenolic compounds, fatty acids, and alkanes); besides, its microbial communities inhibit pathogenic organisms (Sugino Souffront et al. 2022; Kocaman et al. 2024). It also acts as a biostimulant, promoting vegetative growth and helping plants recover from abiotic stresses (Hamed et al. 2022).

Vermicompost has high potential as a germination substrate for horticultural plants, offering a sustainable and environmentally friendly alternative to peat (Zaller 2007; Kinigopoulou et al. 2023; Bustamante et al. 2021). Peat-based substrates have been widely used in horticulture due to their excellent water retention and aeration properties (Kitir et al. 2018). However, peat extraction has serious environmental risks, as peatlands are among the most important carbon sinks on Earth (Harenda et al. 2018). These ecosystems, developed over thousands of years, are being destroyed, releasing stored $CO_2$ and contributing to global warming (Räsänen et al. 2023). With increasing concerns over climate change, it is crucial to find renewable alternatives to maintain soil health and plant growth efficiency, such as vermicompost. Replacing peat with vermicompost allows the agricultural industry to reduce its environmental footprint, protect fragile peatlands, and contribute to efforts toward climate change mitigation (Agarwal et al. 2023; Manh and Wang 2014). The sustainable use of vermicompost in horticulture represents a step toward more eco-friendly agricultural practices while ensuring high-quality plant growth (Rehman et al. 2023).

## 6 Final Remarks

Vermicomposting aligns with the current European soil initiatives, such as the European Green Deal and the Soil Strategy for 2030, which emphasize the importance of sustainable soil management and the transition to a circular economy. Furthermore, vermicomposting contributes to multiple United Nations Sustainable Development Goals (SDGs), particularly those addressing sustainable agriculture, food security, and environmental sustainability.

By transforming organic residues into nutrient-rich vermicompost, this process increases agricultural productivity and food security while contributing to soil health and directly supporting SDG 2 (Zero Hunger). Additionally, vermicomposting reduces the use of irrigation water and reliance on chemical fertilizers, contributing to the achievement of SDG 6 (Clean Water and Sanitation). Moreover, vermicomposting reduces greenhouse gas emissions, supporting SDG 11 (Sustainable Cities and Communities) and enhances carbon sequestration, aligning with SDG 13 (Climate Action). As the synergistic actions of earthworms and microorganisms accelerate organic matter turnover, adopting the sustainable practices of vermicomposting and applying vermicompost to land contribute to SDG 12 (Responsible Consumption and Production) and SDG 15 (Life on Land) by helping the restoration of degraded soils and the promotion of biodiversity.

Integrating vermicomposting into agricultural and residue management practices offers a practical, nature-based solution to meet Europe's targets for sustainable soil management, climate resilience, and waste reduction.

**Funding** The Vine&Wine PT project has received funding from the NextGenerationEU program through Portugal's Recovery and Resilience (PRR) Plan (project no. C644866286–00000011). This work is also supported by National Funds by the Portuguese Foundation for Science and Technology (FCT) under the projects UID/04033/2025: Centre for the Research and Technology of Agro-Environmental and Biological Sciences (https://doi.org/10.54499/UIDB/04033/2020) and LA/P/0126/2020 (https://doi.org/10.54499/LA/P/0126/2020). ENG and CM thank PRR and European Funds NextGenerationEU for their fellowship grant (BI/UTAD/41/2025 and BI/UTAD/30/2024, respectively).

## References

Abad Q, Shafiqi S (2024) Vermicompost: significance and benefits for agriculture. J Res Appl Sci Biotechnol 3:202–207. https://doi.org/10.55544/jrasb.3.2.36

Abbott LK, Macdonald LM, Wong MTF, Webb MJ, Jenkins SN, Farrell M (2018) Potential roles of biological amendments for profitable grain production – a review. Agric Ecosyst Environ 256: 34–50. https://doi.org/10.1016/j.agee.2017.12.021

Adam J, Wellacher M, Azizi F, Loidl A, Zöscher A, Poschacher F, Pomberger R, Tischberger-Aldrian A (2024) Determination and improvement of the quality of separately collected bio-waste from households. Waste Manag Res J Sustain Circ Econ 42:842–854. https://doi.org/10.1177/0734242X241259895

Afshari RT, Seyyedi SM, Mirmiran SM (2023) Clean application of pistachio residues-based vermicompost with γ-aminobutyric acid can alleviate the negative effects of high soil pH on P uptake in saffron (*Crocus sativus* L.). Ind Crops Prod 195:116443. https://doi.org/10.1016/j.indcrop.2023.116443

Agarwal P, Saha S, Hariprasad P (2023) Agro-industrial-residues as potting media: physicochemical and biological characters and their influence on plant growth. Biomass Convers Biorefinery 13:9601–9624. https://doi.org/10.1007/s13399-021-01998-6

Aguiar NO, Olivares FL, Novotny EH, Dobbss LB, Balmori DM, Santos-Júnior LG, Chagas JG, Façanha AR, Canellas LP (2013) Bioactivity of humic acids isolated from vermicomposts at different maturation stages. Plant Soil 362:161–174. https://doi.org/10.1007/s11104-012-1277-5

Ahamad L, Bhat AH, Kumar H, Rana A, Hasan MN, Ahmed I, Ahmed S, Machado RAR, Ameen F (2023) From soil to plant: strengthening carrot defenses against *Meloidogyne incognita* with vermicompost and arbuscular mycorrhizal fungi biofertilizers. Front Microbiol 14:1206217. https://doi.org/10.3389/fmicb.2023.1206217

Ahmad A, Aslam Z, Hussain S, Bibi A, Khaliq A, Javed T, Hussain S, Alotaibi SS, Kalaji HM, Telesiński A, Iwai CB, Kumar U (2022) Rice straw vermicompost enriched with cellulolytic microbes ameliorate the negative effect of drought in wheat through modulating the morpho-physiological attributes. Front Environ Sci 10:902999. https://doi.org/10.3389/fenvs.2022.902999

Ahn HK, Richard TL, Choi HL (2007) Mass and thermal balance during composting of a poultry manure—wood shavings mixture at different aeration rates. Process Biochem 42:215–223. https://doi.org/10.1016/j.procbio.2006.08.005

Aira M, Monroy F, Domínguez J (2007) Microbial biomass governs enzyme activity decay during aging of worm-worked substrates through vermicomposting. J Environ Qual 36:448–452. https://doi.org/10.2134/jeq2006.0262

Aira M, Gómez-Brandón M, González-Porto P, Domínguez J (2011) Selective reduction of the pathogenic load of cow manure in an industrial-scale continuous-feeding vermireactor. Bioresour Technol 102:9633–9637. https://doi.org/10.1016/j.biortech.2011.07.115

Ali U, Sajid N, Khalid A, Riaz L, Rabbani MM, Syed JH, Malik RN (2015) A review on vermicomposting of organic wastes. Environ Prog Sustain Energy 34:1050–1062. https://doi.org/10.1002/ep.12100

Anand K, Sinha PB (2020) Vermitechnology: a solution for agricultural waste. In: Bhat RA, Qadri H, Wani KA, Dar GH, Mehmood MA (eds) Practice, progress, and proficiency in sustainability. IGI Global, pp 273–290

Ascencion LC, Liang W-J, Yen T-B (2015) Control of *Rhizoctonia solani* damping-off disease after soil amendment with dry tissues of *Brassica* results from increase in Actinomycetes population. Biol Control 82:21–30. https://doi.org/10.1016/j.biocontrol.2014.11.010

Avery LM, Booth P, Campbell C, Tompkins D, Hough RL (2012) Prevalence and survival of potential pathogens in source-segregated green waste compost. Sci Total Environ 431:128–138. https://doi.org/10.1016/j.scitotenv.2012.05.020

Azarmi R, Ziveh PS, Satari MR (2008) Effect of vermicompost on growth, yield and nutrition status of tomato (*Lycopersicum esculentum*). Pak J Biol Sci 11:1797–1802. https://doi.org/10.3923/pjbs.2008.1797.1802

Badhwar VK, Singh S, Singh B (2021) Biotransformation of industrial wastes for nutrient rich vermicompost—a review of the bioconversion process by earthworms. In: Singh H, Singh

Cheema PP, Garg P (eds) Sustainable development through engineering innovations. Springer Singapore, Singapore, pp 257–265

Baghbani-Arani A, Modarres-Sanavy SAM, Poureisa M (2021) Improvement the soil physico-chemical properties and fenugreek growth using zeolite and vermicompost under water deficit conditions. J Soil Sci Plant Nutr 21:1213–1228. https://doi.org/10.1007/s42729-021-00434-y

Baral B, Bhattarai N, Shrestha G (2015) Microbial effectiveness through vermicomposting technique for the biological stabilization of solid waste. J Nat Hist Mus 26:126–135. https://doi.org/10.3126/jnhm.v26i0.14136

Bhat SA, Singh J, Vig AP (2018) Earthworms as organic waste managers and biofertilizer producers. Waste Biomass Valoriz 9:1073–1086. https://doi.org/10.1007/s12649-017-9899-8

Blaya J, Marhuenda FC, Pascual JA, Ros M (2016) Microbiota characterization of compost using omics approaches opens new perspectives for phytophthora root rot control. PLoS One 11: e0158048. https://doi.org/10.1371/journal.pone.0158048

Bohacz J (2018) Microbial strategies and biochemical activity during lignocellulosic waste composting in relation to the occurring biothermal phases. J Environ Manag 206:1052–1062. https://doi.org/10.1016/j.jenvman.2017.11.077

Bustamante MA, Gomis MP, Pérez-Murcia MD, Gangi D, Ceglie FG, Paredes C, Pérez-Espinosa A, Bernal MP, Moral R (2021) Use of livestock waste composts as nursery growing media: effect of a washing pre-treatment. Sci Hortic 281:109954. https://doi.org/10.1016/j.scienta.2021.109954

Cáceres R, Coromina N, Malińska K, Martínez-Farré FX, López M, Soliva M, Marfà O (2016) Nitrification during extended co-composting of extreme mixtures of green waste and solid fraction of cattle slurry to obtain growing media. Waste Manag 58:118–125. https://doi.org/10.1016/j.wasman.2016.08.014

Campitelli P, Ceppi S (2008) Chemical, physical and biological compost and vermicompost characterization: a chemometric study. Chemom Intell Lab Syst 90:64–71. https://doi.org/10.1016/j.chemolab.2007.08.001

Cao Y, Tian Y, Wu Q, Li J, Zhu H (2021) Vermicomposting of livestock manure as affected by carbon-rich additives (straw, biochar and nanocarbon): a comprehensive evaluation of earthworm performance, microbial activities, metabolic functions and vermicompost quality. Bioresour Technol 320:124404. https://doi.org/10.1016/j.biortech.2020.124404

Chan YC, Sinha RK, Wang W (2011) Emission of greenhouse gases from home aerobic composting, anaerobic digestion and vermicomposting of household wastes in Brisbane (Australia). Waste Manag Res J Sustain Circ Econ 29:540–548. https://doi.org/10.1177/0734242X10375587

Chiba A, Vitow N, Baum C, Zacher A, Kahle P, Leinweber P, Schloter M, Schulz S (2024) Earthworm activities change phosphorus mobilization and uptake strategies in deep soil layers. Appl Soil Ecol 193:105168. https://doi.org/10.1016/j.apsoil.2023.105168

Danish Toor M, Kizilkaya R, Anwar A, Koleva L, Eldesoky GE (2024) Effects of vermicompost on soil microbiological properties in lettuce rhizosphere: an environmentally friendly approach for sustainable green future. Environ Res 243:117737. https://doi.org/10.1016/j.envres.2023.117737

Domínguez J (2004) State-of-the-art and new perspectives on vermicomposting research. In: Edwards C (ed) Earthworm ecology. CRC Press, pp 401–424

Domínguez J (2018) Earthworms and vermicomposting. In: Ray S (ed) Earthworms – the ecological engineers of soil. InTech

Dominguez J, Edwards C (2010) Relationships between composting and vermicomposting. In: Edwards C, Arancon N, Sherman R (eds) Vermiculture technology. CRC Press, pp 11–25

Domínguez J, Parmelee RW, Edwards CA (2003) Interactions between Eisenia andrei (Oligochaeta) and nematode populations during vermicomposting. Pedobiologia 47:53–60. https://doi.org/10.1078/0031-4056-00169

Domínguez J, Aira M, Gómez-Brandón M (2010) Vermicomposting: earthworms enhance the work of microbes. In: Insam H, Franke-Whittle I, Goberna M (eds) Microbes at work. Springer, Berlin/Heidelberg, pp 93–114

Domínguez J, Aira M, Kolbe AR, Gómez-Brandón M, Pérez-Losada M (2019) Changes in the composition and function of bacterial communities during vermicomposting may explain beneficial properties of vermicompost. Sci Rep 9:9657. https://doi.org/10.1038/s41598-019-46018-w

Domínguez J, Aira M, Crandall KA, Pérez-Losada M (2021) Earthworms drastically change fungal and bacterial communities during vermicomposting of sewage sludge. Sci Rep 11:15556. https://doi.org/10.1038/s41598-021-95099-z

Dores-Silva PR, Landgraf MD, Rezende MOO (2018) Humification process in different kinds of organic residue by composting and vermicomposting: have microbioreactors really accelerated the process? Environ Sci Pollut Res 25:17490–17498. https://doi.org/10.1007/s11356-018-1908-3

Dume B, Hanc A, Svehla P, Michal P, Chane AD, Nigussie A (2022) Vermicomposting technology as a process able to reduce the content of potentially toxic elements in sewage sludge. Agronomy 12:2049. https://doi.org/10.3390/agronomy12092049

Dume B, Hanc A, Svehla P, Michal P, Chane AD, Nigussie A (2023) Composting and vermicomposting of sewage sludge at various C/N ratios: technological feasibility and end-product quality. Ecotoxicol Environ Saf 263:115255. https://doi.org/10.1016/j.ecoenv.2023.115255

Ečimović S, Vrandečić K, Kujavec M, Žulj M, Ćosić J, Velki M (2021) Antifungal activity of earthworm Coelomic fluid obtained from *Eisenia andrei*, *Dendrobaena veneta* and *Allolobophora chlorotica* on six species of Phytopathogenic fungi. Environments 8:102. https://doi.org/10.3390/environments8100102

Edwards CA, Arancon NQ (2022a) The role of earthworms in organic matter and nutrient cycles. In: Biology and ecology of earthworms. Springer, New York, pp 233–274

Edwards CA, Arancon NQ (2022b) The influence of environmental factors on earthworms. In: Biology and ecology of earthworms. Springer, New York, pp 191–232

EL-Mogy MM, Adly MA, Shahein MM, Hassan HA, Mahmoud SO, Abdeldaym EA (2024) Integration of biochar with vermicompost and compost improves agro-physiological properties and nutritional quality of greenhouse sweet pepper. Agronomy 14:2603. https://doi.org/10.3390/agronomy14112603

Eo J, Park K-C (2019) Effect of vermicompost application on root growth and ginsenoside content of Panax ginseng. J Environ Manag 234:458–463. https://doi.org/10.1016/j.jenvman.2018.12.101

Fijalkowski K, Rorat A, Grobelak A, Kacprzak MJ (2017) The presence of contaminations in sewage sludge – the current situation. J Environ Manag 203:1126–1136. https://doi.org/10.1016/j.jenvman.2017.05.068

Fornes F, Mendoza-Hernández D, García-de-la-Fuente R, Abad M, Belda RM (2012) Composting versus vermicomposting: a comparative study of organic matter evolution through straight and combined processes. Bioresour Technol 118:296–305. https://doi.org/10.1016/j.biortech.2012.05.028

Förstner U (2004) Traceability of sediment analysis. TrAC Trends Anal Chem 23:217–236. https://doi.org/10.1016/S0165-9936(04)00312-7

Fracchia L, Dohrmann AB, Martinotti MG, Tebbe CC (2006) Bacterial diversity in a finished compost and vermicompost: differences revealed by cultivation-independent analyses of PCR-amplified 16S rRNA genes. Appl Microbiol Biotechnol 71:942–952. https://doi.org/10.1007/s00253-005-0228-y

Frederickson J, Howell G, Hobson AM (2007) Effect of pre-composting and vermicomposting on compost characteristics. Eur J Soil Biol 43:S320–S326. https://doi.org/10.1016/j.ejsobi.2007.08.032

Ganesh PS, Gajalakshmi S, Abbasi SA (2009) Vermicomposting of the leaf litter of acacia (*Acacia auriculiformis*): possible roles of reactor geometry, polyphenols, and lignin. Bioresour Technol 100:1819–1827. https://doi.org/10.1016/j.biortech.2008.09.051

Garg VK, Gupta R (2009) Vermicomposting of agro-industrial processing waste. In: Singh nee' Nigam P, Pandey A (eds) Biotechnology for agro-industrial residues utilisation. Springer, Dordrecht, pp 431–456

Garg VK, Gupta R (2011) Optimization of cow dung spiked pre-consumer processing vegetable waste for vermicomposting using *Eisenia fetida*. Ecotoxicol Environ Saf 74:19–24. https://doi.org/10.1016/j.ecoenv.2010.09.015

Gazi A, Maity A, Khatua N, Sengupta S, Kundu S, Sarkar T (2024) Effect of vermicompost on soil quality and crop productivity. Int J Agric Ext Soc Dev 7:13–23. https://doi.org/10.33545/26180723.2024.v7.i4Sa.517

Getino-Álvarez M, San-Martin R, Pretzsch H, Pach M, Bravo F, Turrión M-B (2023) Assessing soil C stock and C to N ratio of soil organic matter under mixed pine-beech forests at different scales. Eur J For Res 142:1081–1098. https://doi.org/10.1007/s10342-023-01578-5

Ghahdarijani ARJ, Hoodaji M, Tahmourespour A (2022) Vermicomposting of sewage sludge with organic bulking materials to improve its properties. Environ Monit Assess 194:555. https://doi.org/10.1007/s10661-022-10236-z

Ghandehari Yazdi F, Mokhtari M, Nabi Meibodi M, Sefidkar R, Hatami B, Molavi F, Ghafourzadeh M, Golshiri A, Ebrahimi AA (2024) Bioconversion of cow manure through vermicomposting: effects of tylosin concentration on the weight of worms and manure quality. Sci Rep 14:12575. https://doi.org/10.1038/s41598-024-62839-w

Ghorbani M, Sabour MR, Bidabadi M (2021) Vermicomposting smart closed reactor design and performance assessment by using sewage sludge. Waste Biomass Valoriz 12:6177–6190. https://doi.org/10.1007/s12649-021-01426-w

Gómez Brandón M, Aira M, Kolbe AR, De Andrade N, Pérez-Losada M, Domínguez J (2019) Rapid bacterial community changes during vermicomposting of grape Marc derived from red winemaking. Microorganisms 7:473. https://doi.org/10.3390/microorganisms7100473

Gómez-Brandón M, Domínguez J (2014) Recycling of solid organic wastes through vermicomposting: microbial community changes throughout the process and use of vermicompost as a soil amendment. Crit Rev Environ Sci Technol 44:1289–1312. https://doi.org/10.1080/10643389.2013.763588

Gómez-Brandón M, Aira M, Lores M, Domínguez J (2011a) Changes in microbial community structure and function during vermicomposting of pig slurry. Bioresour Technol 102:4171–4178. https://doi.org/10.1016/j.biortech.2010.12.057

Gómez-Brandón M, Aira M, Lores M, Domínguez J (2011b) Epigeic earthworms exert a bottleneck effect on microbial communities through gut associated processes. PLoS One 6:e24786. https://doi.org/10.1371/journal.pone.0024786

Gómez-Brandón M, Aira M, Santana N, Pérez-Losada M, Domínguez J (2020) Temporal dynamics of bacterial communities in a pilot-scale vermireactor fed with distilled grape Marc. Microorganisms 8:642. https://doi.org/10.3390/microorganisms8050642

Gómez-Roel A, Aira M, Domínguez J (2024) Vermicomposting enhances microbial detoxification of sewage sludge, enabling potential application of the treated product in agroecosystems. Appl Sci 14:7894. https://doi.org/10.3390/app14177894

Gong X, Li S, Sun X, Wang L, Cai L, Zhang J, Wei L (2018) Green waste compost and vermicompost as peat substitutes in growing media for geranium (*Pelargonium zonale* L.) and calendula (*Calendula officinalis* L.). Sci Hortic 236:186–191. https://doi.org/10.1016/j.scienta.2018.03.051

González-Hernández AI, Pérez-Sánchez R, Plaza J, Morales-Corts MR (2022) Compost tea as a sustainable alternative to promote plant growth and resistance against *Rhizoctonia solani* in potato plants. Sci Hortic 300:111090. https://doi.org/10.1016/j.scienta.2022.111090

Grantina-Ievina L, Andersone U, Berkolde-Pīre D, Nikolajeva V, Ievinsh G (2013) Critical tests for determination of microbiological quality and biological activity in commercial vermicompost

samples of different origins. Appl Microbiol Biotechnol 97:10541–10554. https://doi.org/10.1007/s00253-013-4825-x
Gupta AK, Chaudhary A, Panthi B, Chaudhary AK, Gautam E, Badhai S (2022) Vermicomposting. INWASCON Technol Mag 4:29–30. https://doi.org/10.26480/itechmag.04.2022.29.30
Gutiérrez-López M, Moreno G, Trigo D, Juárez E, Jesús JB, Díaz Cosín DJ (2016) The efficiency of earthworm extraction methods is determined by species and soil properties in the Mediterranean communities of Central-Western Spain. Eur J Soil Biol 73:59–68. https://doi.org/10.1016/j.ejsobi.2016.01.005
Hait S, Tare V (2012) Transformation and availability of nutrients and heavy metals during integrated composting–vermicomposting of sewage sludges. Ecotoxicol Environ Saf 79:214–224. https://doi.org/10.1016/j.ecoenv.2012.01.004
Hajam YA, Kumar R, Kumar A (2023) Environmental waste management strategies and vermi transformation for sustainable development. Environ Chall 13:100747. https://doi.org/10.1016/j.envc.2023.100747
Hamed FR, Ismail HEM, Abo El-Maati MF, Desoky EM (2022) Effect of vermicompost-tea and plant extracts on growth, physiological, and biochemical traits of lettuce (*Lactuca sativa* L.). SABRAO J Breed Genet 54:1202–1215. https://doi.org/10.54910/sabrao2022.54.5.21
Hanc A, Hrebeckova T, Kuzel S (2019) Vermicomposting of distillery residues in a vertical-flow windrow system. Waste Biomass Valoriz 10:3647–3657. https://doi.org/10.1007/s12649-019-00671-4
Hanc A, Hrebeckova T, Pliva P, Cajthaml T (2020) Vermicomposting of sludge from a malt house. Waste Manag 118:232–240. https://doi.org/10.1016/j.wasman.2020.08.027
Harenda KM, Lamentowicz M, Samson M, Chojnicki BH (2018) The role of peatlands and their carbon storage function in the context of climate change. In: Zielinski T, Sagan I, Surosz W (eds) Interdisciplinary approaches for sustainable development goals. Springer International Publishing, Cham, pp 169–187
Hasnat M, Alam MA, Khanam M, Binte BI, Kabir MH, Alam MS, Kamal MZU, Rahman GKMM, Haque MM, Rahman MM (2022) Effect of nitrogen fertilizer and biochar on organic matter mineralization and carbon accretion in soil. Sustainability 14:3684. https://doi.org/10.3390/su14063684
He X, Zhang Y, Shen M, Zeng G, Zhou M, Li M (2016) Effect of vermicomposting on concentration and speciation of heavy metals in sewage sludge with additive materials. Bioresour Technol 218:867–873. https://doi.org/10.1016/j.biortech.2016.07.045
Hu X, Zhang T, Tian G, Zhang L, Bian B (2021a) Performance and mechanism of high-speed vermicomposting of dewatered sludge using a new type of laboratory earthworm reactor. Environ Sci Pollut Res 28:26132–26144. https://doi.org/10.1007/s11356-021-12438-3
Hu X, Zhang T, Tian G, Zhang L, Bian B (2021b) Pilot-scale vermicomposting of sewage sludge mixed with mature vermicompost using earthworm reactor of frame composite structure. Sci Total Environ 767:144217. https://doi.org/10.1016/j.scitotenv.2020.144217
Hua Z, Wang Y-H, Cao H-W, Pu L-J, Cui Y-D (2011) Purification of a protein from coelomic fluid of the earthworm *Eisenia foetida* and evaluation of its hemolytic, antibacterial, and antitumor activities. Pharm Biol 49:269–275. https://doi.org/10.3109/13880209.2010.508498
Hughes RJ, Nair J, Ho G (2008) The toxicity of ammonia/ammonium to the vermifiltration wastewater treatment process. Water Sci Technol 58:1215–1220. https://doi.org/10.2166/wst.2008.478
Hussain N, Chatterjee SK, Maiti TK, Goswami L, Das S, Deb U, Bhattacharya SS (2021) Metal induced non-metallothionein protein in earthworm: a new pathway for cadmium detoxification in chloragogenous tissue. J Hazard Mater 401:123357. https://doi.org/10.1016/j.jhazmat.2020.123357
Ingham ER (2005) The compost tea brewing manual, 5th edn. Soil Foodweb
Inglis GD, McAllister TA, Larney FJ, Topp E (2010) Prolonged survival of campylobacter species in bovine manure compost. Appl Environ Microbiol 76:1110–1119. https://doi.org/10.1128/AEM.01902-09

Iticescu C, Georgescu P-L, Arseni M, Rosu A, Timofti M, Carp G, Cioca L-I (2021) Optimal solutions for the use of sewage sludge on agricultural lands. Water 13:585. https://doi.org/10.3390/w13050585

Javoreková S, Kovácsová S, Medo J, Maková J, Petrová J, Hleba L, Košťálová D, Cinkocki R (2019) Soil amended with organic fertilizers as a source of actinomycetes with high potential as potential as biocontrol agents. J Microbiol Biotechnol Food Sci 8:1352–1359. https://doi.org/10.15414/jmbfs.2019.8.6.1352-1359

Jesna M, Livingstone AR (2023) Effect of temperature on regeneration of *Eisenia fetida*. Int J Eng Technol Manag Sci 7:52–55. https://doi.org/10.46647/ijetms.2023.v07i04.011

Jindo K, Martim SA, Navarro EC, Pérez-Alfocea F, Hernandez T, Garcia C, Aguiar NO, Canellas LP (2012) Root growth promotion by humic acids from composted and non-composted urban organic wastes. Plant Soil 353:209–220. https://doi.org/10.1007/s11104-011-1024-3

Kaur T (2020) Vermicomposting: an effective option for recycling organic wastes. In: Kumar Das S (ed) Organic agriculture. IntechOpen

Kauser H, Khwairakpam M (2022) Organic waste management by two-stage composting process to decrease the time required for vermicomposting. Environ Technol Innov 25:102193. https://doi.org/10.1016/j.eti.2021.102193

Khwairakpam M, Bhargava R (2009) Bioconversion of filter mud using vermicomposting employing two exotic and one local earthworm species. Bioresour Technol 100:5846–5852. https://doi.org/10.1016/j.biortech.2009.06.038

Kinigopoulou V, Hatzigiannakis E, Stefanou S, Guitonas A, Oikonomou EK (2023) Utilization of Vermicompost sludge instead of peat in olive tree nurseries in the frame of circular economy and sustainable development. Agri Engineering 5:1630–1643. https://doi.org/10.3390/agriengineering5030101

Kitir N, Yildirim E, Şahin Ü, Turan M, Ekinci M, Ors S, Kul R, Ünlü H, Ünlü H (2018) Peat use in horticulture. In: Topcuoğlu B, Turan M (eds) Peat. InTech

Kiyasudeen K, Ibrahim MH, Ismail SA (2020) Vermicomposting of organic wastes and the production of vermicompost. In: Biovalorisation of wastes to renewable chemicals and biofuels. Elsevier, pp 277–285

Kocaman A, İnci Y, Kıtır N, Turan M, Argın S, Yıldırım E, Giray G, Ersoy N, Güneş A, Katırcıoğlu H, Gürkan B, Bilgili AV, Aydemir ÖE, Akça M (2024) The effect of novel biotechnological vermicompost on tea yield, plant nutrient content, antioxidants, amino acids, and organic acids as an alternative to chemical fertilizers for sustainability. BMC Plant Biol 24:868. https://doi.org/10.1186/s12870-024-05504-8

Koskey G, Avio L, Turrini A, Sbrana C, Bàrberi P (2023) Biostimulatory effect of vermicompost extract enhances soil mycorrhizal activity and selectively improves crop productivity. Plant Soil 484:183–199. https://doi.org/10.1007/s11104-022-05783-w

Kuhad RC, Chandna P, Lata, Singh A (2011) Composting of lignocellulosic waste material for soil amendment. In: Singh A, Parmar N, Kuhad RC (eds) Bioaugmentation, biostimulation and biocontrol. Springer, Berlin/Heidelberg, pp 107–128

Lazcano C, Gómez-Brandón M, Domínguez J (2008) Comparison of the effectiveness of composting and vermicomposting for the biological stabilization of cattle manure. Chemosphere 72:1013–1019. https://doi.org/10.1016/j.chemosphere.2008.04.016

Li L, Wu J, Tian G, Xu Z (2009) Effect of the transit through the gut of earthworm (*Eisenia fetida*) on fractionation of Cu and Zn in pig manure. J Hazard Mater 167:634–640. https://doi.org/10.1016/j.jhazmat.2009.01.013

Liang C, Das KC, McClendon RW (2003) The influence of temperature and moisture contents regimes on the aerobic microbial activity of a biosolids composting blend. Bioresour Technol 86:131–137. https://doi.org/10.1016/S0960-8524(02)00153-0

Liebeke M, Garcia-Perez I, Anderson CJ, Lawlor AJ, Bennett MH, Morris CA, Kille P, Svendsen C, Spurgeon DJ, Bundy JG (2013) Earthworms produce phytochelatins in response to arsenic. PLoS One 8:e81271. https://doi.org/10.1371/journal.pone.0081271

Lim SL, Wu TY, Lim PN, Shak KPY (2015) The use of vermicompost in organic farming: overview, effects on soil and economics: the use of vermicompost in organic farming. J Sci Food Agric 95:1143–1156. https://doi.org/10.1002/jsfa.6849

Lim SL, Lee LH, Wu TY (2016) Sustainability of using composting and vermicomposting technologies for organic solid waste biotransformation: recent overview, greenhouse gases emissions and economic analysis. J Clean Prod 111:262–278. https://doi.org/10.1016/j.jclepro.2015.08.083

Lirikum KLN, Thyug L, Mozhui L (2022) Vermicomposting: an eco-friendly approach for waste management and nutrient enhancement. Trop Ecol 63:325–337. https://doi.org/10.1007/s42965-021-00212-y

Liu X, Zhang J, Wang Q, Shaghaleh H, Chang T, Hamoud YA (2022) Modification of soil physical properties by maize straw biochar and earthworm manure to enhance hydraulic characteristics under greenhouse condition. Sustainability 14:13590. https://doi.org/10.3390/su142013590

Loh T (2005) Vermicomposting of cattle and goat manures by *Eisenia foetida* and their growth and reproduction performance. Bioresour Technol 96:111–114. https://doi.org/10.1016/j.biortech.2003.03.001

Lv B, Xing M, Yang J (2016) Speciation and transformation of heavy metals during vermicomposting of animal manure. Bioresour Technol 209:397–401. https://doi.org/10.1016/j.biortech.2016.03.015

Lv B, Zhang D, Chen Q, Cui Y (2019) Effects of earthworms on nitrogen transformation and the correspond genes (*amoA* and *nirS*) in vermicomposting of sewage sludge and rice straw. Bioresour Technol 287:121428. https://doi.org/10.1016/j.biortech.2019.121428

Ma H, Zhao S, Hou J, Feyissa T, Duan Z, Pan Z, Zhang K, Zhang W (2022) Vermicompost improves physicochemical properties of growing medium and promotes plant growth: a meta-analysis. J Soil Sci Plant Nutr 22:3745–3755. https://doi.org/10.1007/s42729-022-00924-7

Ma L, Shao M, Wang Y, Li T, Jing X, Jia K, Zhang Y (2024) The characteristics and influential factors of earthworm and vermicompost under different land use in a temperate area, China. Forests 15:1389. https://doi.org/10.3390/f15081389

Majlessi M, Eslami A, Najafi Saleh H, Mirshafieean S, Babaii S (2012) Vermicomposting of food waste: assessing the stability and maturity. Iran J Environ Health Sci Eng 9:25. https://doi.org/10.1186/1735-2746-9-25

Manea EE, Bumbac C, Dinu LR, Bumbac M, Nicolescu CM (2024) Composting as a sustainable solution for organic solid waste management: current practices and potential improvements. Sustainability 16:6329. https://doi.org/10.3390/su16156329

Manh VH, Wang CH (2014) Vermicompost as an important component in substrate: effects on seedling quality and growth of muskmelon (*Cucumis melo* L.). APCBEE Procedia 8:32–40. https://doi.org/10.1016/j.apcbee.2014.01.076

Misirlioğlu M, Reynolds JW, Stojanović M, Trakić T, Sekulić J, James SW, Csuzdi C, Decaëns T, Lapied E, Phillips HRP, Cameron EK, Brown GG (2023) Earthworms (Clitellata, Megadrili) of the world: an updated checklist of valid species and families, with notes on their distribution. Zootaxa 5255:417–438. https://doi.org/10.11646/zootaxa.5255.1.33

Mohite DD, Chavan SS, Jadhav VS, Kanase T, Kadam MA, Singh AS (2024) Vermicomposting: a holistic approach for sustainable crop production, nutrient-rich bio fertilizer, and environmental restoration. Discov Sustain 5:60. https://doi.org/10.1007/s43621-024-00245-y

Moinard V, Redondi C, Etiévant V, Savoie A, Duchene D, Pelosi C, Houot S, Capowiez Y (2021) Short- and long-term impacts of anaerobic digestate spreading on earthworms in cropped soils. Appl Soil Ecol 168:104149. https://doi.org/10.1016/j.apsoil.2021.104149

Monroy F, Aira M, Domínguez J (2009) Reduction of total coliform numbers during vermicomposting is caused by short-term direct effects of earthworms on microorganisms and depends on the dose of application of pig slurry. Sci Total Environ 407:5411–5416. https://doi.org/10.1016/j.scitotenv.2009.06.048

Morgan JE, Morgan AJ (1999) The accumulation of metals (Cd, Cu, Pb, Zn and Ca) by two ecologically contrasting earthworm species (*Lumbricus rubellus* and *Aporrectodea caliginosa*):

implications for ecotoxicological testing. Appl Soil Ecol 13:9–20. https://doi.org/10.1016/S0929-1393(99)00012-8
Mupambwa HA, Mnkeni PNS (2018) Optimizing the vermicomposting of organic wastes amended with inorganic materials for production of nutrient-rich organic fertilizers: a review. Environ Sci Pollut Res 25:10577–10595. https://doi.org/10.1007/s11356-018-1328-4
Nair J, Sekiozoic V, Anda M (2006) Effect of pre-composting on vermicomposting of kitchen waste. Bioresour Technol 97:2091–2095. https://doi.org/10.1016/j.biortech.2005.09.020
Nascimento-Gonçalves E, Azevedo T, Lopes H, Sousa JR, Oliveira PA, Roboredo M, Coimbra AM, Morais MC (2024) Vermicomposting as a valorization solution to the winery sector by-products. Agronomy 14:1111. https://doi.org/10.3390/agronomy14061111
Nasiru A, Ibrahim MH, Ismail N (2014) Nitrogen losses in ruminant manure management and use of cattle manure vermicast to improve forage quality. Int J Recycl Org Waste Agric 3:57. https://doi.org/10.1007/s40093-014-0057-z
Nigussie A, Kuyper TW, Bruun S, De Neergaard A (2016) Vermicomposting as a technology for reducing nitrogen losses and greenhouse gas emissions from small-scale composting. J Clean Prod 139:429–439. https://doi.org/10.1016/j.jclepro.2016.08.058
Nleya Y, Young B, Nooraee E, Baroutian S (2023) Opportunities and challenges for anaerobic digestion of farm dairy effluent. Chem Bio Eng Rev 10:924–940. https://doi.org/10.1002/cben.202300021
Nunes N, Ragonezi C, Gouveia CSS, Pinheiro De Carvalho MÂA (2021) Review of sewage sludge as a soil amendment in relation to current international guidelines: a heavy metal perspective. Sustainability 13:2317. https://doi.org/10.3390/su13042317
Oyege I, Balaji Bhaskar MS (2023) Effects of vermicompost on soil and plant health and promoting sustainable agriculture. Soil Syst 7:101. https://doi.org/10.3390/soilsystems7040101
Ozer H, Dede OH, Ozdemir S (2024) Investigation of biomass ash vermicompost as nutrient rich growing media for impatiens. Int J Environ Sci Technol. https://doi.org/10.1007/s13762-024-05964-3
Pant A, Radovich TJK, Hue NV, Arancon NQ (2011) Effects of vermicompost tea (aqueous extract) on Pak Choi yield, quality, and on soil biological properties. Compost Sci Util 19:279–292. https://doi.org/10.1080/1065657X.2011.10737010
Parastesh F, Alikhani HA, Etesami H (2019) Vermicompost enriched with phosphate–solubilizing bacteria provides plant with enough phosphorus in a sequential cropping under calcareous soil conditions. J Clean Prod 221:27–37. https://doi.org/10.1016/j.jclepro.2019.02.234
Pathma J, Sakthivel N (2012) Microbial diversity of vermicompost bacteria that exhibit useful agricultural traits and waste management potential. Springerplus 1:26. https://doi.org/10.1186/2193-1801-1-26
Pathma J, Sakthivel N (2013) Molecular and functional characterization of bacteria isolated from straw and goat manure based vermicompost. Appl Soil Ecol 70:33–47. https://doi.org/10.1016/j.apsoil.2013.03.011
Patra RK, Behera D, Mohapatra KK, Sethi D, Mandal M, Patra AK, Ravindran B (2022) Juxtaposing the quality of compost and vermicompost produced from organic wastes amended with cow dung. Environ Res 214:114119. https://doi.org/10.1016/j.envres.2022.114119
Paul S, Goswami L, Pegu R, Sundar Bhattacharya S (2020) Vermiremediation of cotton textile sludge by *Eudrilus eugeniae*: insight into metal budgeting, chromium speciation, and humic substance interactions. Bioresour Technol 314:123753. https://doi.org/10.1016/j.biortech.2020.123753
Peneva S, Phan Le QN, Munhoz DR, Wrigley O, Macan GPF, Doose H, Amelung W, Braun M (2025) Plastic input and dynamics in industrial composting. Waste Manag 193:283–292. https://doi.org/10.1016/j.wasman.2024.11.043
Peng N, Zhang J, Hu R, Liu S, Liu F, Fan Y, Yang H, Huang J, Ding J, Chen R, Li L, He Z, Wang C (2024) Hidden pathogen risk in mature compost: low optimal growth temperature confers pathogen survival and activity during manure composting. J Hazard Mater 480:136230. https://doi.org/10.1016/j.jhazmat.2024.136230

Peramban E, Ramachandran M, Rayan R, Manoharan V, Sridharan AP, Sarangan P (2023) Effects of physico chemical parameters on biomass produced by using earthworm *Eudrilus Eugeniae*. Int J Curr Sci Res Rev 06. https://doi.org/10.47191/ijcsrr/V6-i4-02

Plavšin I, Velki M, Ečimović S, Vrandečić K, Ćosić J (2017) Inhibitory effect of earthworm coelomic fluid on growth of the plant parasitic fungus *Fusarium oxysporum*. Eur J Soil Biol 78:1–6. https://doi.org/10.1016/j.ejsobi.2016.11.004

Plaza C, Nogales R, Senesi N, Benitez E, Polo A (2008) Organic matter humification by vermicomposting of cattle manure alone and mixed with two-phase olive pomace. Bioresour Technol 99:5085–5089. https://doi.org/10.1016/j.biortech.2007.09.079

Pottipati S, Kundu A, Kalamdhad AS (2022) Process optimization by combining in-vessel composting and vermicomposting of vegetable waste. Bioresour Technol 346:126357. https://doi.org/10.1016/j.biortech.2021.126357

Pundee K, Akeprathumchai S, Tripetchkul S, Salaipeth L (2023) Unveiling the microbial dynamics in vermicomposting with coir pith as earthworm substrate. Heliyon 9:e22945. https://doi.org/10.1016/j.heliyon.2023.e22945

Räsänen A, Albrecht E, Annala M, Aro L, Laine AM, Maanavilja L, Mustajoki J, Ronkanen A-K, Silvan N, Tarvainen O, Tolvanen A (2023) After-use of peat extraction sites – a systematic review of biodiversity, climate, hydrological and social impacts. Sci Total Environ 882:163583. https://doi.org/10.1016/j.scitotenv.2023.163583

Rasschaert G, Van Elst D, Colson L, Herman L, De Carvalho Ferreira HC, Dewulf J, Decrop J, Meirlaen J, Heyndrickx M, Daeseleire E (2020) Antibiotic residues and antibiotic-resistant bacteria in pig slurry used to fertilize agricultural fields. Antibiotics 9:34. https://doi.org/10.3390/antibiotics9010034

Ravindran B, Wong JWC, Selvam A, Sekaran G (2016) Influence of microbial diversity and plant growth hormones in compost and vermicompost from fermented tannery waste. Bioresour Technol 217:200–204. https://doi.org/10.1016/j.biortech.2016.03.032

Ravindran B, Mupambwa HA, Silwana S, Mnkeni PNS (2017) Assessment of nutrient quality, heavy metals and phytotoxic properties of chicken manure on selected commercial vegetable crops. Heliyon 3:e00493. https://doi.org/10.1016/j.heliyon.2017.e00493

Rehman SU, De Castro F, Aprile A, Benedetti M, Fanizzi FP (2023) Vermicompost: enhancing plant growth and combating abiotic and biotic stress. Agronomy 13:1134. https://doi.org/10.3390/agronomy13041134

Rékási M, Ragályi P, Sándor DB, Szabó A, Rivier P-A, Farkas C, Szécsy O, Uzinger N (2023) Effect of composting and vermicomposting on potentially toxic element contents and bioavailability in sewage sludge digestate. Bioresour Technol Rep 21:101307. https://doi.org/10.1016/j.biteb.2022.101307

Rivier P-A, Jamniczky D, Nemes A, Makó A, Barna G, Uzinger N, Rékási M, Farkas C (2022) Short-term effects of compost amendments to soil on soil structure, hydraulic properties, and water regime. J Hydrol Hydromech 70:74–88. https://doi.org/10.2478/johh-2022-0004

Roubalová R, Procházková P, Hanč A, Dvořák J, Bilej M (2020) Mutual interactions of E. andrei earthworm and pathogens during the process of vermicomposting. Environ Sci Pollut Res 27:33429–33437. https://doi.org/10.1007/s11356-019-04329-5

Sahab S, Suhani I, Srivastava V, Chauhan PS, Singh RP, Prasad V (2021) Potential risk assessment of soil salinity to agroecosystem sustainability: current status and management strategies. Sci Total Environ 764:144164. https://doi.org/10.1016/j.scitotenv.2020.144164

Santana NA, Jacques RJS, Antoniolli ZI, Martínez-Cordeiro H, Domínguez J (2020) Changes in the chemical and biological characteristics of grape marc vermicompost during a two-year production period. Appl Soil Ecol 154:103587. https://doi.org/10.1016/j.apsoil.2020.103587

Sethulakshmi K, Ranilakshmi K, Thomas A (2018) Antibacterial and antifungal potentialities of earthworm *Eudrilus eugeniae* paste and Coelomic fluid. Asian J Biol 5:1–7. https://doi.org/10.9734/AJOB/2018/39786

Shah WUH, Yasmeen R, Sarfraz M, Ivascu L (2023) The repercussions of economic growth, industrialization, foreign direct investment, and technology on municipal solid waste: evidence from OECD economies. Sustainability 15:836. https://doi.org/10.3390/su15010836

Shakeel A, Jaggi S, Bhuyan S, Kumar D, Gupta S (2024) Vermicompost: nature's gold. In: Rhizosphere revolution, 1st edn. CRC Press, London, pp 96–115

Sharma K, Garg VK (2020) Conversion of a toxic weed into vermicompost by *Eisenia fetida*: nutrient content and earthworm fecundity. Bioresour Technol Rep 11:100530. https://doi.org/10.1016/j.biteb.2020.100530

Sharma P, Sharma P, Thakur N (2024) Sustainable farming practices and soil health: a pathway to achieving SDGs and future prospects. Discov Sustain 5:250. https://doi.org/10.1007/s43621-024-00447-4

Shober AL, Maguire RO (2018) Manure management. In: Reference module in earth systems and environmental sciences. Elsevier, p B9780124095489091156

Siddiqui Z (2004) Effects of plant growth promoting bacteria and composed organic fertilizers on the reproduction of *Meloidogyne incognita* and tomato growth. Bioresour Technol 95:223–227. https://doi.org/10.1016/j.biortech.2003.01.002

Simsek Ersahin Y, Haktanir K, Yanar Y (2009) Vermicompost suppresses *Rhizoctonia solani* Kühn in cucumber seedlings. J Plant Dis Prot 116:182–188. https://doi.org/10.1007/BF03356308

Singh WR, Kalamdhad AS (2016) Transformation of nutrients and heavy metals during vermicomposting of the invasive green weed *Salvinia natans* using *Eisenia fetida*. Int J Recycl Org Waste Agric 5:205–220. https://doi.org/10.1007/s40093-016-0129-3

Singh S, Sinha RK (2022) Vermicomposting of organic wastes by earthworms: making wealth from waste by converting 'garbage into gold' for farmers. In: Advanced organic waste management. Elsevier, pp 93–120

Singh R, Jiang X, Luo F (2010) Thermal inactivation of heat-shocked *Escherichia coli* O157:H7, Salmonella, and *Listeria monocytogenes* in dairy compost. J Food Prot 73:1633–1640. https://doi.org/10.4315/0362-028X-73.9.1633

Singh J, Laurenti R, Sinha R, Frostell B (2014) Progress and challenges to the global waste management system. Waste Manag Res J Sustain Circ Econ 32:800–812. https://doi.org/10.1177/0734242X14537868

Singh J, Schädler M, Demetrio W, Brown GG, Eisenhauer N (2019) Climate change effects on earthworms – a review. Soil Org 91:113–137. https://doi.org/10.25674/SO91ISS3PP114

Singh T, Kapila R, Kumar M (2024) A minireview on vermicompost and vermiwash as green pesticide for sustainable crop production: approaches, applications, and advancements. JSFA Rep 4:4–10. https://doi.org/10.1002/jsf2.172

Soobhany N (2018) Preliminary evaluation of pathogenic bacteria loading on organic municipal solid waste compost and vermicompost. J Environ Manag 206:763–767. https://doi.org/10.1016/j.jenvman.2017.11.029

Soukiazis E, Proença S (2020) The determinants of waste generation and recycling performance across the Portuguese municipalities – a simultaneous equation approach. Waste Manag 114:321–330. https://doi.org/10.1016/j.wasman.2020.06.039

Srivastava PK, Tiwari GN, Sinha ASK (2024) Enhanced vermicomposting of rice straw and pressmud with biogas slurry employing *Eisenia fetida*: production, characterization, growth, and toxicological risk assessment. J Environ Manag 352:120032. https://doi.org/10.1016/j.jenvman.2024.120032

St. Martin CCG, Dorinvil W, Brathwaite RAI, Ramsubhag A (2012) Effects and relationships of compost type, aeration and brewing time on compost tea properties, efficacy against *Pythium ultimum*, phytotoxicity and potential as a nutrient amendment for seedling production. Biol Agric Hortic 28:185–205. https://doi.org/10.1080/01448765.2012.727667

Stanford K, Hao X, Xu S, McAllister TA, Larney F, Leonard JJ (2009) Effects of age of cattle, turning technology and compost environment on disappearance of bone from mortality compost. Bioresour Technol 100:4417–4422. https://doi.org/10.1016/j.biortech.2008.11.061

Stipniece AA, Vladinovskis V, Daugulis P, Zemite M, Vitola L, Mezule L (2022) Advantages and challenges of composting reactors for household use: smart reactor concept. Sustainability 14: 10030. https://doi.org/10.3390/su141610030

Suárez-Estrella F, Vargas-García C, López MJ, Capel C, Moreno J (2007) Antagonistic activity of bacteria and fungi from horticultural compost against *Fusarium oxysporum f. sp. melonis*. Crop Prot 26:46–53. https://doi.org/10.1016/j.cropro.2006.04.003

Sugino Souffront DK, Salazar-Amoretti D, Jayachandran K (2022) Influence of vermicompost tea on secondary metabolite production in tomato crop. Sci Hortic 301:111135. https://doi.org/10.1016/j.scienta.2022.111135

Sundararasu K (2019) Growth and yield effects of vermicompost and vermicompost rubbish on selected vegetable crop. Int J Curr Microbiol Appl Sci 8:975–984. https://doi.org/10.20546/ijcmas.2019.802.113

Suthar S (2009) Vermicomposting of vegetable-market solid waste using *Eisenia fetida*: impact of bulking material on earthworm growth and decomposition rate. Ecol Eng 35:914–920. https://doi.org/10.1016/j.ecoleng.2008.12.019

Suthar S, Singh S (2008) Feasibility of vermicomposting in biostabilization of sludge from a distillery industry. Sci Total Environ 394:237–243. https://doi.org/10.1016/j.scitotenv.2008.02.005

Swati A, Hait S (2017) Fate and bioavailability of heavy metals during vermicomposting of various organic wastes—a review. Process Saf Environ Prot 109:30–45. https://doi.org/10.1016/j.psep.2017.03.031

Szczech M, Smolińska U (2001) Comparison of Suppressiveness of Vermicomposts produced from animal manures and sewage sludge against *Phytophthora nicotianae* Breda de Haan var. *nicotianae*. J Phytopathol 149:77–82. https://doi.org/10.1046/j.1439-0434.2001.00586.x

Thirunavukkarasu A, Sivashankar R, Nithya R, Sathya AB, Priyadharshini V, Kumar BP, Muthuveni M, Krishnamoorthy S (2023) Sustainable organic waste management using vermicomposting: a critical review on the prevailing research gaps and opportunities. Environ Sci Process Impacts 25:364–381. https://doi.org/10.1039/D2EM00324D

Tian M, Yu R, Guo S, Yang W, Liu S, Du H, Liang J, Zhang X (2024) Effect of vermicompost application on the soil microbial community structure and fruit quality in melon (*Cucumis melo*). Agronomy 14:2536. https://doi.org/10.3390/agronomy14112536

Tognetti C, Laos F, Mazzarino MJ, Hernández MT (2005) Composting vs. vermicomposting: a comparison of end product quality. Compost Sci Util 13:6–13. https://doi.org/10.1080/1065657X.2005.10702212

Toor MD, Basit A, Okorie B, Nath D, Din MMU, Kumar Verma P, Saleem S, Ullah I, Yousef HN, Mohamed HI (2024) Earthworms as catalysts for climate-resilient agriculture: enhancing food security and water management in the face of climate change. Water Air Soil Pollut 235:779. https://doi.org/10.1007/s11270-024-07576-6

Tyagi J, Ahmad S, Malik M (2022) Nitrogenous fertilizers: impact on environment sustainability, mitigation strategies, and challenges. Int J Environ Sci Technol 19:11649–11672. https://doi.org/10.1007/s13762-022-04027-9

Usmani Z, Kumar V, Gupta P, Gupta G, Rani R, Chandra A (2019) Enhanced soil fertility, plant growth promotion and microbial enzymatic activities of vermicomposted fly ash. Sci Rep 9: 10455. https://doi.org/10.1038/s41598-019-46821-5

Vuković A, Velki M, Ečimović S, Vuković R, Štolfa Čamagajevac I, Lončarić Z (2021) Vermicomposting—facts, benefits and knowledge gaps. Agronomy 11:1952. https://doi.org/10.3390/agronomy11101952

Vyas P, Sharma S, Gupta J (2022) Vermicomposting with microbial amendment: implications for bioremediation of industrial and agricultural waste. Biotechnologia 103:203–215. https://doi.org/10.5114/bta.2022.116213

Wang L, Zhang Y, Lian J, Chao J, Gao Y, Yang F, Zhang L (2013) Impact of fly ash and phosphatic rock on metal stabilization and bioavailability during sewage sludge vermicomposting. Bioresour Technol 136:281–287. https://doi.org/10.1016/j.biortech.2013.03.039

Wang Y, Yuan Z, Tang Y (2021) Enhancing food security and environmental sustainability: a critical review of food loss and waste management. Resour Environ Sustain 4:100023. https://doi.org/10.1016/j.resenv.2021.100023

Wang F, Yao W, Zhang W, Miao L, Wang Y, Zhang H, Ding Y, Zhu W (2022) Humic acid characterization and heavy metal behaviour during vermicomposting of pig manure amended with$^{13}$ C-labelled rice straw. Waste Manag Res J Sustain Circ Econ 40:736–744. https://doi.org/10.1177/0734242X211035943

Wang F, Zhang Y, Su Y, Wu D, Xie B (2024a) Pollutant control and nutrient recovery of organic solid waste by earthworms: mechanism and agricultural benefits of vermicomposting. J Environ Chem Eng 12:112610. https://doi.org/10.1016/j.jece.2024.112610

Wang F, Pan T, Fu D, Fotidis IA, Moulogianni C, Yan Y, Singh RP (2024b) Pilot-scale membrane-covered composting of food waste: initial moisture, mature compost addition, aeration time and rate. Sci Total Environ 926:171797. https://doi.org/10.1016/j.scitotenv.2024.171797

Wang F, Xiang L, Sze-Yin Leung K, Elsner M, Zhang Y, Guo Y, Pan B, Sun H, An T, Ying G, Brooks BW, Hou D, Helbling DE, Sun J, Qiu H, Vogel TM, Zhang W, Gao Y, Simpson MJ, Luo Y, Chang SX, Su G, Wong BM, Fu T-M, Zhu D, Jobst KJ, Ge C, Coulon F, Harindintwali JD, Zeng X, Wang H, Fu Y, Wei Z, Lohmann R, Chen C, Song Y, Sanchez-Cid C, Wang Y, El-Naggar A, Yao Y, Huang Y, Cheuk-Fung Law J, Gu C, Shen H, Gao Y, Qin C, Li H, Zhang T, Corcoll N, Liu M, Alessi DS, Li H, Brandt KK, Pico Y, Gu C, Guo J, Su J, Corvini P, Ye M, Rocha-Santos T, He H, Yang Y, Tong M, Zhang W, Suanon F, Brahushi F, Wang Z, Hashsham SA, Virta M, Yuan Q, Jiang G, Tremblay LA, Bu Q, Wu J, Peijnenburg W, Topp E, Cao X, Jiang X, Zheng M, Zhang T, Luo Y, Zhu L, Li X, Barceló D, Chen J, Xing B, Amelung W, Cai Z, Naidu R, Shen Q, Pawliszyn J, Zhu Y, Schaeffer A, Rillig MC, Wu F, Yu G, Tiedje JM (2024c) Emerging contaminants: a one health perspective. The Innovation 5: 100612. https://doi.org/10.1016/j.xinn.2024.100612

Wei X, Xie B, Wan C, Song R, Zhong W, Xin S, Song K (2024) Enhancing soil health and plant growth through microbial fertilizers: mechanisms, benefits, and sustainable agricultural practices. Agronomy 14:609. https://doi.org/10.3390/agronomy14030609

Wu Q, Wan W (2023) Insight into application of phosphate-solubilizing bacteria promoting phosphorus availability during chicken manure composting. Bioresour Technol 373:128707. https://doi.org/10.1016/j.biortech.2023.128707

Xu P, Shu L, Li Y, Zhou S, Zhang G, Wu Y, Yang Z (2023) Pretreatment and composting technology of agricultural organic waste for sustainable agricultural development. Heliyon 9: e16311. https://doi.org/10.1016/j.heliyon.2023.e16311

Yadav A, Garg VK (2011) Industrial wastes and sludges management by vermicomposting. Rev Environ Sci Biotechnol 10:243–276. https://doi.org/10.1007/s11157-011-9242-y

Yatoo AM, Ali MN, Baba ZA, Hassan B (2021) Sustainable management of diseases and pests in crops by vermicompost and vermicompost tea. A review. Agron Sustain Dev 41:7. https://doi.org/10.1007/s13593-020-00657-w

Yatoo AM, Ali MN, Baba ZA, Alsohim AS, Muthukumaran M, Sayyed RZ (2024) Effect of macrophyte biomass-based vermicompost and vermicompost tea on plant growth, productivity, and biocontrol of Fusarium wilt disease in tomato. Biocatal Agric Biotechnol 60:103320. https://doi.org/10.1016/j.bcab.2024.103320

Yu X, Li X, Ren C, Wang J, Wang C, Zou Y, Wang X, Li G, Li Q (2022) Co-composting with cow dung and subsequent vermicomposting improve compost quality of spent mushroom. Bioresour Technol 358:127386. https://doi.org/10.1016/j.biortech.2022.127386

Zaller JG (2007) Vermicompost as a substitute for peat in potting media: effects on germination, biomass allocation, yields and fruit quality of three tomato varieties. Sci Hortic 112:191–199. https://doi.org/10.1016/j.scienta.2006.12.023

Zhang B-G, Li G-T, Shen T-S, Wang J-K, Sun Z (2000) Changes in microbial biomass C, N, and P and enzyme activities in soil incubated with the earthworms *Metaphire guillelmi* or *Eisenia fetida*. Soil Biol Biochem 32:2055–2062. https://doi.org/10.1016/S0038-0717(00)00111-5

Zhang H, Tan SN, Teo CH, Yew YR, Ge L, Chen X, Yong JWH (2015) Analysis of phytohormones in vermicompost using a novel combinative sample preparation strategy of ultrasound-assisted extraction and solid-phase extraction coupled with liquid chromatography–tandem mass spectrometry. Talanta 139:189–197. https://doi.org/10.1016/j.talanta.2015.02.052

Zhao H-T, Li T-P, Zhang Y, Hu J, Bai Y-C, Shan Y-H, Ke F (2017) Effects of vermicompost amendment as a basal fertilizer on soil properties and cucumber yield and quality under continuous cropping conditions in a greenhouse. J Soils Sediments 17:2718–2730. https://doi.org/10.1007/s11368-017-1744-y

Zhao F, Zhang Y, Dong W, Zhang Y, Zhang G, Sun Z, Yang L (2019) Vermicompost can suppress *Fusarium oxysporum f. sp. lycopersici* via generation of beneficial bacteria in a long-term tomato monoculture soil. Plant Soil 440:491–505. https://doi.org/10.1007/s11104-019-04104-y

Zhao X, Lu H, Xin H, Shi Q, Li D, Liu L, Sun Z (2024) Improve Avermectin fermentation residue decomposition efficiency, bacterial community and mature compost quality by inoculating mature compost. Waste Biomass Valoriz 15:3785–3802. https://doi.org/10.1007/s12649-024-02420-8

Zuhair R, Moustafa YTA, Mustafa NSA, El-Dahshouri MF, Zhang L, Ageba MF (2022) Efficacy of amended vermicompost for bio-control of root knot nematode (RKN) *Meloidogyne incognita* infesting tomato in Egypt. Environ Technol Innov 27:102397. https://doi.org/10.1016/j.eti.2022.102397

Zuo Y, Zhang J, Zhao R, Dai H, Zhang Z (2018) Application of vermicompost improves strawberry growth and quality through increased photosynthesis rate, free radical scavenging and soil enzymatic activity. Sci Hortic 233:132–140. https://doi.org/10.1016/j.scienta.2018.01.023

# Greenhouse Gas Emissions During Composting Under Field Conditions and Their Impact on Environmental Health

**Pooja L. R., Renu Singh, Mamta Bisht, Manoj Shrivastava, and H. M. Akshay Kumar**

## 1 Introduction

Composting is widely used to manage organic waste yet presents a paradox. The practice contributes significantly to waste diminution and nutrient recycling but, at the same time, produces greenhouse gases (GHGs) like methane ($CH_4$), nitrous oxide ($N_2O$), and carbon dioxide ($CO_2$) (Fig. 1). Among them, $CH_4$ emits the largest amount of global warming potential (GWP) during the composting of yard waste, whereas $N_2O$ is the principal gaseous emission with respect to nitrogen-rich substrates. Despite this, however, the environmental footprint remains a concern. It is reported that the contribution of GHG emissions by composting form 0.01–0.06% of the global sources (Andersen 2010). The ecological implications of composting are highly variable, depending on several factors such as the type of organic waste being composted, the composting technology in use, management techniques, and the ultimate use of the compost. For example, whereas yard waste composting primarily emits $CH_4$, nitrogen-rich wastes are more susceptible to emissions of $N_2O$. Nevertheless, the implementation of efficient management strategies, including enhanced aeration, moisture regulation, and pretreatment techniques like anaerobic digestion, can significantly reduce these emissions. Moreover, cost-effective biological

Pooja L. R. (✉) · R. Singh · M. Shrivastava
Division of Environment Science, ICAR- Indian Agricultural Research Institute, New Delhi, Delhi, India

M. Bisht
Division of Environment Science, ICAR- Indian Agricultural Research Institute, New Delhi, Delhi, India

ICAR-Indian Institute of Water Management, Bhubaneswar, India

H. M. Akshay Kumar
Division of Plant Pathology, ICAR-Indian Agricultural Research Institute, Dhemaji, Assam, India



R. K. Singh et al. (eds.), *Composting of Farm Residues by Nature-Based Solutions*,
https://doi.org/10.1007/978-3-032-12255-1_4

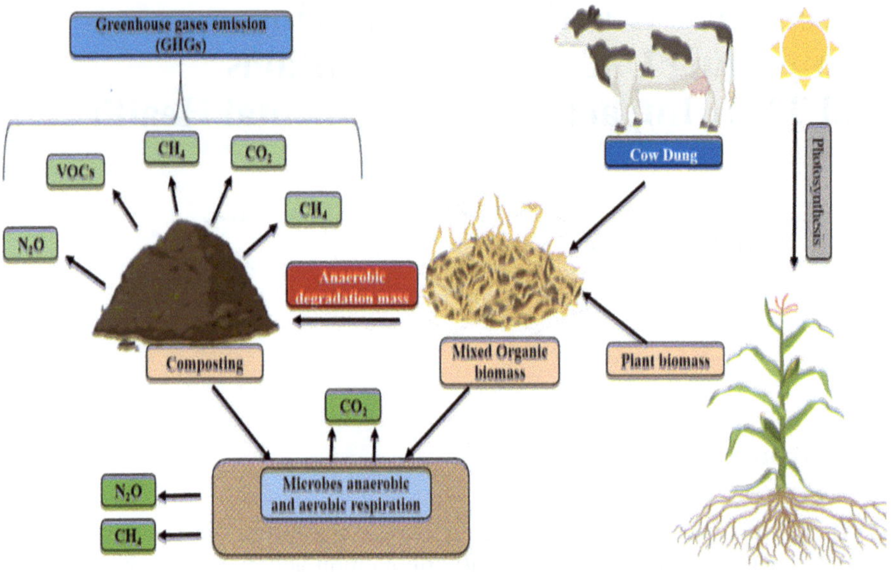


**Fig. 1** Greenhouse gas emissions from the composting process

technologies, such as biofiltration, have demonstrated potential in minimizing off-gas emissions, thereby further improving the sustainability of composting methods. The measurement of GHG emissions at each step of the composting process is important for developing reliable emission factors, which are important for conducting environmental impact assessments and for evaluating different waste treatment alternatives. Tools like Life Cycle Assessment have been instrumental in examining the trade-offs and overall sustainability of composting operations. In addition, the implementation of best practices such as efficient pile management and advanced off-gas cleaning systems can drastically reduce emissions, making composting a viable strategy to reduce GHGs associated with organic waste. Composting, despite its challenges, is increasingly being recognized as a sustainable waste management approach. It helps divert organic waste from landfills and contributes to soil health and agricultural productivity through the application of nutrient-rich compost.

As research and technological breakthroughs continue to enhance the methods of composting, its contribution to reducing climate change and circular economy principles will continue to grow. This chapter examines the dichotomous characteristics of composting, emphasizing its possible ecological advantages while concurrently outlining approaches to mitigate its shortcomings.

## 2 Greenhouse Gases in Composting: Sources, Mechanisms, and Implications

### *2.1 Methane Emissions: The Silent Contributor to Global Warming*

Methane ($CH_4$) is a potent greenhouse gas that plays a significant role in global warming, although its atmospheric concentration is less than carbon dioxide ($CO_2$). Its warming potential is strikingly high, approximately 80 times greater than $CO_2$ over a 20-year period. Methane emissions originate from diverse sources such as natural gas production, livestock farming, waste management, and melting permafrost (Thakur and Solanki 2022). The most alarming emission from melting permafrost is methane, which creates a positive feedback loop where increased emissions enhance global warming, thereby accelerating permafrost thaw and further methane release. Moreover, economic activities like beef production, oil and natural gas extraction, and biomass energy production have been considered the main drivers of methane emission intensity. Cross-national studies indicate that economic development and foreign direct investment in such sectors increase methane emissions, while environmental policies and state-led environmentalism can help reduce their impact. Since methane accounts for a major contribution to global warming, its abatement is imperative to curtail climate change. Reducing the emissions of methane would, therefore, require integrating policies that address natural as well as anthropogenic sources through focused practices at sectoral levels. That is a lesson for which international cooperation and innovation may be necessary.

#### 2.1.1 Biochemical Pathways Leading to Methane Production

The three distinct pathways for the process of methanogenesis are *hydrogenotrophic*, *aceticlastic,* and *methylotrophic* methanogenesis. These rely on different substrates and biochemical mechanisms and show the metabolic diversity of methanogenic archaea in forming methane.

(a) Hydrogenotrophic Methanogenesis

In hydrogenotrophic methanogenesis, carbon dioxide is reduced to methane by hydrogen or formate acting as an electron donor. This process is the dominant pathway in environments such as coal beds, aquatic systems, and neotropical tank bromeliads, where $H_2$ and $CO_2$ are in abundance. A defining characteristic of this process is the reliance on electron bifurcation for energy conservation, an example of which is the organism *Methanococcus maripaludis*. The occurrence of this pathway is susceptible to environmental conditions such as temperature and substrate availability and forms a central part of the methane cycle in a variety of ecosystems. In addition, the use of biogas upgrading is of particular interest, considering that supplementing $H_2$ would improve the methane yield as well as the microbial community efficiency under thermophilic conditions.

(b) Aceticlastic Methanogenesis

The aceticlastic pathway produces the greatest proportion of methane in acetate-rich environments by cleaving acetate into methane and $CO_2$. Specific methanogens, such as *Methanosarcina acetivorans*, use this pathway, which is very dominant in environments such as wetlands and wastewater treatment systems that contain accumulated acetate. The process is sensitive to environmental variables, such as ammonia levels and temperature, with thermophilic conditions favoring aceticlastic activity in specific settings.

(c) Methylotrophic Methanogenesis

Methylotrophic methanogenesis is a pathway that uses one-carbon compounds, such as methanol and methylamines, as substrates to produce methane. This pathway is highly prevalent in saline habitats and other environments where compatible solutes degrade into methylated compounds. Unlike the other pathways, methylotrophic methanogenesis exhibits exceptional biochemical versatility, incorporating unconventional substrates like methylated sulfur compounds, methoxylated aromatics, and tertiary amines. These special metabolic capabilities are sustained by new enzymes and cofactors that highlight the biochemical flexibility of methanogens. Recent work has also revealed this pathway in archaeal lineages outside of the *Euryarchaeota* and in the *Verstraetearchaeota*, which adds to the ecological and evolutionary relevance it is gaining. The different characteristics of hydrogenotrophic, aceticlastic, and methylotrophic methanogenesis point out their complementary roles in methane production across different ecosystems.

Hydrogenotrophic and aceticlastic pathways are predominant in conventional anoxic environments, but methylotrophic methanogenesis assumes a greater role in saline and specialized habitats. Collectively, these pathways shape the global methane budget, providing critical insights for climate change mitigation, renewable energy development, and biogeochemical modeling. Understanding the interplay between these pathways helps us better manage methane emissions and harness microbial processes to achieve sustainable solutions.

### 2.1.2 Factors Driving Methane Emissions in Composting

Methane ($CH_4$) emissions from composting are the result of a dynamic interaction of feedstock characteristics, technology choice, and operational practices. A clear understanding of these factors is necessary to minimize environmental impacts while maximizing the benefits of composting as a sustainable waste management strategy.

The type of organic material significantly influences methane production. Nitrogen-rich feedstocks, such as food waste and manure, often produce more nitrous oxide ($N_2O$) due to microbial nitrogen transformations, whereas carbon-rich feedstocks like yard and garden waste are more likely to emit methane, especially under anaerobic conditions. The carbon-to-nitrogen (C:N) ratio is a critical determinant; ratios that are too low favor excessive nitrogen loss, and more critical ratios may create an environment with anaerobic pockets, which favor the production of $CH_4$. Seasonal factors also play a role in this process because temperatures and humidity can vary during changing weather conditions, thereby influencing the microbial activity and aeration in compost piles. There are also operational patterns that include working hours and time for turning events that determine the oxygen diffusion and rates of decomposition and, therefore, methane emissions (Edjabou and Scheutz 2023). Open windrow composting system directly impacts the levels of methane emission. Open windrow composting systems, which are usually applied for garden and park wastes, yield lower methane emissions because their design is to ensure appropriate aeration. However, when anaerobic digestion integrates with open windrow composting, the residual organic materials from digestion can provide significant methane if they end up in anaerobic conditions during the subsequent composting phases. Despite this potential drawback, anaerobic digestion is a valuable pretreatment step that reduces the emissions of nitrous oxide ($N_2O$), ammonia ($NH_3$), and volatile organic compounds (VOCs), if managed appropriately.

The digestate from anaerobic digestion can improve nutrient recycling with minimized methane release if it is properly integrated into aerobic composting systems. Advanced composting systems with built-in aeration controls, biofilters, or closed environments also minimize methane emissions by keeping the process aerobic.

### Environmental and Operational Factors

Environmental parameters such as oxygen availability, moisture levels, and temperature have threshold effects on methane generation. Composting piles lacking sufficient oxygen often develop anaerobic microenvironments that favor the growth of methane-producing archaea. Moisture content exceeding 60% can exacerbate this problem by impeding air diffusion, while excessively dry conditions slow microbial decomposition. Another factor regulating emission rates is temperature. This is because microbial activity is highly sensitive to temperatures. High thermophilic temperatures of above 55 °C in the early stages of composting improve organic matter breakdown, but, at the same time, anaerobic pockets may develop if not checked. Methane emissions tend to be highly pronounced within the first 3 weeks since there is significant microbial action during this period (Vergara and Silver 2019). Operational best practices, such as frequent turning, appropriate pile sizing, and maintenance of optimal moisture levels, are critical to reduce methane emissions. Enforcing forced aeration systems will further enhance oxygen penetration, thereby suppressing anaerobic conditions. Other developing technologies include biochar or microbial inoculants designed to increase aerobic microbial activity and mitigate methane emissions. Mitigation measures for methane emissions from the composting process include the optimization of the feedstock C:N ratio, effective aeration, effective pile management, and other preprocessing methods, like anaerobic digestion to stabilize the feedstock prior to composting. Biofiltration is another emerging technology that offers significant potential for mitigating methane emissions while conserving both ecological and economic advantages. By controlling these factors that drive methane emissions, composting remains a sustainable and environmentally friendly solution for organic waste management, contributing to the global efforts to reduce greenhouse gas emissions.

### 2.1.3 Quantifying Methane's Role in the Composting Process

Emissions of methane ($CH_4$), a potent greenhouse gas, from composting have been studied in great detail to understand their sources, magnitudes, and mitigation strategies. Emission rates from composting facilities vary widely, with reported values ranging from 8.0 to 42.5 kg $CH_4$ $h^{-1}$ (Edjabou and Scheutz 2023) and from 1.3 to 10.7 kg $CH_4$ $h^{-1}$ (Hrad et al. 2014). A multitude of factors affect these differences, including variations in feedstock, technological implementation related to composting, actual operations, and seasonal impacts, among other factors. Some other operational factors are related to windrow turning. Several investigations demonstrate that turning windrows dramatically enhances methane emissions. It tends to be elevated when they are turned since the gases trapped in an anaerobic state are temporarily released through compost. High-nitrogen, high-moisture feedstocks are particularly prone to generating methane, although emissions can be mitigated by optimizing feedstock mixtures to improve aeration and reduce anaerobic conditions. Despite the emissions generated during composting, the process can contribute to net greenhouse gas reductions by diverting organic waste from the anaerobic conditions in landfills, where methane generation would be significantly higher. Understanding and formulating effective mitigation strategies toward the environmental impacts of composting facilities would require an accurate quantification of methane emissions. Measurement techniques have been quite diverse, with significant differences in accuracy and applicability. Small-scale methods include static flux chambers and funnel methods, which are widely applied but tend to underestimate the methane emissions due to the failure in capturing spatial and temporal variability. Dynamic plume methods, often combined with Fourier transform infrared (FTIR) spectroscopy, have been more robust for emission estimation. These techniques are more accurate, as they take into account the differences in gas dispersion and atmospheric conditions. Tracer gas dispersion and inverse dispersion methods have also been used for the determination of emissions from commercial composting facilities, with information about process-level emissions and their temporal dynamics. Emerging technologies, such as airborne imaging spectrometry, have further improved the ability to detect and quantify methane emissions from large-scale operations, including composting facilities and landfills. Advanced techniques can capture emission patterns in real time and with a high resolution that traditional methods or biogas generation models often miss. Such methodologies are invaluable for improving emission inventories and informing policy decisions on waste management and greenhouse gas mitigation.

## 2.2 *Nitrous Oxide Emissions: The Potent Greenhouse Gas from Nitrogen-Rich Wastes*

Nitrous oxide is a potent greenhouse gas emitted during wastewater treatment, contributing greatly to climate change. The $N_2O$ emissions in wastewater treatment plants are influenced by several operational parameters, such as low dissolved oxygen, increased nitrite concentrations, and low chemical oxygen demand (COD)/N ratios. The main microbial pathways leading to $N_2O$ production are hydroxylamine oxidation, nitrifier denitrification, and heterotrophic denitrification. While $N_2O$ is an environmental concern, research has been conducted recently about its potential as an energy source through decomposition and recovery processes. In order to reduce $N_2O$ emissions, the optimization of operational parameters, selection of proper N-removal processes, and development of validated dynamic models are important (Massara et al. 2017).

### 2.2.1 Nitrous Oxide Emissions from Composting: Processes, Drivers, and Global Implications

Nitrous oxide is a potent greenhouse gas with a global warming potential approximately 300 times greater than carbon dioxide ($CO_2$). It is one of the major by-products of nitrogen transformations during composting. The production of $N_2O$ in composting systems is mainly associated with nitrification and denitrification processes, which depend on the availability of nitrogen (N), carbon (C), oxygen, moisture, and temperature. Ammonia-oxidizing bacteria and archaea play critical roles in these processes, contributing to $N_2O$ generation through their metabolic pathways (Prosser et al. 2020). Effective aeration and balanced feedstock composition can mitigate anaerobic conditions that favor $N_2O$ emissions. Environmental conditions act as key triggers for $N_2O$ emissions during composting. Oxygen concentration is particularly critical, as limited oxygen availability promotes denitrification, increasing the $N_2O$:$N_2$ ratio. Seasonal temperature and moisture fluctuations also affect microbial activity, with higher emissions frequently associated with warm, moist conditions. Composting N-rich materials, such as food and animal waste, generally results in higher emissions of $N_2O$ compared to C-rich materials like yard trimmings.

While composting accounts for emissions of $N_2O$, this contribution to the global warming problem is significantly smaller compared to other waste management practices. Composting helps cut down the total greenhouse gas emission by keeping organic waste from landfills, which typically experience low oxygen conditions, under which the production of $CH_4$ is more prolific. It is, therefore, crucial to optimize practices in composting in ways that limit $N_2O$ emissions, such as incorporating nitrification inhibitors or biochar amendments and minimizing moisture and oxygen levels. The broader global warming potential of $N_2O$ is still controversial. Although some studies highlight its significant role in climate change and stratospheric ozone depletion, recent research findings indicate that its warming impact may be less serious than previously estimated. Advances in nitrogen management, especially in agriculture, offer significant opportunities for emission reductions. On the other hand, tight regulations on $N_2O$ emissions might jeopardize food security. Thus, the implementation of mitigation strategies needs to be balanced. Denitrification in composting necessitates an integrated approach toward nitrogen transformations, microbial dynamics, and environmental triggers to minimize $N_2O$ emission. Future research in this area should focus on optimizing $N_2O$-to-$N_2$ conversion during denitrification and innovative ways of minimizing emissions while preserving the quality of the compost.

## *2.3 Ammonia ($NH_3$) and Volatile Organic Compounds (VOCs): The Overlooked Air Pollutants*

### 2.3.1 Ammonia Emissions: Sources, Mechanisms, and Environmental Impacts

Emissions of ammonia ($NH_3$) are primarily from agricultural sources, especially manure from livestock and urea-based fertilizers, which account for around 75% of the global ammonia emissions. Nonagricultural sources include human activities, natural vegetation, wild animals, biomass burning, and industrial processes, among others, which contribute to ammonia emissions but to a smaller extent. Livestock-related emissions are 81% of the total in Europe, and major contributions come from farm animal wastes. Minor sources such as coal combustion, waste incineration, and sewage treatment plants are also found to be important in global emissions, according to emerging research. Ammonia formation mainly occurs through the hydrolysis of urea, catalyzed by the enzyme urease. This process is sensitive to environmental conditions like pH, temperature, and moisture content, in which pH and temperature play a crucial role in determining ammonia flux. Various mitigation measures have been developed to reduce the emissions of ammonia. These include the application of urease inhibitors (enhanced-efficiency fertilizers) and biofertilizers, with reported average mitigation efficiencies of 54% and 50.1%, respectively (Li et al. 2023).

Ammonia emissions from agriculture have considerable environmental impacts, including soil acidification, loss of biodiversity, and the formation of particulate matter. In Scotland, ammonia emissions are variable with livestock density and management practices and are higher in lowland areas. These result in nitrogen deposition, often exceeding critical levels. Accurate models of emission, taking into account manure management and weather conditions, have to be developed in the creation of policies that tend to harmonize environmental protection and agricultural productivity.

### 2.3.2 The Role of VOCs in Air Quality and Odor Formation

Composting of municipal solid waste produces a range of VOCs, which cause air pollution and may lead to health issues. More than 100 VOCs have been identified in composting processes, and terpenes are the major group (Sánchez-Monedero et al. 2018). VOC concentrations are highest during the primary fermentation and tipping stages of composting, while the early stages of composting, especially the first 2 weeks, produce the most intense emissions. Food waste composting typically results in higher VOC and odor concentrations compared to other materials, such as sewage sludge or digested organic waste. Although individual VOCs are not usually a health hazard, their cumulative effects from mixtures resulting from composting are a source of concern (Nie et al. 2018). These VOCs are aromatic compounds, sulfur-containing organics, and terpenes. In addition, odor-causing VOCs include ammonia, methanethiol, and sulfides, which are particularly prevalent during the thermophilic phase of composting. VOC emission and odor correlation is, therefore, essential for upgrading composting management practices to reduce environmental impacts. Aeration strategies and biofiltration are typically used for VOC emission control. The latter is more promising since treatment efficiencies of up to 90% have been observed for VOCs (Dhamodharan et al. 2019). However, there is still a challenge to determine the optimal operational parameters as well as the mechanisms behind VOC biodegradation, which are necessary to adequately manage emissions from composting facilities. Therefore, better VOC and odor-producing substance management is important to minimize the environmental and health effects of composting operations.

### 2.3.3 Interaction of $NH_3$ and VOCs with Greenhouse Gases

Composting is an effective process of waste management. However, it has been associated with emitting GHGs, such as $CH_4$, $N_2O$, and $CO_2$, as a result of decomposition activity in organic matter. These compounds are ammonia ($NH_3$) and VOCs; however, they interact very complicatedly with the GHG dynamics. The indirect pathway for $N_2O$ production is also possible when deposited materials enter an ecosystem through nitrification and denitrification resulting from volatilization of $NH_3$ produced from the microbial degradation of substrates that are rich in nitrogen. Microbially Induced Chemistry may produce other precursors to secondary pollutants, such as VOCs like alcohols, ketones, and terpenes, which can affect the atmospheric chemistry. These interactions are controlled by microbial activity, aeration, temperature, and moisture levels. Increasing aeration will enhance the production of $CH_4$ and $N_2O$ but reduce the production of $CH_4$. Composting in the new styles like adding biochar, managing aeration, and controlling pH minimizes the $NH_3$ emission through adsorption by biochar, optimization of the microbial activities, and suppression of the VOCs' release during composting. The inclusion of gas capture technologies in the utilization of microbial inoculants may further optimize composting sustainability. Low-emission strategies, therefore, need to understand the dynamics between $NH_3$ and VOC within composting systems. Reducing these interactions is significant to minimize the environmental impact and improve air quality by way of being part of the climate change mitigation measures.

# 3 Factors Influencing GHG Emissions During Composting: An In-Depth Analysis

## *3.1 Feedstock Mix*

Composting is widely used to recycle organic wastes, with varying levels of greenhouse gas emissions depending upon the feedstock mix and the conditions used for processing. There are different methane generation potentials, when landfilled or lagooned, for feedstocks, including municipal solid waste, biosolids, and animal manures. High-nitrogen, high-moisture materials generally produce the most GHG emissions during composting. Among various materials, food waste and chicken litter are responsible for the highest GHG emissions (Zeng et al. 2023). However, composting can also lead to GHG savings through methane avoidance and carbon sequestration in soil. Overall, the GHG impact of composting can vary from significant savings to a net load, depending on factors like the type of waste, technology, and the final use of the compost. Optimizing feedstock selection and mixing is critical in reducing GHG emissions during composting.

(a) *Carbon-to-Nitrogen (C:N) Ratio*

The carbon-to-nitrogen (C:N) ratio plays an important role in managing the GHG emissions produced during composting. An optimal C:N ratio of 25–30 helps minimize emissions and improve the compost quality. Very low ratios tend to enhance $N_2O$ emissions, while higher C:N ratios often decrease overall GHG emissions. Availability of nitrogen affects the microbial activity, decomposition rate, and gaseous loss that occurs during the process. Aeration and turning frequency are also important; higher turning frequencies and optimum C:N ratios reduce the nitrogen loss. Proper management of these factors will minimize methane ($CH_4$) and hydrogen sulfide ($H_2S$) emissions, which will enhance environmentally friendly composting.

(b) *Diversity in Organic Material Types*

The same organic materials emit different kinds of GHGs during composting. For example, emissions from yard waste composting are mainly methane, but those from nitrogen-rich materials, such as animal manure, have more $N_2O$. Emissions vary with different factors such as energy sources, operating conditions, soil types, and application methods. Effective pile management and aeration can reduce the production of methane, though forced aeration may increase ammonia emissions. Mitigation techniques, such as biofiltration and the use of biochar, can reduce emissions but not universally for all materials. Emission factors should also be evaluated on a case-by-case basis since the production level of GHGs varies due to the differences in parameters of the composting scenarios studied.

### 3.2 Impact of Decomposability, Moisture Content, and Nutrient Concentration

Moisture content, aeration, and turning frequency are the greatest factors that affect the emission of GHGs in a composting system. For example, the use of decomposers in composting goat manure has been proven to reduce methane emissions without the loss of nutrient content (Puastuti et al. 2021). Water content affects microbial activity in a composting process, as it influences oxygen diffusion; and consequently, it reflects its effect on the entire composting process. The ideal moisture content for composting ranges between 50% and 60%, although other studies have indicated a range of 50–70%. Moisture levels below 45% or above 65% may inhibit the composting process (Shan et al. 2021). Higher moisture (>64%) will cause anaerobic conditions that produce odors and higher sulfate concentrations, whereas low moisture (<64%) promotes aerobic degradation. Biological drying, metabolic production of water, and changing compaction and porosity over time make it difficult to maintain and achieve an optimal moisture content. A water activity of 0.94 is recommended to find the lower moisture limit. Air flow and potential clogging in the compost layer influence the upper limit. It is known that the moisture content varies depending on the compost mix and the stages of the process. Moisture management is critical in minimizing $CH_4$ and $N_2O$ emissions. In cattle feedlots, maintaining the soil moisture at 0.2–0.4 g $H_2O$ $g^{-1}$ dry matter optimizes the environmental impact (Miller and Berry 2005).

### 3.3 Environmental Conditions

The main environmental factor that drives microbial activity during composting is temperature. There is optimal microbial activity at 35–50 °C, though the activity declines sharply above 55–60 °C. Other factors that determine decomposition include the C:N ratio, moisture, pH, and oxygen content (Nemet et al. 2021). As decomposition proceeds, pH, protein content, and total organic content mirror the microbial activity. Microbial community composition changes over the course of the process, with bacteria dominating early in the process, followed by fungi later in the process.

Temperature control is critical in composting to ensure that optimal microbial activity and pathogen inactivation are achieved. Temperatures above 60 °C suppress microbial communities, whereas low temperatures enhance decomposition. An appropriate temperature balance will prevent conditions that may hamper the efficiency of composting by balancing microbial activity, rate of decomposition, and pathogen reduction. pH is one of the significant factors in microbial dynamics during composting. Compost pH is initially acidic but tends to be neutralized as decomposition progresses. The pH, moisture content, and total carbon affect microbial community succession (Ge et al. 2022). Microbial diversity is positively correlated with the compost pH, nitrogen content, C:N ratio, and protease activity. Enzyme activities, especially those of lignocellulolytic enzymes, play an important role in the biodegradation of organic matter.

### 3.4 Oxygen and Aeration

The availability of oxygen, which is very crucial for aerobic respiration, impacts composting efficiency and GHG emissions. Aeration rates that are high decompose materials at faster rates and also decrease methane emissions. The measurement of oxygen concentration is critical in maintaining aerobic conditions since low concentrations of oxygen facilitate anaerobic degradation, thereby enhancing the production of methane (Ermolaev et al. 2019). Forcing aeration can increase oxygen availability but will introduce some operational complications affecting the emission profiles.

All factors that influence GHG emissions in composting have to be managed to optimize the environmental outcome. Feedstock composition, moisture content, and environmental conditions all have an impact on microbial activity and the balance between aerobic and anaerobic processes. Understanding the interplay between these factors allows us to design composting systems that would minimize GHG emissions while reducing organic waste and enhancing sustainable practices in waste management.

## 4 Mitigation Strategies: Advanced Techniques for Reducing GHG Emissions

### 4.1 Optimizing Composting Conditions: Best Practices for Emission Control

Temperature and moisture management is essential to minimize greenhouse gas emissions during composting. Studies suggest maintaining thermophilic conditions of 50–60 °C for better microbial activity, enhanced organic matter degradation, and inhibition of methane production. Moisture management in the optimal range of 50–60% prevents the formation of anaerobic zones, which could cause methane and nitrous oxide emissions. Hyperthermophilic composting, a newer method, has shown promise in reducing composting time while enhancing organic matter decomposition and lowering GHG emissions by increasing temperatures through microbial inoculation. Aeration is crucial for promoting aerobic decomposition, reducing anaerobic hotspots, and controlling methane and ammonia release. Recent innovations include forced aeration systems and the strategic addition of bulking agents to maintain porosity and oxygen flow (Table 1). Balanced moisture management complements aeration, thus preventing waterlogging and creating anaerobic conditions; this forms the primary source of nitrous oxide emission. Turning and mixing compost piles improve aeration, distribute moisture, and minimize compaction. All of these are critical for maintaining aerobic conditions. It has been shown that frequent turning—ideally once per week during active composting—should reduce ammonia and nitrous oxide emissions and ensure uniform decomposition. Automated and mechanized turning technologies are increasingly used to enhance the control and efficiency of the process. These practices underpin the necessity of precision in managing parameters in composting to mitigate GHG emissions and ensure good-quality compost suitable for agriculture.

## 4.2 *Use of Additives: Innovations in Emission Reduction*

Composting organic waste can produce GHGs and lead to nitrogen loss, but additives can mitigate these issues. Various additives have been shown to reduce GHG emissions during composting, with physical additives like biochar and zeolite being particularly effective for $N_2O$ reduction (Table 2). Biochar addition during composting has shown promising results in mitigating greenhouse gas emissions and improving compost quality. Studies have demonstrated that biochar can significantly reduce methane ($CH_4$) emissions by up to 58% and reduce various air pollutants by 61–70% during dairy manure composting (Harrison et al. 2024). The effectiveness of biochar in reducing emissions depends on factors such as application rate, raw material type, and pyrolysis temperature. While higher biochar application rates (20%) may increase $CH_4$ emissions, they can decrease $N_2O$ and $NH_3$ emissions, resulting in a net reduction of global warming potential. Biochar addition at lower rates (3–5%) can accelerate organic matter degradation, enhance N mineralization, and improve physical properties of the compost mixture without significantly impacting gaseous emissions (Gao et al. 2023). These findings suggest that biochar-composting could be an effective strategy for reducing the environmental impacts of manure management.

Zeolite can decrease $CH_4$ emissions by up to 88.45% and nitrogen loss by 28.80% in sewage sludge composting (Wang et al. 2017). Additives can also improve the compost quality, reduce heavy metal mobility, and increase plant-available nutrients. Acidic additives such as apple pomace, citric acid, and magnesium hydrogen

**Table 1** The effects of process conditions on GHGs and odors during composting

| Feedstock | Measure | Impact on gaseous emissions (relativ0e to control) | | References |
|---|---|---|---|---|
| | | $CH_4$ | $N_2O$ | |
| Cow manure, wheat straw | Functional membrane–covered composting | – | ↑ | Xiong et al. (2022) |
| Chicken manure, mushroom residue, cron stalk, bran | Semipermeable membrane–covered composting | ↓ | ↓ | Sun et al. (2022) |
| Cow manure | Semipermeable membrane–covered composting | ↓ | ↓ | Fang et al. (2021) |
| Kitchen waste, garden waste | Aeration intensity (0.48 L/kg/min) | ↓ | ↓ | Xu et al. (2021) |
| Cow manure, cornstalk | Negative pressure aeration (0.75 L/min/kg$^2$) | ↑ | ↓ | Wang et al. (2018) |
| Chicken manure, mushroom residue | Intermittent aeration (10 min on–30 min off) | ↑ | ↑ | Ma et al. (2020) |
| Chicken manure, rice husk | Electric field–assisted composting (2 V DC) | ↓ | ↓ | Li et al. (2022) |
| Chicken manure, mature compost, rice husk, dewatered sewage sludge | Electric field–assisted composting (2 V DC) | ↓ | ↓ | Tang et al. (2019) |

↑ increase (no detailed data), ↓ decrease (no detailed data)

phosphate have shown promise in improving nitrogen conservation and reducing GHG emissions during sewage sludge composting (Pan et al. 2018). However, some acidic additives like elemental sulfur and phosphoric acid may increase $N_2O$ emissions and salinity while reducing compost maturity. Emerging additives have the potential to significantly enhance composting processes and outcomes. Organic, inorganic, and biological additives can modify the physical, chemical, and microbial properties of compost, improving its final quality. These additives can affect key parameters like temperature, pH, and moisture, influencing the composting process and reducing greenhouse gas emissions. Specific additives, such as silage and flue gas desulfurization gypsum, have been shown to accelerate the composting of green waste, producing high-quality compost in shorter periods (Yin et al. 2021). The use of additives, particularly in combination with vermicomposting, can increase plant-available nutrients while decreasing nitrogen leaching and heavy metal mobility. To maximize sustainability, co-composting and co-vermicomposting strategies should be optimized locally, integrating the resulting amendments into a circular economy approach for agricultural systems.

**Table 2** The effects of different additives on GHGs and odors during composting

| Feedstock | Additive | Impact on gaseous emissions (relative to control) | | | Note | References |
|---|---|---|---|---|---|---|
| | | $CH_4$ | $N_2O$ | Volatile sulfur compounds (VSCs) | | |
| Chicken manure, rice husk | Electric field–assisted composting (2 V) + 10% biochar | – 70% | – 30% | ↓ | Lasted for 30 days; promoted temperature rise; increased electrical conductivity | Li et al. (2022) |
| Pig manure, apple sawdust | 5% biochar +10% sugarcane bagasse | – 63% | ↓ | –36% | Lasted for 60 days; similar temperature variation | Wang et al. (2022) |
| Sewage sludge, straw | 5% bamboo biochar +0.3% aerobic microorganism agent | – 44% | ↓ | – | Lasted for 29 days; promoted temperature rise | Xue et al. (2019) |
| Dewatered sewage sludge, wheat straw | 12% wheat straw biochar +4% calcium-bentonite | ↓ | ↓ | ↑ | Lasted for 43 days; promoted temperature rise | Awasthi et al. (2018) |
| Pig manure, cornstalk | 0.2% dicyandiamide +6.6% dihydrogen phosphate | – 39% | – 36% | –18% | Lasted for 40 days; similar temperature variation; initial C/N: 17 | Yang et al. (2022) |
| Pig manure, cornstalk | 0.6% dicyandiamide +6.6% dihydrogen phosphate | – 33% | – 25% | –21% | Lasted for 40 days; promoted temperature rise; increased electrical conductivity | Yang et al. (2022) |
| Food waste composting | Mature compost and sulfur powder | ↓ | – 37% | – | Lasted for 15 days | Xiong et al. (2023) |
| Pig manure, sawdust | 0.6% dicyandiamide +0.5% thiophosphoryl triamide | ↓ | – 64% | – | Lasted for 20 days; similar temperature variation | Jiang et al. (2021) |
| Sewage sludge, cornstalk | 2% dicyandiamide +2.5% dihydrogen phosphate | ↓ | – 86% | – | Lasted for 35 days; later temperature peak; reduced $CH_4$ emissions by more than 50% | Zhao et al. (2020) |
| Pig manure, cornstalk | 10% chicken manure | ↓ | – 88% | – | Lasted for 49 days; promoted temperature rise; $H_2S$: –35%, $Me_2S$: –100%, $Me_2S_2$: –63% | Liu et al. (2020) |
| Chicken manure, wheat straw | 5% manure compost | – 59% | – 9% | –41% | Lasted for 42 days; similar temperature variations | Chen et al. (2020) |

(continued)

**Table 2** (continued)

| Feedstock | Additive | Impact on gaseous emissions (relative to control) | | | Note | References |
|---|---|---|---|---|---|---|
| | | $CH_4$ | $N_2O$ | Volatile sulfur compounds (VSCs) | | |
| Kitchen waste, cornstalk | 10% manure compost | – 45% | – 74% | – | Lasted for 35 days; promoted temperature rise | Yang et al. (2019) |
| Sewage sludge, straw | 0.3% aerobic microorganism agent | ↓ | ↓ | | Lasted for 29 days; promoted temperature rise | Xue et al. (2021) |
| Kitchen waste, garden waste | 0.9% aerobic microbial inoculant | ↓ | ↓ | ↑ | Lasted for 35 days; promoted temperature rise | Gao et al. (2021) |
| Sewage sludge, rice husk | 5% thermotolerant sulfide-oxidizing compound bacterial consortium (w/w) | ↓ | ↓ | ↓ | Lasted for 22 days; promoted temperature rise; reduced $H_2S$ by 49% | Chen et al. (2022) |
| Sewage sludge, spent mushroom substrate | 5% thermophilic nitrifying bacteria (v/v) | | | ↓ | Lasted for 20 days; promoted temperature rise; initial C/N: 16 | Zhao et al. (2020) |
| Sewage sludge, wheat straw | 10% Ca-bentonite | – 86% | – 81% | ↑ | Lasted for 56 days; promoted temperature rise; Ca-bentonite led to higher $NH_3$ emissions | Wang et al. (2017) |
| Poultry manure, sugarcane straw, mature compost | 10% poultry litter biochar | ↓ | ↓ | | Lasted for 60 days; similar temperature variation | Agyarko-Mintah et al. (2017) |
| Chicken manure composting | Wheat straw | – 55.9% | – 9% | – | | Chen et al. (2022) |
| Biochar | Wood vinegar | – 62% | – 35% | – | | Wang et al. (2022) |

↑ increase (no detailed data), ↓ decrease (no detailed data)

## 4.3 *Improved Aeration Techniques: Striking a Balance Between Aerobic Efficiency and $NH_3$ Control*

### 4.3.1 Forced Aeration: Mechanisms and Benefits

The forced aeration systems have a higher performance in terms of the promotion of microbial growth, enzyme activity, and contaminant degradation compared with manual turning. It can maintain higher optimal composting temperatures of around 50–65 °C for even very harsh winter climates. High degrees of composting, as well as higher nitrogen conversion into organic forms, are also obtained with forced aeration. However, a comparative study found that passive aeration methods may maintain higher temperatures for longer periods than forced aeration. Forced aeration offers several advantages over passive or natural aeration methods. It ensures a more reliable oxygen supply and heat removal, which reduces the need for turning and space. Compared to the turned windrow system, this improves better growth of mesophilic and thermophilic microorganisms in cellulase activity as well as nicotine degradation in the composting of tobacco waste. Another variation of forced aeration, called passive aeration, displayed better performance in maintaining a high temperature compared with natural aeration, conserving nitrogen content and resulting in a reduction in the composting time.

### 4.3.2 Passive Aeration Systems: Low-Cost Solutions with a High Impact

Passive aeration systems are a cost-effective means of composting, especially for small-scale operations. These systems can match the performance of active aeration methods but with reduced capital and operating costs. Passive aeration with vertical pipes is found to be more efficient than that with horizontal pipes in terms of delivering air to the compost piles to reach the thermophilic temperatures that will kill pathogens. A practical analytical model has been developed to predict passive airflow in compost, taking into account physical characteristics and temperature. Although both active and passive aeration systems work well in household composting, active aeration may be preferred because of shorter maturation periods and reduced odor and fly nuisance. Overall, passive aeration systems are very promising solutions for composting at a low cost and high impact, especially in small farms, where nitrogen loss can be minimized.

### 4.3.3 Integrated Aeration Strategies: Combining Techniques for Optimal Results

The results of studies have shown that integrated aeration strategies in composting significantly influence the efficiency and quality of the composting process. A few studies demonstrated that different aeration methods, including one-directional airflow and its recirculation, resulted in distinct temperature and moisture profiles, and one-directional airflow resulted in the highest decomposition rates. The combination of pile turning with natural ventilation improved the compost quality while reducing energy consumption and emissions. Alkoaik (2019) observed that a continuous aeration–rotation process significantly reduced the active composting phase to only 4.5 days, improving temperature and nitrogen content control. Additional modeling showed that one-way airflow resulted in the highest temperature gradients, while backflow improved moisture control. Collectively, these studies imply that combining various aeration methods can improve the efficiency of composting and product quality.

## 4.4 *Effective Pile Management: Strategic Approaches to Minimize Emissions*

Proper pile management can therefore reduce the emissions generated by composting. In this process, the most significant gases are $CH_4$ and $N_2O$. Mixing piles leads to an increase of approximately 20% in overall GHG emissions. Compared to static piles, it is evident that the initial three weeks of composting are highly significant because they encompass a large proportion of both $CH_4$ and $CO_2$ emissions (Vergara and Silver 2019). Maintaining adequate oxygen levels and appropriate moisture content can help decrease methanogenesis. The addition of bulking agents can greatly reduce $N_2O$ and $CH_4$ emissions by 53% and 71%, respectively. Turned composting systems have the potential to decrease GHGs, and covering or compacting piles can greatly reduce $NH_3$ emissions (Pardo et al. 2015). However, such strategies may result in pollution swapping and, therefore, require holistic approaches in the development of sustainable solid waste management systems.

### 4.4.1 Oxygen Monitoring and Its Role in Reducing Methane Production

Oxygen monitoring plays an essential role in the optimization of composting processes as well as the reduction in methane production. Oxygen monitoring equipment allows one to make real-time changes to aeration strategies while holding the oxygen level optimal at any given time. An oxygen concentration greater than 14% can minimize hydrogen sulfide generation by an aerobic compost pile. Methane production usually happens in the anaerobic layer of compost piles, yet it can be oxidized through the upper layers, avoiding the release of methane when it is in the maturity phase. Low oxygen percentages of 1–2.5% may even prolong the mesophilic phase and decrease microbial respiration, thus delaying ammonia release and reducing its output at higher oxygen percentages, for example, 16%. However, at low oxygen percentages during the thermophilic phase, anaerobic conditions may still generate methane. Therefore, maintaining adequate oxygen levels throughout the composting process is essential for minimizing methane formation and optimizing the overall composting efficiency.

#### 4.4.2 Frequency and Methods of Pile Turning: Enhancing Decomposition and Aeration

Composting of agricultural and municipal biowaste is a good method of waste management and nutrient recycling. The frequency of turning is very important in terms of efficiency in composting and the quality of the final product. More frequent turning, such as every 2–4 days, tends to generate higher temperatures and ensure faster removal of moisture and greater organic matter reduction compared to less frequent or no turning (Stegenta et al. 2019). However, this can increase the loss of nitrogen. Pile size and pile configuration also impact the composting process. Large piles may limit oxygen penetration and temperature distribution. Seasonal variation impacts composting dynamics, with winter piles having lower carbon monoxide production. Optimal turning strategies may include more frequent turning in the initial stages and less as the process progresses. Proper composting management yields a final product with lower moisture contents, higher nutrient concentrations, and higher stability.

#### 4.4.3 Real-Time Monitoring Systems for Dynamic Pile Management

Real-time monitoring systems for composting processes have been developed to optimize waste management and produce high-quality compost. Such systems use wireless sensor networks to measure key parameters like temperature, moisture, and electrical conductivity at multiple points in compost piles (Park et al. 2019). Intelligent monitoring devices make continuous data collection and visualization possible, allowing users to follow the progress of composting through real-time databases or mobile applications. The implementation of these technologies allows timely interventions, such as dumping and aeration, to enhance the inactivation of pathogens and the retention of nutrients. Sensor placement is an important aspect because measurements in a single compost vessel vary significantly. Advanced monitoring systems can also monitor other factors like pH and levels of compost. These technologies, therefore, reduce the need for manual intervention but improve overall efficiency and hygiene in composting.

## 5 Environmental Health Impacts: Assessing the Broader Implications of Composting Emissions

### *5.1 The Positive Contribution of Composting to Environmental Health*

Composting is an environmentally friendly process that converts organic wastes into nutrient-rich soil amendments, providing many environmental benefits. It improves soil health in terms of its physical, chemical, and biological properties, with increased organic matter, water-holding capacity, and microbial activity. Compost application may result in carbon sequestration; up to 40% of applied carbon is retained in the short term. This decreases the application of chemical fertilizers since compost is rich in all the necessary nutrients, including nitrogen, phosphorus, and potassium. In addition, the application of compost enhances crop yield by up to 52% and the stability of soil aggregates by 29–63% and reduces soil bulk density by 0.7–20%. These results add to better-quality soil, lower soil erosion, and increased productivity in agriculture. Therefore, composting becomes a sound strategy for waste management as well as environmental protection.

### 5.1.1 Reducing Landfill-Related GHG Emissions: A Quantitative Perspective

Composting organic waste provides a good opportunity for minimizing GHG emissions compared to landfilling. Studies have shown that composting can reduce GHG emissions by up to 66.72% in comparison to conventional waste management practices (Wong et al. 2017). The most significant source of GHG reduction arises from the avoidance of methane generation in landfills. Composting one ton of municipal solid waste is estimated to reduce methane emissions by approximately 0.2–0.3 tons. However, the global warming factor of composting varies widely, ranging from −900 to 300 kg $CO_2$ equivalents per ton of wet waste, depending on factors such as waste composition, the technology used, and compost application (Nguyen Thanh 2012). The greatest GHG savings are achieved when compost is used as a substitute for peat in growth media production. These findings highlight composting as a viable strategy for sustainable organic waste management and GHG mitigation.

### 5.1.2 Enhancing Soil Health Through Compost Application

Compost application was proven to significantly improve the health of the soil and increase crop yields. It enhanced the soil structure, its water-holding capacity, and nutrient availability (Wright et al. 2022). Research has found that compost increases the carbon content in the soil organic matter, cation exchange capacity, and microbial diversity. Field experiments have shown that compost application increased corn yield by two to four times and improved the soil moisture content (Matsui et al. 2016). For tomato and cabbage crops, compost stimulated growth, yield, and product quality. Of course, compost also improves the efficiency of nitrogen use if applied with chemical fertilizers. Although both drum composting and vermicomposting are efficient, vermicompost is less contaminated with heavy metals. Thus, compost application is sustainable in improving soil fertility and recycling nutrients and could potentially minimize nutrient leaching.

### 5.1.3 Carbon Sequestration Potential of Compost: Long-Term Benefits

Significant potential of long-term carbon sequestration and climate change mitigation exists in compost amendments to grasslands and rangelands. Research has revealed that the application of compost may enhance soil carbon stocks with long-term greenhouse gas benefits that persist for decades. Climate mitigation potential of compost amendments varies from 63 to 158 g $CO_2$-eq/m$^2$/year, depending on the time considered. In ornamental horticulture, partially substituting peat-based growth media with compost or biochar could potentially store up to three million tons of $CO_2$ equivalent annually (Alvarez-Vega et al. 2022). Long-term modeling studies predict that compost and biosolids can significantly increase soil organic carbon stocks compared to anaerobic digestate and nitrogen fertilizer, with crop rotation further enhancing carbon sequestration (Badewa et al. 2023). They emphasize the potential role of compost in sustainable land management and climate change mitigation.

## 5.2 *Potential Negative Impacts: Unintended Consequences of Composting Emissions*

Composting is positive for organic waste management but results in some negative environmental impacts due to gaseous emissions. The main concerns include the emission of greenhouse gases ($CO_2$, $CH_4$, and $N_2O$) and air pollutants ($NH_3$ and VOCs). $CH_4$ is the largest contributor to global warming potential in the case of yard waste composting, while $N_2O$ dominates in the case of nitrogen-rich wastes. Emissions are characterized by waste characteristics, pretreatment processes, and conditions in composting. Effective pile management and aeration are essential in achieving minimal $CH_4$ emissions, especially in the first three weeks of composting. Life Cycle Assessment studies illustrate the dependency of system boundaries and functional units by showing that in-vessel and home composting are considered the most favorable options among all composting methods. Emissions notwithstanding, composting can still have a net negative life cycle emission when taking into consideration avoided emissions and sinks.

# 6 Conclusion

Sustainable composting forms a significant part of diminishing GHG emissions while ensuring the improvement of soil health and agricultural productivity. The benefits of field-based composting are immense, yet handling the emissions of methane, nitrous oxide, and ammonia during the process is challenging. However, through nature-based solutions, such as optimal temperature and moisture control, precise management of aeration, and the incorporation of biochar or microbial inoculants, significant emission reductions can be achieved, with good compost quality. Innovative strategies, including hyperthermophilic composting and automated turning systems, offer opportunities to improve the efficiency and sustainability of composting processes. Such approaches, therefore, are not only likely to reduce the organic waste management footprint on the environment but also contribute to the realization of circular agriculture through the recycling of nutrients back to the soil. Adding emission-monitoring and mitigation technologies ensures that composting aligns with global climate change goals as well as supports environmental health. By embracing these best practices, composting can become an important solution for managing farm residues, mitigating GHG emissions, and promoting sustainable agricultural systems. Efforts should be devoted to further research, encouraging technology use, and policy support for optimizing composting practices under field conditions, ensuring a resilient and sustainable agricultural landscape.

**Competing Interests** The authors have no conflicts of interest to declare that are relevant to the content of this chapter.

## References

Alkoaik FN (2019) Integrating aeration and rotation processes to accelerate composting of agricultural residues. PLoS One 14:e0220343. https://doi.org/10.1371/journal.pone.0220343

Alvarez-Vega C, Streitwieser DA, Álvarez-Barreto J, Leon-Reyes A (2022) Characterization of low-density polyethylene greenhouse films during the composting of rose residues. SN Appl Sci 4:275. https://doi.org/10.1007/s42452-022-05153-7

Andersen JK (2010) Composting of organic waste: quantification and assessment of greenhouse gas emissions. Technical University of Denmark

Awasthi MK, Wang Q, Chen H et al (2018) In-vessel co-composting of biosolid: focusing on mitigation of greenhouse gases emissions and nutrients conservation. Renew Energy 129:814–823. https://doi.org/10.1016/j.renene.2018.06.066

Badewa EA, Yeung CC, Whalen JK, Oelbermann M (2023) Compost and biosolids increase long-term soil organic carbon stocks. Can J Soil Sci 103:483–492. https://doi.org/10.1139/CJSS-2022-0120

Chen H, Awasthi SK, Liu T et al (2020) Effects of microbial culture and chicken manure biochar on compost maturity and greenhouse gas emissions during chicken manure composting. J Hazard Mater 389:121908. https://doi.org/10.1016/j.jhazmat.2019.121908

Chen L, Li W, Zhao Y et al (2022) Effects of compound bacterial agent on gaseous emissions and compost maturity during sewage sludge composting. J Clean Prod 366:133015. https://doi.org/10.1016/j.jclepro.2022.133015

Dhamodharan K, Varma VS, Veluchamy C et al (2019) Emission of volatile organic compounds from composting: a review on assessment, treatment and perspectives. Sci Total Environ 695:133725. https://doi.org/10.1016/j.scitotenv.2019.133725

Edjabou ME, Scheutz C (2023) Quantification of–and determining factors affecting–methane emissions from composting plants. Waste Manag 170:287–296. https://doi.org/10.1016/j.wasman.2023.06.002

Ermolaev E, Sundberg C, Pell M et al (2019) Effects of moisture on emissions of methane, nitrous oxide, and carbon dioxide from food and garden waste composting. J Clean Prod 240:118165. https://doi.org/10.1016/j.jclepro.2019.118165

Fang C, Yin H, Han L et al (2021) Effects of semi-permeable membrane covering coupled with intermittent aeration on gas emissions during aerobic composting from the solid fraction of dairy manure at industrial scale. Waste Manag 131:1–9. https://doi.org/10.1016/j.wasman.2021.07.018

Gao S, Harrison BP, Thao T et al (2023) Biochar co-compost improves nitrogen retention and reduces carbon emissions in a winter wheat cropping system. GCB Bioenergy 15:462–477. https://doi.org/10.1111/gcbb.13024

Gao X, Xu Z, Li Y et al (2021) Bacterial dynamics for gaseous emission and humification in bio-augmented composting of kitchen waste. Sci Total Environ 801:149640. https://doi.org/10.1016/j.scitotenv.2021.149640

Ge M, Shen Y, Ding J et al (2022) New insight into the impact of moisture content and pH on dissolved organic matter and microbial dynamics during cattle manure composting. Bioresour Technol 344:126236. https://doi.org/10.1016/j.biortech.2021.126236

Harrison BP, Gao S, Thao T et al (2024) Methane and nitrous oxide emissions during biochar-composting are driven by biochar application rate and aggregate formation. GCB Bioenergy 16: e13121. https://doi.org/10.1111/gcbb.13121

Hrad M, Binner E, Piringer M, Huber-Humer M (2014) Quantification of methane emissions from full-scale open windrow composting of biowaste using an inverse dispersion technique. Waste Manag 34:2445–2453. https://doi.org/10.1016/j.wasman.2014.08.004

Jiang J, Yu D, Wang Y et al (2021) Use of additives in composting informed by experience from agriculture: effects of nitrogen fertilizer synergists on gaseous nitrogen emissions and corresponding genes (amoA and nirS). Bioresour Technol 319:124127. https://doi.org/10.1016/j.biortech.2020.124127

Li L, Zhao C, Wang X et al (2023) Effects of nitrification and urease inhibitors on ammonia-oxidizing microorganisms, denitrifying bacteria, and greenhouse gas emissions in greenhouse vegetable fields. Environ Res 237:116781. https://doi.org/10.1016/j.envres.2023.116781

Li X, Zhao Y, Xu A et al (2022) Conductive biochar promotes oxygen utilization to inhibit greenhouse gas emissions during electric field-assisted aerobic composting. Sci Total Environ 842:156929. https://doi.org/10.1016/j.scitotenv.2022.156929

Liu Y, Ma R, Li D et al (2020) Effects of calcium magnesium phosphate fertilizer, biochar, and spent mushroom substrate on compost maturity and gaseous emissions during pig manure composting. J Environ Manag 267:110649. https://doi.org/10.1016/j.jenvman.2020.110649

Ma S, Xiong J, Cui R et al (2020) Effects of intermittent aeration on greenhouse gas emissions and bacterial community succession during large-scale membrane-covered aerobic composting. J Clean Prod 266:121551. https://doi.org/10.1016/j.jclepro.2020.121551

Massara TM, Malamis S, Guisasola A et al (2017) A review on nitrous oxide (N2O) emissions during biological nutrient removal from municipal wastewater and sludge reject water. Sci Total Environ 596:106–123. https://doi.org/10.1016/j.scitotenv.2017.04.081

Matsui N, Nakata K, Cornelius C, Macdonald M (2016) Chemical properties and fermentation behavior of the composts prepared by three composting methods in Malawi. Afr J Agric Res 11: 2213–2223. https://doi.org/10.5897/AJAR2015.10485

Miller DN, Berry ED (2005) Cattle feedlot soil moisture and manure content: I. Impacts on greenhouse gases, odor compounds, nitrogen losses, and dust. J Environ Qual 34:644–655. https://doi.org/10.2134/jeq2005.0644

Nemet F, Perić K, Lončarić Z (2021) Microbiological activities in the composting process–a review. Columella 8:41–53. https://doi.org/10.18380/SZIE.COLUM.2021.8.2.41

Nguyen Thanh P (2012) Greenhouse gas emissions from composting and anaerobic digestion plants (Doctoral dissertation, Universitäts- und Landesbibliothek Bonn)

Nie E, Zheng G, Shao Z et al (2018) Emission characteristics and health risk assessment of volatile organic compounds produced during municipal solid waste composting. Waste Manag 79:188–195. https://doi.org/10.1016/j.wasman.2018.06.004

Pan J, Cai H, Zhang Z et al (2018) Comparative evaluation of the use of acidic additives on sewage sludge composting quality improvement, nitrogen conservation, and greenhouse gas reduction. Bioresour Technol 270:467–475. https://doi.org/10.1016/j.biortech.2018.09.044

Pardo G, Moral R, Aguilera E, Del Prado A (2015) Gaseous emissions from management of solid waste: a systematic review. Glob Chang Biol 21:1313–1327. https://doi.org/10.1111/gcb.12806

Park J, Cho KH, Ligaray M, Choi MJ (2019) Organic matter composition of manure and its potential impact on plant growth. Sustainability 11:2346. https://doi.org/10.3390/su11082346

Prosser JI, Hink L, Gubry-Rangin C, Nicol GW (2020) Nitrous oxide production by ammonia oxidizers: physiological diversity, niche differentiation and potential mitigation strategies. Glob Chang Biol 26:103–118. https://doi.org/10.1111/gcb.14877

Puastuti W, Yulistiani D, Widiawati Y (2021) Nutrient content and greenhouse gas emissions of goat manure compost processed without and with decomposer. IOP Conf Ser Earth Environ Sci 648:012113. https://doi.org/10.1088/1755-1315/648/1/012113

Sánchez-Monedero MA, Fernández-Hernández A, Higashikawa FS, Cayuela ML (2018) Relationships between emitted volatile organic compounds and their concentration in the pile during municipal solid waste composting. Waste Manag 79:179–187. https://doi.org/10.1016/j.wasman.2018.07.008

Shan G, Li W, Gao Y et al (2021) Additives for reducing nitrogen loss during composting: a review. J Clean Prod 307:127308. https://doi.org/10.1016/j.jclepro.2021.127308

Stegenta S, Sobieraj K, Pilarski G et al (2019) Analysis of the spatial and temporal distribution of process gases within municipal biowaste compost. Sustainability 11:2340. https://doi.org/10.3390/su11082340

Sun B, Bai Z, Li Y et al (2022) Emission mitigation of CH4 and N2O during semi-permeable membrane covered hyperthermophilic aerobic composting of livestock manure. J Clean Prod 379:134850. https://doi.org/10.1016/j.jclepro.2022.134850

Tang J, Li X, Zhao W et al (2019) Electric field induces electron flow to simultaneously enhance the maturity of aerobic composting and mitigate greenhouse gas emissions. Bioresour Technol 279:234–242. https://doi.org/10.1016/j.biortech.2019.01.123

Thakur S, Solanki H (2022) Role of methane in climate change and options for mitigation–a brief review. Int Assoc Biol Comput Dig 1:275–281

Vergara SE, Silver WL (2019) Greenhouse gas emissions from windrow composting of organic wastes: patterns and emissions factors. Environ Res Lett 14:124027. https://doi.org/10.1088/1748-9326/ab55f4

Wang M, Awasthi MK, Wang Q et al (2017) Comparison of additives amendment for mitigation of greenhouse gases and ammonia emission during sewage sludge co-composting based on correlation analysis. Bioresour Technol 243:520–527. https://doi.org/10.1016/j.biortech.2017.06.135

Wang X, Liu X, Wang Z et al (2022) Greenhouse gas reduction and nitrogen conservation during manure composting by combining biochar with wood vinegar. J Environ Manag 324:116349. https://doi.org/10.1016/j.jenvman.2022.116349

Wang X, Bai Z, Yao Y et al (2018) Composting with negative pressure aeration for the mitigation of ammonia emissions and global warming potential. J Clean Prod 195:448–457. https://doi.org/10.1016/j.jclepro.2018.05.204

Wong JH, Ho CS, Mansor NNA, Lee CT (2017) Mitigation of greenhouse gases emission through food waste composting and replacement of chemical fertiliser. Chem Eng Trans 56:367–372. https://doi.org/10.3303/CET1756062

Wright J, Kenner S, Lingwall B (2022) Utilization of compost as a soil amendment to increase soil health and to improve crop yields. Open J Soil Sci 12:216–224. https://doi.org/10.4236/ojss.2022.126012

Xiong J, Su Y, He X et al (2022) Effects of functional-membrane covering technique on nitrogen succession during aerobic composting: metabolic pathways, functional enzymes, and functional genes. Bioresour Technol 354:127205. https://doi.org/10.1016/j.biortech.2022.127205

Xiong S, Liu Y, Zhang H et al (2023) Effects of chemical additives and mature compost on reducing nitrogen loss during food waste composting. Environ Sci Pollut Res 30:39000–39011. https://doi.org/10.1007/s11356-022-24752-5

Xu Z, Qi C, Zhang L et al (2021) Bacterial dynamics and functions for gaseous emissions and humification in response to aeration intensities during kitchen waste composting. Bioresour Technol 337:125369. https://doi.org/10.1016/j.biortech.2021.125369

Xue S, Zhou L, Zhong M et al (2021) Bacterial agents affected bacterial community structure to mitigate greenhouse gas emissions during sewage sludge composting. Bioresour Technol 337: 125397. https://doi.org/10.1016/j.biortech.2021.125397

Yang F, Li Y, Han Y et al (2019) Performance of mature compost to control gaseous emissions in kitchen waste composting. Sci Total Environ 657:262–269. https://doi.org/10.1016/j.scitotenv.2018.12.011

Yang Y, Yin Z, Li L et al (2022) Effects of dicyandiamide, phosphogypsum and superphosphate on greenhouse gas emissions during pig manure composting. Sci Total Environ 846:157487. https://doi.org/10.1016/j.scitotenv.2022.157487

Yin Y, Yang C, Li M et al (2021) Research progress and prospects for using biochar to mitigate greenhouse gas emissions during composting: a review. Sci Total Environ 798:149294. https://doi.org/10.1016/j.scitotenv.2021.149294

Zeng J, Michel FC Jr, Huang G (2023) Comparison and evaluation of GHG emissions during simulated thermophilic composting of different municipal and agricultural feedstocks. Int J Environ Res Public Health 20:3002. https://doi.org/10.3390/ijerph20043002

Zhao Y, Li W, Chen L et al (2020) Effect of enriched thermotolerant nitrifying bacteria inoculation on reducing nitrogen loss during sewage sludge composting. Bioresour Technol 311:123461. https://doi.org/10.1016/j.biortech.2020.123461

# Composting Techniques for Poultry Industry–Based Organic Residues and Their Sustainable Application to Improve Soil Nitrogen Content

**Catarina Medeiros, Tiago Azevedo, Elisabete Nascimento-Gonçalves, Maria Cristina Morais, João Ricardo Sousa, Paula A. Oliveira, Marta Roboredo, and Ana M. Coimbra**

## 1 Introduction

The poultry industry plays a crucial role in global food security, serving as a significant source of affordable and high-quality protein for billions of people worldwide (Castro et al. 2023). In response to the growing global demand for protein, the industry has rapidly expanded (Agyarko-Mintah et al. 2017), leading to substantial increases in poultry meat production (Fig. 1) and per capita consumption over recent decades.

This growth has led to the generation of considerable amounts of poultry industry residues, including litter, manure, feathers, carcasses, and by-products from slaughterhouse and hatchery (Fig. 2) (Karuppannan et al. 2021; Rahman et al. 2022). These waste materials pose serious environmental challenges, such as the emission of greenhouse gases (GHGs), namely ammonia ($NH_3$) and nitrous oxide ($N_2O$), and other pollutants, contributing to soil, air, and water pollution (Gržinić et al. 2023;

Authors Catarina Medeiros, Tiago Azevedo and Elisabete Nascimento-Gonçalves have equally contributed to this chapter.

C. Medeiros · T. Azevedo
Centre for the Research and Technology of Agro-Environmental and Biological Sciences (CITAB), University of Trás-os-Montes and Alto Douro (UTAD), Vila Real, Portugal

E. Nascimento-Gonçalves · M. C. Morais · J. R. Sousa · P. A. Oliveira (✉) · M. Roboredo
A. M. Coimbra
Centre for the Research and Technology of Agro-Environmental and Biological Sciences (CITAB), University of Trás-os-Montes and Alto Douro (UTAD), Vila Real, Portugal

Institute for Innovation, Capacity Building and Sustainability of Agri-Food Production (Inov4Agro), UTAD, Vila Real, Portugal
e-mail: pamo@utad.pt

 
R. K. Singh et al. (eds.), *Composting of Farm Residues by Nature-Based Solutions*,
https://doi.org/10.1007/978-3-032-12255-1_5

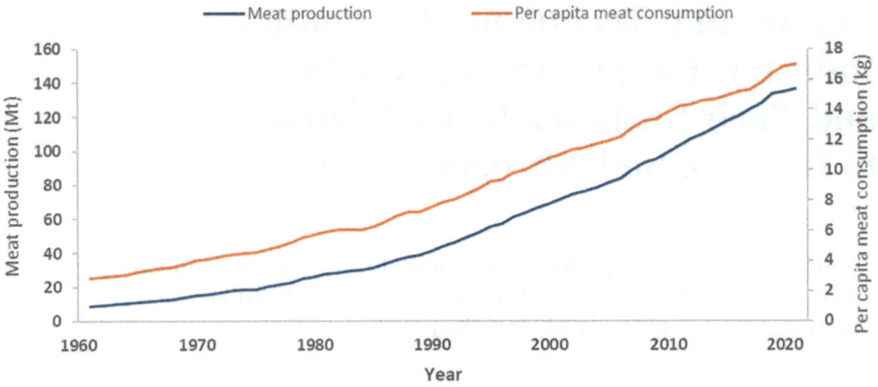


**Fig. 1** World poultry meat production and per capita poultry meat consumption. (Adapted from Food and Agriculture Organization of the United Nations, 2023)

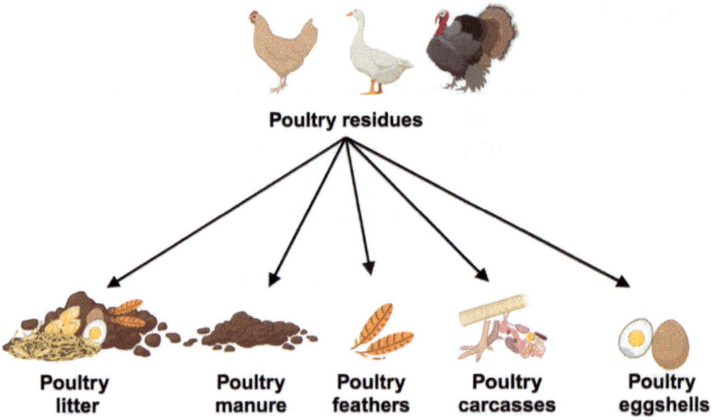


**Fig. 2** Poultry industry residues

Richa et al. 2020) and, ultimately, to exacerbating climate change (Zhang et al. 2023). Additionally, poultry industry residues are also associated with potential health risks due to the presence and spread of pathogens (Gržinić et al. 2023; Mozhiarasi and Natarajan 2025). Despite these hazards, poultry wastes are rich in valuable nutrients and can serve as raw materials for producing value-added products with various applications (Karuppannan et al. 2021).

Composting has emerged as a sustainable and promising strategy to transform these materials into N-rich organic fertilizers, which are essential for soil fertility and plant productivity (Richa et al. 2020), thus reducing dependence on chemical fertilizers (Mironiuk et al. 2023).

The main goal of this chapter is to provide a comprehensive overview of the properties of poultry industry residues (e.g., litter, manure, carcasses, and eggshells), their potential for generating value-added products, and the composting methods for managing them, with a focus on their application to improve soil N content.

## 2 Characterization of Poultry Industry Organic Residues

The poultry industry generates a considerable volume of organic residues that pose significant environmental concerns (Zhang et al. 2023). From feathers and manure to eggshells and processing by-products, these residues present both environmental concerns and unexplored opportunities. Understanding the characteristics of these residues is essential for optimizing their use in composting processes and ensuring their sustainable application as soil amendments. This subchapter examines the primary waste streams generated by the poultry industry, their characteristics, and the environmental challenges associated with their disposal.

### 2.1 *Poultry Litter and Manure*

Poultry litter and manure are the primary by-products of poultry farming. Poultry litter is a mixture of poultry excreta, spilled feed, and bedding materials, while poultry manure is defined as the excreted product itself, not including any bedding material (Vaishnav et al. 2023). Due to the variability in bird species, feed, production system, and handling methods, among other factors, the composition, quantity, and physical form (in the case of manure, which can range from liquid to solid) are highly variable (Mohamed Amanullah et al. 2010). Globally, poultry manure production is substantial, with annual outputs of approximately 20,708 million tons, including 11,514 million tons in Asia and 2039 million tons in Europe (Jurgutis et al. 2020).

Poultry litter and manure are rich in macronutrients and micronutrients essential for soil fertility. Their typical composition includes high contents of C (29.15–45.30%), N (3.27–6.42%), P (1.5–3.5%), S (0.36–1.45%), and K (1.5–3%) (Mohamed Amanullah et al. 2010; Baniasadi et al. 2016; Mante and Agblevor 2011; Ro et al. 2010; Hussein et al. 2017; Yıldız et al. 2019). Regarding N, this nutrient is predominantly present in organic forms (70–80%), derived from proteins and amino acids (Khodadadi et al. 2024). Thus, the nutrient composition, coupled with the organic matter content, makes poultry litter and manure potentially valuable for composting and agronomic use.

Despite these attributes, several environmental problems can be associated with poultry litter and manure, particularly when they are improperly managed, and their disposal requires careful consideration. Direct land application, which is one of the most frequent forms of discarding poultry litter and manure (Singh et al. 2014), requires particular attention. First, the addition of mineral salts to poultry feed can lead to high levels of sodium and potassium salts, potentially causing soil salinity and impairing plant growth (Li-Xian et al. 2007). Second, poultry litter and manure are also significant sources of greenhouse gases, including methane ($CH_4$), carbon dioxide ($CO_2$), and nitrous oxide ($N_2O$), primarily emitted during storage and land application (Moore et al. 2011; Rodhe and Karlsson 2002). Methane is released through the anaerobic decomposition of organic matter, while nitrous oxide is generated by the microbial nitrification and denitrification processes in soils. Ammonia volatilization, a frequent problem with poultry manure, indirectly contributes to $N_2O$ formation and atmospheric N deposition (Swelum et al. 2021). Third, these residues may present nutrient imbalances, such as the N-to-P ratio. In fact, the typical nitrogen-to-phosphorus (N:P) ratio of 2:1 found in poultry litter often results in the overapplication of P, relative to crop requirements (Staver and Brinsfield 2001). Surpluses of N and P can leach into water bodies, contributing to the eutrophication of aquatic systems, algal blooms, and hypoxic zones, which then affect fish habitats (Tiwari and Pal 2022). Furthermore, the direct application of these residues to agricultural land can be problematic, given their high content in contaminants and their association with the potential for pathogen spread. Antibiotics and metals can adversely affect plant growth and other nontarget organisms (Ko et al. 2008; Narciso et al. 2023), while the spread of bacteria (*Salmonella*, *Campylobacter jejuni*, and *Listeria monocytogenes*) (Chen and Jiang 2014) and viruses (e.g., avian influenza) (Figueroa et al. 2021) poses a threat to human and animal health (Fatoba et al. 2022).

To mitigate these environmental issues, sustainable management practices for poultry litter and manure have been developed and recommended:

1. Anaerobic digestion, a biological process that breaks down organic matter in the absence of oxygen, producing biogas, a renewable energy source (Paranhos et al. 2023; Mansour et al. 2024)
2. Composting, a controlled aerobic decomposition, producing stable, pathogen-free organic matter that can improve soil health while minimizing odor and N losses (Limoges et al. 2022; Okada et al. 2024)
3. Pyrolysis, the transformation of biomass under high temperatures and low-oxygen conditions, producing biochar, a stable carbon-rich material suitable for soil improvement (Pachón Gómez et al. 2023)

In aquaculture of plankton-eating fish, treated poultry manure has also been explored as a nutrient source to promote algal growth, aiming to reduce costs associated with fish feed (Cavalcanti et al. 2021).

## 2.2 Poultry Feathers

Poultry feathers represent a major waste by-product in the global poultry industry, presenting considerable environmental concerns. Feathers are a by-product of poultry processing, generated when birds are slaughtered, scalded in hot water, and mechanically plucked using specialized feather-removal equipment (Shung et al. 2023). This industry produces billions of birds annually, with feathers accounting for approximately 10% of a bird's total weight (Ozdemir and Yetilmezsoy 2020). For example, the world poultry chicken production generates an estimated 8.5 million tons of feathers annually (Reddy and Santosh 2016), which must be promptly managed to prevent decay, odor, and contamination of processing facilities.

Feathers are primarily composed of keratin (90%), a highly durable structural protein formed by the extensive cross-linking of cysteine bonds, which makes feathers challenging to degrade (Sypka et al. 2021). This durability demands intensive transformation processes to convert feathers into useful products (Tesfaye et al. 2017a). Feathers contain a high proportion of nonessential amino acids (Latshaw et al. 1994) and are chemically rich in C (64.47%), N (10.41%), and S (2.64%) (Tesfaye et al. 2017b). Therefore, their disposal presents significant challenges and raises environmental concerns, albeit with great potential in terms of alternative applications. Currently, a large proportion of feathers is directed to landfills, contributing to landfill overload due to their lengthy decomposition time (Tesfaye et al. 2017a). Furthermore, improper disposal can contribute to groundwater contamination through the release of harmful leachates or runoff, negatively affecting soil and aquatic ecosystems. Feather incineration is another disposal alternative, but it contributes to the release of greenhouse gases and other atmospheric pollutants (Stingone and Wing 2011). Additionally, poultry feathers, like manure and litter, can harbor pathogens, including specific bacterial strains (e.g., *Escherichia coli* and *Salmonella* spp.), with risks to human, animal, and environmental health under the One Health framework (Franke-Whittle and Insam 2013; Dréano et al. 2021). Therefore, strict monitoring and innovative waste management strategies are essential to mitigate these challenges while exploring sustainable solutions for repurposing this abundant and N-rich source.

In recent years, several strategies have been developed to recycle this residue into sustainable products. On the one hand, keratin can be extracted from feathers and transformed into porous sponges, films, microfibers, and hydrogels (for medical applications and food packaging) or used in the development of environmentally friendly biomaterials (Wang et al. 2025). On the other hand, feather hydrolysates, produced through high-pressure and high-temperature processes, enzymatic treatments, or keratinolytic microbial action, have shown potential as N-rich fertilizers, either in their raw form or after undergoing vermicomposting (Shahiful Hizam et al. 2024; Gupta et al. 2023). These hydrolysates also hold potential as animal feed ingredients due to their protein content (Fakhfakh et al. 2011) and in biofuel production (Xia et al. 2012), among others (Callegaro et al. 2019). Moreover, feathers can be used as adsorbents for the removal of wastewater contaminants, including metal ions, dyes, and antibiotics (Okoya et al. 2020; Solís-Moreno et al. 2021; Li et al. 2019; Mittal et al. 2013).

## 2.3 Poultry Carcasses

Another significant residue generated by the poultry industry is carcasses, resulting from mortality during rearing, waste from slaughterhouses, culled birds, and unsold or expired products. These carcass residues encompass a variety of components, such as legs, feet, blood, bone fragments, cartilage, skin, trachea, viscera, and combs or wattles (Bravo et al. 2023). Due to their diverse origins, these residues exhibit considerable organic composition variability. Nevertheless, they are consistently characterized by a high organic content, primarily due to their richness in proteins, fats, and other organic molecules (Baéza et al. 2022; Moukazis et al. 2018). However, quantifying the exact volume of these residues is challenging due to differences in production scales and practices. It is estimated that approximately one billion tons of bone, feather, offal, and blood are obtained annually from broiler and fowl production in the United Kingdom (McGauran et al. 2021). Studies have reported that slaughterhouse residues are rich in C (54.41%), N (9.45%), and S (6.76%), highlighting their potential as an N source, with an estimated yield of 22.4 kg of N per 1000 poultry individuals processed (Yoon et al. 2014).

The rapid biodegradability of these residues presents several challenges in terms of their use, including odor emissions and the proliferation of pathogenic microorganisms, such as *Salmonella* sp., *Staphylococcus* sp., and *Clostridium* sp., if not managed rapidly and appropriately. Moreover, poultry carcasses may contain several contaminants, such as metals, drugs, and other chemical additives introduced through feed, either for nutritional or pharmaceutical purposes (Haapapuro et al. 1997). While burial and landfill deposition can impair soil and water quality due to N and P leaching, open-air burning and incineration release greenhouse gases and N oxides ($NO_x$) (Izydorczyk et al. 2022). In Europe, due to the strict European Union regulations, these practices have been largely replaced by alternative processes such as controlled incineration and alkaline hydrolysis (Gwyther et al. 2011). However, in developing countries, many rural slaughterhouses continue to dispose of untreated waste due to inadequate infrastructural facilities and governance (Bhunia et al. 2022). Despite advancements in waste management, many current disposal practices still fail to incorporate sustainable treatment technologies or adopt cleaner production protocols, which are essential for mitigating both environmental and public health risks (Chiarelotto et al. 2021).

Poultry carcass residues are rich in N and have a low carbon-to-nitrogen (C:N) ratio, ranging from 7 to 10 (Kuboń et al. 2024). This imbalance can hinder certain valorization processes, such as anaerobic co-digestion. Therefore, to enhance biogas yield, low-N materials, such as fruit and vegetable agro-industry residues, are often co-digested with poultry slaughterhouse residues (Samadi et al. 2024). Similarly, composting poultry carcass residues also requires the addition of C-rich materials such as coconut fiber, agricultural residues, cardboard, and wood dust that allow their effective degradation (Bharathy et al. 2012; Asses et al. 2019). Besides, these residues also hold high potential for fertilizer purposes; for example, acid solubilization, followed by neutralization with potassium hydroxide, can result in nutrient-rich liquid fertilizers (Izydorczyk et al. 2022).

Finally, research has shown the suitability of slaughterhouse residues for biodiesel and hydrogen production (Chowdhury et al. 2022) and their use, after high-temperature rendering, as feed ingredient for monogastric animal (Prabakaran and Valavan 2021; Wang et al. 2024).

### *2.4 Poultry Eggshells*

The Food and Agriculture Organization (FAO) records indicate a global egg production of more than 86.4 million tons, just for chicken eggs, in 2021 (FAOSTAT 2025). This substantial egg production is responsible for significant quantities of eggshell residues, particularly in the hatchery and processing sectors. Although the exact volume of eggshell residues is difficult to quantify, a considerable portion is disposed of in landfills without any pretreatment, which is recognized as a major environmental issue, as improper disposal in non-designated areas can lead to health hazards over time, including fungal growth and the attraction of pests, such as rats (Waheed et al. 2020). Eggshells originate from two main sources, accidental breakage during production and the egg products market. The latter has seen notable growth in recent years due to rising consumer demand for processed egg products, such as liquid, frozen, concentrated, or powdered eggs (Quina et al. 2017). Eggshell residues consist predominantly of calcium carbonate (96% of the total mass) (Singh et al. 2021), along with organic matter (such as collagen and glycoproteins) and trace minerals like Mg and P (Laca et al. 2017). While eggshells from industrial processes are often employed in agriculture to neutralize acidic soils (Oliveira et al. 2013), their potential economic and environmental value remains largely unexploited. This opens opportunities to explore sustainable solutions for adding value to eggshell materials.

Thus, in recent years, there has been a growing interest in repurposing eggshells as a natural biomaterial due to their abundance, high calcium carbonate content, and unique microstructure (Wilson 2017). Studies have shown a wide range of applications for eggshell-derived materials, including animal feed additives and human nutritional supplements, catalysts for the production of biodiesel, natural adsorbents for removing heavy metals and organic dyes from wastewater, soil fertilizers, bone substitutes, and drug delivery carriers in biomedical applications (Rahman et al. 2014; Yadav et al. 2018; Laohavisuti et al. 2021; Odetoye et al. 2021).

A summary of the main advantages and challenges associated with poultry industry residues is presented in Table 1.

**Table 1** Advantages and challenges of poultry industry residues

| Poultry residues | Advantages | Challenges |
|---|---|---|
| Litter and manure | High content of macro- (N, P, and K) and micronutrients, constituting a promising organic fertilizer that can reduce dependence on mineral fertilizers | Nutrient leaching, water contamination, and pathogen spread, affecting soil, human, and animal health |
| Feathers | Rich in keratin and nonessential amino acids, with potential animal and industrial uses | Difficult to degrade, requiring complex processing. Improper disposal can lead to soil, water, and air contamination |
| Carcasses | Rich in proteins, fats, and other organic molecules for various applications | Rapid biodegradability causes unpleasant odors and pathogen risks. Requires strict regulations |
| Eggshells | Rich in calcium carbonate, making them useful for soil pH adjustment | Improper disposal can lead to environmental pollution |

# 3 Composting Processes and Techniques

Composting is widely recognized as a sustainable and effective method for managing large amounts of organic residues, including those generated by the poultry industry. This process offers both environmental and economic advantages, including waste valorization, pathogen reduction, and odor control (Richa et al. 2020; Bist et al. 2024). By transforming poultry residues into nutrient-rich compost, composting aligns with the principles of sustainable waste management and resource recovery, being a critical strategy within a circular economy. To ensure effectiveness, it is imperative to identify optimal composting conditions that support microbial activity, maintain a balanced C:N ratio (25–35), and regulate critical parameters, such as moisture content (40–60%), oxygen levels, and temperature (Azim et al. 2018). Composting is an aerobic biological process wherein microorganisms decompose organic material, typically conducted either in windrows/piles or in-vessel (Irfan et al. 2020; Brodie et al. 2000). The process develops through four main stages, which often overlap rather than occur sequentially: mesophilic, thermophilic, cooling, and maturation (Belyaeva and Haynes 2009). During the mesophilic stage, bacteria, fungi, and actinomycetes actively metabolize substrates. As temperatures rise to approximately 54–66 °C (thermophilic phase), thermophilic bacteria, such as *Bacillus* sp., dominate the decomposition process, and pathogens are eliminated. Subsequently, the cooling phase involves the renewed activity of mesophilic microorganisms to further decompose complex organic matter, and the process concludes with the maturation phase, when microbial activity stabilizes and humification creates a stable nutrient-rich compost.

Composting offers numerous advantages, such as residue sanitization, mass and volume reduction, and cost-effective, nutrient-rich compost production (Asses et al. 2019). However, poultry residue composting, due to their specific chemical composition, often requires co-composting with carbon-rich materials like sawdust or wood chips, which help achieve optimal C:N ratios and enhance microbial activity and structural porosity (Mozhiarasi and Natarajan 2025; Adhikari et al. 2008; Qasim et al. 2018). However, when composting livestock by-products, careful site location consideration is required since these residues can attract pests, such as mosquitoes, cockroaches, rats, and flies, which may spread diseases to nearby residential areas (Mozhiarasi and Natarajan 2025). Additionally, leachates from poorly managed composting sites may increase groundwater contamination risk. Therefore, risk mitigation measures are required, including strategic site selection, residue containment, and effective process control.

Apart from traditional composting, vermicomposting, an alternative composting strategy that incorporates earthworms to enhance compost quality and accelerate nutrient release, has been gaining attention in recent years (Nascimento-Gonçalves et al. 2024; Lirikum et al. 2022). The earthworm species commonly used include *Eisenia fetida*, *Eudrilus eugeniae*, *Perionyx sansibaricus*, *Pontoscolex corethrurus*, *Megascolex chinensis*, and *Lampito mauritii* (Waqas et al. 2023). Earthworm activity helps fragment and aerate organic material and enrich microbial communities, resulting in composts with improved nutrient profiles (Soobhany 2019). Vermicomposting is particularly useful for producing high-value compost, suitable for agricultural applications; however, earthworms are highly sensitive to the high ammonia and salt contents of poultry residues (Kaur 2020). Poultry residue pre-composting or C-rich material incorporation are strategies that can help reduce these problems (Niedzialkoski et al. 2021; Srivastava et al. 2023).

## 4 Soil Application and Effect on Nitrogen Availability

The application of poultry residues should be timed so that nutrient availability aligns with optimal conditions for plant uptake, maximizing yield and nutrient absorption while minimizing potential nutrient losses to the environment (Sistani et al. 2010; Adekiya and Agbede 2017). In particular, application timing is a key factor influencing the efficiency of N use by crops (de la Casa et al. 2013; Jn-Baptiste et al. 2013). For example, in warm regions, spring applications of poultry manure and poultry litter resulted in higher levels of available N in the soil compared to fall applications (Jn-Baptiste et al. 2013). Yet fall applications have been found to be as effective as the spring ones for corn grain yield in regions with cold and mild winters, such as the Midwestern United States and Kentucky (Ruiz Diaz and Sawyer 2008).

Besides application timing, other factors also influence N availability, namely the form of poultry residues and the application method. In the case of poultry litter, the application in narrow subsurface bands resulted in better cotton performance compared to surface broadcasting, as the latter practice exposes litter nutrients to volatilization losses (Tewolde et al. 2009). Similarly, in the case of poultry manure, soil incorporation increased soil organic matter, as well as soil and leaf N, P, K, Ca, and Mg contents, resulting in higher tomato growth and yield, compared to surface broadcasting (Adekiya and Agbede 2017). Regarding corn and soybean production, the incorporation of poultry litter through conventional tillage resulted in higher corn yields compared to no-till and strip-tillage methods, while in soybean yield, no significant differences were observed (Watts and Torbert 2011).

When considering the form in which poultry residues are applied to soils, composting and vermicomposting have become the most widely adopted practices (Manea et al. 2024; Martins et al. 2024). During composting and vermicomposting, bulking agents can still be incorporated to improve N content, although significant differences in the amount and type of nutrients may occur, depending on the bulking agents used (Kaur et al. 2023; Alarefee et al. 2021; Vicentin et al. 2021).

Furthermore, the strategic combination of composted poultry residues with amendments like biochar, lignite, and zeolite offers a sustainable pathway to improve soil N content, boost crop yields, reduce dependence on chemical fertilizers, and mitigate environmental impacts, thus supporting the advancement of eco-friendly agricultural practices (Agyarko-Mintah et al. 2017; Cao et al. 2020; Prasai et al. 2018). Biochar, as a highly porous pyrolysis product of plant and animal waste biomass with a high C content, is frequently added as a conditioner to improve compost stability and quality (Zhang et al. 2021). Oxidation of biochar develops negatively charged functional groups, improving cation exchange, particularly for $NH_4^+$. This allows biochar to adsorb $NH_4^+$, reducing $NH_3$ emissions during composting (Agyarko-Mintah et al. 2017; Zhang et al. 2021). It has been shown that the addition of biochar during the composting process promotes rapid organic matter degradation, accelerates stabilization and detoxification, and also enhances N mineralization (ammonification and nitrification) (Sánchez-García et al. 2015). Lignite, also known as brown coal, is widely used as a natural adsorbent, influencing microbial communities by increasing the water-holding capacity, decreasing the bulk density, improving aeration, retaining nutrients, and buffering pH (Cao et al. 2020; Costello et al. 2021). It also inhibits urease activity, which hydrolyzes urea to $NH_3$, thereby reducing its emissions (Cao et al. 2020). Zeolites, crystalline, hydrated aluminosilicates of alkali and alkaline earth cations, with infinite three-dimensional structures, have the ability to lose and gain water reversibly and exchange cations (Mumpton and Fishman 1977). The application of zeolite during poultry residue management has shown significant potential, such as N retention, emission reduction, and the improvement of composting processes and the biodegradability of organic matter (Turan 2009).

To enhance efficiency, reduce processing time, and improve the quality of composting products, the use of microbial consortia, including strains such as *Bacillus subtilis*, *Azotobacter chroococcum*, *Actinomadura keratinilytica*, and *Phanerochaete chrysosporium*, has gained increasing attention (Greff et al. 2022; Habbeche et al. 2014). These microorganisms contribute to the accelerated degradation of organic matter and improved nutrient availability in poultry compost, particularly through the action of keratinase-producing microbes that facilitate efficient keratin breakdown. Additionally, microbial bioaugmentation, particularly through the introduction of ammonifying, nitrifying, nitrite-oxidizing, and N-fixing bacteria, has been shown to significantly reduce $NH_3$ emissions (15.6–59.4%) and mitigate N losses (1.87–23.21%) during chicken manure composting (Li et al. 2021).

Therefore, careful consideration of all these factors is essential to optimize the benefits of poultry residue application to agricultural soils. In summary, the effectiveness of poultry residues in enhancing crop productivity depends on their nutrient availability, which is influenced by soil properties, agricultural management strategies, application rate, timing, and method (Lin et al. 2018). Table 2 presents an overview of examples of poultry residue applications to soil, highlighting the different methods and main results.

## 5 Impact on Agricultural and Environmental Sustainability

N losses are a significant challenge when composting N-rich residues, like the ones resulting from the poultry industry (Chen et al. 2023). These losses occur through $NH_3$ volatilization, leaching, and incomplete denitrification and are influenced by temperature, pH, and moisture levels (Hoang et al. 2022). Traditional composting often results in substantial N losses, mainly during the thermophilic phase, where high temperatures and alkaline conditions favor $NH_3$ volatilization (Chowdhury et al. 2014; Oviedo-Ocaña et al. 2021; Skrzypczak et al. 2024). Excessive moisture can also lead to N losses in the form of $NO_3^-$ leaching, especially in systems with poor drainage or inadequate control over moisture content (Kacprzak et al. 2023). Under anaerobic conditions, when aeration during composting is insufficient, denitrification of $NO_3^-$ or $NO_2^-$ results in the release of nitrous oxides ($N_2O$) (Oviedo-Ocaña et al. 2021). All these losses reduce the N content of the compost, compromising its fertilizer value. On the other hand, vermicomposting minimizes N losses due to the absence of high-temperature phases and pH fluctuations (Yadav and Garg 2011). In addition, increased microbial activity during vermicomposting stabilizes N by enhancing nitrification and humification, thereby converting labile N into more stable organic compounds (Nigussie et al. 2016).

By replacing synthetic fertilizers with compost and vermicompost, farmers can lower production costs while mitigating the negative environmental impacts associated with synthetic fertilizer use, such as soil degradation, water pollution from nutrient runoff, and greenhouse gas emissions from fertilizer manufacturing (Wang et al. 2018; Köninger et al. 2021). Moreover, the use of vermicompost aligns with regenerative agriculture and agroecological practices by enhancing soil fertility, promoting microbial diversity, and improving soil structure, thereby supporting sustainable farming systems (Pereira et al. 2023; Manzoor et al. 2024; Oyege and Balaji Bhaskar 2023).

**Table 2** Examples of poultry residues' effects on soil N levels and soil characteristics

| Method | Type of residue | Main results | Refs. |
|---|---|---|---|
| Direct application | Poultry manure | Duplicated total N uptake compared to control<br>Increased agronomic efficiency (87.3%) compared to no amendment | Kaur et al. (2023) |
| Composting | Poultry manure | Increased organic matter (0.68–0.76%)<br>Enhanced available soil NPK | Farhad et al. (2018) |
| | Poultry manure + biochar | Reduced $NH_3$ emissions (30–44%) | Janczak et al. (2017) |
| | Poultry manure + two types of biochar | Reduced $NH_3$ emissions (50–82% with wheat stalk biochar and 86–97% with rice husk biochar) | Zhang et al. (2021) |
| | Poultry litter + biochar | Reduced $NH_3$ emissions (64%)<br>Decreased total N losses (52%)<br>Promoted faster decomposition of poultry litter | Steiner et al. (2010) |
| | Poultry litter + biochar | Reduced $NH_3$ emissions<br>Decreased $NH_3$ toxicity to microorganisms<br>Enhanced N retention | Agyarko-Mintah et al. (2017) |
| | Poultry litter + biochar | Improved N stability<br>Reduced leaching<br>Promoted gradual mineralization | Alarefee et al. (2021) |
| | Poultry litter + lignite | Enhanced nitrification and denitrification<br>Reduced $NH_3$ emissions (39.3% and 50.2%) | Cao et al. (2024) |
| | Poultry litter + lignite | Increased N retention (25%)<br>Boosted mineral N (60%) | Cao et al. (2020) |
| | Poultry litter + natural zeolite | Reduced N losses (16%) | Turan (2009) |
| | Poultry litter + natural zeolite | Decreased global warming potential (21.9%)<br>Increased N content in the final compost (27.25%) | Peng et al. (2019) |
| | Poultry feather + *Streptomyces* sp. | Increased total N content (15.41%) and available N (1.51%) compared to untreated feathers | Rasel et al. (2020) |
| | Eggshell | Favored nitrification, leading to accumulation of $NO_3^-$ | Soares et al. (2017) |
| Composting and vermicomposting | Poultry litter | Greater macronutrient concentrations and lower C:N ratio in vermicompost compared to compost | Joshi et al. (2020) |
| | Poultry litter | Decreased N-$NH_4^+$ levels | Elisabet Masin et al. (2020) |
| | Chicken manure and feathers | Higher N content due to an optimal C/N balance in vermicompost containing chicken manure, enhanced by the presence | Shahiful Hizam et al. (2024) |

(continued)

**Table 2** (continued)

| Method | Type of residue | Main results | Refs. |
|---|---|---|---|
| | | of earthworms and feather-degrading bacteria | |
| | Hydrolyzed chicken feathers | Increased $NO_3^-$ (79.6%) and reduced $NH_4^+$ (91%) in vermicomposting<br>Increased total N (42.34–61.39%) in traditional composting<br>Decreased C:N ratio (61.70–67.87%) | Dume et al. (2022) |

## 6 Challenges and Future Prospects

The rapid growth of poultry production in recent years has increased environmental and public health challenges due to the volume of residues generated, including litter, manure, feathers, carcasses, and eggshells. From a circular economy perspective, these residues, characterized by their diverse properties and nutritive values, offer promising opportunities for sustainable agriculture, particularly through their potential to enhance soil N content and reduce dependence on chemical fertilizers.

Effective management of poultry industry residues through composting not only mitigates environmental impacts but also contributes to enhancing soil fertility and health and improving plant growth. This review describes the characteristics of poultry residues, the environmental effects of their direct application to soil, and the advantages of composting as a sustainable management strategy. Furthermore, it highlights significant knowledge gaps such as the need to optimize composting processes by testing different amendments, aiming for N retention maximization, greenhouse gas emission and soil contamination reduction, pathogenic organisms' elimination, and chemical N fertilizer dependence reduction. Currently, most studies dealing with composting of poultry industry residues are focused on manure. Thus, conducting and optimizing composting assays using other poultry residues, particularly those more resistant to decomposition (e.g., feathers), should be considered in further studies. In addition, comprehensive economic and environmental assessments are necessary to ensure the wide use of composting. Regulatory frameworks and policy guidelines ensure compost quality, pathogen limits, and permissible levels of metals or other environmental contaminants in composted poultry industry residues. Furthermore, innovative applications of poultry industry residues, for example, energy recovery from carcasses or industrial uses for feathers and eggshells, should be explored to maximize their value and minimize their environmental footprint. Addressing these challenges will unlock the full potential of poultry industry residues, in particular, in advancing sustainable farming methods, contributing to long-term soil fertility and health and ecosystem resilience.

Recent advancements in composting focus on enhancing efficiency, reducing processing time, and improving compost quality. The application of biochar and microbial consortia has been employed to accelerate organic matter degradation and enhance nutrient availability, particularly in keratin decomposition. Moreover, microbial bioaugmentation, through the introduction of ammonifying, nitrifying, nitrite-oxidizing, and N-fixing bacteria, can significantly reduce $NH_3$ emissions and mitigate N losses during poultry manure composting.

**Acknowledgments/Funding** The Vine&Wine PT project has received funding from the NextGenerationEU program through Portugal's Recovery and Resilience (PRR) Plan (project no. C644866286–00000011). This work is also supported by National Funds by the Portuguese Foundation for Science and Technology (FCT) under the projects UID/04033/2025: Centro de Investigação e de Tecnologias Agroambientais e Biológicas and LA/P/0126/2020 (https://doi.org/10.54499/LA/P/0126/2020). ENG and CM thank PRR and European Funds NextGenerationEU for their fellowship grant (BI/UTAD/41/2025 and BI/UTAD/30/2024).

## References

Adekiya AO, Agbede TM (2017) Effect of methods and time of poultry manure application on soil and leaf nutrient concentrations, growth and fruit yield of tomato (*Lycopersicon esculentum* Mill). J Saudi Soc Agric Sci 16:383–388. https://doi.org/10.1016/j.jssas.2016.01.006

Adhikari BK, Barrington S, Martinez J, King S (2008) Characterization of food waste and bulking agents for composting. Waste Manag 28:795–804. https://doi.org/10.1016/j.wasman.2007.08.018

Agyarko-Mintah E, Cowie A, Van Zwieten L, Singh BP, Smillie R, Harden S, Fornasier F (2017) Biochar lowers ammonia emission and improves nitrogen retention in poultry litter composting. Waste Manag 61:129–137. https://doi.org/10.1016/j.wasman.2016.12.009

Alarefee HA, Ishak CF, Karam DS, Othman R (2021) Efficiency of Rice husk biochar with poultry litter co-composts in Oxisols for improving soil Physico-chemical properties and enhancing maize performance. Agronomy 11:2409. https://doi.org/10.3390/agronomy11122409

Asses N, Farhat W, Hamdi M, Bouallagui H (2019) Large scale composting of poultry slaughterhouse processing waste: microbial removal and agricultural biofertilizer application. Process Saf Environ Prot 124:128–136. https://doi.org/10.1016/j.psep.2019.02.004

Azim K, Soudi B, Boukhari S, Perissol C, Roussos S, Thami Alami I (2018) Composting parameters and compost quality: a literature review. Org Agric 8:141–158. https://doi.org/10.1007/s13165-017-0180-z

Baéza E, Guillier L, Petracci M (2022) Review: production factors affecting poultry carcass and meat quality attributes. Animal 16(Suppl 1):100331. https://doi.org/10.1016/j.animal.2021.100331

Baniasadi M, Tugnoli A, Conti R, Torri C, Fabbri D, Cozzani V (2016) Waste to energy valorization of poultry litter by slow pyrolysis. Renew Energy 90:458–468. https://doi.org/10.1016/j.renene.2016.01.018

Belyaeva ON, Haynes RJ (2009) Chemical, microbial and physical properties of manufactured soils produced by co-composting municipal green waste with coal fly ash. Bioresour Technol 100:5203–5209. https://doi.org/10.1016/j.biortech.2009.05.032

Bharathy N, Sakthivadivu R, Sivakumar K, Saravanakumar V (2012) Disposal and utilization of broiler slaughter waste by composting. Vet World 5:359. https://doi.org/10.5455/vetworld.2012.359-361

Bhunia S, Bhowmik A, Mukherjee J (2022) Chapter 26: Waste management of rural slaughterhouses in developing countries. In: Hussain C, Hait S (eds) Advanced organic waste management. Elsevier, pp 425–449

Bist RB, Bist K, Poudel S, Subedi D, Yang X, Paneru B, Mani S, Wang D, Chai L (2024) Sustainable poultry farming practices: a critical review of current strategies and future prospects. Poult Sci 103:104295. https://doi.org/10.1016/j.psj.2024.104295

Bravo FI, Calvo E, López-Villalba RA, Torres-Fuentes C, Muguerza B, García-Ruiz A, Morales D (2023) Valorization of chicken slaughterhouse byproducts to obtain antihypertensive peptides. Nutrients 15:457. https://doi.org/10.3390/nu15020457

Brodie HL, Carr, Lewis E, Condon P (2000) A comparison of static pile and turned windrow methods for poultry litter compost production. Compost Sci Util 8:178–189. https://doi.org/10.1080/1065657X.2000.10701990

Callegaro K, Brandelli A, Daroit DJ (2019) Beyond plucking: feathers bioprocessing into valuable protein hydrolysates. Waste Manag 95:399–415. https://doi.org/10.1016/j.wasman.2019.06.040

Cao Y, Bai M, Han B, Impraim R, Butterly C, Hu H, He J, Chen D (2020) Enhanced nitrogen retention by lignite during poultry litter composting. J Clean Prod 277:122422. https://doi.org/10.1016/j.jclepro.2020.122422

Cao Y, Bai M, Han B, Butterly C, Hu H, He J, Griffith DWT, Chen D (2024) $NH_3$ and greenhouse gas emissions during co-composting of lignite and poultry wastes and the following amendment of co-composted products in soil. Environ Technol 45:5794–5807. https://doi.org/10.1080/09593330.2024.2306799

Castro FLS, Chai L, Arango J, Owens CM, Smith PA, Reichelt S, DuBois C, Menconi A (2023) Poultry industry paradigms: connecting the dots. J Appl Poult Res 32:100310. https://doi.org/10.1016/j.japr.2022.100310

Cavalcanti L, Gouveia É, Souza E, Maud J (2021) Effect of poultry litter as an organic fertilizer on water quality, parasitic abundance, and growth of nile tilapia. Bol Inst Pesca. https://doi.org/10.20950/1678-2305/bip.2021.47.e622

Chen Z, Jiang X (2014) Microbiological safety of chicken litter or chicken litter-based organic fertilizers: a review. Agriculture 4:1–29. https://doi.org/10.3390/agriculture4010001

Chen J, Jin C, Sun S, Yang D, He Y, Gan P, Nalume WG, Ma Y, He W, Li G (2023) Recognizing the challenges of composting: critical strategies for control, recycling, and valorization of nitrogen loss. Resour Conserv Recycl 198:107172. https://doi.org/10.1016/j.resconrec.2023.107172

Chiarelotto M, Restrepo JCPS, Lorin HEF, Damaceno FM (2021) Composting organic waste from the broiler production chain: a perspective for the circular economy. J Clean Prod 329:129717. https://doi.org/10.1016/j.jclepro.2021.129717

Chowdhury MA, de Neergaard A, Jensen LS (2014) Potential of aeration flow rate and bio-char addition to reduce greenhouse gas and ammonia emissions during manure composting. Chemosphere 97:16–25. https://doi.org/10.1016/j.chemosphere.2013.10.030

Chowdhury MW, Nabi MN, Arefin MA, Rashid F, Islam MT, Gudimetla P, Muyeen SM (2022) Recycling slaughterhouse wastes into potential energy and hydrogen sources: an approach for the future sustainable energy. Bioresour Technol Rep 19:101133. https://doi.org/10.1016/j.biteb.2022.101133

Costello B, Chen D, Han B, Zhang W, Butterly CR (2021) Lignite, dewatered lignite and modified subbituminous coal reduce nitrogen loss from broiler litter. Waste Manag 136:113–121. https://doi.org/10.1016/j.wasman.2021.10.005

de la Casa A, Ovando G, Bressanini L, Martínez J (2013) Aquacrop model calibration in potato and its use to estimate yield variability under field conditions. Atmos Clim Sci 3:397–407. https://doi.org/10.4236/acs.2013.33041

Dréano E, Laurentie M, Hurtaud-Pessel D, Mompelat S (2021) Multi–class analysis of 30 antimicrobial residues in poultry feathers by liquid chromatography tandem mass spectrometry. Food Addit Contam Part A 38:1701–1716. https://doi.org/10.1080/19440049.2021.1942561

Dume B, Hanc A, Svehla P, Michal P, Solcova O, Chane AD, Nigussie A (2022) Nutrient recovery and changes in enzyme activity during vermicomposting of hydrolysed chicken feather residue. Environ Technol:1–15. https://doi.org/10.1080/09593330.2022.2147451

Elisabet Masin C, Fernandez ME, Lescano MR, Zalazar CS (2020) Bioconversion of agro-industrial wastes: combined compost and vermicompost processes using *Eisenia fetida* for stabilization of poultry litter. IJROWA 9. https://doi.org/10.30486/ijrowa.2020.1885790.1011

Fakhfakh N, Ktari N, Haddar A, Mnif IH, Dahmen I, Nasri M (2011) Total solubilisation of the chicken feathers by fermentation with a keratinolytic bacterium, *Bacillus pumilus* A1, and the production of protein hydrolysate with high antioxidative activity. Process Biochem 46:1731–1737. https://doi.org/10.1016/j.procbio.2011.05.023

FAOSTAT. https://www.fao.org/faostat/en/#data/QCL. Accessed 18 Mar 2025

Farhad W, Cheema MA, Hammad HM, Saleem MF, Fahad S, Abbas F, Khosa I, Bakhat HF (2018) Influence of composted poultry manure and irrigation regimes on some morpho-physiology parameters of maize under semiarid environments. Environ Sci Pollut Res 25:19918–19931. https://doi.org/10.1007/s11356-018-2125-9

Fatoba DO, Amoako DG, Abia ALK, Essack SY (2022) Transmission of antibiotic-resistant Escherichia coli from chicken litter to agricultural soil. Front Environ Sci 9. https://doi.org/10.3389/fenvs.2021.751732

Figueroa A, Derksen T, Biswas S, Nazmi A, Rejmanek D, Crossley B, Pandey P, Gallardo RA (2021) Persistence of low and highly pathogenic avian influenza virus in reused poultry litter, effects of litter amendment use, and composting temperatures. J Appl Poult Res 30:100096. https://doi.org/10.1016/j.japr.2020.09.011

Franke-Whittle IH, Insam H (2013) Treatment alternatives of slaughterhouse wastes, and their effect on the inactivation of different pathogens: a review. Crit Rev Microbiol 39:139–151. https://doi.org/10.3109/1040841X.2012.694410

Greff B, Szigeti J, Nagy Á, Lakatos E, Varga L (2022) Influence of microbial inoculants on co-composting of lignocellulosic crop residues with farm animal manure: a review. J Environ Manag 302:114088. https://doi.org/10.1016/j.jenvman.2021.114088

Gržinić G, Piotrowicz-Cieślak A, Klimkowicz-Pawlas A, Górny RL, Ławniczek-Wałczyk A, Piechowicz L, Olkowska E, Potrykus M, Tankiewicz M, Krupka M, Siebielec G, Wolska L (2023) Intensive poultry farming: a review of the impact on the environment and human health. Sci Total Environ 858:160014. https://doi.org/10.1016/j.scitotenv.2022.160014

Gupta S, Sharma S, Aich A, Verma AK, Bhuyar P, Nadda AK, Mulla SI, Kalia S (2023) Chicken feather waste hydrolysate as a potential biofertilizer for environmental sustainability in organic agriculture management. Waste Biomass Valor 14:2783–2799. https://doi.org/10.1007/s12649-023-02123-6

Gwyther CL, Williams AP, Golyshin PN, Edwards-Jones G, Jones DL (2011) The environmental and biosecurity characteristics of livestock carcass disposal methods: a review. Waste Manag 31:767–778. https://doi.org/10.1016/j.wasman.2010.12.005

Haapapuro ER, Barnard ND, Simon M (1997) Review–animal waste used as livestock feed: dangers to human health. Prev Med 26:599–602. https://doi.org/10.1006/pmed.1997.0220

Habbeche A, Saoudi B, Jaouadi B, Haberra S, Kerouaz B, Boudelaa M, Badis A, Ladjama A (2014) Purification and biochemical characterization of a detergent-stable keratinase from a newly thermophilic actinomycete *Actinomadura keratinilytica* strain Cpt29 isolated from poultry compost. J Biosci Bioeng 117:413–421. https://doi.org/10.1016/j.jbiosc.2013.09.006

Hoang HG, Thuy BTP, Lin C, Vo D-VN, Tran HT, Bahari MB, Le VG, Vu CT (2022) The nitrogen cycle and mitigation strategies for nitrogen loss during organic waste composting: a review. Chemosphere 300:134514. https://doi.org/10.1016/j.chemosphere.2022.134514

Hussein MS, Burra KG, Amano RS, Gupta AK (2017) Temperature and gasifying media effects on chicken manure pyrolysis and gasification. Fuel 202:36–45. https://doi.org/10.1016/j.fuel.2017.04.017

Irfan M, Mehmood S, Mahmud A, Anjum AA (2020) An assessment of chemical and microbiological properties of different types of poultry waste compost prepared by bin and windrow composting system. Braz J Poult Sci 22:eRBCA. https://doi.org/10.1590/1806-9061-2020-1278

Izydorczyk G, Mikula K, Skrzypczak D, Witek-Krowiak A, Mironiuk M, Furman K, Gramza M, Moustakas K, Chojnacka K (2022) Valorization of poultry slaughterhouse waste for fertilizer purposes as an alternative for thermal utilization methods. J Hazard Mater 424:127328. https://doi.org/10.1016/j.jhazmat.2021.127328

Janczak D, Malińska K, Czekała W, Cáceres R, Lewicki A, Dach J (2017) Biochar to reduce ammonia emissions in gaseous and liquid phase during composting of poultry manure with wheat straw. Waste Manag 66:36–45. https://doi.org/10.1016/j.wasman.2017.04.033

Jn-Baptiste M, Sistani KR, Tewolde H (2013) Poultry litter time and method of application effects on corn yield. Soil Sci 178:109. https://doi.org/10.1097/SS.0b013e31828f5fa0

Joshi TN, Nepali DB, Sah R, Bhattarai T, Midmore DJ (2020) A comparison of composting and vermicomposting for the disposal of poultry waste. Anim Prod Sci 60:986. https://doi.org/10.1071/AN17177

Jurgutis L, Slepetiene A, Volungevicius J, Amaleviciute-Volunge K (2020) Biogas production from chicken manure at different organic loading rates in a mesophilic full scale anaerobic digestion plant. Biomass Bioenergy 141:105693. https://doi.org/10.1016/j.biombioe.2020.105693

Kacprzak M, Malińska K, Grosser A, Szołtysek JS, Wystalska K, Dróżdż D, Jasińska A, Meers E (2023) Cycles of carbon, nitrogen and phosphorus in poultry manure management technologies – environmental aspects. Crit Rev Environ Sci Technol 53:914–938. https://doi.org/10.1080/10643389.2022.2096983

Karuppannan SK, Dowlath MJH, Raiyaan GID, Rajadesingu S, Arunachalam KD (2021) Chapter 5: Application of poultry industry waste in producing value-added products—a review. In: Hussain CM (ed) Concepts of advanced zero waste tools. Elsevier, pp 91–121

Kaur T (2020) Vermicomposting: an effective option for recycling organic wastes. In: Organic agriculture. IntechOpen

Kaur P, Saini KS, Sharma S, Kaur J, Bhatt R, Alamri S, Alfagham AT, Hussain S (2023) Increasing the efficiency of the rice–wheat cropping system through integrated nutrient management. Sustainability 15:12694. https://doi.org/10.3390/su151712694

Khodadadi M, Masoumi A, Sadeghi M (2024) Drying, a practical technology for reduction of poultry litter (environmental) pollution: methods and their effects on important parameters. Poult Sci 103:104277. https://doi.org/10.1016/j.psj.2024.104277

Ko HJ, Kim KY, Kim HT, Kim CN, Umeda M (2008) Evaluation of maturity parameters and heavy metal contents in composts made from animal manure. Waste Manag 28:813–820. https://doi.org/10.1016/j.wasman.2007.05.010

Köninger J, Lugato E, Panagos P, Kochupillai M, Orgiazzi A, Briones MJI (2021) Manure management and soil biodiversity: towards more sustainable food systems in the EU. Agric Syst 194:103251. https://doi.org/10.1016/j.agsy.2021.103251

Kuboń M, Komorowska M, Niemiec M, Sikora J, Szeląg-Sikora A, Olech E, Molik E, Gajda J (2024) The impact of biochar additives and fat-emulsifying substances on the efficiency of the slaughterhouse waste Biogasing process. Energies 17:3065. https://doi.org/10.3390/en17133065

Laca A, Laca A, Díaz M (2017) Eggshell waste as catalyst: a review. J Environ Manag 197:351–359. https://doi.org/10.1016/j.jenvman.2017.03.088

Laohavisuti N, Boonchom B, Boonmee W, Chaiseeda K, Seesanong S (2021) Simple recycling of biowaste eggshells to various calcium phosphates for specific industries. Sci Rep 11:15143. https://doi.org/10.1038/s41598-021-94643-1

Latshaw JD, Musharaf N, Retrum R (1994) Processing of feather meal to maximize its nutritional value for poultry. Anim Feed Sci Technol 47:179–188. https://doi.org/10.1016/0377-8401(94)90122-8

Li H, Hu J, Wang X, An L (2019) Development of a bio-inspired photo-recyclable feather carbon adsorbent towards removal of amoxicillin residue in aqueous solutions. Chem Eng J 373:1380–1388. https://doi.org/10.1016/j.cej.2019.03.160

Li M-X, He X-S, Tang J, Li X, Zhao R, Tao Y-Q, Wang C, Qiu Z-P (2021) Influence of moisture content on chicken manure stabilization during microbial agent-enhanced composting. Chemosphere 264:128549. https://doi.org/10.1016/j.chemosphere.2020.128549

Limoges MA, Neher DA, Weicht TR, Millner PD, Sharma M, Donnelly C (2022) Differential survival of Escherichia coli and listeria spp. in northeastern U.S. soils amended with dairy manure compost, poultry litter compost, and heat-treated poultry pellets and fate in raw edible radish crops. J Food Prot 85:1708–1715. https://doi.org/10.4315/JFP-21-261

Lin Y, Watts DB, van Santen E, Cao G (2018) Influence of poultry litter on crop productivity under different field conditions: a meta-analysis. Agron J 110:807–818. https://doi.org/10.2134/agronj2017.09.0513

Lirikum KLN, Thyug L, Mozhui L (2022) Vermicomposting: an eco-friendly approach for waste management and nutrient enhancement. Trop Ecol 63:325–337. https://doi.org/10.1007/s42965-021-00212-y

Li-Xian Y, Guo-Liang L, Shi-Hua T, Gavin S, Zhao-Huan H (2007) Salinity of animal manure and potential risk of secondary soil salinization through successive manure application. Sci Total Environ 383:106–114. https://doi.org/10.1016/j.scitotenv.2007.05.027

Manea EE, Bumbac C, Dinu LR, Bumbac M, Nicolescu CM (2024) Composting as a sustainable solution for organic solid waste management: current practices and potential improvements. Sustainability 16:6329. https://doi.org/10.3390/su16156329

Mansour M-N, Lendormi T, Drévillon L, Naji A, Louka N, Maroun RG, Hobaika Z, Lanoisellé J-L (2024) Influence of substrate/inoculum ratio, inoculum source and ammonia inhibition on anaerobic digestion of poultry waste. Environ Technol 45:1894–1907. https://doi.org/10.1080/09593330.2022.2157754

Mante OD, Agblevor FA (2011) Parametric study on the pyrolysis of manure and wood shavings. Biomass Bioenergy 35:4417–4425. https://doi.org/10.1016/j.biombioe.2011.08.017

Manzoor A, Naveed MS, Ali RMA, Naseer MA, UL-Hussan M, Saqib M, Hussain S, Farooq M (2024) Vermicompost: a potential organic fertilizer for sustainable vegetable cultivation. Sci Hortic 336:113443. https://doi.org/10.1016/j.scienta.2024.113443

Martins GL, Pellegrinetti TA, de Freitas AS, Rocha GS, Monteiro GGTN, Mendes LW, Tsai SM, Oliveira FC, Regitano JB (2024) Differential effects of fresh and composted organic wastes on soil bacterial community and maize growth. Appl Soil Ecol 204:105715. https://doi.org/10.1016/j.apsoil.2024.105715

McGauran T, Dunne N, Smyth BM, Cunningham E (2021) Feasibility of the use of poultry waste as polymer additives and implications for energy, cost and carbon. J Clean Prod 291:125948. https://doi.org/10.1016/j.jclepro.2021.125948

Mironiuk M, Samoraj M, Witek-Krowiak A, Górecki H, Moustakas K, Chojnacka K (2023) Processing of nuisance animal waste into agricultural products. Environ Pollut 319:120924. https://doi.org/10.1016/j.envpol.2022.120924

Mittal A, Thakur V, Gajbe V (2013) Adsorptive removal of toxic azo dye Amido Black 10B by hen feather. Environ Sci Pollut Res 20:260–269. https://doi.org/10.1007/s11356-012-0843-y

Mohamed Amanullah M, Sekar S, Muthukrishnan P (2010) Prospects and potential of poultry manure. Asian J Plant Sci 9:172–182. https://doi.org/10.3923/ajps.2010.172.182

Moore PA, Miles D, Burns R, Pote D, Berg K, Choi IH (2011) Ammonia emission factors from broiler litter in barns, in storage, and after land application. J Environ Qual 40:1395–1404. https://doi.org/10.2134/jeq2009.0383

Moukazis I, Pellera F-M, Gidarakos E (2018) Slaughterhouse by-products treatment using anaerobic digestion. Waste Manag 71:652–662. https://doi.org/10.1016/j.wasman.2017.07.009

Mozhiarasi V, Natarajan TS (2025) Slaughterhouse and poultry wastes: management practices, feedstocks for renewable energy production, and recovery of value added products. Biomass Conv Bioref 15:1705–1728. https://doi.org/10.1007/s13399-022-02352-0

Mumpton FA, Fishman PH (1977) The application of natural zeolites in animal science and aquaculture. J Anim Sci 45:1188–1203. https://doi.org/10.2527/jas1977.4551188x

Narciso A, Barra Caracciolo A, De Carolis C (2023) Overview of direct and indirect effects of antibiotics on terrestrial organisms. Antibiotics 12:1471. https://doi.org/10.3390/antibiotics12091471

Nascimento-Gonçalves E, Azevedo T, Lopes H, Sousa JR, Oliveira PA, Roboredo M, Coimbra AM, Morais MC (2024) Vermicomposting as a valorization solution to the winery sector by-products. Agronomy 14:1111. https://doi.org/10.3390/agronomy14061111

Niedzialkoski RK, Marostica R, Damaceno FM, Costa LA d M, Costa MSS d M (2021) Combination of biological processes for agro-industrial poultry waste management: effects on vermicomposting and anaerobic digestion. J Environ Manag 297:113127. https://doi.org/10.1016/j.jenvman.2021.113127

Nigussie A, Kuijper T, Bruun S, de Neergaard A (2016) Vermicomposting as a technology for reducing nitrogen losses and greenhouse gas emissions from small-scale composting. J Clean Prod 139:429–439. https://doi.org/10.1016/j.jclepro.2016.08.058

Odetoye TE, Agu JO, Ajala EO (2021) Biodiesel production from poultry wastes: waste chicken fat and eggshell. J Environ Chem Eng 9:105654. https://doi.org/10.1016/j.jece.2021.105654

Okada E, Young BJ, Pérez DJ, Pellegrini MC, Carciochi WD, Lavallén CM, Ponce AG, Dopchiz MC, Hernández Guijarro K, Franco M d R, Di Martino AM, Rizzo PF (2024) Effect of on-farm poultry litter composting processes on physicochemical, biological, and toxicological parameters and reduction of antibiotics and antibiotic-resistant *Escherichia coli*. Waste Manag 174: 310–319. https://doi.org/10.1016/j.wasman.2023.12.005

Okoya AA, Ochor NO, Akinyele AB, Olaiya OO (2020) The use of chicken feather waste as an adsorbent for crude oil clean up from polluted water. J Agric Ecol Res Int:43–53. https://doi.org/10.9734/jaeri/2020/v21i330136

Oliveira DA, Benelli P, Amante ER (2013) A literature review on adding value to solid residues: egg shells. J Clean Prod 46:42–47. https://doi.org/10.1016/j.jclepro.2012.09.045

Oviedo-Ocaña ER, Hernández-Gómez AM, Ríos M, Portela A, Sánchez-Torres V, Domínguez I, Komilis D (2021) A comparison of two-stage and traditional co-composting of green waste and food waste amended with phosphate rock and sawdust. Sustainability 13:1109. https://doi.org/10.3390/su13031109

Oyege I, Balaji Bhaskar MS (2023) Effects of vermicompost on soil and plant health and promoting sustainable agriculture. Soil Syst 7:101. https://doi.org/10.3390/soilsystems7040101

Ozdemir S, Yetilmezsoy K (2020) A mini literature review on sustainable management of poultry abattoir wastes. J Mater Cycles Waste Manag 22:11–21. https://doi.org/10.1007/s10163-019-00934-1

Pachón Gómez EM, Domínguez RE, López DA, Téllez JF, Marino MD, Almada N, Gange JM, Moyano EL (2023) Chicken litter: a waste or a source of chemicals? Fast pyrolysis and hydrothermal conversion as alternatives in the valorisation of poultry waste. J Anal Appl Pyrolysis 169:105796. https://doi.org/10.1016/j.jaap.2022.105796

Paranhos AG d O, Pereira AR, Coelho LDN, Silva S d Q, de Aquino SF (2023) Hydrothermal pre-treatment followed by anaerobic digestion for the removal of tylosin and antibiotic resistance agents from poultry litter. Environ Sci Pollut Res 30:42443–42455. https://doi.org/10.1007/s11356-023-25329-6

Peng S, Li H, Xu Q, Lin X, Wang Y (2019) Addition of zeolite and superphosphate to windrow composting of chicken manure improves fertilizer efficiency and reduces greenhouse gas emission. Environ Sci Pollut Res 26:36845–36856. https://doi.org/10.1007/s11356-019-06544-6

Pereira MMA, Moraes LC, Mogollón MCT, Borja CJF, Duarte M, Buttrós VHT, Luz JMQ, Pasqual M, Dória J (2023) Cultivating biodiversity to harvest sustainability: vermicomposting and inoculation of microorganisms for soil preservation and resilience. Agronomy 13:103. https://doi.org/10.3390/agronomy13010103

Prabakaran R, Valavan SE (2021) Wealth from poultry waste: an overview. Worlds Poult Sci J 77: 389–401. https://doi.org/10.1080/00439339.2021.1901557

Prasai TP, Walsh KB, Midmore DJ, Jones BEH, Bhattarai SP (2018) Manure from biochar, bentonite and zeolite feed supplemented poultry: moisture retention and granulation properties. J Environ Manag 216:82–88. https://doi.org/10.1016/j.jenvman.2017.08.040

Qasim W, Lee MH, Moon BE, Okyere FG, Khan F, Nafees M, Kim HT (2018) Composting of chicken manure with a mixture of sawdust and wood shavings under forced aeration in a closed reactor system. Int J Recycl Org Waste Agric 7:261–267. https://doi.org/10.1007/s40093-018-0212-z

Quina MJ, Soares MAR, Quinta-Ferreira R (2017) Applications of industrial eggshell as a valuable anthropogenic resource. Resour Conserv Recycl 123:176–186. https://doi.org/10.1016/j.resconrec.2016.09.027

Rahman MM, Netravali AN, Tiimob BJ, Rangari VK (2014) Bioderived "Green" composite from soy protein and eggshell Nanopowder. ACS Sustain Chem Eng 2:2329–2337. https://doi.org/10.1021/sc5003193

Rahman MM, Hassan A, Hossain I, Jahangir MMR, Chowdhury EH, Parvin R (2022) Current state of poultry waste management practices in Bangladesh, environmental concerns, and future recommendations. J Adv Vet Anim Res 9:490–500. https://doi.org/10.5455/javar.2022.i618

Rasel MNH, Halder M, Joardar JC (2020) Biodegradation of poultry feather using *Streptomyces* sp. for nitrogen and its effect on growth and yield of okra. Int J Recycl Org Waste Agric 9. https://doi.org/10.30486/ijrowa.2020.1898655.1064

Reddy N, Santosh MS (2016) Chapter 14: Recovery and applications of feather proteins. In: Singh Dhillon G (ed) Protein byproducts. Academic, pp 255–274

Richa, Kumar V, Singh J, Sharma N (2020) Poultry manure and poultry waste management: a review. Int J Curr Microbiol Appl Sci 9:3483–3495. https://doi.org/10.20546/ijcmas.2020.906.410

Ro KS, Cantrell KB, Hunt PG (2010) High-temperature pyrolysis of blended animal manures for producing renewable energy and value-added biochar. Ind Eng Chem Res 49:10125–10131. https://doi.org/10.1021/ie101155m

Rodhe L, Karlsson S (2002) SE—structures and environment: ammonia emissions from broiler manure — influence of storage and spreading method. Biosyst Eng 82:455–462. https://doi.org/10.1006/bioe.2002.0081

Ruiz Diaz DA, Sawyer JE (2008) Plant-available nitrogen from poultry manure as affected by time of application. Agron J 100:1318–1326. https://doi.org/10.2134/agronj2008.0010

Samadi MT, Leili M, Rahmani A, Moradi S, Godini K (2024) Anaerobic co-digestion using poultry slaughterhouse and vegetable wastes to enhance biogas yield: effect of different C/N ratios. Biomass Conv Bioref 14:28303–28311. https://doi.org/10.1007/s13399-022-03501-1

Sánchez-García M, Alburquerque JA, Sánchez-Monedero MA, Roig A, Cayuela ML (2015) Biochar accelerates organic matter degradation and enhances N mineralisation during composting of poultry manure without a relevant impact on gas emissions. Bioresour Technol 192:272–279. https://doi.org/10.1016/j.biortech.2015.05.003

Shahiful Hizam N, Omar N, Sabri N, Samsuddin N, Zainol Abidin M (2024) Evaluation of physicochemical properties and nutrient composition of chicken feather Vermicompost for sustainable waste management system. IOP Conf Ser: Earth Environ Sci 1426:012014. https://doi.org/10.1088/1755-1315/1426/1/012014

Shung C-C, Hsin K-Y, Tan F-J, Chen S-E (2023) The Defeathering effect by scalding in chickens follows their intrinsic dermal Histologies. Animals 13:2584. https://doi.org/10.3390/ani13162584

Singh S, Dutta S, Inamdar S (2014) Land application of poultry manure and its influence on spectrofluorometric characteristics of dissolved organic matter. Agric Ecosyst Environ 193:25–36. https://doi.org/10.1016/j.agee.2014.04.019

Singh A, Kelkar N, Natarajan K, Selvaraj S (2021) Review on the extraction of calcium supplements from eggshells to combat waste generation and chronic calcium deficiency. Environ Sci Pollut Res 28:46985–46998. https://doi.org/10.1007/s11356-021-15158-w

Sistani KR, Adeli A, Tewolde H (2010) Apparent use efficiency of nitrogen and phosphorus from litter applied to bermudagrass. Commun Soil Sci Plant Anal 41:1873–1884. https://doi.org/10.1080/00103624.2010.492444

Skrzypczak D, Trzaska K, Gil F, Izydorczyk G, Chojnacka K (2024) Guidelines for efficient nitrogen preservation in sewage sludge-based fertilizers. Sci Total Environ 947:174460. https://doi.org/10.1016/j.scitotenv.2024.174460

Soares MAR, Quina MJ, Reis MS, Quinta-Ferreira R (2017) Assessment of co-composting process with high load of an inorganic industrial waste. Waste Manag 59:80–89. https://doi.org/10.1016/j.wasman.2016.09.044

Solís-Moreno CA, Cervantes-González E, Saavedra-Leos MZ (2021) Use and treatment of chicken feathers as a natural adsorbent for the removal of copper in aqueous solution. J Environ Health Sci Eng 19:707–720. https://doi.org/10.1007/s40201-021-00639-4

Soobhany N (2019) Insight into the recovery of nutrients from organic solid waste through biochemical conversion processes for fertilizer production: a review. J Clean Prod 241: 118413. https://doi.org/10.1016/j.jclepro.2019.118413

Srivastava PK, Singh A, Kumari S, Arora S, Choubey AK, Sinha ASK (2023) Production and characterization of sustainable vermimanure derived from poultry litter and rice straw using tiger worm *Eisenia fetida*. Bioresour Technol 369:128377. https://doi.org/10.1016/j.biortech.2022.128377

Staver KW, Brinsfield RB (2001) Agriculture and water quality on the Maryland eastern shore: where do we go from here?: long-term solutions to accelerated eutrophication must provide mechanisms for redistributing nutrients flowing into concentrated animal-producing regions. Bioscience 51:859–868. https://doi.org/10.1641/0006-3568(2001)051[0859:AAWQOT]2.0.CO;2

Steiner C, Das KC, Melear N, Lakly D (2010) Reducing nitrogen loss during poultry litter composting using biochar. J Environ Qual 39:1236–1242. https://doi.org/10.2134/jeq2009.0337

Stingone JA, Wing S (2011) Poultry litter incineration as a source of energy: reviewing the potential for impacts on environmental health and justice. New Solut 21:27–42. https://doi.org/10.2190/NS.21.1.g

Swelum AA, El-Saadony MT, Abd El-Hack ME, Abo Ghanima MM, Shukry M, Alhotan RA, Hussein EOS, Suliman GM, Ba-Awadh H, Ammari AA, Taha AE, El-Tarabily KA (2021) Ammonia emissions in poultry houses and microbial nitrification as a promising reduction strategy. Sci Total Environ 781:146978. https://doi.org/10.1016/j.scitotenv.2021.146978

Sypka M, Jodłowska I, Białkowska AM (2021) Keratinases as versatile enzymatic tools for sustainable development. Biomolecules 11:1900. https://doi.org/10.3390/biom11121900

Tesfaye T, Sithole B, Ramjugernath D (2017a) Valorisation of chicken feathers: a review on recycling and recovery route—current status and future prospects. Clean Techn Environ Policy 19:2363–2378. https://doi.org/10.1007/s10098-017-1443-9

Tesfaye T, Sithole B, Ramjugernath D, Chunilall V (2017b) Valorisation of chicken feathers: characterisation of chemical properties. Waste Manag 68:626–635. https://doi.org/10.1016/j.wasman.2017.06.050

Tewolde H, Armstrong S, Way TR, Rowe DE, Sistani KR (2009) Cotton response to poultry litter applied by subsurface banding relative to surface broadcasting. Soil Sci Soc Am J 73:384–389. https://doi.org/10.2136/sssaj2008.0127

Tiwari AK, Pal DB (2022) Chapter 11: nutrients contamination and eutrophication in the river ecosystem. In: Madhav S, Kanhaiya S, Srivastav A, Singh V, Singh P (eds) Ecological significance of river ecosystems. Elsevier, pp 203–216

Turan NG (2009) Nitrogen availability in composted poultry litter using natural amendments. Waste Manag Res 27:19–24. https://doi.org/10.1177/0734242X07087993

Vaishnav S, Saini T, Chauhan A, Gaur GK, Tiwari R, Dutt T, Tarafdar A (2023) Livestock and poultry farm wastewater treatment and its valorization for generating value-added products: recent updates and way forward. Bioresour Technol 382:129170. https://doi.org/10.1016/j.biortech.2023.129170

Vicentin R, Masín CE, Lescano MR, Zalazar CS (2021) Poultry litter stabilization by two-stage composting-vermicomposting process: environmental, energetic and economic performance. Chemosphere 281:130872. https://doi.org/10.1016/j.chemosphere.2021.130872

Waheed M, Yousaf M, Shehzad A, Inam-Ur-Raheem M, Khan MKI, Khan MR, Ahmad N, Abdullah ARM (2020) Channelling eggshell waste to valuable and utilizable products: a

comprehensive review. Trends Food Sci Technol 106:78–90. https://doi.org/10.1016/j.tifs.2020.10.009
Wang Y, Zhu Y, Zhang S, Wang Y (2018) What could promote farmers to replace chemical fertilizers with organic fertilizers? J Clean Prod 199:882–890. https://doi.org/10.1016/j.jclepro.2018.07.222
Wang S, Wei Z, Wang L (2024) Improving slaughterhouse byproducts utilization via anaerobic digestion, composting, and rendering. Renew Sust Energ Rev 189:113881. https://doi.org/10.1016/j.rser.2023.113881
Wang Z, Xiao N, Guo S, Liu X, Liu C, Ai M (2025) Unlocking the potential of keratin: a comprehensive exploration from extraction and structural properties to cross-disciplinary applications. J Agric Food Chem 73:1014–1037. https://doi.org/10.1021/acs.jafc.4c07102
Waqas M, Hashim S, Humphries UW, Ahmad S, Noor R, Shoaib M, Naseem A, Hlaing PT, Lin HA (2023) Composting processes for agricultural waste management: a comprehensive review. Processes 11:731. https://doi.org/10.3390/pr11030731
Watts DB, Torbert HA (2011) Long-term tillage and poultry litter impacts on soybean and corn grain yield. Agron J 103:1479–1486. https://doi.org/10.2134/agronj2011.0073
Wilson PB (2017) Recent advances in avian egg science: a review. Poult Sci 96:3747–3754. https://doi.org/10.3382/ps/pex187
Xia Y, Massé DI, McAllister TA, Beaulieu C, Ungerfeld E (2012) Anaerobic digestion of chicken feather with swine manure or slaughterhouse sludge for biogas production. Waste Manag 32: 404–409. https://doi.org/10.1016/j.wasman.2011.10.024
Yadav A, Garg VK (2011) Industrial wastes and sludges management by vermicomposting. Rev Environ Sci Biotechnol 10:243–276. https://doi.org/10.1007/s11157-011-9242-y
Yadav T, Mungray AA, Mungray AK (2018) Generation of $TiO_2$ nanoparticle-based acacia saturated eggshell bio-composite for pathogen removal. Environ Nanotechnol Monitor Manag 9:50–57. https://doi.org/10.1016/j.enmm.2017.12.002
Yıldız Z, Kaya N, Topcu Y, Uzun H (2019) Pyrolysis and optimization of chicken manure wastes in fluidized bed reactor: $CO_2$ capture in activated bio-chars. Process Saf Environ Prot 130:297–305. https://doi.org/10.1016/j.psep.2019.08.011
Yoon Y-M, Kim S-H, Oh S-Y, Kim C-H (2014) Potential of anaerobic digestion for material recovery and energy production in waste biomass from a poultry slaughterhouse. Waste Manag 34:204–209. https://doi.org/10.1016/j.wasman.2013.09.020
Zhang F, Wei Z, Wang JJ (2021) Integrated application effects of biochar and plant residue on ammonia loss, heavy metal immobilization, and estrogen dissipation during the composting of poultry manure. Waste Manag 131:117–125. https://doi.org/10.1016/j.wasman.2021.05.037
Zhang L, Ren J, Bai W (2023) A review of poultry waste-to-wealth: technological Progress, modeling and simulation studies, and economic- environmental and social sustainability. Sustainability 15:5620. https://doi.org/10.3390/su15075620

# Exploring the Organic Residual Matter from Cattle Industries and Microbial Dynamics in Composting

**Oyindamola John Samson, Doyinsola Amos Samson, Jamiu Oluwatosin Adeyemi, Hafeez Aderinsayo Adekola, and Ismail Babatunde Onajobi**

## 1 Introduction

The cattle industry is a significant contributor to organic residual waste, primarily in the form of manure, bedding materials, feed leftovers, and wastewater. These organic residuals are nutrient-rich, containing essential elements like nitrogen, phosphorus, and potassium, making them valuable for soil enrichment (Riaz et al. 2023). However, if improperly managed, these wastes pose environmental and public health risks, including water contamination, greenhouse gas emissions, and the spread of pathogens. The sheer volume of waste generated globally explains the need for effective strategies to mitigate its impact while harnessing its potential benefits for sustainable agriculture (de Oliveira et al. 2023).

Sustainable waste management in the cattle industry is essential for reducing its environmental footprint. Effective practices not only address the challenges of waste disposal but also align with global efforts to mitigate climate change (Koul et al. 2022). By transforming waste into useful products, sustainable waste management promotes resource efficiency and reduces dependency on chemical fertilizers.

O. J. Samson (✉)
Department of Microbiology and Biotechnology, Ajayi Crowther University, Oyo, Nigeria

Department of Microbiology, Olabisi Onabanjo University, Ago-Iwoye, Nigeria

D. A. Samson
Department of Agricultural Economics and Farm Management, Federal University of Agriculture, Abeokuta, Nigeria

J. O. Adeyemi · I. B. Onajobi
Department of Microbiology, Olabisi Onabanjo University, Ago-Iwoye, Nigeria

H. A. Adekola
Department of Molecular Biology and Biotechnology, Nigerian Institute of Medical Research, Lagos, Nigeria

 
R. K. Singh et al. (eds.), *Composting of Farm Residues by Nature-Based Solutions*,
https://doi.org/10.1007/978-3-032-12255-1_6

Furthermore, proper management minimizes the release of methane and ammonia, thereby reducing air pollution and contributing to healthier ecosystems. It also helps prevent the leaching of nutrients into water bodies, which can lead to eutrophication and biodiversity loss (Hanif et al. 2022).

Composting has emerged as a vital strategy for managing cattle industry waste sustainably, offering a biological process to recycle organic matter (Yuvaraj et al. 2021). This process converts waste into stable, humus-like compost that enriches the soil with nutrients and improves its structure. By facilitating the breakdown of organic materials through microbial activity, composting not only reduces the bulk of waste but also eliminates harmful pathogens and weed seeds. The end product is an eco-friendly alternative to synthetic fertilizers, promoting soil health while reducing environmental risks associated with untreated waste disposal (Bhunia et al. 2021).

In the context of agricultural sustainability, composting plays a pivotal role by enhancing soil fertility and fostering long-term productivity. Compost improves soil organic matter content, enhances water retention, and promotes beneficial microbial activity, all of which contribute to resilient agricultural systems (Rastogi et al. 2023). Additionally, it aids in carbon sequestration by stabilizing organic carbon in the soil, making it a crucial component in combating climate change. The use of compost derived from cattle industry waste supports a circular economy, ensuring that agricultural by-products are reintegrated into the production cycle instead of becoming pollutants (Raihan 2023).

The integration of composting into cattle waste management systems represents a sustainable approach to addressing environmental challenges while advancing agricultural productivity. It aligns with global sustainability goals by reducing waste, promoting soil health, and mitigating climate impacts (Manea et al. 2024). By recognizing the value of organic residuals and adopting composting technologies, the cattle industry can transition from a waste generator to a contributor to sustainable agriculture and environmental stewardship. Thus, this chapter aims to explore the biogeochemical and microbial processes governing the composting of organic waste from cattle industries.

## 2 Characteristics of Organic Residual Matter from Cattle Industries

The cattle industry generates diverse types of organic residual matter, which can be broadly categorized into manure, bedding materials, feed leftovers, and wastewater (Marlina et al. 2024). Manure is the most significant component, comprising a mix of feces and urine. Bedding materials, such as straw, wood shavings, or sawdust, are used for animal comfort and absorbency but are discarded once saturated. Feed leftovers include uneaten portions of hay, silage, or grain, while wastewater originates from cleaning operations in barns and feedlots. Each type of waste varies in

**Table 1** Characteristics and environmental challenges of organic residuals from cattle industries

| Waste type | Components | Nutrient profile | Environmental challenges | References |
|---|---|---|---|---|
| Manure | Feces and urine | High in N, P, and K | Methane emissions, water contamination, and pathogen spread | Carpio et al. (2021) |
| Bedding materials | Straw, sawdust, and wood shavings | Carbon-rich, low in nutrients | Excess bulk and slow decomposition if unmanaged | Wang et al. (2023a, b) |
| Feed leftovers | Uneaten hay, silage, and grains | Proteins, carbohydrates, and micronutrients | Attract pests and potential nutrient leaching | Koul et al. (2022) |
| Wastewater | Cleaning runoff and dissolved organic matter | Dissolved nutrients and trace organics | Eutrophication and groundwater contamination | Marlina et al. (2024) |

composition and volume depending on the farm's size, feeding practices, and waste management systems (Koul et al. 2022).

The chemical composition of these wastes reflects their potential as a resource for soil enrichment. Manure is rich in macronutrients like nitrogen (N), phosphorus (P), and potassium (K), essential for plant growth. Bedding materials, while lower in nutrients, contribute carbon-rich organic matter that balances the carbon-to-nitrogen (C/N) ratio during composting. Feed leftovers may contain residual proteins, carbohydrates, and micronutrients, while wastewater often carries dissolved nutrients and organic compounds. The nutrient profile of these wastes depends on factors such as the cattle's diet, feed efficiency, and waste handling practices, making characterization critical for effective recycling (Yahaya et al. 2023).

Improper disposal of organic residuals from cattle industries poses significant environmental challenges. Direct discharge of manure and wastewater into water bodies can cause nutrient loading, leading to eutrophication and loss of aquatic biodiversity. Decomposing manure emits methane and nitrous oxide, potent greenhouse gases contributing to climate change. Accumulated waste in feedlots or storage facilities can result in odor issues and attract pests, creating public health concerns. Additionally, the presence of pathogens, antibiotic residues, and heavy metals in these wastes threatens soil and water quality if untreated or poorly managed (Sahu and Arya 2024).

Systematically understanding the characteristics of each waste type is essential. For example, optimizing the C/N ratio is critical for efficient composting, which balances nutrient release and minimizes greenhouse gas emissions. Similarly, treating wastewater to remove excess nutrients before reuse or discharge ensures environmental safety. Incorporating feed leftovers into composting systems not only recycles nutrients but also reduces waste volume. These strategies highlight the need for integrated management practices to harness the benefits of organic residuals while mitigating risks (Table 1).

## 3 Composting as a Waste Management Strategy

Composting is a biological process that converts organic waste into stable, nutrient-rich compost through the activity of microorganisms under controlled aerobic conditions. The primary principles of composting include maintaining adequate aeration, moisture, and a balanced carbon-to-nitrogen (C/N) ratio to support microbial activity. These conditions allow microorganisms to break down complex organic materials into simpler compounds, generating heat as a byproduct and producing a humus-like material suitable for soil amendment. Effective composting requires regular monitoring to ensure the process progresses efficiently and yields high-quality compost (Ezeah et al. 2023).

Composting cattle industry waste offers numerous environmental and agricultural benefits. It reduces the bulk and environmental risks associated with untreated waste by stabilizing nutrients and eliminating harmful pathogens. The resulting compost enriches soil fertility by providing essential nutrients and improving soil structure, water retention, and microbial activity. Additionally, composting mitigates the release of greenhouse gases like methane, which are typically emitted from anaerobic decomposition in unmanaged waste piles. By recycling organic residues into valuable compost, the cattle industry contributes to sustainable waste management and resource efficiency (Wei et al. 2024).

The composting process involves three key stages: the mesophilic, thermophilic, and curing phases. During the *mesophilic phase*, moderate temperatures (25–40 °C) encourage the activity of mesophilic microorganisms that initiate the decomposition of readily degradable organic matter (Bher et al. 2022). This is followed by the *thermophilic phase*, where temperatures rise to 45–70 °C due to the intense microbial activity of thermophilic bacteria and fungi. This phase is crucial for breaking down complex materials like lignin and cellulose and for killing pathogens and weed seeds. Finally, in the *curing phase*, temperatures decline, and mesophilic microorganisms reestablish themselves to complete the stabilization of the compost, ensuring a mature and nutrient-balanced end product (Yin et al. 2023).

The composting process operates as a continuous cycle, transforming waste into resources while promoting environmental sustainability. Organic residues are collected and prepared, then undergo controlled decomposition in the mesophilic and thermophilic stages, followed by stabilization in the curing phase. The finished compost is applied to agricultural soils, enriching fertility and closing the loop in waste management. This cyclical approach not only addresses the disposal challenges of cattle industry waste but also enhances soil health and supports sustainable agricultural practices (Azis et al. 2023).

The microbial dynamics in composting begin with the collection and preparation of organic waste, where moisture, aeration, and the carbon-to-nitrogen ratio are adjusted (Fig. 1). In the *mesophilic phase*, microorganisms like *Escherichia* and *Bacillus* initiate decomposition. This transitions into the *thermophilic phase*, where high temperatures break down complex materials and suppress pathogens. Finally,

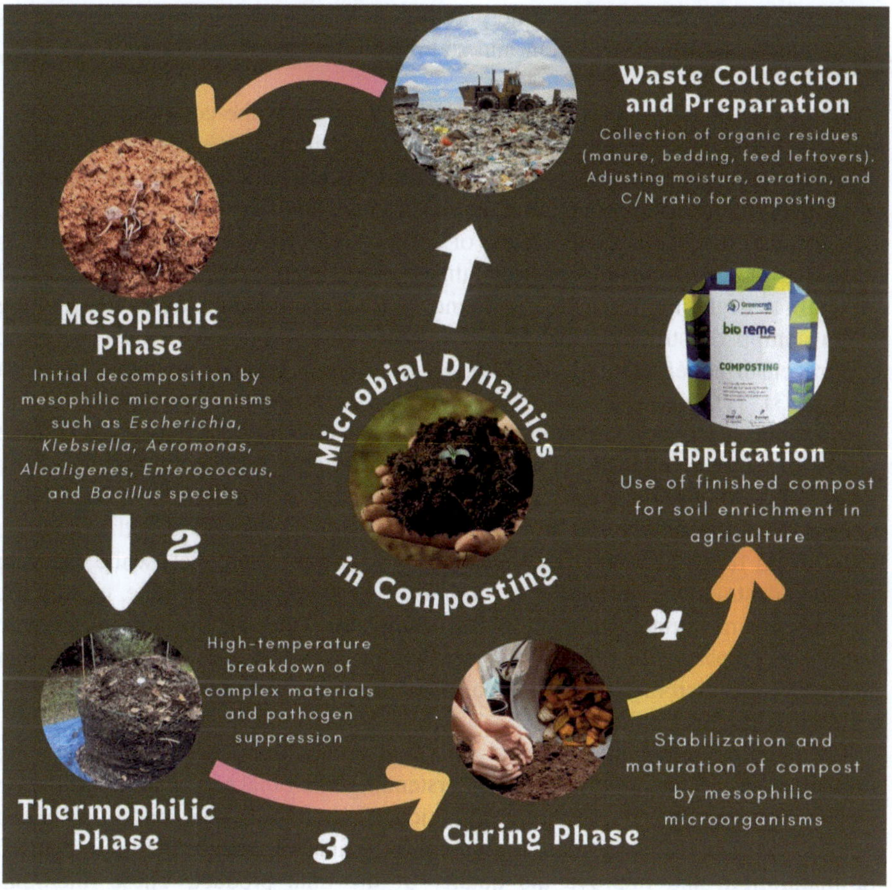


**Fig. 1** Composting cycle for cattle industry waste

during the *curing phase*, the compost stabilizes with mesophilic microbes, becoming mature and ready for application as a nutrient-rich soil enhancer in agriculture.

## 4 Microbial Dynamics in the Composting Process

Microorganisms are the primary drivers of organic matter degradation during composting, transforming complex organic compounds into simpler, stable forms. These microorganisms break down carbohydrates, proteins, lipids, and lignin into carbon dioxide, water, and nutrient-rich humus. Their metabolic activity generates heat, which creates a self-sustaining environment for the composting process. The efficiency of composting largely depends on maintaining optimal conditions, such as temperature, moisture, aeration, and the carbon-to-nitrogen (C/N) ratio, to support microbial activity. The dynamic interplay between different microbial communities ensures the complete degradation and stabilization of organic matter (Wang et al. 2023a, b).

Major microbial groups play distinct roles throughout the composting process. *Bacteria*, being the most abundant, initiate decomposition by breaking down readily degradable compounds such as sugars and proteins (Kumari and Das 2023). *Fungi* are critical for decomposing more complex materials like cellulose and lignin, making them essential in later stages of composting (Yao et al. 2023). *Actinomycetes*, a group of filamentous bacteria, thrive in moderate-to-high temperatures and are vital for degrading resistant compounds, such as chitin and waxes, contributing to the compost's structural stability. Together, these groups create a synergistic system, ensuring efficient decomposition of diverse organic materials (Sharma et al. 2023).

Microbial succession is a hallmark of the composting process, marked by distinct phases of activity aligned with temperature changes. In the *mesophilic phase* (25–40 °C), mesophilic bacteria dominate, breaking down simple organic compounds. As temperatures rise in the *thermophilic phase* (45–70 °C), thermophilic bacteria and fungi take over, decomposing more complex materials while ensuring pathogen destruction. During the *curing phase*, as temperatures decline, mesophilic microorganisms reestablish themselves, completing the stabilization of organic matter and preparing the compost for agricultural use. This succession ensures that each group of microorganisms optimally contributes to different stages of decomposition (Rillig et al. 2024).

The dynamic interaction of these microbial groups throughout the composting process highlights the importance of understanding microbial ecology for optimizing composting practices. By fostering conditions favorable to beneficial microbes at each phase, composting systems can maximize efficiency, reduce the time required for maturation, and improve the quality of the final product. These microbial dynamics also aid in designing strategies to minimize undesirable effects such as odors and greenhouse gas emissions, promoting sustainable waste management. This systematic understanding of microbial dynamics can guide the development of enhanced composting systems that are both efficient and environmentally friendly (Table 2).

## 5 Factors Influencing Microbial Activity and Compost Quality

Temperature is one of the most critical factors influencing microbial activity and the quality of compost. Different microbial communities dominate at various temperature ranges during the composting process. In the *mesophilic phase* (25–40 °C), mesophilic microbes thrive, initiating the breakdown of simple organic compounds. As the process progresses into the *thermophilic phase* (45–70 °C), thermophilic

**Table 2** Microbial groups and their roles in composting phases

| Microbial group | Role in composting | Dominant phase | Activities | References |
|---|---|---|---|---|
| Mesophilic bacteria | Degradation of simple organic compounds | Mesophilic phase (25–40 °C) | Breakdown of sugars, proteins, and other soluble compounds | Yao et al. (2023) |
| Thermophilic bacteria | Degradation of complex organic matter | Thermophilic phase (45–70 °C) | Breakdown of cellulose, hemicellulose, and pathogen suppression | Kumari and Das (2023) |
| Fungi | Decomposition of lignin and complex fibers | Thermophilic and curing phases | Breakdown of lignin, cellulose, and resistant organic materials | Wang et al. (2023a, b) |
| Actinomycetes | Degradation of recalcitrant compounds | Curing phase | Breakdown of waxes, chitin, and stabilization of humus | Rillig et al. (2024) |

microorganisms take over, degrading more complex materials such as cellulose and lignin. This high-temperature phase also eliminates pathogens and weed seeds, ensuring a safe and hygienic end product. If temperatures exceed 70 °C, microbial activity can decline, emphasizing the need to regulate heat through aeration and turning (Zhou et al. 2023).

Moisture content and aeration are equally vital for maintaining an active microbial population. Microorganisms require water for metabolic processes, and an optimal moisture level of 40–60% ensures microbial survival without creating anaerobic conditions. Excessive moisture can lead to poor aeration, encouraging anaerobic microbes that produce odors and slow decomposition. Aeration, achieved through periodic turning or forced aeration systems, provides oxygen needed for aerobic microbes, prevents excessive heat buildup, and ensures even decomposition throughout the composting pile. Balancing moisture and aeration is key to producing high-quality compost efficiently.

The carbon-to-nitrogen (C/N) ratio plays a pivotal role in composting, as it influences microbial growth and activity. A balanced C/N ratio, typically ranging from 25:1 to 30:1, provides microbes with the necessary carbon for energy and nitrogen for protein synthesis. High-carbon materials (e.g., straw, sawdust) serve as energy sources, while nitrogen-rich materials (e.g., manure, green plant residues) promote microbial proliferation. An imbalance, such as too much carbon, slows the process, while excess nitrogen leads to ammonia emissions and odors. Adjusting the mix of materials to achieve the optimal C-N ratio ensures efficient decomposition and nutrient retention in the final compost (Lalremruati and Devi 2023).

The use of additives or inoculants can enhance microbial activity and compost quality. Additives, such as lime or biochar, help regulate pH and improve aeration, while microbial inoculants introduce beneficial microorganisms to accelerate decomposition and nutrient cycling. These inoculants, often derived from compost or commercial microbial formulations, ensure rapid colonization of the pile and enhance the breakdown of resistant organic compounds. In some cases, enzymes or

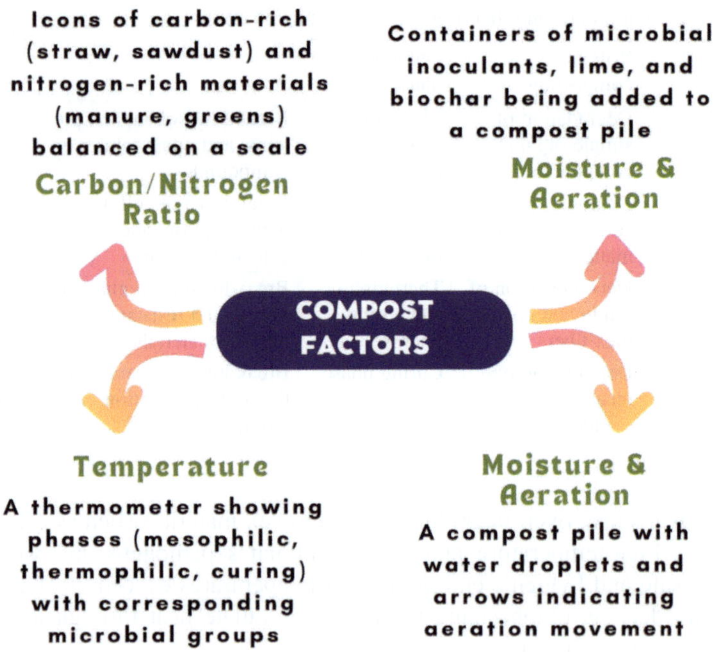


**Fig. 2** Factors influencing microbial activity and compost quality

nutrient supplements are added to address specific deficiencies in the composting mix. Proper use of additives and inoculants can significantly reduce composting time and improve the nutritional value of the final product (Ansari et al. 2023).

The key factors influencing composting include the *carbon/nitrogen ratio*, which balances carbon-rich materials (e.g., straw, sawdust) with nitrogen-rich ones (e.g., manure, greens) to optimize microbial activity (Fig. 2). *Moisture and aeration* are critical, requiring adequate water content and airflow for decomposition. *Temperature* governs composting phases (mesophilic, thermophilic, and curing), each driven by specific microbial groups to ensure effective organic matter breakdown and nutrient stabilization.

## 6 Challenges in Composting Cattle Industry Waste

Odor management and the emission of greenhouse gases are significant challenges in the composting of cattle industry waste. The anaerobic breakdown of organic matter, often caused by inadequate aeration or excessive moisture, can produce foul-smelling gases such as ammonia and hydrogen sulfide. These odors not only affect the environment but also create issues for nearby communities. Additionally, composting cattle waste emits greenhouse gases like methane and nitrous oxide, particularly during the early stages or in poorly managed systems. Optimizing aeration, moisture levels, and carbon-to-nitrogen (C/N) ratios is critical for minimizing these emissions while ensuring efficient composting (Andraskar et al. 2021).

Pathogen survival in cattle waste poses another critical challenge. Manure and other cattle industry residues often harbor pathogens such as *Escherichia coli*, *Salmonella*, and *Clostridium* species, which can persist if composting conditions are inadequate. The thermophilic phase, characterized by temperatures of 55–70 °C, is essential for pathogen suppression, as sustained high temperatures effectively kill harmful microorganisms. However, if the composting process is inconsistent or does not reach the required temperature, pathogens may survive, rendering the compost unsafe for agricultural use. Proper management practices, including frequent turning and monitoring of temperature, are vital for ensuring pathogen-free compost (Alegbeleye and Sant'Ana 2020).

The presence of antibiotic residues and heavy metals in cattle waste adds complexity to composting. Antibiotics used in livestock can remain in manure, affecting microbial activity and potentially leading to the development of antibiotic-resistant bacteria in the compost. Similarly, heavy metals from feed additives or other sources can accumulate in the compost, posing risks to soil and plant health when applied to farmland. Strategies such as pretreatment of waste, careful selection of feed additives, and the use of biochar or other amendments can help immobilize these contaminants, reducing their impact and ensuring safer compost products (Zalewska et al. 2021).

These challenges require a holistic approach to composting cattle industry waste. Integrated management practices that include proper aeration, monitoring of temperature and moisture, and the use of amendments to mitigate contaminants can significantly enhance the quality and safety of compost. Additionally, advances in microbial inoculants and enzyme technologies can improve degradation efficiency and pathogen suppression. The cattle industry can adopt sustainable composting practices that minimize environmental risks while maximizing agricultural benefits (Antonangelo et al. 2021; Manea et al. 2024). When these strategies are systematically implemented, as shown in Table 3, composting cattle industry waste can become a safe, efficient, and environmentally friendly process.

## 7 Advances in Enhancing Composting Efficiency

Advancements in microbial inoculants and enzyme additives have significantly improved composting efficiency. Microbial inoculants, consisting of specialized bacteria, fungi, and actinomycetes, are designed to accelerate the decomposition of organic matter and enhance nutrient cycling. These formulations often include thermophilic microorganisms capable of thriving in high-temperature conditions, optimizing degradation during the thermophilic phase. Enzyme additives, such as cellulases, proteases, and lipases, further aid in breaking down complex organic compounds like lignin, proteins, and fats. The combined use of these

**Table 3** Challenges and mitigation strategies in composting cattle industry waste

| Challenge | Mechanisms | Mitigation strategies | References |
|---|---|---|---|
| Odor and greenhouse gases | Emission of ammonia, methane, and nitrous oxide due to anaerobic conditions or improper aeration | Maintain optimal aeration, adjust C/N ratio, and use biofilters or additives like biochar for odor control | Gao et al. (2023); O'Neill et al. (2022) |
| Pathogen survival | Persistence of harmful microorganisms in manure and residues | Ensure sustained thermophilic phase (55–70 °C), regular turning, and monitoring of temperature | Madejón et al. (2022) |
| Antibiotic residues and metals | Residual antibiotics and heavy metals in waste affecting microbial activity and potentially contaminating soil | Pretreatment of waste, use of biochar or other amendments to immobilize contaminants, and careful feed selection | Saravanan et al. (2023) |

biotechnological innovations not only speeds up composting but also enhances the nutrient profile and stability of the final compost product (Silva et al. 2022).

Advances in aeration technologies and composting system designs have also contributed to greater efficiency. Forced aeration systems, including perforated pipes and blowers, provide a consistent oxygen supply, maintaining aerobic conditions and reducing odors. Static pile composting systems equipped with forced aeration reduce the need for frequent turning, saving time and labor. Additionally, in-vessel composting systems, which operate in enclosed environments, enable precise control over temperature, moisture, and aeration, ensuring uniform decomposition. These innovative designs improve process control, reduce environmental impacts, and are particularly useful for large-scale operations such as those in the cattle industry (Nguyen et al. 2024).

Monitoring and prediction tools have revolutionized the ability to manage composting processes in real time. Sensors embedded in compost piles can measure key parameters like temperature, moisture content, and oxygen levels, providing instant feedback for adjustments. Advanced predictive modeling software analyzes these data points, forecasting compost maturity and highlighting potential inefficiencies. Internet of Things (IoT) technologies further enhance monitoring by integrating sensor data into centralized systems accessible remotely, allowing operators to optimize composting processes proactively. These tools ensure consistent quality and reduce waste by enabling precise, data-driven management practices (Chamara et al. 2022).

The integration of these advances has transformed composting into a more efficient and sustainable process. The use of microbial inoculants and enzymes accelerates decomposition and improves compost quality, while modern aeration and system designs minimize labor and environmental challenges. Meanwhile, monitoring tools ensure real-time optimization and consistency, making composting a viable solution for organic waste management on both small and industrial scales. As research and technology continue to evolve, these innovations are likely to drive further improvements in composting efficiency and sustainability (Cao et al. 2023).

## 8 Applications and Benefits of Compost from Cattle Waste

Compost derived from cattle waste plays a pivotal role in improving soil health and enhancing fertility. The nutrient-rich organic matter in compost enriches soil with essential macronutrients like nitrogen, phosphorus, and potassium, as well as micronutrients vital for plant growth. Furthermore, compost enhances soil structure by improving its water retention capacity, aeration, and aggregate stability, which are crucial for root development and microbial activity. The organic matter content also boosts soil microbial diversity, promoting nutrient cycling and overall soil health. Regular application of compost from cattle waste transforms degraded soils into fertile ground for sustainable agriculture (Walia and Kaur 2024).

The use of compost significantly impacts crop productivity and environmental sustainability. With the steady release of nutrients, compost reduces the dependence on synthetic fertilizers, which can be costly and environmentally harmful. This steady nutrient supply enhances crop yields while minimizing the risks of nutrient leaching and groundwater contamination. Additionally, compost application reduces soil erosion and improves the resilience of crops to drought and other environmental stresses. This dual benefit of enhancing productivity while preserving the environment underscores the importance of compost in modern sustainable farming practices (Shaji et al. 2021).

Compost also plays a critical role in carbon sequestration, contributing to climate change mitigation. The organic carbon in compost binds with soil particles, forming stable complexes that store carbon for extended periods. This process reduces the amount of carbon dioxide released into the atmosphere, mitigating greenhouse gas emissions. Additionally, using compost as a soil amendment can indirectly reduce methane and nitrous oxide emissions by decreasing the reliance on untreated manure disposal and synthetic fertilizers. By incorporating compost into agricultural practices, farmers contribute to global efforts to combat climate change (Bhattacharyya et al. 2022).

The environmental benefits of compost extend beyond the farm. Using cattle waste compost helps with waste diversion, reducing the burden on landfills and minimizing the associated methane emissions. Furthermore, the recycling of organic residues into valuable compost aligns with circular economy principles, transforming waste into a resource. This practice promotes sustainable waste management systems and exemplifies how agricultural industries can adopt eco-friendly practices to address environmental challenges (Dou et al. 2024).

Incorporating compost from cattle waste into agricultural systems delivers a comprehensive suite of benefits, from improving soil health and crop productivity to supporting climate change mitigation efforts. This valuable resource will help farmers not only enhance their productivity but also contribute to a more sustainable and resilient agricultural ecosystem. The long-term benefits of adopting composting practices for cattle waste highlight its importance as a cornerstone of sustainable agriculture. The transformative potential of compost from cattle waste in fostering sustainable agricultural and environmental practices (Table 4).

**Table 4** Applications and benefits of compost from cattle waste

| Application | Description | Benefits | References |
|---|---|---|---|
| Soil health improvement | Enrichment with nutrients and organic matter and improved soil structure | Enhanced fertility, water retention, aeration, and microbial activity | Sivaram et al. (2023) |
| Crop productivity | Steady nutrient supply for crops | Higher yields, reduced reliance on synthetic fertilizers, and better resilience to environmental stresses | Sarma et al. (2024) |
| Environmental sustainability | Reduction in nutrient leaching, erosion control | Minimized groundwater contamination and improved ecosystem health | Ekemezie and Digitemie (2024) |
| Carbon sequestration | Stabilization of organic carbon in soils | Mitigation of greenhouse gas emissions and contribution to climate change efforts | Harmsen et al. (2023) |
| Waste management | Conversion of organic residues into valuable compost | Reduced landfill burden, methane emission reduction, and alignment with circular economy principles | Lakhouit (2024) |

# 9 Conclusion

Composting cattle waste offers a transformative approach to addressing agricultural and environmental challenges. By converting nutrient-rich organic residues into high-quality compost, this practice not only enhances soil health and fertility but also boosts crop productivity while reducing dependency on synthetic fertilizers. Composting promotes environmental sustainability by mitigating nutrient leaching, preventing soil erosion, and improving ecosystem resilience. Beyond immediate agricultural benefits, compost from cattle waste plays a critical role in combating climate change. Its ability to sequester carbon in soils and reduce greenhouse gas emissions highlights its significance as a sustainable waste management strategy. Moreover, by diverting organic waste from landfills and aligning with circular economy principles, composting contributes to waste reduction and resource recovery. The adoption of composting practices underscores the potential for the cattle industry to move toward more eco-friendly and sustainable systems. This integration of waste management and agricultural productivity ensures that resources are utilized efficiently while minimizing environmental footprints. Compost derived from cattle waste offers multifaceted benefits, making it a cornerstone of sustainable agricultural practices and climate change mitigation strategies. By harnessing its potential, farmers and industries alike can drive progress toward a greener, more resilient future.

## References

Alegbeleye OO, Sant'Ana AS (2020) Manure-borne pathogens as an important source of water contamination: an update on the dynamics of pathogen survival/transport as well as practical risk mitigation strategies. Int J Hyg Environ Health 227:113524. https://doi.org/10.1016/j.ijheh.2020.113524

Andraskar J, Yadav S, Kapley A (2021) Challenges and control strategies of odor emission from composting operation. Appl Biochem Biotechnol 193:2331–2356. https://doi.org/10.1007/s12010-021-03490-3

Ansari SA, Shakeel A, Sawarkar R, Maddalwar S, Khan D, Singh L (2023) Additive facilitated co-composting of lignocellulosic biomass waste, approach towards minimizing greenhouse gas emissions: an up to date review. Environ Res 224:115529. https://doi.org/10.1016/j.envres.2023.115529

Antonangelo JA, Sun X, Zhang H (2021) The roles of co-composted biochar (COMBI) in improving soil quality, crop productivity, and toxic metal amelioration. J Environ Manag 277:111443. https://doi.org/10.1016/j.jenvman.2020.111443

Azis FA, Choo M, Suhaimi H, Abas PE (2023) The effect of initial carbon to nitrogen ratio on kitchen waste composting maturity. Sustainability 15(7):6191. https://doi.org/10.3390/su15076191

Bhattacharyya SS, Ros GH, Furtak K, Iqbal HM, Parra-Saldívar R (2022) Soil carbon sequestration–an interplay between soil microbial community and soil organic matter dynamics. Sci Total Environ 815:152928. https://doi.org/10.1016/j.scitotenv.2022.152928

Bher A, Mayekar PC, Auras RA, Schvezov CE (2022) Biodegradation of biodegradable polymers in mesophilic aerobic environments. Int J Mol Sci 23(20):12165. https://doi.org/10.3390/ijms232012165

Bhunia S, Bhowmik A, Mallick R, Mukherjee J (2021) Agronomic efficiency of animal-derived organic fertilizers and their effects on biology and fertility of soil: a review. Agronomy 11(5):823. https://doi.org/10.3390/agronomy11050823

Cao X, Williams PN, Zhan Y, Coughlin SA, McGrath JW, Chin JP, Xu Y (2023) Municipal solid waste compost: global trends and biogeochemical cycling. Soil Environ Health 1(4):100038. https://doi.org/10.1016/j.seh.2023.100038

Carpio MJ, Sánchez-Martín MJ, Rodríguez-Cruz MS, Marín-Benito JM (2021) Effect of organic residues on pesticide behavior in soils: a review of laboratory research. Environments 8(4):32. https://doi.org/10.3390/environments8040032

Chamara N, Islam MD, Bai GF, Shi Y, Ge Y (2022) Ag-IoT for crop and environment monitoring: past, present, and future. Agric Syst 203:103497. https://doi.org/10.1016/j.agsy.2022.103497

de Oliveira RT, Ghobakhloo M, Figueira S (2023) Industry 4.0 towards social and environmental sustainability in multinationals: enabling circular economy, organizational social practices, and corporate purpose. J Clean Prod 139712. https://doi.org/10.1016/j.jclepro.2023.139712

Dou Z, Dierenfeld ES, Wang X, Chen X, Shurson GC (2024) A critical analysis of challenges and opportunities for upcycling food waste to animal feed to reduce climate and resource burdens. Resour Conserv Recycl 203:107418. https://doi.org/10.1016/j.resconrec.2024.107418

Ekemezie IO, Digitemie WN (2024) Climate change mitigation strategies in the oil & gas sector: a review of practices and impact. Eng Sci Technol J 5(3):935–948. https://doi.org/10.51594/estj.v5i3.948

Ezeah C, Osuagwu JC, Adesiyan IM (2023) Design principles of sustainable composting and recovery of value from compost. In: Current developments in biotechnology and bioengineering. Elsevier, pp 143–168. https://doi.org/10.1016/B978-0-323-91874-9.00008-5

Gao J, Xu Z, Daniel CT (2023) Biochar-supported odor control. In: Biochar applications for wastewater treatment, pp 243–271. https://doi.org/10.1002/9781119764403.ch15

Hanif S, Lateef M, Hussain K, Hyder S, Usman B, Zaman K, Asif M (2022) Controlling air pollution by lowering methane emissions, conserving natural resources, and slowing urbanization in a panel of selected Asian economies. PLoS One 17(8):e0271387

Harmsen M, Tabak C, Höglund-Isaksson L, Humpenöder F, Purohit P, van Vuuren D (2023) Uncertainty in non-CO2 greenhouse gas mitigation contributes to ambiguity in global climate policy feasibility. Nat Commun 14(1):2949. https://doi.org/10.1038/s41467-023-38577-4

Koul B, Yakoob M, Shah MP (2022) Agricultural waste management strategies for environmental sustainability. Environ Res 206:112285. https://doi.org/10.1016/j.envres.2021.112285

Kumari S, Das S (2023) Bacterial enzymatic degradation of recalcitrant organic pollutants: catabolic pathways and genetic regulations. Environ Sci Pollut Res 30(33):79676–79705. https://doi.org/10.1007/s11356-023-28130-7

Lakhouit A (2024) Mitigating landfill emissions strategies for effective waste management in Tabuk. Cleaner Waste Syst 100187. https://doi.org/10.1016/j.clwas.2024.100187

Lalremruati M, Devi AS (2023) Duration of composting and changes in temperature, pH and C/N ratio during composting: a review. Agric Rev 44(3):350–356. https://doi.org/10.18805/ag.R-2197

Madejón E, Panettieri M, Madejón P, Perez-de-Mora A (2022) Composting as sustainable managing option for seaweed blooms on recreational beaches. Waste Biomass Valorization 13(2): 863–875. https://doi.org/10.1007/s12649-021-01548-1

Manea EE, Bumbac C, Dinu LR, Bumbac M, Nicolescu CM (2024) Composting as a sustainable solution for organic solid waste management: current practices and potential improvements. Sustainability 16(15):6329. https://doi.org/10.3390/su16156329

Marlina ET, Hidayati YA, Harlia E, Ismail N (2024) 7 Biological waste management hazardous of animal husbandry and food processing industry. In: Solid waste management: challenges and recent solutions, p 132

Nguyen NT, Vo TS, Tran-Nguyen PL, Nguyen MN, Matsuhashi R, Kim K, Vo TTBC (2024) A comprehensive review of aeration and wastewater treatment. Aquaculture 741113. https://doi.org/10.1016/j.aquaculture.2024.741113

O'Neill T, Rynk R, Black G, Bokowa AF (2022) Odor management and community relations. In: The composting handbook. Academic Press, pp 549–625. https://doi.org/10.1016/B978-0-323-85602-7.00019-4

Raihan A (2023) A review of climate change mitigation and agriculture sustainability through soil carbon sequestration. J Agric Sustain Environ 2997:271X. https://doi.org/10.56556/jase.v2i2.783

Rastogi M, Verma S, Kumar S, Bharti S, Kumar G, Azam K, Singh V (2023) Soil health and sustainability in the age of organic amendments: a review. Int J Environ Clim Change 13(10): 2088–2102. https://doi.org/10.9734/ijecc/2023/v13i102870

Riaz S, Mehmood R, Kareem HA, Sadia H (2023) Livestock waste, types, sources, pollution potential, and country-wise comparisons. In: Waste problems and management in developing countries. Apple Academic Press, pp 205–231

Rillig MC, Kim SW, Zhu YG (2024) The soil plastisphere. Nat Rev Microbiol 22(2):64–74. https://doi.org/10.1038/s41579-023-00967-2

Sahu S, Arya SK (2024) Mitigation of greenhouse gas emissions in agriculture and food processing through sustainable management practices for climate change. In: Advances and technology development in greenhouse gases: emission, capture and conversion. Elsevier, pp 71–96. https://doi.org/10.1016/B978-0-443-19231-9.00010-7

Saravanan A, Swaminaathan P, Kumar PS, Yaashikaa PR, Kamalesh R, Rangasamy G (2023) A comprehensive review on immobilized microbes-biochar and their environmental remediation: mechanism, challenges and future perspectives. Environ Res 116723. https://doi.org/10.1016/j.envres.2023.116723

Sarma HH, Borah SK, Dutta N, Sultana N, Nath H, Das BC (2024) Innovative approaches for climate-resilient farming: strategies against environmental shifts and climate change. Int J Environ Clim Change 14(9):217–241. https://doi.org/10.9734/ijecc/2024/v14i94407

Shaji H, Chandran V, Mathew L (2021) Organic fertilizers as a route to controlled release of nutrients. In: Controlled release fertilizers for sustainable agriculture. Academic Press, pp 231–245. https://doi.org/10.1016/B978-0-12-819555-0.00013-3

Sharma A, Arya SK, Singh J, Kapoor B, Bhatti JS, Suttee A, Singh G (2023) Prospects of chitinase in sustainable farming and modern biotechnology: an update on recent progress and challenges. Biotechnol Genet Eng Rev 40(1):310–340. https://doi.org/10.1080/02648725.2023.2183593

Silva GDC, Kitano IT, Ribeiro IADF, Lacava PT (2022) The potential use of actinomycetes as microbial inoculants and biopesticides in agriculture. Front Soil Sci 2:833181. https://doi.org/10.3389/fsoil.2022.833181

Sivaram AK, Abinandan S, Chen C, Venkateswartlu K, Megharaj M (2023) Microbial inoculant carriers: soil health improvement and moisture retention in sustainable agriculture. Adv Agron 180:35–91. https://doi.org/10.1016/bs.agron.2023.03.001

Walia SS, Kaur T (2024) Beneficial role of vermicompost: nutrient content in vermicompost and success stories. In: Earthworms and vermicomposting: species, procedures and crop application. Springer, Singapore, pp 135–146. https://doi.org/10.1007/978-981-99-8953-9_13

Wang P, Han S, Lin Y (2023a) Role of microbes and microbial dynamics during composting. In: Current developments in biotechnology and bioengineering. Elsevier, pp 169–220

Wang Z, Hu Y, Wang S, Wu G, Zhan X (2023b) A critical review on dry anaerobic digestion of organic waste: characteristics, operational conditions, and improvement strategies. Renew Sust Energ Rev 176:113208. https://doi.org/10.1016/j.rser.2023.113208

Wei X, Xie B, Wan C, Song R, Zhong W, Xin S, Song K (2024) Enhancing soil health and plant growth through microbial fertilizers: mechanisms, benefits, and sustainable agricultural practices. Agronomy 14(3):609. https://doi.org/10.3390/agronomy14030609

Yahaya SM, Mahmud AA, Abdullahi M, Haruna A (2023) Recent advances in the chemistry of nitrogen, phosphorus and potassium as fertilizers in soil: a review. Pedosphere 33(3):385–406. https://doi.org/10.1016/j.pedsph.2022.07.012

Yao W, Cai D, Huang F, Mohamed TA, Li P, Qiao X, Wu J (2023) Promoting lignin exploitability in compost: a cooperative microbial depolymerization mechanism. Process Saf Environ Prot 174:856–868. https://doi.org/10.1016/j.psep.2023.05.003

Yin J, Xie M, Yu X, Feng H, Wang M, Zhang Y, Chen T (2023) A review of the definition, influencing factors, and mechanisms of rapid composting of organic waste. Environ Pollut 123125. https://doi.org/10.1016/j.envpol.2023.123125

Yuvaraj A, Thangaraj R, Ravindran B, Chang SW, Karmegam N (2021) Centrality of cattle solid wastes in vermicomposting technology–A cleaner resource recovery and biowaste recycling option for agricultural and environmental sustainability. Environ Pollut 268:115688. https://doi.org/10.1016/j.envpol.2020.115688

Zalewska M, Błażejewska A, Czapko A, Popowska M (2021) Antibiotics and antibiotic resistance genes in animal manure–consequences of its application in agriculture. Front Microbiol 12: 610656. https://doi.org/10.3389/fmicb.2021.610656

Zhou SP, Tang SQ, Ke X, Zhou HY, Zou SP, Xue YP, Zheng YG (2023) Hyperthermophilic pretreatment significantly accelerates thermophilic composting humification through improving bacterial communities and promoting microbial cooperation. Bioresour Technol 385:129467. https://doi.org/10.1016/j.biortech.2023.129467

# Evolution of Modern Methods to Understand the Microbial Community and Metabolic Dynamics of Compost

**Alexandros Phokas, Panayiota Pissaridou, Johana Rodosthenous, Christiana Hadjimichael, and Michalis Omirou**

## 1 Introduction

Understanding microbial communities in compost is crucial for maintaining soil health and promoting sustainable agricultural practices. Compost not only serves as an organic amendment but also acts as a reservoir for diverse microbial communities that can significantly influence soil properties and ecosystem functioning. Microbial communities play a central role in the composting process due to their substantial impact on organic matter decomposition, nutrient cycling, and the establishment of beneficial relationships among microorganisms and the surrounding environment. Primarily, microorganisms facilitate the breakdown of organic materials during composting, which is crucial for converting waste into nutrient-rich compost. Various studies have demonstrated that bacteria, fungi, and archaea are the primary players in this biodegradation process (Aguilar-Paredes et al. 2023; Zang et al. 2018). For instance, *Actinobacteria*, *Clostridia*, and *Bacilli* are notably active during composting, contributing significantly to the degradation of plant biomass (Antunes et al. 2016). The metabolic activities of these microorganisms help release nutrients such as nitrogen and phosphorus bound in organic material, making them available for plant uptake (Li et al. 2019a).

The dynamics of microbial communities throughout the different phases of composting also highlight their essential role. Firstly, the microbial taxa involved in composting are incredibly diverse, encompassing bacteria, archaea, and fungi with differing metabolic pathways and environmental tolerances (Antunes et al. 2016). For instance, the composting process transitions through several phases—mesophilic, thermophilic, and curing—each characterised by distinct microbial communities (Araujo et al. 2021). For example, during the thermophilic phase,

A. Phokas · P. Pissaridou · J. Rodosthenous · C. Hadjimichael · M. Omirou (✉)
Agrobiotechnology Section, Agricultural Research Institute, Nicosia, Cyprus
e-mail: momirou@ari.moa.gov.cy



R. K. Singh et al. (eds.), *Composting of Farm Residues by Nature-Based Solutions*,
https://doi.org/10.1007/978-3-032-12255-1_7

which occurs at elevated temperatures, specific thermophilic microorganisms dominate due to their capacity to thrive in extreme conditions (Araujo et al. 2021; Danon et al. 2008). These microorganisms play a crucial role in breaking down complex materials, such as lignocellulose, thereby accelerating the composting process (López et al. 2021).

Traditional methods, such as culture-based techniques and microscopy, have contributed valuable insights into microbial taxonomy and functional roles in compost, enabling the isolation and characterisation of specific organisms (Ryckeboer et al. 2003; Anastasi et al. 2005; Green et al. 2004). However, their major limitation lies in detecting only a small fraction of the microbial community, thereby overlooking unculturable species that are critical to compost dynamics (Hill et al. 2000).

High-throughput sequencing technologies enable comprehensive characterisation of microbial diversity, including both bacterial and fungal populations, far beyond the limited scope of traditional culture-based methods (Neher et al. 2013). More importantly, these technologies have evolved to not only reveal taxonomic profiles but also to infer and quantify metabolic functions within microbial communities (Fig. 1). Approaches such as metagenomics, metatranscriptomics, and metaproteomics have progressively enabled researchers to move from the question of "who is there?" to "what are they doing?"—capturing functional potential, gene expression, and protein activity.

Understanding the metabolic dynamics of compost microbiomes is now a research priority, as this functional insight is more directly tied to compost quality, maturation, and environmental impact than taxonomic data alone. Functional omics approaches are particularly valuable in revealing microbial contributions to critical processes, such as organic matter decomposition, nutrient mineralisation, methane and ammonia emissions, and the fate of antibiotic resistance genes.

This chapter traces the evolution of methods used to study microbial communities in compost, from traditional techniques to cutting-edge omics approaches. It aims to highlight how integrating taxonomic and functional insights through modern molecular tools can improve our understanding and management of composting systems.

## 2 Metagenomics

Metagenomics refers to the direct genetic analysis of complex microbial communities, including uncultured microorganisms, allowing for a comprehensive exploration of microbial diversity and functional potential (Jindal 2020). In recent years, Next-Generation Sequencing (NGS) technologies have been widely applied in metagenomic studies to analyse and characterise microbial communities in environmental samples. These technologies include metabarcoding and shotgun sequencing, which offer complementary approaches to studying microbial diversity and function. While metabarcoding focuses on sequencing specific marker genes to assess

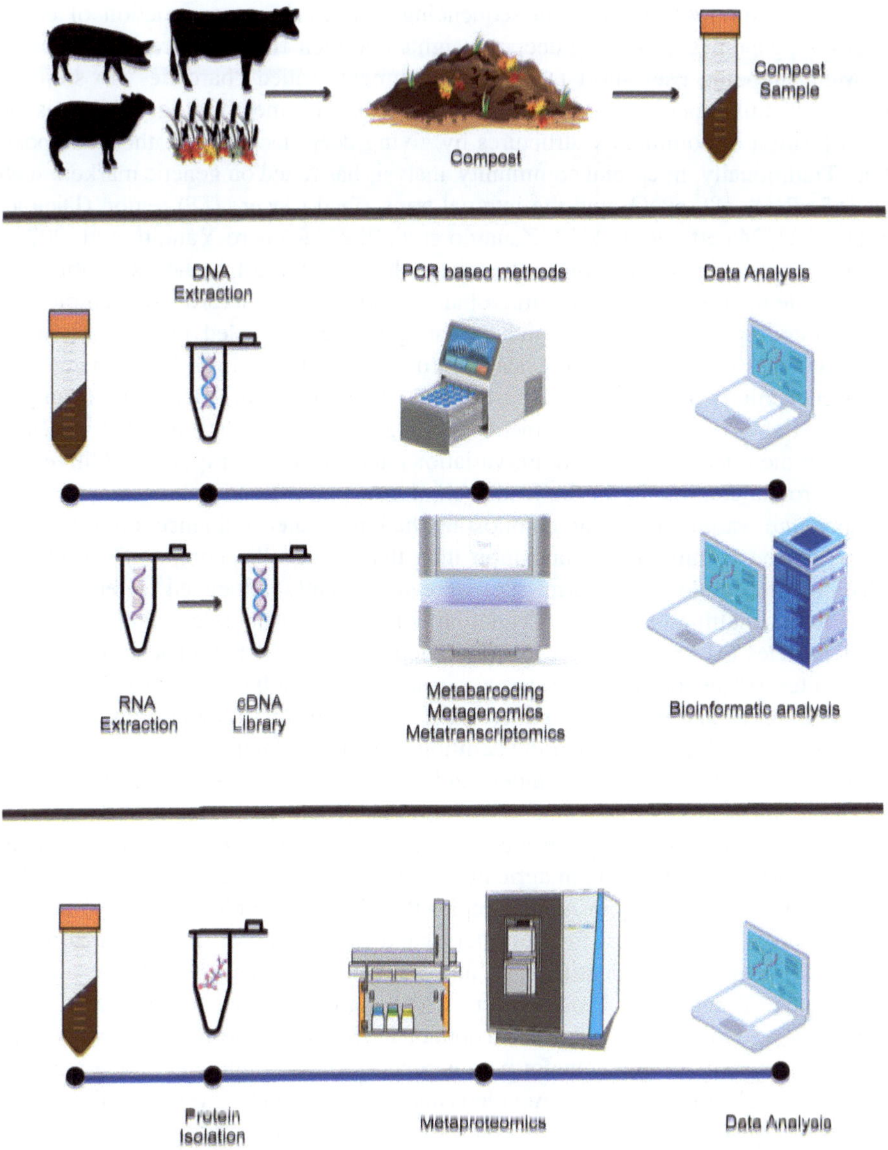


**Fig. 1** Overview of multi-omics approaches used to analyze microbial communities and functional activity in compost. Three primary molecular approaches are illustrated: (1) DNA-based methods, including PCR for the quantification of functional and taxonomic marker genes; (2) RNA-based methods, involving RNA extraction and cDNA library preparation for metatranscriptomics and metabarcoding to assess gene expression and community composition; and (3) Protein-based methods, where proteins are isolated for metaproteomics to identify actively expressed proteins. Each approach concludes with bioinformatic or statistical data analysis to interpret the microbial composition and functional potential or activity within the compost ecosystem

taxonomic composition, shotgun sequencing enables the reconstruction of entire microbial genomes, providing deeper insights into their functional capabilities.

Metabarcoding uses short DNA/RNA fragments called "barcodes" to simultaneously identify many taxa within the same sample. This methodology facilitates the identification of community structures by giving deep insights into their composition. Traditionally, microbial community analysis has relied on genetic markers such as *16S rRNA, 18S rRNA*, and the internal transcribed spacer (*ITS*) region (Djemiel et al. 2022; Maretto et al. 2023; Zanardo et al. 2023; Romero-Yahuitl et al. 2024). One of the primary applications of metabarcoding in compost studies is its ability to capture temporal changes in microbial community structure. By employing metabarcoding techniques, researchers can generate a detailed map of microbial diversity at various composting stages, correlating these shifts with changes in physicochemical parameters (Meng et al. 2019; Song et al. 2023). For example, Chang et al. (2019) used DNA metabarcoding of the *16S rRNA* and *ITS* genes to evaluate the effects of temperature variations and metallic compound additives on the microbial community in the feedstock of cow manure and corn stalks. Overall, temperature variation during composting had a greater influence on microbial diversity and community composition than that of metallic compound additives. Phyla of Bacteroidetes and Firmicutes were significantly influenced by temperature variation, resulting from the variation of three distinctive families within Bacteroidetes (Marinilabiaceae, Porphyromonadaceae, and Flavobacteriaceae) and Firmicutes (Planococcaceae, Aerococcaceae, and Carnobacteriaceae). Moreover, temperature variations during composting had a greater influence on microbial α-diversity and β-diversity than the addition of metallic compounds.

Mang et al. (2022) sought to understand the microbiome of a slurry cow digestate produced on a farm (ADG) and of a more stabilised digestate-derived compost (DdC) obtained after 50 days of a composting process to determine their potential uses as organic amendments in agriculture using NGS sequencing of the *16S rRNA* gene. They have detected in ADG samples that the top five phyla were Firmicutes, Bacteroidetes, Tenericutes, Proteobacteria, and Euryarchaeota. In comparison, the most abundant phyla identified in all the DdC samples were Bacteroidetes, Firmicutes, Tenericutes, Proteobacteria, and Actinobacteria. Additionally, they have observed differences in bacterial abundance related to the composting process. At the beginning of the transformation process, they found that Bacteroidales, Bacillales, Clostridiales, and Acholepasmatales had higher relative abundance values in digestate samples than in compost. Other orders, such as Flavobacteriales, Campylobacterales, Pseudomonadales, and Actinomycetales, were found to be more abundant at the end of the process, and thus were more prevalent in the DdC than in the ADG samples (Mang et al. 2022).

Ma et al. (2024) and Guerreiro et al. (2024) studied how biochar can affect cow manure aerobic composting. They have observed that biochar supplementation decreases the diversity of microbial populations (Ma et al. 2024; Guerreiro et al. 2024). The addition of volcanic rock has also been shown to improve the richness and diversity of microorganisms during cow manure composting, mainly due to the ability of the volcanic rock to improve aerobic composting (Ma et al. 2024). The composition of cow manure changes during the composting process. *Flavobacterium*, *Acinetobacter*, and *Solibacillus* were the dominant bacteria in the initial phase. *Flavobacterium* was the dominant bacterial genus in the thermophilic (day 1) and cooling (day 15) periods. *Flavobacterium* is known to promote the breakdown of cellulose, hemicellulose, and lignin during composting, playing a key role in this process. Additionally, *Thermobacillus* and *Geobacillus* were also recorded, which are known to degrade organic matter and promote the condensation and aromatisation of humus (Ma et al. 2024).

Metabarcoding is also a powerful tool for comprehensively analysing microbial interactions within compost. These approaches offer valuable insights into microbial complexity, enabling the identification of key functional groups involved in nutrient cycling and overall compost stability (Sun et al. 2023). Also, recent studies have expanded the scope to include additional targets such as antibiotic resistance genes (ARG) to assess the potential risks associated with compost application (Fan et al. 2023; J. Zhang et al. 2023a; Zhou et al. 2024).

Despite these advancements, the abundance and diversity of the microbial community that harbours functional genes associated with key biochemical cycles remain underutilised in compost metabarcoding studies. For example, the diversity of microbial communities involved in nitrogen transformation, such as nitrite oxidoreductase (*nxrA*), bacterial and archaeal ammonia monooxygenase (*amoA*), periplasmic nitrate reductase (*napA*), membrane-bound nitrate reductase (*narG*), nitrite reductase (*nirK*, *nirS*), nitric oxide reductase (*norB*) and nitrous oxide reductase (*nosZ*), has yet to be widely explored in compost microbiome studies. However, functional gene metabarcoding has been successfully applied in soil and water studies, particularly targeting *amoA* and comammox genes (Abell et al. 2012; Huang et al. 2018; Vasileiadis et al. 2018; Liu et al. 2021; Papadopoulou et al. 2024). These approaches allow researchers to identify specific microbial groups based on functional potential rather than taxonomic classification. For example, the use of *amoA* allow for elucidating the mechanism of action of SDZ and/or Cu on soil nitrification in black soils (Liu et al. 2021). Applying similar functional gene analyses to composting could enhance our knowledge regarding the microbial roles during different composting stages. By tracking functional diversity, researchers can address key questions related to organic matter degradation, nutrient cycling, and the impact of compost amendments on microbial processes. More recently, in multiple disciplines, the use of a multi-amplicon DNA metabarcoding approach has enabled the application of large-scale sequencing as a surveillance technology in the field at a lower cost (Francis et al. 2024a, b; Leray et al. 2016; Papadopoulou et al. 2024). This approach, along with other sequencing methodologies, enhances the characterisation of microbial communities in compost samples by providing detailed insights into community structure and functional associations. Additionally, its cost-effectiveness facilitates a deeper understanding of microbial contributions to compost maturation and biological characterisation, as well as their potential role in soil after compost application.

Shotgun sequencing has emerged as a powerful tool for studying microbial communities in complex environments, including compost. Unlike metabarcoding, which targets specific marker genes (e.g., *16S rRNA, 18S rRNA*, or *ITS*) to profile microbial diversity, shotgun sequencing sequences all the DNA in a sample (Tessler et al. 2017). This untargeted approach provides a more comprehensive understanding of microbial taxonomy, functional potential, and metabolic interactions within an ecosystem. For example, Eckstrom and Barlow (2019) used shotgun metagenomics to profile bacterial species in food waste-derived compost, demonstrating its sensitivity in detecting diverse microbial taxa and tracking their relative abundances throughout the composting process. Additionally, Antunes et al. (2016) demonstrated that whole-metagenome shotgun sequencing enables an in-depth analysis of microbial community structure and dynamics during thermophilic composting. By classifying coding sequences using orthologous groups, the study provided insights into microbial functions linked to various metabolic processes, highlighting how different taxa contribute to composting functionality. This level of detail is crucial for identifying microbial roles in nutrient cycling and organic matter degradation—key processes in composting. It is particularly useful for assessing bacterial diversity and quantifying microbial abundance, including unculturable microorganisms that are otherwise difficult or impossible to analyse (Quince et al. 2017; Deng et al. 2020).

Additionally, the metabolic capabilities of microbial communities in compost can be revealed through shotgun sequencing. For example, shotgun sequencing has been extensively used to identify ARG during the composting process of animal manure (Keenum et al. 2021) and to monitor their persistence and distribution after manure application in soil (Yang et al. 2020; Wang et al. 2023; Abdugheni et al. 2023; Zhang et al. 2023b). Recently, through shotgun sequencing, Liao et al. (2019) highlighted the presence and persistence of ARGs in food waste composting, where composting significantly alters microbial community composition but does not necessarily eliminate ARGs. Despite high composting temperatures reducing the abundance of some ARG-associated bacterial taxa, ARGs persist through horizontal gene transfer (HGT) and physicochemical interactions, such as correlations with MGEs, pH, $NO_3^-$, and moisture. Understanding HGT through shotgun sequencing enables researchers to more accurately assess the risks associated with composting waste that may harbour resistant bacteria.

Additionally, shotgun sequencing provides valuable insights into the metabolic functions of compost microbiomes, which are associated with nutrient cycling. By identifying genes that encode carbohydrate-active enzymes (CAZymes), it reveals microbial contributions to lignocellulose breakdown, which comprises cellulose, hemicellulose, and lignin. Santos-Pereira et al. (2023) employed metagenomic analysis to identify a diverse array of CAZymes in compost, with a particular focus on glycoside hydrolases (GHs), including cellulases and hemicellulases, which play a crucial role in driving plant polysaccharide hydrolysis (Santos-Pereira et al. 2023). Heiss-Blanquet et al. (2016) further demonstrated that composting conditions, particularly the thermophilic phase, enrich cellulolytic bacteria, highlighting the role of shotgun sequencing in linking microbial taxa to enzymatic functions (Heiss-Blanquet et al. 2016). Shotgun sequencing has also played a crucial role in lignin degradation. Wang et al. (2016a) identified *Actinobacteriota* as key producers of lignin-degrading enzymes, including lignin peroxidases, which accelerate composting. Similarly, Montella et al. (2017) identified genes encoding glycoside hydrolases essential for lignocellulose breakdown, thereby reinforcing the importance of metagenomics in understanding microbially driven composting processes. A recent study by Xu et al. (2024) used metagenomic approaches to investigate the microbial sources and sinks of $N_2O$ during pig and cattle manure composting. A total of 105 MAGs that contained *narBGHI, napAB, nirABDKS, norBC, nosZ, nrfAH, nasAB, amoA*, and *nxrAB* genes were identified from the metagenomic data. Overall, during the experiment, the $NO_3^-$ content decreased to an undetectable level at the beginning of composting and then gradually increased at the cooling and mature stages. Although data analysis revealed a gradual increase in the $NO_3^-$ content, the $N_2O$ emission weakened at the mature composting stage, possibly due to increased microbial reduction by *nosZ*. During the mesophilic and thermophilic stages of compost maturation, the relevant bacteria primarily belonged to the classes *Bacillota*, *Pseudomonadota*, and *Actinomycetota*. The dropping temperature during compost maturation's cooling and mature stages allowed the Proteobacteria and Bacteroidetes to gradually become the dominant phyla (Xu et al. 2024).

Despite their great impact on environmental microbiology and ecology, metabarcoding and shotgun metagenomics have some limitations and are not panaceas. For example, while metabarcoding sequencing can provide a rapid and cost-effective means of determining the taxonomic composition of a microbial community, it has well-recognised limitations associated with low taxonomic resolution, PCR amplification biases and low functional insights (Janda and Abbott 2007; Edgar 2017; Links et al. 2018). The use of shotgun sequencing improves the resolution of microbial diversity and provides functional information; however, it is more expensive and can be less feasible for large-scale studies. In addition, the complexity of the resulting data necessitates substantial bioinformatics resources and expertise for proper analysis, which can be a significant barrier for many research teams (Quince et al. 2017). Therefore, integrating metagenomics with complementary approaches such as metatranscriptomics, metaproteomics, and stable isotope probing can provide a more comprehensive understanding of microbial activity, functional potential, and metabolic dynamics in compost ecosystems.

## 3 Metatranscriptomics

Metatranscriptomics is a high-throughput sequencing and analysis of RNA transcripts in complex microbial communities, enabling the quantification of gene expression and metabolic activities of these communities under specific environmental or physiological conditions (Giannoukos et al. 2012). This approach provides a dynamic snapshot of microbial function via gene expression within the microbiome community rather than the mere presence of potentially active genes, as in the case of metagenomic data (Franzosa et al. 2014). Furthermore, it can highlight the metabolic dynamics (such as nutrient cycling, decomposition, and enzymatic activity) in samples, including temporal changes during experimental procedures. In detail, metatranscriptomics involves the isolation of total RNA from environmental samples, typically followed by the removal of ribosomal RNA (rRNA) to enrich for messenger RNA (mRNA), which is subsequently reverse transcribed into complementary DNA (cDNA) for sequencing (Giannoukos et al. 2012). Advancements in sequencing technologies and computational pipelines have significantly improved the feasibility of these analyses, making it possible to analyse complex datasets that include the RNA profiles of microorganisms inhabiting diverse environments, such as the human gut, marine ecosystems, and soil microbiomes (Westreich et al. 2018; Hatch et al. 2019; Peng et al. 2023).

The implementation of metatranscriptomics in composts is crucial for understanding the active microbial processes that drive organic matter decomposition, nutrient cycling, and greenhouse gas emissions. Compost is a material that involves complex microbial successions that regulate the breakdown of organic matter. For example, a recent study showed that the turning procedure had a strong impact on the lignocellulosic biomass deconstruction that occurs synergistically and sequentially, with hemicellulose being degraded preferentially to cellulose and lignin (Antunes et al. 2016). In the same study, metatranscriptomics revealed that lignocellulose degradation in a thermophilic composting process is primarily or even exclusively the result of bacterial enzymatic activity. Thus, lignocellulolytic enzymes of bacterial origin may be of special interest in industrial applications in which fungal enzymes cannot be used, In addition, Zhang et al. (2020a) demonstrated that the expression of β-glucosidase genes fluctuates during composting due to the dynamic response of the active microbial community, which regulates β-glucosidase gene expression to adapt to temperature variations.

Furthermore, Li et al. (2021a) used metatranscriptomics to investigate the mechanism of bioaugmentation adopted in a continuous anaerobic process treating cow manure. This study revealed that bioaugmentation enriched key bacterial genera, including *Clostridium, Cellvibrio, Cellulomonas, Bacillus,* and *Fibrobacter*, which enhanced cellulase activity and facilitated the breakdown of complex organic matter. Moreover, an increase in *Bathyarchaeia* abundance was noticed, an archaeal group known for its role in methanogenesis. The presence of *Bathyarchaeia* was associated with increased methyl-coenzyme M reductase activity, accelerating methane production and improving the efficiency of anaerobic digestion (Li et al. 2021a).

As observed in the metagenomics literature, most studies on livestock manure fertiliser implementation have focused on the resistome, antibiotic resistance that can be transferred to soil. The integration of metabarcoding and metatranscriptomics substantially improves our understanding of how different compost procedures and treatments affect both the presence and expression of antibiotic resistance genes (Zhao et al. 2025).

Metatranscriptomics faces several limitations that impact its effectiveness in studying microbial communities. One major challenge is RNA instability, as mRNA degrades rapidly, requiring careful sample collection, preservation, and processing to prevent information loss. Additionally, extracting high-quality RNA from biological or environmental samples, such as faeces, is challenging due to contamination and degradation issues. A significant obstacle is that a large proportion of extracted RNA consists of rRNA, which can dramatically reduce mRNA coverage, thereby limiting the functional insights gained from transcriptomic studies. Although rRNA depletion methods have been developed to mitigate this issue (Peano et al. 2013), they add complexity and cost to the process. Metatranscriptomics is also costly and technically demanding, requiring RNA enrichment and specialised sequencing technologies, making large-scale studies less feasible.

Another critical challenge is the lack of comprehensive reference databases, which limits the accurate functional annotation of transcripts, particularly from uncultured microbial species. Moreover, bioinformatics challenges arise due to the complexity of distinguishing microbial from host transcripts, handling non-coding RNAs, and interpreting gene expression levels. While commercial enrichment kits are available to help separate microbial RNA from host RNA, computational methods can also be used when a reference genome for the host is available, as demonstrated by Pérez-Losada et al. (2015) in their study of host–pathogen interactions in the human airway microbiome.

Additionally, defining a clear threshold for gene expression remains difficult due to the highly dynamic nature of gene expression in microbial communities. Another limitation is that mRNA abundance does not always correlate with protein production, potentially leading to misinterpretation or overestimation of metabolic activity. Finally, variability in data analysis pipelines introduces bias, as different bioinformatics tools apply distinct filtering criteria and stringency levels (Li et al. 2025).

Even though metatranscriptomics can provide many insights into the microbiome structures and communities, it is best to be coupled with metagenomics to better understand the microbial communities at a genetic and activity level. The value of combining metagenome and metatranscriptome datasets is that the assembly becomes more representative of the community.

Despite these challenges, continued advancements in sequencing technologies, rRNA depletion strategies, and computational tools are expected to enhance the accuracy, feasibility, and overall application of metatranscriptomics in microbial ecology and environmental studies.

## 4 Metaproteomics

Metaproteomics is an emerging field in microbiome research that focuses on the large-scale characterisation of proteins expressed by microbial communities in environmental samples. By identifying and quantifying proteins, this approach provides direct insights into active metabolic processes and ecological interactions, offering a functional perspective beyond the genetic potential assessed by metagenomics (Hettich et al. 2013). Identification and quantification of proteins synthesised by the compost microbiome using metaproteomics can offer more accurate information on physiological activity and translational regulation taking place during composting (Wang et al. 2016b). Proteins are also more stable molecules relative to RNA and do not require enrichment relative to DNA; therefore, their handling and processing are much easier (Zhang et al. 2018). Furthermore, in contrast to metagenomic data, which are derived from both living and dead microbial cells, proteomic data reflect only metabolically active cells, providing direct evidence of metabolic dynamics (Zhang et al. 2016).

Metaproteomics methods can be broadly classified into two categories: the bottom-up and the top-down approaches. In the bottom-up approach, proteins are extracted, digested into peptides, and analysed using liquid chromatography coupled with mass spectrometry (LC-MS) or tandem MS (LC-MS/MS), while in the top-down approach, protein digestion is bypassed and intact proteins are identified via LC-MS (Abiraami et al. 2020). The bottom-up approach, also known as shotgun proteomics, is the most common method in microbial metaproteomics (Abiraami et al. 2020). It typically involves protease digestion, and may include prior protein separation using one-dimensional or two-dimensional polyacrylamide gel electrophoresis (Wang et al. 2016b). While gel-free approaches can increase the number of detectable proteins, additional strategies are often required to enhance the identification of insoluble membrane proteins (Wang et al. 2016b).

Shotgun metaproteomics is generally more practical and widely used in complex environmental studies than the top-down approach, since peptides are smaller molecules, more soluble, easier to ionise and more readily detected by mass spectrometry (Amunugama et al. 2013).

However, a key limitation of shotgun proteomics is protein interference, since several peptides can be shared among different proteins or protein families, leading to incomplete protein coverage (Po and Eyers 2023). In contrast, the top-down approach, which directly analyses intact proteins without proteolytic digestion into peptides, provides greater protein coverage and deeper insights into posttranslational modifications and alternative splicing (Melby et al. 2021). Despite these advantages, top-down proteomics has seen limited application in compost microbial dynamics due to its technical complexity and analytical challenges (Abiraami et al. 2020).

Both approaches face additional challenges, including high costs, particularly when multiple replicates are required, reliance on incomplete protein sequence databases, potential inefficiencies in protein extraction and purification due to organic acid interference, and a bias towards highly abundant proteins, which may skew the characterisation of microbial communities and metabolic functions(Wang et al. 2016b). Despite these limitations, metaproteomics remains a powerful tool, as it provides a direct measure of functional microbial activity, offering insights beyond those obtained from metagenomics or metatranscriptomics.

In recent decades, metaproteomics has been applied to diverse environmental samples, including marine water, soil and activated sludge, and more recently to livestock compost, to elucidate microbial diversity, physiological functions and ecological interactions (Wang et al. 2016b). Zhang et al. (2018) used metaproteomics to study protein expression in chicken manure compost, identifying *Bacillus* and *Lentibacillus* as key degraders, secreting a wide variety of proteases.

Additionally, maize straw supplementation enhanced carbohydrate metabolism while inhibiting the growth of Proteobacteria, a phylum that includes many pathogenic bacteria. Using metaproteomics on municipal solid waste samples, Liu et al. (2015) showed that fungi primarily facilitated cellulose degradation during the early stages of composting, while bacterial communities dominated during the thermophilic phase, suggesting shifts in microbial abundance over time. Furthermore, Lam et al. (2021) studied the metaproteome of chicken manure during anaerobic digestion for biogas production to characterise and identify the sources of methanogenic substrates, concluding that ammonia inhibition favoured hydrogenotrophic methanogenesis. Similarly, Lin et al. (2016) used metaproteomics to investigate methane metabolism in a thermophilic swine manure digester and proposed future biotechnological directions, such as manipulating methanotroph abundance to enhance methane production.

Metaproteomics is often integrated with other omics approaches to gain a more comprehensive understanding of metabolic activity and community dynamics. For example, Zhang et al. (2020b) used metagenomics, real-time PCR, and metaproteomics to analyse dairy cow manure under different composting strategies. Their findings revealed that antibiotic resistance was activated in the manure, as indicated by the presence of heat shock proteins associated with antibiotic-induced stress and metallo-β-lactamase family proteins, which degrade β-lactam antibiotics. In addition, composting led to lower microbial diversity and a reduction of antimicrobial resistance genes, highlighting its potential to mitigate antimicrobial resistance gene risks. Similarly, Zhang et al. (2016) combined metaproteomics and metagenomics to study maize straw compost supplemented with cow manure and demonstrated that hemicellulose degradation in maize straw was primarily driven by fungal genera, such as *Thermomyces* and *Aspergillus*, while lignocellulose degradation was carried out by bacterial genera, such as *Thermobifida*.

Although metaproteomics has been increasingly applied to analyse metabolic activity in environmental samples, its use in livestock composting remains limited. However, advancements in metaproteomics for other matrices, such as municipal waste and soil, along with continuous improvements in protein sequence databases, are expected to expand its application in livestock composting. This progress will enable a more comprehensive characterisation of microbial metabolic dynamics and ecological interactions.

## 5 Polymerase Chain Reaction (PCR)-Based Methods

PCR-based methods can provide an understanding of microbial community and metabolic dynamics of compost without the need for microbial cultivation, which can be time-consuming or unfeasible for many microorganisms (Cao et al. 2020; Mirabile et al. 2024). These techniques amplify and quantify microbial target gene fragments that serve as indicators of microbial abundance and metabolic activity (Cao et al. 2020). Target fragments are derived from genomic DNA, providing information about microbial functional gene abundance and thus community dynamics, or cDNA, synthesised from RNA via reverse transcription, providing insights into microbial gene expression and metabolic activity (Wang et al. 2021).

Real-time PCR (RT-PCR) is currently the benchmark method for the quantification of nucleic acids (Quan et al. 2018). It measures target PCR product accumulation in real time by detecting a fluorescent signal at the end of each amplification cycle (Park et al. 2021). Quantification is achieved by comparing fluorescence signals to a standard curve generated from samples with known target gene copy numbers (Quan et al. 2018). Fluorescence reporters commonly employed for gene quantification are either SYBR Green I, which is a universal double-stranded DNA intercalating dye, or TaqMan probes, which bind to specific regions of target genes (Cao et al. 2020).

The majority of applications of qPCR to study composting include the characterisation of functional genes involved in nitrogen transformation, such as *nxrA* and bacterial and archaeal *amoA*, regulating nitrification, *napA*, *narG*, *nirK*, *nirS*, *norB* and *nosZ*, regulating denitrification, and nitrogenase iron protein (*nifH*), regulating nitrogen fixation (e.g. Sun et al. 2019; He et al. 2021; Cao et al. 2024). Other marker genes examined more sporadically include those of the sulfur cycle, such as adenosine-5′-phosphosulfate reductase (*aprA*) (Chen et al. 2020), genes involved in carbon sequestration, such as ribulose-1,5-bisphosphate carboxylase/oxygenase genes (*cbbL* and *cbbM*) (Li et al. 2020b) and more commonly bacterial or archaeal *16S rRNA* genes (Zhong et al. 2020) used to estimate total numbers of each microbial kingdom.

Sampling at different timepoints during the composting process and performing qPCR can provide important insights into the temporal patterns of microbial metabolism. For example, Zhong et al. (2020) employed qPCR of nitrogen cycle functional genes and concluded that during the thermophilic and cooling phases of dairy cow manure composting, denitrification and nitrogen fixation predominate, while nitrification is more prevalent during the maturation phase. Furthermore, carrying out different treatments on the compost and investigating the effects on microbial community abundance is another common approach in which qPCR is employed. Treatments may be chemical, such as the addition of biochar (e.g. Chen et al. 2020) or biological, such as the addition of lignocellulose-degrading microbes (e.g. Yu et al. 2020), and are often coupled with characterisation of gene abundance regulating nitrogen transformation. For example, He et al. (2021) treated pig manure compost with powder bamboo biochar and demonstrated that nitrous oxide and ammonia emissions were reduced, most likely due to changes in the abundance of denitrifying genes. Cao et al. (2024) applied *Bacillus* and *Saccharomyces* to livestock manure compost and showed that nitrogen transformation was facilitated, likely due to decreased abundance of denitrifying genes and increased abundance of nitrogen fixation and nitrification genes.

A very informative approach is to combine qPCR data with metagenomic data or data from physicochemical analyses, such as temperature, pH, nitrate and ammonium nitrogen measurements, to deduce how microbial community abundance correlates with and possibly regulates other variables during composting. For example, Cao et al. (2024) used correlation analysis and showed that the abundance of the denitrifying genes *narG*, *nirS* and *nosZ* was positively correlated with livestock compost total nitrogen content, while the abundance of *nifH* and *nxrA*, responsible for nitrogen fixation and nitrite oxidation, respectively, was positively correlated with the abundance of specific bacterial genera identified through metagenomic sequencing. Chen et al. (2020) used correlation and network analysis to correlate gene abundance data from qPCR with high-throughput DNA sequencing data and inferred the host microorganisms of each gene regulating nitrogen transformation. Path analysis has also been employed to understand how gene abundance quantified by qPCR affects physicochemical parameters directly or indirectly during composting. For example, Sun et al. (2019) used path analysis and showed that total nitrogen transformation was driven by the synergistic actions of nitrification and denitrification genes during cattle manure and rice straw co-composting. Such approaches, involving the combination of data from multifactorial analyses, can be very useful, as they can direct us to study more closely specific relationships that drive bacterial community and metabolic activity dynamics and characterise them more comprehensively.

Despite being the benchmark method for nucleic acid quantification, RT-PCR comes with a big limitation: the requirement for the generation of a standard curve, which can be a serious source of variation and uncertainty in quantification (Tiwari et al. 2022). This limitation has been overcome with the development of digital PCR (dPCR), which is currently recognised as the most precise method for absolute quantification of functional gene abundance and expression levels, acting as a reliable tool for tracking microbial biodegradation (Cao et al. 2020).

dPCR differs from RT-PCR in that the reaction mixture is split into thousands of partitions, generated through water–oil emulsion or microfluidic chips, in which amplification takes place during thermal cycling (Tiwari et al. 2022). Thus, dPCR is generally classified as either droplet-based (ddPCR), where partitions are generated through water-oil emulsion, or chip-based (cdPCR), where partitions are generated physically via chip microwells (Cao et al. 2020). Fluorescence is measured at the end of the reaction, and partitions are categorised as positive or negative based on a fluorescence threshold determined algorithmically or set by the user (Tiwari et al. 2022). Poisson statistics are used to estimate gene copy numbers based on the endpoint fluorescence measured in each partition (Park et al. 2021).

Among the biggest advantages of dPCR over RT-PCR are higher estimation accuracy, higher analytical sensitivity and decreased susceptibility to PCR inhibitors, due to reaction partitioning (Sancha Dominguez et al. 2024). In addition, the lack of necessity for a standard curve makes dPCR a more accurate and efficient method, especially when analysing a large number of samples. The absence of standard curve generation enhances the versatility of the method, making quantification results more comparable between different laboratories, as well as surpassing the requirement for reference materials, which may be difficult to find or generate (Park et al. 2021; Sancha Dominguez et al. 2024). In spite of these advantages of dPCR, RT-PCR can be superior in some cases, as amplicons can be retrieved for sequencing, quantitative dynamic range can be higher (up to 8 $\log_{10}$), technology is better established and more widely used, and running expenses can be relatively lower (Tiwari et al. 2022). However, to understand microbial community and metabolic dynamics of compost, a high level of accuracy and reliability is required during quantification, and this can be better achieved via dPCR (Cao et al. 2020).

dPCR has been used extensively in different matrices, such as soil, manure, water and food, to quantify total bacterial abundance with *16S rRNA* primers, as well as specific bacterial family abundance using specific primers, demonstrating the superiority of dPCR over RT-PCR (Cavé et al. 2016; Wang et al. 2021). The most widely used application of dPCR regarding the study of compost microbial dynamics is for the tracking of antibiotic resistance marker genes (Park et al. 2021), such as *sul*, *tet*, *bla* and *qnr* conferring resistance to sulfonamides, tetracyclines, β-lactam and quinolones, respectively (e.g. Cavé et al. 2016; Duan et al. 2019; Esperón et al. 2020). In some studies, such as in Jauregi et al. (2023), compost underwent different treatments, and then the abundance of antibiotic resistance genes was quantified following compost application as a fertiliser. However, the majority of these types of studies are mainly focused on monitoring the extent to which the human population and the environment are exposed to antibiotic-resistant bacteria, rather than investigating the dynamics of different microbial populations and their metabolic activity during composting. There is only a limited number of studies where dPCR is used to investigate microbial population dynamics over several weeks of composting, and these concentrate only on a single species or on a single phylogenetic group of microorganisms. For example, Li et al. (2020a) quantified the abundance of *Lactobacillus plantarum* in sheep manure compost over an eight-week period using ddPCR and found that the species population abundance decreased at the later stages of composting, presumably due to elevated temperature.

The use of dPCR to study metabolic dynamics related to the nitrogen cycle during composting holds huge potential, as significant progress has already been made using agricultural soil samples. A ddPCR assay using the universal double-stranded DNA intercalating dye EvaGreen has been developed to quantify the abundance of all bacteria via *16S rRNA* gene targeting, as well as the most commonly studied genes regulating the nitrogen cycle: bacterial and archaeal *amoA*, regulating nitrification, and *nirK, nirS, nosZI* and *nosZII*, regulating denitrification (Voegel et al. 2020). In addition, an EvaGreen ddPCR assay has also been developed and applied to lake water samples to target *amoA* genes from Comammox bacteria, which perform complete oxidation of ammonia into nitrates (Harringer and Alfreider 2021). dPCR assays employing highly specific TaqMan probes have also been developed, targeting *16S rRNA* and *ITS* genes to quantify total bacterial and total fungal abundance, respectively (Wang et al. 2022), bacterial *amoA* (Dong et al. 2014) and *nosZI* genes (Johnson et al. 2024). However, these assays have been developed using soil DNA samples, and their application to compost samples remains largely understudied. The modification and application of those dPCR protocols to compost samples has the potential to revolutionise the study and understanding of microbial community and metabolic dynamics. Especially if coupled with multiplexing and/or RNA extraction, dPCR can provide quick and accurate results regarding the metabolism of nitrogen, sulfur, carbon and many other physiological pathways during composting. These data can then be combined with compost chemical and physicochemical data to provide deeper insights into the biological processes and their interplay with environmental parameters during composting, such as the characterisation of the driving forces of greenhouse gas emissions.

## 6 Stable Isotope Probing

Stable isotope probing (SIP) is a molecular technique widely used in microbial ecology to identify which microorganisms actively metabolise compounds labelled with heavy isotopes of key elements, mainly $^{13}C$ and $^{15}N$ (Xiong et al. 2024). $^{13}C$ is commonly used to track the metabolism of methanol, methane, acetate, toluene, glucose, pesticides, lignin and organic solvents, while $^{15}N$ is used to track ammonium, nitrogen-containing pollutants, antibiotics and herbicides (Madsen 2006; Kim et al. 2023). Depending on the molecular biomarker that the isotope is incorporated into, such as DNA, RNA or protein, the method can be coupled with an omics or other suitable technique to investigate the flow of the isotope within the compost or any other environmental matrix (Jameson et al. 2017).

DNA- and RNA-SIP are the most preferred methods due to the availability of sequencing resources; however, both methods require the addition of a high amount of substrate to separate labelled from unlabelled nucleic acids during ultracentrifugation (Xiong et al. 2024). This can be a source of bias, as high substrate enrichment levels may not reflect natural conditions. In addition, bias may arise due to cross-feeding, whereby nucleic acids of non-target microbes may be labelled due to substrate degradation, especially when long incubation times are needed (Kim et al. 2023). Nevertheless, these methods are still preferred as they can provide valuable insights into the metabolic capabilities of microorganisms in an affordable and reliable manner (Jameson et al. 2017).

Protein-based SIP, involving metaproteomics, is also widely used, enabling the accurate identification of labelled peptides by MS, despite the challenges of metaproteomics approaches discussed earlier (Jehmlich et al. 2016). Compared to DNA-SIP, incorporation of the stable isotope into proteins does not require cell replication, which decreases the incubation time required and thus the risk of cross-feeding (Jehmlich et al. 2016). In addition, as with any metaproteomics approach, more insights can be gained into the metabolic activity of microorganisms compared to nucleic acid-based approaches.

In the past decade, $^{13}C$ labelling coupled with metagenomics, metatranscriptomics or metaproteomics has been used to study carbon metabolism in environmental samples, including compost. For example, Hanajima et al. (2019) used $^{13}C$ glucose to label dead bacterial cells in dairy cow compost and, by applying DNA- and RNA-SIP, identified *Sphingobium* sp., myxobacterial lineages and *Bacillales* as assimilators of dead bacterial biomass. In a similar study, DNA- and RNA-SIP, involving $^{13}C$ glucose labelling of dead bacterial cells in aerated pig manure slurry, revealed that the predominant dead biomass assimilators were *Acholeplasma* spp. (Hanajima et al. 2015).

In addition, protein-SIP involving $^{13}C$ acetate has been used in other systems, such as anaerobic digesters of livestock manure. In a study by Mosbæk et al. (2016) involving protein-SIP, the identified peptides were mapped onto a binned microbial metagenome, and *Methanosarcina*, *Methanoculleus* and *Clostridia* spp. were identified as acetate-consuming microorganisms. Furthermore, the emission of $^{13}C$-labelled carbon dioxide and methane and the presence of *Clostridia* spp., which harbour the formyltetrahydrofolate synthetase (*FTFHS*) gene, suggested that the mechanism of acetate turnover utilised by these microorganisms was syntrophic oxidation coupled with hydrogenotrophic methanogenesis. Similarly, Sun et al. (2017) used DNA-SIP to track the metabolism of $^{13}C$ acetate and $^{13}C$ lactate during thermophilic anaerobic digestion of cow manure. Employment of metagenomics suggested *Clostridium, Hydrogenophaga, Fervidobacterium, Spirochaeta, Limnohabitans*, and *Rhodococcus* as acetate oxidisers, and *Anaeromyxobacter, Desulfobulbus, Syntrophus, Cystobacterineae,* and *Azospira* as lactate consumers. Interestingly, cluster analysis indicated that acetate and lactate oxidation could be linked to nitrate and/or nitrite reduction (Sun et al. 2017).

$^{15}N$ labelling has also been used extensively in environmental samples, such as compost, to study nitrogen transformation, particularly the substrates responsible for $N_2O$ fluxes (Clough et al. 2013). Labelling has frequently been achieved directly, by supplementing the compost with $^{15}N$ chemical fertiliser or indirectly, by applying $^{15}N$ fertiliser to cultivated crops on which livestock are fed (Li et al. 2021b). In both approaches, there is a risk of heterogeneous labelling due to the complex composition of compost (Li et al. 2021b).

In the case of $^{15}N$ labelling in compost, coupling with omics techniques is not yet a widespread approach. There are some studies where DNA-SIP with $^{15}N$ labelling has been used; for example, Li et al. (2019b) used $^{15}N$ ammonium dihydrogen phosphate and showed that nitrogen transformation occurred mainly via the anammox pathway, but this study was conducted on wastewater. DNA-SIP with $^{15}N$ labelling has also been used on soil samples, as in Reed et al. (2025), where application of $^{15}N$ labelled ammonium sulfate as fertiliser on agricultural soils enabled the elucidation of microbial genera responsible for nitrogen assimilation.

In compost, $^{15}N$ labelling is more commonly coupled with MS approaches, whereby changes in $^{15}N$ to $^{14}N$ isotope ratio ($\delta^{15}N$) are measured. Nishida et al. (2008) used indirect $^{15}N$ labelling to study nitrogen flow during the application of cattle manure compost to rice paddy fields over a three-year period. Employing isotope-ratio MS to measure the $\delta^{15}N$, they showed that 7% of nitrogen was incorporated into rice, 66–69% into soil, and the remaining was unrecovered. In addition, Lim et al. (2010) compared the $\delta^{15}N$ of synthetic fertiliser, livestock manure slurry, and composted livestock manure and concluded that faster ammonia volatilisation during composting resulted in a higher $\delta^{15}N$ in livestock compost. In another study, Li et al. (2021b) used indirect labelling by fertilising fodder crops with $^{15}N$ ammonium sulfate and investigated the time required to achieve homogeneous labelling when glucose is provided as a labile carbon source. They found that after 45 days of treatment, labelling was homogeneous in the different nitrogen pools: active, stable and recalcitrant. This work paves the way for future studies using $^{15}N$ labelling in compost, as it proposes how to eliminate the bias of heterogeneous labelling.

## 7 Conclusion

In conclusion, SIP-based methods, coupled with omics or MS techniques, have been widely used in the past decade to elucidate community and metabolic dynamics in complex microbial ecosystems. Yet, very few studies have applied these methods to livestock compost samples, which highlights a significant research gap in this area. With the increasing availability of isotopically labelled compounds and the improvement of genomic, transcriptomic and proteomic sequence databases, it is expected that the use of SIP in studying the microbiome of compost will become more prevalent in the near future (Jameson et al. 2017). The next step will be to capitalise on the knowledge acquired to facilitate the mitigation of greenhouse gas emissions or enhance the biodegradation of contaminants (Kim et al. 2023).

## References

Abiraami TV, Singh S, Nain L (2020) Soil metaproteomics as a tool for monitoring functional microbial communities: promises and challenges. Rev Environ Sci Biotechnol 19(1):73–102). Springer. https://doi.org/10.1007/s11157-019-09519-8

Abdugheni R, Li L, Yang ZN, Huang Y, Fang BZ, Shurigin V, Mohamad OAA, Liu YH, Li WJ (2023) Microbial risks caused by livestock excrement: current research status and prospects. Microorganisms 11:1897. https://doi.org/10.3390/MICROORGANISMS11081897

Abell GCJ, Robert SS, Frampton DMF, Volkman JK, Rizwi F, Csontos J, Bodrossy L (2012) High-throughput analysis of ammonia oxidiser community composition via a novel, amoA-based functional gene array. PLoS One 7:e51542. https://doi.org/10.1371/JOURNAL.PONE.0051542

Aguilar-Paredes A, Valdés G, Araneda N, Valdebenito E, Hansen F, Nuti M (2023) Microbial community in the composting process and its positive impact on the soil biota in sustainable agriculture. Agronomy 13(2):542. https://doi.org/10.3390/agronomy13020542

Amunugama R, Jones R, Ford M, Allen D (2013) Bottom-up mass spectrometry–based proteomics as an investigative analytical tool for discovery and quantification of proteins in biological samples. Adv Wound Care 2(9):549–557. https://doi.org/10.1089/wound.2012.0384

Anastasi A, Varese GC, Marchisio VF (2005) Isolation and identification of fungal communities in compost and vermicompost. Mycologia 97:33–44

Antunes LP, Martins LF, Pereira RV, Thomas AM, Barbosa D, Lemos LN, Silva GMM, Moura LMS, Epamino GWC, Digiampietri LA, Lombardi KC, Ramos PL, Quaggio RB, De Oliveira JCF, Pascon RC, Da Cruz JB, Da Silva AM, Setubal JC (2016) Microbial community structure and dynamics in thermophilic composting viewed through metagenomics and metatranscriptomics. Sci Rep 6:38915. https://doi.org/10.1038/SREP38915

Araujo ASF, de Pereira APA, Antunes JEL, Oliveira LMdS, de Melo WJ, Rocha SMB, do Amorim MR, Araujo FF, Melo VMM, Mendes LW (2021) Dynamics of bacterial and archaeal communities along the composting of tannery sludge sludge. Environ Sci Pollut Res 28 (45):64295–64306. https://doi.org/10.1007/s11356-021-15585-9

Cao Y, Yu M, Dong G, Chen B, Zhang B (2020) Digital PCR as an emerging tool for monitoring of microbial biodegradation. Molecules 25(3) MDPI AG. https://doi.org/10.3390/molecules25030706

Cao R, Huang Y, Li R, Li K, Ren Z, Wu J (2024) Regulation of nitrogen transformation and microbial community by inoculation during livestock manure composting. Environ Microbiol Rep 16(2). https://doi.org/10.1111/1758-2229.13256

Cavé L, Brothier E, Abrouk D, Bouda PS, Hien E, Nazaret S (2016) Efficiency and sensitivity of the digital droplet PCR for the quantification of antibiotic resistance genes in soils and organic residues. Appl Microbiol Biotechnol 100(24):10597–10608. https://doi.org/10.1007/s00253-016-7950-5

Chang R, Yao Y, Cao W, Wang J, Wang X, Chen Q (2019) Effects of composting and carbon based materials on carbon and nitrogen loss in the arable land utilization of cow manure and corn stalks. J Environ Manag, 233:283–290. https://doi.org/10.1016/j.jenvman.2018.12.021

Chen M, Huang Y, Wang C, Gao H (2020) The conversion of organic nitrogen by functional bacteria determines the end-result of ammonia in compost. Bioresour Technol 299. https://doi.org/10.1016/j.biortech.2019.122599

Clough TJ, Müller C, Laughlin RJ (2013) Using stable isotopes to follow excreta N dynamics and $N_2O$ emissions in animal production systems. Animal 7 Suppl 2:418–426. https://doi.org/10.1017/S1751731113000773

Danon M, Franke-Whittle IH, Insam H, Chen Y, Hadar Y (2008) Molecular analysis of bacterial community succession during prolonged compost curing. FEMS Microbiol Ecol 65(1): 133–144. https://doi.org/10.1111/j.1574-6941.2008.00506.x

Deng X, Achari A, Federman S, Yu G, Somasekar S, Bártolo I, Yagi S, Mbala-Kingebeni P, Kapetshi J, Ahuka-Mundeke S, Muyembe-Tamfum JJ, Ahmed AA, Ganesh V, Tamhankar M, Patterson JL, Ndembi N, Mbanya D, Kaptue L, McArthur C, Muñoz-Medina JE, Gonzalez-Bonilla CR, López S, Arias CF, Arevalo S, Miller S, Stone M, Busch M, Hsieh K, Messenger S, Wadford DA, Rodgers M, Cloherty G, Faria NR, Thézé J, Pybus OG, Neto Z, Morais J, Taveira N, Hackett JR, Chiu CY (2020) Metagenomic sequencing with spiked primer enrichment for viral diagnostics and genomic surveillance. Nat Microbiol 5:443–454. https://doi.org/10.1038/s41564-019-0637-9

Djemiel C, Maron PA, Terrat S, Dequiedt S, Cottin A, Ranjard L (2022) Inferring microbiota functions from taxonomic genes: a review. Gigascience 11:giab090. https://doi.org/10.1093/GIGASCIENCE/GIAB090

Dong L, Meng Y, Wang J, Liu Y (2014) Evaluation of droplet digital PCR for characterizing plasmid reference material used for quantifying ammonia oxidizers and denitrifiers. Anal Bioanal Chem 406(6):1701–1712. https://doi.org/10.1007/s00216-013-7546-1

Duan M, Zhang Y, Zhou B, Wang Q, Gu J, Liu G, Qin Z, Li Z (2019) Changes in antibiotic resistance genes and mobile genetic elements during cattle manure composting after inoculation with Bacillus subtilis. Bioresour Technol 292. https://doi.org/10.1016/j.biortech.2019.122011

Eckstrom K, Barlow JW (2019) Resistome metagenomics from plate to farm: the resistome and microbial composition during food waste feeding and composting on a Vermont poultry farm. PLoS One 14:e0219807. https://doi.org/10.1371/JOURNAL.PONE.0219807

Edgar RC (2017) Accuracy of microbial community diversity estimated by closed- and open-reference OTUs. PeerJ 2017:e3889. https://doi.org/10.7717/PEERJ.3889/SUPP-2

Esperón F, Albero B, Ugarte-Ruíz M, Domínguez L, Carballo M, Tadeo JL, del Mar Delgado M, Moreno MÁ, de la Torre A (2020) Assessing the benefits of composting poultry manure in reducing antimicrobial residues, pathogenic bacteria, and antimicrobial resistance genes: a field-scale study. Environ Sci Pollut Res 27(22):27738–27749. https://doi.org/10.1007/s11356-020-09097-1

Fan Q, Zhang J, Shi H, Chang S, Hou F (2023) Metagenomic profiles of yak and cattle manure resistomes in different feeding patterns before and after composting. Appl Environ Microbiol 89. https://doi.org/10.1128/aem.00645-23

Francis EK, Antonopoulos A, Westman ME, McKay-Demeler J, Laing R, Šlapeta J (2024a) A mixed amplicon metabarcoding and sequencing approach for surveillance of drug resistance to levamisole and benzimidazole in Haemonchus spp. Int J Parasitol 54:55–64. https://doi.org/10.1016/J.IJPARA.2023.07.002

Francis EK, Antonopoulos A, Westman ME, McKay-Demeler J, Laing R, Šlapeta J (2024b) A mixed amplicon metabarcoding and sequencing approach for surveillance of drug resistance to levamisole and benzimidazole in Haemonchus spp. Int J Parasitol 54:55–64. https://doi.org/10.1016/j.ijpara.2023.07.002

Franzosa EA, Morgan XC, Segata N, Waldron L, Reyes J, Earl AM, Giannoukos G, Boylan MR, Ciulla D, Gevers D, Izard J, Garrett WS, Chan AT, Huttenhower C (2014) Relating the metatranscriptome and metagenome of the human gut. Proc Natl Acad Sci USA 111:E2329. https://doi.org/10.1073/PNAS.1319284111/-/DCSUPPLEMENTAL

Giannoukos G, Ciulla DM, Huang K, Haas BJ, Izard J, Levin JZ, Livny J, Earl AM, Gevers D, Ward DV, Nusbaum C, Birren BW, Gnirke A (2012) Efficient and robust RNA-seq process for cultured bacteria and complex community transcriptomes. Genome Biol 13:1–13. https://doi.org/10.1186/GB-2012-13-3-R23/FIGURES/5

Green SJ, Michel FC, Hadar Y, Minz D (2004) Similarity of bacterial communities in sawdust- and straw-amended cow manure composts. FEMS Microbiol Lett 233:115–123

Guerreiro JF, Pires AJ, Nunes M, Esteves A, Chambel L, Pascoal P, Pereira M, Fangueiro D, Tavares L, Dias R, Bexiga R, Oliveira M (2024) Biochar supplementation affects the microbiome of recycled manure solids for cow bedding: a metagenomic analysis. J Dairy Sci. https://doi.org/10.3168/jds.2024-25616

Hanajima D, Aoyagi T, Hori T (2015) Survival of free-living *Acholeplasma* in aerated pig manure slurry revealed by $^{13}C$-labeled bacterial biomass probing. Front Microbiol 6(OCT). https://doi.org/10.3389/fmicb.2015.01206

Hanajima D, Aoyagi T, Hori T (2019) Dead bacterial biomass-assimilating bacterial populations in compost revealed by high-sensitivity stable isotope probing. Environ Int 133. https://doi.org/10.1016/j.envint.2019.105235

Harringer M, Alfreider A (2021) Primer evaluation and development of a droplet digital PCR protocol targeting amoA genes for the quantification of Comammox in lakes. Sci Rep 11(1). https://doi.org/10.1038/s41598-021-82613-6

Hatch A, Horne J, Toma R, Twibell BL, Somerville KM, Pelle B, Canfield KP, Genkin M, Banavar G, Perlina A, Messier H, Klitgord N, Vuyisich M (2019) A robust metatranscriptomic technology for population-scale studies of diet, gut microbiome, and human health. Int J Genom 2019:1718741. https://doi.org/10.1155/2019/1718741

He X, Yin H, Fang C, Xiong J, Han L, Yang Z, Huang G (2021) Metagenomic and q-PCR analysis reveals the effect of powder bamboo biochar on nitrous oxide and ammonia emissions during aerobic composting. Bioresour Technol 323. https://doi.org/10.1016/j.biortech.2020.124567

Heiss-Blanquet S, Fayolle-Guichard F, Lombard V, Hébert A, Coutinho PM, Groppi A, Barre A, Henrissat B (2016) Composting-like conditions are more efficient for enrichment and diversity of organisms containing cellulase-encoding genes than submerged cultures. PLoS One 11: e0167216. https://doi.org/10.1371/JOURNAL.PONE.0167216

Hettich RL, Pan C, Chourey K, Giannone RJ (2013) Metaproteomics: harnessing the power of high performance mass spectrometry to identify the suite of proteins that control metabolic activities in microbial communities. Anal Chem 85(9):4203–4214. https://doi.org/10.1021/ac303053e

Hill GT, Mitkowski NA, Aldrich-Wolfe L, Emele LR, Jurkonie DD et al (2000) Methods for assessing the composition and diversity of soil microbial communities. Appl Soil Ecol 15:25–36

Huang Z, Jiang Y, Song X, Hallerman E, Peng L, Dong D, Ma T, Zhai J, Li W (2018) Ammonia-oxidizing bacteria and archaea within biofilters of a commercial recirculating marine aquaculture system. AMB Express 8:1–12. https://doi.org/10.1186/S13568-018-0551-1/FIGURES/3

Jameson E, Taubert M, Coyotzi S, Chen Y, Eyice Ö, Schäfer H, Murrell JC, Neufeld JD, Dumont MG (2017) DNA-, RNA-, and protein-based stable-isotope probing for high-throughput biomarker analysis of active microorganisms. In: Methods in molecular biology, vol 1539. Humana Press Inc, pp 57–74. https://doi.org/10.1007/978-1-4939-6691-2_5

Janda JM, Abbott SL (2007) 16S rRNA gene sequencing for bacterial identification in the diagnostic laboratory: pluses, perils, and pitfalls. J Clin Microbiol 45:2761. https://doi.org/10.1128/JCM.01228-07

Jauregi L, González A, Garbisu C, Epelde L (2023) Organic amendment treatments for antimicrobial resistance and mobile element genes risk reduction in soil-crop systems. Sci Rep 13(1). https://doi.org/10.1038/s41598-023-27840-9

Jehmlich N, Vogt C, Lünsmann V, Richnow HH, von Bergen M (2016) Protein-SIP in environmental studies. Curr Opin Biotechnol 41:26–33). Elsevier Ltd. https://doi.org/10.1016/j.copbio.2016.04.010

Jindal S (2020) Microbes in soil and their metagenomics. In: Microbial diversity, interventions and scope, pp 85–96. https://doi.org/10.1007/978-981-15-4099-8_5/TABLES/2

Johnson FE, Lerch RN, Motavalli PP, Veum KS, Scharf PC (2024) Comparative analysis of three next-generation sequencing techniques to measure nosZ gene abundance in Missouri claypan soils. Environ Res 249. https://doi.org/10.1016/j.envres.2024.118346

Kim J, Hwangbo M, Shih CH, Chu KH (2023) Advances and perspectives of using stable isotope probing (SIP)-based technologies in contaminant biodegradation. Water Res X 20. https://doi.org/10.1016/j.wroa.2023.100187

Keenum I, Williams RK, Ray P, Garner ED, Knowlton KF, Pruden A (2021) Combined effects of composting and antibiotic administration on cattle manure–borne antibiotic resistance genes. Microbiome 9:1–16. https://doi.org/10.1186/S40168-021-01006-Z/FIGURES/7

Lam TK, Yang JT, Lai SJ, Liang SY, Wu SH (2021) Meta-proteomics analysis of microbial ecosystem during the anaerobic digestion of chicken manure in biogas production farm. Bioresour Technol Rep 13. https://doi.org/10.1016/j.biteb.2021.100643

Leray M, Haenel Q, Bourlat SJ (2016) Preparation of amplicon libraries for metabarcoding of marine eukaryotes using illumina MiSeq: the adapter ligation method. In: Methods in molecular biology. Humana Press Inc., pp 209–218

Li N, Zeng W, Guo Y, Li C, Ma C, Peng Y (2019b) Nitrogen-associated niche characteristics and bacterial community estimated by $^{15}$N-DNA-stable isotope probing in one-stage partial nitritation/anammox process with different ammonium loading. J Environ Manag 247:603–612. https://doi.org/10.1016/j.jenvman.2019.06.110

Li W, Liu Y, Hou Q, Huang W, Zheng H, Gao X, Yu J, Kwok LY, Zhang H, Sun Z (2020a) Lactobacillus plantarum improves the efficiency of sheep manure composting and the quality of the final product. Bioresour Technol 297. https://doi.org/10.1016/j.biortech.2019.122456

Li M, Xu J, Jiang Z, Li Q (2020b) Molecular understanding of autotrophic CO2-fixing bacterial communities in composting based on RuBisCO genes analysis. J Biotechnol 320:36–43. https://doi.org/10.1016/j.jbiotec.2020.06.010

Li Y, Zhao J, Zhang Z (2021a) Implementing metatranscriptomics to unveil the mechanism of bioaugmentation adopted in a continuous anaerobic process treating cow manure. Bioresour Technol 330:124962. https://doi.org/10.1016/J.BIORTECH.2021.124962

Li C, Ding S, Du C, He Y, Ma Z, Li G, Sun Z, Abhilash C, Abdelhamid MT, Zhang H, Zebib B, Hatano R (2021b) The transformation dynamics and homogeneity of different N fractions in compost following glucose addition. https://doi.org/10.3390/agriculture

Li Z, Yang Y, Xia Y, Wu T, Zhu J, Wang Z et al (2019a) The succession pattern of bacterial diversity in compost using pig manure mixed with wood chips analyzed by 16s rRNA gene analysis. https://doi.org/10.1101/674069

Li W, Fan Q, Yang Y, Xiao X, Li J, Zhang Y (2025) Benchmarking and optimizing qualitative and quantitative pipelines in environmental metatranscriptomics using mixture controlling experiments. ISME Commun 5(1). https://doi.org/10.1093/ismeco/ycaf090

Liao H, Friman VP, Geisen S, Zhao Q, Cui P, Lu X, Chen Z, Yu Z, Zhou S (2019) Horizontal genetransfer and shifts in linked bacterial community composition are associated with maintenance of antibioticresistance genes during food waste composting. Sci Total Environ, 660:841–850. https://doi.org/10.1016/j.scitotenv.2018.12.353

Lim S-S, Lee S-M, Lee S-H, Choi W-J (2010) Nitrogen isotope compositions of synthetic fertilizer, raw livestock manure slurry, and composted livestock manure. Korean J Soil Sci Fert 43(4)

Lin YW, Tuan NN, Huang SL (2016) Metaproteomic analysis of the microbial community present in a thermophilic swine manure digester to allow functional characterization: a case study. Int Biodeterior Biodegrad 115:64–73. https://doi.org/10.1016/j.ibiod.2016.06.013

Links MG, Dumonceaux TJ, McCarthy L, Hemmingsen SM, Topp E, Town JR (2018) CaptureSeq: hybridization-based enrichment of *cpn60* gene fragments reveals the community structures of synthetic and natural microbial ecosystems. https://doi.org/10.1101/492116

Liu D, Li M, Xi B, Zhao Y, Wei Z, Song C, Zhu C (2015) Metaproteomics reveals major microbial players and their biodegradation functions in a large-scale aerobic composting plant. Microb Biotechnol 8(6):950–960. https://doi.org/10.1111/1751-7915.12290

Liu X, Shao Y, Dong Y, Dong M, Xu Z, Hu X, Liu A (2021) Response of ammonia-oxidizing archaea and bacteria to sulfadiazine and copper and their interaction in black soils. Environ Sci Pollut Res 28:11357–11368. https://doi.org/10.1007/S11356-020-11356-0/FIGURES/7

López M, Jurado M, López-González JA, Estrella-González MJ, Martínez-Gallardo M, Toribio AJ et al (2021) Characterization of thermophilic lignocellulolytic microorganisms in composting. Front Microbiol 12. https://doi.org/10.3389/fmicb.2021.697480

Ma S, Shen Y, Ding J, Cheng H, Zhou H, Ge M, Wang J, Cheng Q, Zhang D, Zhang Y, Xu P, Zhang P (2024) Effects of biochar and volcanic rock addition on humification and microbial community during aerobic composting of cow manure. Bioresour Technol 391. https://doi.org/10.1016/j.biortech.2023.129973

Madsen EL (2006) The use of stable isotope probing techniques in bioreactor and field studies on bioremediation. Curr Opin Biotechnol 17(1):92–97. https://doi.org/10.1016/j.copbio.2005.12.004

Mang SM, Trotta V, Scopa A, Camele I (2022) Metagenomic analysis of bacterial community structure and dynamics of a digestate and a more stabilized digestate-derived compost from agricultural waste. PRO 10. https://doi.org/10.3390/pr10020379

Maretto L, Deb S, Ravi S, Della Lucia MC, Borella M, Campagna G, Squartini A, Concheri G, Nardi S, Stevanato P (2023) 16S metabarcoding, total soil DNA content, and functional bacterial genes quantification to characterize soils under long-term organic and conventional farming systems. Chem Biol Technol Agric 10:1–15. https://doi.org/10.1186/S40538-023-00450-3/TABLES/4

Melby JA, Roberts DS, Larson EJ, Brown KA, Bayne EF, Jin S, Ge Y (2021) Novel strategies to address the challenges in top-down proteomics. J Am Soc Mass Spectrom 32(6):1278–1294). American Chemical Society. https://doi.org/10.1021/jasms.1c00099

Meng Q, Yang W, Men M, Bello A, Xu X, Xu B, Deng L, Jiang X, Sheng S, Wu X, Han Y, Zhu H (2019) Microbial community succession and response to environmental variables during cow manure and corn straw composting. Front Microbiol 10:529. https://doi.org/10.3389/FMICB.2019.00529/FULL

Mirabile A, Sangiorgio G, Bonacci PG, Bivona D, Nicitra E, Bonomo C, Bongiorno D, Stefani S, Musso N (2024) Advancing pathogen identification: the role of digital PCR in enhancing diagnostic power in different settings. Diagnostics 14(15) Multidisciplinary Digital Publishing Institute (MDPI). https://doi.org/10.3390/diagnostics14151598

Montella S, Ventorino V, Lombard V, Henrissat B, Pepe O, Faraco V (2017) Discovery of genes coding for carbohydrate-active enzyme by metagenomic analysis of lignocellulosic biomasses. Sci Rep 7:1–15. https://doi.org/10.1038/srep42623

Mosbæk F, Kjeldal H, Mulat DG, Albertsen M, Ward AJ, Feilberg A, Nielsen JL (2016) Identification of syntrophic acetate-oxidizing bacteria in anaerobic digesters by combined protein-based stable isotope probing and metagenomics. ISME J 10(10):2405–2418. https://doi.org/10.1038/ismej.2016.39

Neher DA, Weicht TR, Bates ST, Leff JW, Fierer N (2013) Changes in bacterial and fungal communities across compost recipes, preparation methods, and composting times. PLoS One 8(11):e79512. https://doi.org/10.1371/journal.pone.0079512

Nishida M, Sumida H, Kato N (2008) Fate of nitrogen derived from 15N-labeled cattle manure compost applied to a paddy field in the cool climate region of Japan. Soil Sci Plant Nutr 54(3): 459–466. https://doi.org/10.1111/j.1747-0765.2008.00255.x

Papadopoulou ES, Bachtsevani E, Katsoula A, Charami C, Lampronikou E, Vasileiadis S, Karpouzas DG (2024) Nitrification inhibitors impose distinct effects on comammox bacteria and canonical ammonia oxidizers under high N fertilization regimes. Appl Soil Ecol 199. https://doi.org/10.1016/j.apsoil.2024.105417

Park S, Rana A, Sung W, Munir M (2021) Competitiveness of quantitative polymerase chain reaction (qPCR) and droplet digital polymerase chain reaction (ddPCR) technologies, with a particular focus on detection of antibiotic resistance genes (ARGs). Appl Microbiol 1(3): 426–444). Multidisciplinary Digital Publishing Institute (MDPI). https://doi.org/10.3390/applmicrobiol1030028

Peano C, Pietrelli A, Consolandi C, Rossi E, Petiti L, Tagliabue L, De Bellis G, Landini P (2013) An efficient rRNA removal method for RNA sequencing in GC-rich bacteria

Peng J, Zhou X, Rensing C, Liesack W, Zhu YG (2023) Soil microbial ecology through the lens of metatranscriptomics. Soil Ecol Lett 6:1–15. https://doi.org/10.1007/S42832-023-0217-Z

Pérez-Losada M, Castro-Nallar E, Bendall ML, Freishtat RJ, Crandall KA (2015) Dual transcriptomic profiling of host and microbiota during health and disease in pediatric asthma. PLoS One 10:e0131819. https://doi.org/10.1371/JOURNAL.PONE.0131819

Po A, Eyers CE (2023) Top-down proteomics and the challenges of true proteoform characterization. J Proteome Res 22(12):3663–3675). American Chemical Society. https://doi.org/10.1021/acs.jproteome.3c00416

Quan PL, Sauzade M, Brouzes E (2018) DPCR: a technology review. Sensors (Basel) 18(4) MDPI AG. https://doi.org/10.3390/s18041271

Quince C, Walker AW, Simpson JT, Loman NJ, Segata N (2017) Shotgun metagenomics, from sampling to analysis. Nat Biotechnol 35:833–844. https://doi.org/10.1038/nbt.3935

Reed K, Dang C, Walkup J, Purcell A, Hungate B, Morrissey E (2025) Comparing field and lab quantitative stable isotope probing for nitrogen assimilation in soil microbes. Appl Environ Microbiol. https://doi.org/10.1128/aem.01849-24

Romero-Yahuitl V, Zarco-González KE, Toriz-Nava AL, Hernández M, Velázquez-Fernández JB, Navarro-Noya YE, Luna-Guido M, Dendooven L (2024) The archaeal and bacterial community structure in composted cow manures is defined by the original populations: a shotgun metagenomic approach. Front Microbiol 15. https://doi.org/10.3389/fmicb.2024.1425548

Ryckeboer J, Mergaert J, Vaes K, Klammer S, de Clercq D et al (2003) A survey of bacteria and fungi occurring during composting and self-heating processes. Ann Microbiol 53:349–410

Sancha Dominguez L, Cotos Suárez A, Sánchez Ledesma M, Muñoz Bellido JL (2024) Present and future applications of digital PCR in infectious diseases diagnosis. Diagnostics 14-(9) Multidisciplinary Digital Publishing Institute (MDPI). https://doi.org/10.3390/diagnostics14090931

Santos-Pereira C, Sousa J, Costa ÂMA, Santos AO, Rito T, Soares P, Franco-Duarte R, Silvério SC, Rodrigues LR (2023) Functional and sequence-based metagenomics to uncover carbohydrate-degrading enzymes from composting samples. Appl Microbiol Biotechnol 107:5379. https://doi.org/10.1007/S00253-023-12627-9

Song X, Lu C, Luo J, Gong X, Guo D, Ma Y (2023) Matured compost amendment improves compost nutrient content by changing the bacterial community during the composting of Chinese herb residues. Front Microbiol 14:1146546. https://doi.org/10.3389/FMICB.2023.1146546/FULL

Sun W, Krumins V, Dong Y, Gao P, Ma C, Hu M, Li B, Xia B, He Z, Xiong S (2017) A combination of stable isotope probing, illumina sequencing, and co-occurrence network to investigate thermophilic acetate- and lactate-utilizing bacteria. Microb Ecol 75(1):113–122. https://doi.org/10.1007/s00248-017-1017-8

Sun Y, Sheng S, Jiang X, Bello A, Wu X, Meng Q, Deng L, Xu X, Li H (2019) Genetic associations as indices for assessing nitrogen transformation processes in co-composting of cattle manure and rice straw. Bioresour Technol 291. https://doi.org/10.1016/j.biortech.2019.121815

Sun N, Fan B, Yang F, Zhao L, Wang M (2023) Effects of adding corn steep liquor on bacterial community composition and carbon and nitrogen transformation during spent mushroom substrate composting. BMC Microbiol 23:1–13. https://doi.org/10.1186/S12866-023-02894-X/FIGURES/6

Tessler M, Neumann JS, Afshinnekoo E, Pineda M, Hersch R, Velho LFM, Segovia BT, Lansac-Toha FA, Lemke M, Desalle R, Mason CE, Brugler MR (2017) Large-scale differences in microbial biodiversity discovery between 16S amplicon and shotgun sequencing. Sci Rep 7: 6589. https://doi.org/10.1038/S41598-017-06665-3

Tiwari A, Ahmed W, Oikarinen S, Sherchan SP, Heikinheimo A, Jiang G, Simpson SL, Greaves J, Bivins A (2022) Application of digital PCR for public health-related water quality monitoring. Sci Total Environ 837). Elsevier B.V. https://doi.org/10.1016/j.scitotenv.2022.155663

Vasileiadis S, Puglisi E, Papadopoulou ES, Pertile G, Suciu N, Pappolla RA, Tourna M, Karas PA, Papadimitriou F, Kasiotakis A, Ipsilanti N, Ferrarini A, Sulowicz S, Fornasier F, Menkissoglu-Spiroudi U, Nicol GW, Trevisan M, Karpouzas DG (2018) Blame it on the metabolite: 3,5-dichloroaniline rather than the parent compound is responsible for the decreasing diversity and function of soil microorganisms. Appl Environ Microbiol 84:e01536–e01518. https://doi.org/10.1128/AEM.01536-18

Voegel TM, Larrabee MM, Nelson LM (2020) Development of droplet digital PCR assays to quantify genes involved in nitrification and denitrification, comparison with quantitative real-time PCR and validation of assays in vineyard soil. Can J Microbiol 67(2). https://doi.org/10.1139/cjm-2020-0033

Wang C, Dong D, Wang H, Müller K, Qin Y, Wang H, Wu W (2016a) Metagenomic analysis of microbial consortia enriched from compost: new insights into the role of Actinobacteria in lignocellulose decomposition. Biotechnol Biofuels 9:1–17. https://doi.org/10.1186/S13068-016-0440-2/FIGURES/5

Wang DZ, Kong LF, Li YY, Xie ZX (2016b) Environmental microbial community proteomics: status, challenges and perspectives. Int J Mol Sci 17(8) MDPI AG. https://doi.org/10.3390/ijms17081275

Wang X, Howe S, Deng F, Zhao J (2021) Current applications of absolute bacterial quantification in microbiome studies and decision-making regarding different biological questions. Microorganisms 9(9). https://doi.org/10.3390/microorganisms9091797

Wang D, Wang S, Du X, He Q, Liu Y, Wang Z, Feng K, Li Y, Deng Y (2022) ddPCR surpasses classical qPCR technology in quantitating bacteria and fungi in the environment. Mol Ecol Resour 22(7):2587–2598. https://doi.org/10.1111/1755-0998.13644

Wang H, Qi JF, Qin R, Ding K, Graham DW, Zhu YG (2023) Intensified livestock farming increases antibiotic resistance genotypes and phenotypes in animal feces. Commun Earth Environ 4:1–12. https://doi.org/10.1038/s43247-023-00790-w

Westreich ST, Treiber ML, Mills DA, Korf I, Lemay DG (2018) SAMSA2: a standalone metatranscriptome analysis pipeline. BMC Bioinformatics 19:1–11. https://doi.org/10.1186/S12859-018-2189-Z/TABLES/2

Xiong Y, Mueller RS, Feng S, Guo X, Pan C (2024) Proteomic stable isotope probing with an upgraded Sipros algorithm for improved identification and quantification of isotopically labeled proteins. Microbiome 12(1). https://doi.org/10.1186/s40168-024-01866-1

Xu Z, Gao X, Li G, Nghiem LD, Luo W, Zhang F (2024) Microbial sources and sinks of nitrous oxide during organic waste composting. Environ Sci Technol 58:7367–7379. https://doi.org/10.1021/acs.est.3c10341

Yang Y, Ashworth AJ, DeBruyn JM, Durso LM, Savin M, Cook K, Moore PA, Owens PR (2020) Antimicrobial resistant gene prevalence in soils due to animal manure deposition and long-term pasture management. PeerJ 8. https://doi.org/10.7717/peerj.10258

Yu J, Gu J, Wang X, Guo H, Wang J, Lei L, Dai X, Zhao W (2020) Effects of inoculation with lignocellulose-degrading microorganisms on nitrogen conversion and denitrifying bacterial community during aerobic composting. Bioresour Technol 313. https://doi.org/10.1016/j.biortech.2020.123664

Zanardo M, Giannattasio M, Sablok G, Pindo M, La Porta N, Lorenzetti M, Noro C, Stevanato P, Concheri G, Squartini A (2023) Metabarcoding analysis of the bacterial and fungal communities during the maturation of preparation 500, used in biodynamic agriculture, suggests a rational link between horn and manure. Environ Dev Sustain 26:29891–29909. https://doi.org/10.1007/S10668-023-03144-W/FIGURES/7

Zang X, Liu M, Fan Y, Xu J, Xu X, Li H (2018) The structural and functional contributions of β-glucosidase-producing microbial communities to cellulose degradation in composting. Biotechnol Biofuels 11(1). https://doi.org/10.1186/s13068-018-1045-8

Zhang L, Zhang H, Wang Z, Chen G, Wang L (2016) Dynamic changes of the dominant functioning microbial community in the compost of a 90-$m^3$ aerobic solid state fermentor revealed by integrated meta-omics. Bioresour Technol 203:1–10. https://doi.org/10.1016/j.biortech.2015.12.040

Zhang L, Li L, Pan X, Shi Z, Feng X, Gong B, Li J, Wang L (2018) Enhanced growth and activities of the dominant functional microbiota of chicken manure composts in the presence of maize straw. Front Microbiol 9(MAY). https://doi.org/10.3389/fmicb.2018.01131

Zhang X, Ma B, Liu J, Chen X, Li S, Su E, Gao L, Li H (2020a) β-Glucosidase genes differentially expressed during composting. Biotechnol Biofuels 13:1–11. https://doi.org/10.1186/S13068-020-01813-W/FIGURES/5

Zhang L, Li L, Sha G, Liu C, Wang Z, Wang L (2020b) Aerobic composting as an effective cow manure management strategy for reducing the dissemination of antibiotic resistance genes: an integrated meta-omics study. J Hazard Mater 386. https://doi.org/10.1016/j.jhazmat.2019.121895

Zhang J, Yue Z, Ding C, Zhou Z, Zhang T, Wang X (2023a) Metagenomic binning analyses of pig manure composting reveal potential antibiotic-degrading bacteria and their risk of antibiotic resistance genes. Bioresour Technol 371:128540. https://doi.org/10.1016/j.biortech.2022.128540

Zhang Y, Kitazumi A, Liao Y-T, de los Reyes BG, Wu VCH (2023b) Metagenomic investigation reveals bacteriophage-mediated horizontal transfer of antibiotic resistance genes in microbial communities of an organic agricultural ecosystem. Microbiol Spectr 11. https://doi.org/10.1128/SPECTRUM.00226-23/SUPPL_FILE/SPECTRUM.00226-23-S0001.DOCX

Zhao E, Li Y, Zhang J, Geng B (2025) A review on the degradation of antibiotic resistance genes during composting of livestock manure. In Toxics 13:8. Multidisciplinary Digital Publishing Institute (MDPI). https://doi.org/10.3390/toxics13080667

Zhong XZ, Zeng Y, Wang SP, Sun ZY, Tang YQ, Kida K (2020) Insight into the microbiology of nitrogen cycle in the dairy manure composting process revealed by combining high-throughput sequencing and quantitative PCR. Bioresour Technol 301. https://doi.org/10.1016/j.biortech.2020.122760

Zhou Z, Keiblinger KM, Huang Y, Bhople P, Shi X, Yang S, Yu F, Liu D (2024) Virome and metagenomic sequencing reveal the impact of microbial inoculants on suppressions of antibiotic resistome and viruses during co-composting. J Hazard Mater 477:135355. https://doi.org/10.1016/j.jhazmat.2024.135355

# Composting: Improved Crop Performance and Resilience for Environmentally Friendly Practices

**Sana Bashir, Tasmia Bashir, Ali Abdur Rehman, and Mudassar Khan**

## 1 Introduction

Composting is a biological process that converts organic materials into a nutrient-rich soil amendment, mitigating environmental issues and promoting sustainable agriculture. This process involves the decomposition of organic waste, including agricultural, market, and kitchen waste, by a diverse community of microorganisms (Chen et al. 2016; Maina et al. 2017). Effective composting is influenced by factors such as temperature, pH, moisture content, and carbon-to-nitrogen ratio, with two primary types of composting existing: aerobic and anaerobic (Fig. 1). The benefits of composting are multifaceted, encompassing waste reduction and diversion from landfills, improved soil quality and fertility, enhanced water retention, reduced irrigation needs, weed suppression, and attraction of beneficial insects (Soobhany 2019; Oliveira et al. 2017; Malinska et al. 2017; Papafilippaki et al. 2015). Furthermore, composting has significant potential to mitigate climate change by reducing methane emissions, sequestering carbon in soil, and improving crop yields, thereby reducing land clearing and associated carbon dioxide emissions (Bher et al. 2023).

S. Bashir (✉)
Centre for the Research and Technology of Agro-Environmental and Biological Sciences, CITAB, Inov4Agro, Universidade de Trás os Montes e Alto Douro, UTAD, Quinta de Prados, Vila Real, Portugal

T. Bashir
Department of Plant Sciences, Quaid-i-Azam University, Islamabad, Pakistan

A. A. Rehman
Department of Electrical Engineering, COMSATS University, Islamabad, Pakistan

M. Khan
Centre for the Research and Technology of Agro-Environmental and Biological Sciences (CITAB), Department of Agronomy, University of Trás-os-Montes and Alto Douro (UTAD), Quinta de Prados, Vila Real, Portugal



R. K. Singh et al. (eds.), *Composting of Farm Residues by Nature-Based Solutions*, https://doi.org/10.1007/978-3-032-12255-1_8

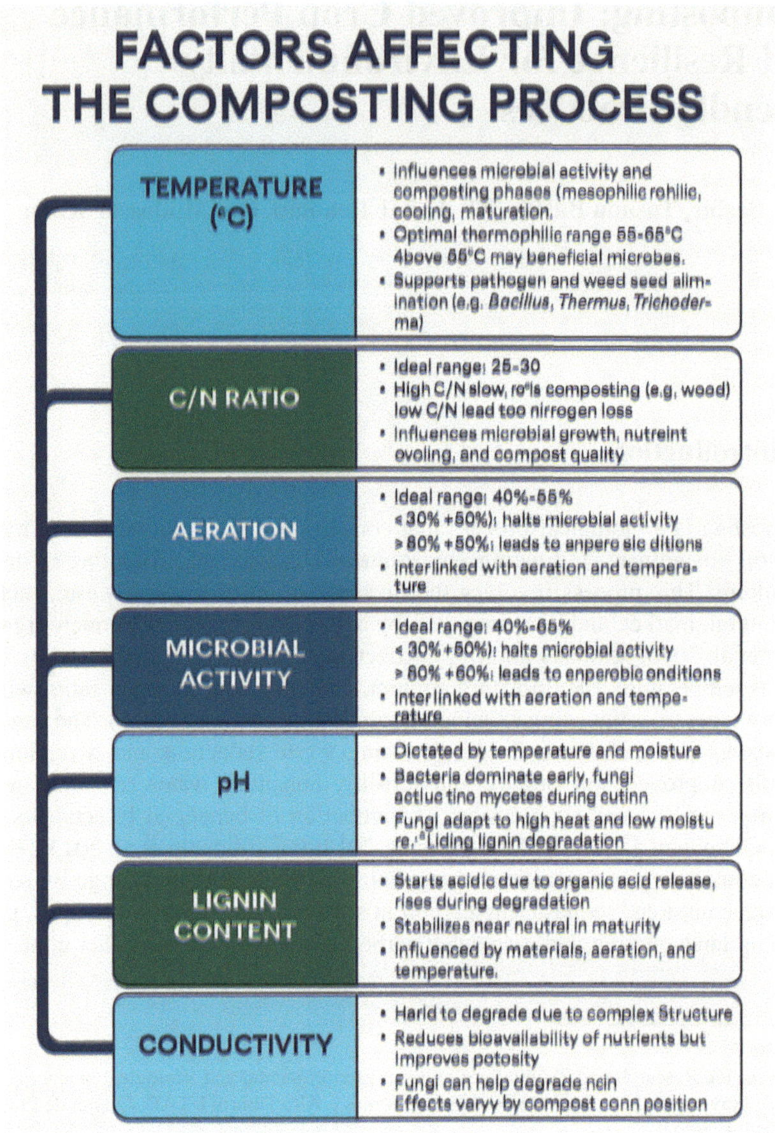


**Fig. 1** Factors affecting the composting progress

By adopting composting practices, individuals and communities can contribute to a more sustainable future, promoting ecosystem services, supporting plant growth, and ensuring environmental sustainability (Bagagiolo et al. 2022).

The exponential escalation in solid waste generation, primarily driven by population growth, urbanization, and economic development, has necessitated the implementation of sustainable waste management strategies worldwide (Sultana et al.

2021). Reuse and recycling have emerged as preferred approaches within integrated solid waste management frameworks, aligning with circular economy principles that prioritize resource efficiency and minimize environmental degradation (Lelicińska-Serafiń et al. 2023; Dede et al. 2023). Composting technology has emerged as a viable solution for converting organic waste into a nutrient-rich compost product, mitigating environmental pressures and promoting agricultural sustainability (Hashim et al. 2022; Külcü and Yaldiz 2014; Ajaweed et al. 2022). This biochemical process recovers valuable nutrients, reduces pathogenic microorganisms, and produces a fertilizer alternative to inorganic chemicals, thereby enhancing soil organic carbon and nitrogen, improving soil aggregation, and supporting long-term fertility and productivity (Meng et al. 2017; Gonawala and Jardosh 2018; Nozhevnikova et al. 2019). Compost characterization reveals high organic matter content and abundance of macro- and micronutrients, positively impacting soil biological and physicochemical properties, such as improved water retention, aeration, and thermal regulation (Azim et al. 2018; Qasim et al. 2018; FAO 2021). Furthermore, compost application has gained recognition globally as an eco-friendly substitute for synthetic fertilizers, offering significant economic savings without compromising crop yields (Drózdz et al. 2020; Lin et al. 2018; Chen et al. 2022; Turcanu 2018). The integration of composting technologies into agricultural practices also contributes to climate change mitigation by reducing greenhouse gas emissions, sequestering carbon in soil, and promoting ecosystem services (Rolewicz-Kalińska et al. 2020; Lin et al. 2022). By adopting composting technologies, communities can promote resource efficiency, reduce environmental degradation, foster resilient agricultural systems, and ensure sustainable development.

## 2 Compositing Technologies

Composting is a biological process that converts organic waste into a nutrient-rich soil amendment through the action of microorganisms, including bacteria, fungi, and actinomycetes, in the presence of oxygen. This ancient technique, employed by pre-Columbian Indians and ancient Egyptians, has evolved significantly over the past four decades, with scientific research highlighting its benefits. Composting involves the breakdown of organic matter into humus, a stable and fertile matrix that enhances soil fertility and structure (Drózdz et al. 2020).

The composting process requires careful management of factors such as moisture content, porosity, temperature, oxygen levels, and carbon-to-nitrogen ratio (Zukowska et al. 2019; Avidov et al. 2017) (Table 1). Microbial activity drives the decomposition and humification of organic matter, transforming it into a nutritious soil amendment. Composting technologies have advanced, with innovative processes developed for large-scale and medium-scale farmers. However, small-scale farmers face challenges due to high-tech equipment requirements (Kalamdhad et al. 2009; Gai et al. 2019; Choudhary et al. 2018).

**Table 1** Factors involved in the composting process

| Key parameters | Effect on composting | Impact on compost quality |
|---|---|---|
| **Temperature** | Optimizes microbial activity and aids decomposition | Influences temperature stability and affects maturity of compost |
| **Oxygen** | Supports microbial respiration and essential for decomposition | Vital for controlling microbial growth and decomposition |
| **Moisture** | Ensures moisture availability for microbes and proper breakdown | Ensures that decomposition progresses efficiently |
| **Porosity** | Affects airflow and microbial activity | Important for air circulation and microbial efficiency |
| **Carbon-to-nitrogen ratio** | Balances decomposition rate and ensuring proper breakdown | Ensures balanced microbial action and critical for humus formation |

Key parameters controlling the composting process include temperature, oxygen, and moisture. Optimal conditions ensure efficient microbial activity, resulting in a stable, mature, and contamination-free compost product (Martin and Brathwaite 2012). Composting is increasingly recognized as an efficient, cost-effective, and environmentally friendly solid waste management strategy, particularly for organic waste.

The process involves complex biochemical transformations, necessitating careful preparation of the input mixture to ensure product quality and environmental safety (Ren et al. 2018). Critical parameters for process control include temperature, oxygen, moisture, porosity, and carbon-to-nitrogen ratio. Successful composting operations rely on precise management of these factors to produce high-quality compost.

Composting offers numerous benefits, including:

1. Nutrient recovery and recycling
2. Soil fertility enhancement
3. Improved soil structure
4. Reduced environmental pollution
5. Climate change mitigation
6. Cost-effective waste management

## 2.1 Decomposition

Composting is a complex, multi-stage process involving microbial decomposition of organic waste. The process undergoes four stages: mesophilic, thermophilic, cooling, and maturation, which can occur concurrently or consecutively, influenced by factors such as mixture composition, water content, air circulation, and microbiological composition (Gonawala and Jardosh 2018). During the mesophilic phase, microorganisms rapidly metabolize carbon-rich substrates at temperatures between 15–40 °C, producing heat. As the process progresses to the thermophilic phase, temperature increases, favoring thermophilic bacteria, and carbon compounds are produced (Misra et al. 2003).

The cooling phase is characterized by the breakdown of complex structures by discrete waste composting fungi, while actinomycetes form humic compounds via condensation processes. Finally, the maturation phase is dominated by aerobic bacteria, resulting in lower oxygen uptake rates and temperatures below 25 °C. Macrofauna and soil organisms enter, metabolizing phytotoxic chemicals and suppressing plant diseases (Nozhevnikova et al. 2019).

Composting offers numerous benefits, including nutrient recovery and recycling, soil fertility enhancement, improved soil structure, reduced environmental pollution, climate change mitigation, and cost-effective waste management. Additionally, composting contributes to bioremediation, weed control, plant disease control, pollution anticipation, and erosion management (Hassan et al. 2023).

Composting technologies can be categorized into aerobic and anaerobic processes. Aerobic composting requires oxygen, resulting in rapid decomposition, heat production, and pathogen elimination. In contrast, anaerobic composting produces methane, carbon dioxide, and organic acids. Optimal composting conditions require careful management of moisture content, carbon-to-nitrogen ratio, oxygen supply, temperature control, and aeration regime.

Effective composting systems involve three primary phases: preparation of the mixture, bio-oxidative phase, and maturation phase. The differences in composting technologies lie in the operation of the bio-oxidative phase, influencing technical and administrative decisions. By understanding these processes, we can harness the potential of composting to promote resource efficiency, reduce environmental degradation, and support resilient agricultural systems (Nozhevnikova et al. 2019).

Overall, composting provides a sustainable solution for organic waste management, offering numerous environmental, economic, and social benefits (Pergola et al. 2018). By adopting composting technologies, we can mitigate climate change, improve soil health, and ensure a more sustainable future.

## 2.2 *Discrete Waste Decomposition (DW)*

Discrete waste composting is a controlled process that converts organic waste into a nutrient-rich fertilizer, mitigating environmental pollution and promoting sustainable agriculture (Ventorino et al. 2019; Sayara et al. 2020). Compost, the end product, is an assortment of essential plant nutrients, including nitrogen, phosphorus, potassium, sulfur, carbon, magnesium, and trace elements. The composting process involves complex biochemical transformations, influenced by factors such as moisture content, temperature, oxygen levels, and carbon-to-nitrogen ratio (Trupiano et al. 2017). Effective composting requires careful management of processing parameters and organic feedstocks, determining the chemical features of the compost, including C/N ratio and pH. By combining these components optimally, nutrient-rich compost substrates can be produced, replacing commercial mineral fertilizers in agriculture (Pane et al. 2014; Coelho et al. 2019; Pane et al. 2019). Discrete waste composting offers numerous benefits, including bioremediation, weed control, plant disease control, pollution anticipation, erosion management, and increased soil biodiversity (Uyizeye et al. 2019; Pose-Juan et al. 2017; Manirakiza and Seker 2020). This sustainable waste management strategy supports food safety, reduces environmental risks associated with synthetic fertilizers, and fosters resilient agricultural systems.

## 2.3 Passive Compositing

Passive composting, typically conducted in windrows or piles, involves forming a mixture of raw materials into a manageable structure that requires periodic stirring to maintain porosity. Aeration is achieved through natural air circulation within the pile, necessitating a smaller scale to ensure adequate oxygenation (Pergola et al. 2018). This method is relatively inexpensive and suitable for small farms with limited spatial constraints. Windrow composting, a specific type of passive composting, involves arranging raw materials in long, narrow piles that are periodically mechanically turned to facilitate aeration. However, relying solely on turning can lead to inconsistent oxygen levels, resulting in decreased microbial activity. Oxygen depletion can occur rapidly, often within an hour of turning, highlighting the need for frequent pile manipulation. Moreover, pile size poses a significant challenge, as heights exceeding 3 meters can hinder effective aeration, leading to technical and financial complexities (Hassan et al. 2023).

## 2.4 Aerated Static Pile Compositing (ASP)

The Aerated Static Pile (ASP) method utilizes blowers to inject air into the compost pile, enhancing oxygenation and cooling through positive pressure. This approach accelerates composting and minimizes spatial requirements. To ensure uniform air distribution, a porous base layer, such as wood chips or straw, is essential, while a top layer of finished compost or sawdust absorbs odors, retains moisture, and regulates heat (Keng et al. 2023). The ASP system alternates between bottom blowing and suction aeration, maintaining consistent temperature and moisture levels throughout the pile. Pile height, typically ranging from 150 to 245 cm, depends on factors like material porosity, weather conditions, and equipment length. Winter composting benefits from increased height to retain heat. A 15-cm layer of finished compost or bulking agent on top prevents surface drying, insulates against heat loss, and filters ammonia and odors (Michel et al. 2022).

Aerated static piles come in two configurations: individual and extended piles. Individual piles are triangular, approximately 300–490 cm wide, with the aeration pipe running lengthwise under the ridge. Suitable for intermittent raw material availability, these piles require careful raw material selection and initial mixing to prevent poor air distribution (Grosu et al. 2023). Static pile composting employs two aeration methods: suction and pressure systems. Suction systems draw air through the pile, collecting it in the aeration pipe for easy odor filtration. Pressure systems force air into the pile, producing greater airflow but making odor control more challenging (Rasapoor et al. 2009). Pressure systems are preferred for temperature control due to their enhanced cooling capabilities.

## 2.5 *Crop Residue Waste Composting*

The global agricultural sector generates substantial waste, necessitating effective management strategies to mitigate environmental, health, and aesthetic concerns. The escalating volume of agricultural waste exacerbates these issues, underscoring the need for research into eco-friendly disposal methods. Composting has emerged as a viable, cost-effective, and environmentally sustainable solution, transforming biodegradable wastes into valuable resources (Gajalakshmi and Abbasi 2008). This process combines organic materials like wood shavings, pine needles, dry leaves, sawdust, and coir pith to create durable humus.

However, decomposing lignin-rich plant products poses challenges. To overcome this, lime is added to accelerate the breakdown process. A ratio of 5 kg of lime per 1000 kg of plant materials is typically used to produce high-quality compost (Clay et al. 2019). When mixed with water, lime forms a semi-solid substance or dry powder, enhancing humification by reducing lignin structure and boosting humus content (Sarkar and Chourasia 2017; Huang and Hall 2017). By adopting suitable processes and quality control measures, various crop leftovers can be converted into valuable compost substrates, supporting sustainable agricultural practices.

## 2.6 *Municipal Solid Waste Composting (MSW)*

The rapid growth of municipal solid waste (MSW) due to urbanization, industrialization, and population increase has become a significant concern globally, particularly in Pakistan (Waqas et al. 2023). Composting has emerged as a viable solution, leveraging the high organic content of MSW to produce valuable soil conditioners. This approach not only reduces waste disposal volumes but also mitigates pollution potential. Modern MSW management composting systems involve processes like sorting, contaminant separation, sizing, mixing, and biological decomposition (Ramachandra et al. 2018; Ayilara et al. 2020).

In Asia, countries like India have successfully implemented large-scale composting facilities, treating up to 500 metric tons of MSW daily. By 2008, composting accounted for 9% of India's MSW treatment. Other countries, such as Bangladesh and the Maldives, have also adopted composting as a small-scale industry (Visvanathan and Trankler 2003; Glawe et al. 2005; Sharholy et al. 2008). MSW comprises diverse materials, primarily from residential, commercial, and garden sources, with biodegradable organic waste being the predominant fraction. This includes kitchen waste, yard trimmings, and other materials susceptible to recuperative treatment through anaerobic digestion, composting, or combined technologies.

Organic waste streams exhibit substantial variability in macronutrient concentrations, including nitrogen, phosphorus, potassium, calcium, magnesium, and sulfur. Factors like initial waste composition, animal farming strategies, waste treatment processes, and additives influence nutrient concentrations and availability in the final compost (Sivaramanan 2015; Gunaruwan and Gunasekara 2016). Effective composting requires identifying the optimal carbon-to-nitrogen ratio and addressing potential contaminants. Nutrient content analysis can be performed using classical chemical methods or alternative techniques like X-ray fluorescence spectrometry and near-infrared spectrometry.

However, composting also raises concerns about potentially toxic substances (PTS) in MSW, including heavy metals, micro- and nanoplastics, persistent organic pollutants, pharmaceuticals, and personal care products. Adopting proper collection strategies and measures to mitigate PTS occurrence in composting feedstock is crucial (Hussain et al. 2020). Generally, composting offers a valuable solution for MSW management, converting organic waste into nutrient-rich soil amendments while reducing waste disposal volumes and pollution potential.

## 2.7 *Biomedical Waste Compositing (BMW)*

The improper management of biomedical waste (BMW) poses significant environmental and health risks. In Pakistan, hospitals generate approximately 2.07 kg of BMW per bed per day, emphasizing the need for safe disposal techniques. Composting offers a sustainable solution, and research has identified neem and tobacco extracts as cost-effective degradation agents for organic BMW, suitable for small local farmers (Dinesh et al. 2010). These extracts can convert organic BMW into valuable fertilizers.

Previous studies recommend treating BMW with 5% sodium hypochlorite (NaOCl) at the disposal site, followed by mixing with cow dung slurry to initiate decomposition. Vermicomposting (VC) can further enhance the process, utilizing epigeal worm species to break down the waste. This approach provides an energy-efficient and sustainable method for eliminating and recycling hazardous BMW (Steen and Brooks 2015; Patil et al. 2019). Conventional composting is a widely used technique for managing various types of waste. Effective composting requires careful consideration of factors such as waste composition, moisture content, and decomposition processes. By adopting suitable composting methods, BMW can be transformed into valuable resources, mitigating environmental and health risks associated with improper disposal (Manzoor and Sharma 2019).

# 3 Conventional Compositing

Over the past few decades, conventional composting processes, including vermicomposting (VC), aerobic composting (AC), and anaerobic composting (AnC), have gained widespread acceptance globally. Numerous studies have demonstrated that agricultural waste management through composting significantly enhances soil texture and structure, yielding multiple benefits for the environment and agricultural productivity. Researchers have focused on optimizing composting techniques to improve bioavailability, consistency, and economic viability, while minimizing environmental impacts (Waqas et al. 2023). This has led to advancements in approaches and green development. Traditional composting methods possess distinct features that influence the resulting compost consistency, which is crucial for effective agricultural applications.

## *3.1 Vermicompositing (VC)*

Vermicomposting (VC) is a composting method that utilizes earthworms to break down organic waste. This process occurs in a controlled environment, such as a bin or tub, with a perforated bottom to facilitate compost tea infiltration (Kapoor et al. 2015). The setup typically consists of layers, including cow manure, worms (e.g., *Eudrilus eugineae/Eisenia Fetida*), and shredded kitchen waste (Huang et al. 2020). Daily water addition supports worm survival and multiplication, and the leached water can be used as compost tea, a liquid fertilizer rich in nutrients. Earthworms can decompose various organic materials, including sewage sludge, cattle farm waste, poultry waste, bagasse, industrial waste, and residential waste (Sanasam and Talukdar 2017). Studies have shown that species like *Eisenia fetida, Eudrilus eugineae*, and Lampito Mauritii are effective for VC. These worms thrive in optimal conditions, such as temperatures between 25 and 30°C, moisture content of 60–75%, and pH levels between 6.5 and 7.5 (Sinha et al. 2008; Meena and Renu 2009; Zhao et al. 2018).

Research has demonstrated that VC enhances nutrient content, reducing C/N and C/P ratios. For instance, a study in Jodhpur, India, found that *Eisenia fetida* produced nutrient-rich compost more efficiently than Lampito Mauritii (Loh et al. 2005) Similarly, combining earthworms with microorganisms (N-fixing, K-fixing, and P-solubilizing bacteria) minimized composting time and increased nutrient enrichment (Hrebecková et al. 2019; Huang et al. 2020). Vermicomposting offers numerous benefits, including improved soil texture and water retention, reduced environmental pollution, and resource conservation (Plaza et al. 2008; Padmavathiamma et al. 2008). The resulting vermicompost contains essential nutrients like calcium, magnesium, manganese, copper, iron, and zinc, as well as beneficial microbes that support plant health and resilience (Ghosh et al. 1999; Pramanik et al. 2007; Gupta and Garg 2008; Yadav et al. 2010). By leveraging farm resources, vermicomposting can promote sustainable soil management and environmental protection.

Vermicomposting parameters vary depending on the type of waste utilized. Studies have reported distinct ranges for factors such as C/N ratio, pH, and electrical conductivity (EC) across different waste materials, including sludge from tanneries and cattle dung, newspaper and sawdust, distillery industry sludge, household waste, and wastewater treatment plant sludge (Garg and Kaushik 2005; Sen and Chandra 2007; Yadav and Garg 2009; Subramanian et al. 2010). Research in Kerala, India, demonstrated the efficacy of Eudrillus eugineae in synthesizing nutrient-enriched compost when inoculated with N-fixing, K-fixing, and P-solubilizing bacteria (Busato et al. 2012). This vermicompost fertilizer accelerated root growth, increased nutrient uptake, and enhanced total yield. Similarly, peppermint mill sludge, solid pulp, and cow dung fed to *Eudrilus eugineae* reduced pollution and converted waste into value-added products, with enhanced N and P content and reduced C/N ratio (Biruntha et al. 2020).

Experiments using four different feedstocks (seaweed, sugarcane trash, coir pith, and vegetable waste) and *Eudrilus eugineae* showed decreased pH, organic matter content, total organic carbon, and cellulose and lignin levels after 50 days of composting. Conversely, nutrient levels increased, indicating vermicompost development. Adding cow dung to the mixture enriched the final compost's nutrient content. Comparative studies revealed that *Eudrilus eugineae* outperformed *Eisenia fetida* and local worms in forming nutrient-enriched products, achieving higher reproduction and growth rates (Tripathi and Bhardwaj 2004; Yadav and Garg 2011; Das et al. 2016). Combining worms, bacteria, and organic waste generated high-quality vermicompost, highlighting the benefits of synergistic relationships in vermicomposting systems.

## 3.2 *Aerobic (In-Vessel) Compositing (AC)*

Aerobic composting (AC) is a degradation process that utilizes microorganisms and oxygen to break down organic matter (OM). This process occurs in the open atmosphere, typically in piles or windrows. To facilitate rapid decomposition, green and brown materials are shredded to 2–3 cm or smaller sizes. The shredded material is then arranged in a pile with specific moisture content, and frequent turnings ensure proper mixing and aeration for microorganism survival (Kulikowska et al. 2015). Studies have demonstrated the effectiveness of AC in various settings. For instance, a mixture of poultry waste, feed waste, wood chips, and feathers was composted under aerated piles, resulting in significant changes during compost formation, including temperature increases due to mesophilic and thermophilic microorganisms. Additionally, composted poultry litter showed higher OM levels than uncomposted litter, reducing soil susceptibility to loss.

Comparative studies between aerated and turned windrows revealed similar compost characteristics, but turned windrows achieved higher temperatures and humification rates. Mechanical turning methods are preferred for large-scale applications, as they achieve higher temperatures and convert waste into valuable resources efficiently (Kalemelawa et al. 2012). Research using olive husks, chicken manure, wheat straw, and bamboo biochar has shown that adding biochar enhances porosity, stabilizes composting rates, and accelerates the process (Baeta-Hall et al. 2005; Cayuela et al. 2006; Liu et al. 2017). Biochar also reduces greenhouse gas emissions ($CO_2$, $N_2O$, $NH_3$, and $CH_4$) and improves compost quality (Tiquia and Tam 1998; O'Connor et al. 2018; Głąb et al. 2018; Zhang et al. 2020).

In-vessel composting offers advantages over windrow or pile composting, including reduced bioaerosols, better waste management, and control over leachate. Experiments using green waste, paper waste, and biosolids in bins demonstrated better results with maximum temperature achievement and humification (Roberts et al. 2007; Avidov et al. 2018). Sleeve composting has also been explored, utilizing green waste and sewage sludge. Maintaining oxygen levels through perforated pipes and controlling moisture content resulted in negligible harmful bacteria and a non-toxic final product (Liang et al. 2003).

Closed bioreactors (in-vessel composting) have been evaluated for various physicochemical parameters, demonstrating satisfactory final compost within 12 days. Correlations between nitrates, phosphates, and other parameters were observed. Comparative analyses of different aeration methods, including windrow and forced-air composting, revealed similar results (Brodie et al. 2000; Solano et al. 2001; Díaz et al. 2002; Makan et al. 2020). However, windrow composting achieves higher temperatures, albeit with complicated temperature control and labor-intensive manual turning. Mechanical turning saves time and reduces costs, producing superior compost in terms of humification, OM, and particle size.

## *3.3 Anerobic Composting (AnC)*

Anaerobic composting (AnC) is a microorganism-driven process that decomposes organic matter (OM) without oxygen utilization. This two-step process requires high moisture levels and nitrogen-rich materials, such as animal manure, food waste, and sewage sludge (Smet et al. 1999) (Table 2). AnC produces compost tea, a nutrient-rich liquid fertilizer, and prevents environmental pollution by utilizing generated methane as an energy source (Liu et al. 2019; Zhang et al. 2017). AnC has been applied to various feedstocks, including kitchen waste, fly ash, crop deposits, garden and animal waste, municipal waste, sawdust, pig manure, and sewage sludge (Zorpas and Loizidou 2008; Arrigoni et al. 2018). This process reduces greenhouse gas emissions, as it occurs in an oxygen-free environment (Xu et al. 2020). However, AnC has drawbacks, such as longer processing times, potential presence of harmful pathogens like E. coli and Salmonella, and odor production.

Two-stage composting is an innovative approach combining AnC and Aerobic Composting (AC) to enhance product quality, process speed, and environmental benefits. AnC serves as primary composting, followed by AC as secondary

**Table 2** Composting methods

| Composting method | Key characteristics | Benefits | Limitations |
|---|---|---|---|
| **Discrete waste composting (DW)** | Controlled process; nutrient-rich output; substitutes synthetic fertilizers | Bioremediation, disease/ weed control, pollution control, soil biodiversity | Requires optimal input and processing control |
| **Passive composting** | Windrow/pile-based; natural aeration; low-cost but manual turning needed | Cost-effective, simple, ideal for small farms | Inconsistent oxygen levels, rapid oxygen depletion |
| **Aerated static pile (ASP)** | Uses blowers for air; porous base; top layer absorbs odor and retains moisture | Fast decomposition, space-saving, good odor management | High setup cost, equipment and careful planning required |
| **Crop residue composting** | Lignin-rich materials; lime enhances decomposition | Enhances soil structure, turns waste into humus-rich compost | Lime requirement; lignin decomposition is slow |
| **Municipal solid waste composting** | High organic content; large-scale treatment; requires sorting and decontamination | Reduces MSW volume, nutrient recycling, sustainable urban waste management | Risk of toxic substances; composition variability |
| **Biomedical waste composting (BMW)** | Neem/tobacco extracts; NaOCl pre-treatment; uses cow dung and vermicomposting | Safe disposal, nutrient-rich fertilizer, energy-efficient | Needs pre-treatment, controlled environment |
| **Vermicomposting (VC)** | Earthworm-based; compost + compost tea production; climate-sensitive setup | Nutrient-rich compost, improved soil structure, liquid fertilizer benefit | Requires worm care, sensitive to moisture/temp changes |
| **Aerobic composting (AC)** | Oxygen-dependent microbes; requires regular turning; temperature-regulated process | Fast composting, reduces pathogens, improves OM content | Needs frequent maintenance and proper material balance |
| **Anaerobic composting (AnC)** | No oxygen; methane production; handles wet waste; slower process | Energy recovery (methane), nutrient-rich tea, reduced GHG emissions | Odors, slow rate, pathogen risk, high moisture requirement |

composting. Research has explored this method using various materials, including sewage sludge, grass, and rape straw (Obersky et al. 2018; Li et al. 2018; Meng et al. 2020). The entire process takes approximately 217 days, with 10 days in a bioreactor for OM oxidation and waste sanitation, followed by 207 days in windrow composting for maturation. Recent studies on two-stage composting focus on increasing process reliability and product consistency. While AnC has benefits, its limitations, such as longer processing times and potential pathogen presence, are mitigated by the subsequent AC stage (Troy et al. 2013; Mandpe et al. 2019). This integrated approach offers a promising solution for efficient and sustainable waste management.

# 4 Process of Compositing

The composting process for municipal organic waste encompasses multiple stages, focusing on effective feedstock management and optimal biochemical conversion. Initially, feedstock management involves the collection, sorting, screening, preprocessing, and mixing of organic waste. This critical step ensures the removal of contaminants, reduction of particle size, and creation of a homogeneous mixture (Hsu and Lo 1999; Niwagaba et al. 2009).

Subsequently, the processed feedstock undergoes biochemical conversion, comprising composting and maturation phases. Composting involves microorganism-driven decomposition, while maturation allows for further stabilization and refinement of the compost. This two-phase process transforms organic waste into a stable, nutrient-rich final product, suitable for agricultural and environmental applications.

## *4.1 Collection of Organic Material*

The initial step in organic waste management involves collecting organic materials, including food scraps, yard trimmings, and agricultural wastes, from households and business communities. Source separation, where waste is segregated at the point of generation, offers significant advantages over mechanical sorting. This approach minimizes contamination risks from plastics, metals, and personal care products, resulting in a cleaner feedstock.

The benefits of source separation include reduced processing requirements, improved carbon-to-nitrogen ratios, and optimized microbial activity. This, in turn, facilitates faster decomposition rates, lower ammonia emissions, and enhanced compost quality (Gaspar et al. 2023). By separating organic waste at the source, waste management systems can produce higher-quality compost, reducing environmental impacts and promoting sustainable waste management practices.

## *4.2 Sort and Screen*

Sorting and screening are crucial steps in preparing organic waste for composting. These procedures utilize mechanical systems, including shredders, trommels, and magnetic separators, to remove inorganic contaminants like plastics and metals. Advanced technologies, such as optical sensors (infrared and color sensors), enhance efficiency by identifying and separating waste paper, plastics, and recyclables (Rahman et al. 2014).

Recent innovations have introduced robotic systems leveraging artificial intelligence algorithms for automated waste sorting. These cutting-edge solutions aim to minimize human labor and boost sorting accuracy through robust detection and manipulation of waste items. By integrating these technologies, waste management facilities can optimize feedstock quality, streamlining the composting process and producing higher-quality end products (Kiyokawa et al. 2021).

## *4.3 Preprocessing of Organic Wastes and Bulking Agents*

Preprocessing of organic wastes and bulking agents is essential for optimal composting. Shredding or crushing operations reduce particle size, increase surface area, and create void spaces within the compost pile. This facilitates improved gas exchange, particularly oxygen diffusion, supporting the growth of thermophilic microorganisms that drive the decomposition of organic carbon (Singh et al. 2023).

## *4.4 Feedstock Management*

Effective composting relies heavily on proper feedstock management and mixing to achieve a balanced carbon-to-nitrogen (C/N) ratio, ideally between 25% and 35%. This balance is crucial for microbial growth and decomposition. To achieve this balance, carbon-rich "brown" materials like dry leaves, straw, sawdust, and paper are combined with nitrogen-rich "green" materials such as grass clippings, food scraps, manure, and coffee grounds (Wei et al. 2017; Reimer et al. 2020; Kulikowska and Bernat 2023; Nordahl et al. 2023).

The C/N ratio significantly impacts composting efficiency. A low ratio can lead to increased greenhouse gas emissions and ammonia release due to nitrogen mineralization. Conversely, a high ratio can result in reduced biological activity, slower degradation rates, inadequate heat generation, and prolonged composting periods (Lin et al. 2018; Gaspar et al. 2023). Adjusting the C/N ratio ensures optimal microbial activity, efficient decomposition, and minimized environmental impacts (Singh et al. 2023; Zhan et al. 2023).

## *4.5 Compositing*

Composting involves blending organic matter into a controlled mixture, which is then formed into outdoor windrows or placed in enclosed vessels for in-vessel composting. Microorganisms, including bacteria and fungi, utilize oxygen to break down organic matter within the compost pile. This decomposition process generates heat, triggering a thermophilic phase characterized by temperatures between 45 °C and 55 °C, occasionally reaching up to 80 °C. The elevated temperatures sanitize the final product by destroying pathogens and weed seeds (Zhao et al. 2022; Nordahl et al. 2023).

Maintaining optimal moisture levels, typically between 50% and 60%, is crucial for supporting microbial activity and preserving the physical properties of the compost pile. Regular turning of the compost pile ensures proper aeration and moisture distribution. This dual approach facilitates efficient decomposition, promotes healthy microbial growth, and yields a high-quality, sanitized compost product (Gaspar et al. 2023).

# 5 Factors Affecting the Process of Compositing

## 5.1 *Temperature (°C)*

Temperature plays a crucial role in the composting process, influencing microbial composition, density, and metabolic activity. Microorganisms require specific temperature ranges for optimal activity. The compost pile's center can reach temperatures of 140 °F (60 °C) due to microbial activity. If temperatures don't rise, anaerobic conditions and rotting occur. The composting process involves four temperature phases: mesophilic (40 °C), thermophilic (60 °C), cooling (40 °C), and maturation. Thermophilic temperatures (55 °C and above) eliminate pathogens, weed seeds, and fly larvae. Compost managers maintain temperatures below 65 °C to prevent bacterial die-off and slowed decomposition (Nozhevnikova et al. 2019).

During the active phase, microbial populations multiply, breaking down organic matter and generating heat. Temperatures rise from mesophilic to thermophilic ranges. In the cooling stage, temperatures drop as energy sources dwindle, and mesophilic bacteria grow. The maturation phase involves humification, adding value to the compost. Various composting techniques exist, including static aerated pile composting, turned composting, passive composting, and in-vessel composting. The choice of technique depends on factors like capital cost, labor cost, time, and land availability (Sayara et al. 2020).

Microorganisms, primarily from the phyla Firmicutes, Proteobacteria, Bacteroidetes, and Actinobacteria, participate in composting under varying temperature conditions. Mesophilic bacteria and fungi initiate decomposition, while thermophiles take over as temperatures rise. Specific microorganisms, such as Bacillus and Thermus species, play key roles in different stages.

Actinomycetes and bacteria from Enterobacteriales and Pseudomonadales promote further decomposition during the cooling and maturation stages. Microscopic fungi, like Aspergillus and Thermomyces, contribute to lignin breakdown. Filamentous fungi, such as Trichoderma and Chaetomium, degrade macromolecular substrates. Temperature, nutrient availability, and nitrogen accessibility influence fungal growth (Hassan et al. 2023). Hyphal development enables filamentous fungi to colonize solid substrates and utilize resources efficiently. Only specific fungi can break down lignin. Viruses and minute eukaryotes, including protozoa, may be present in compost material. Pathogenic viruses decrease significantly during the thermophilic stage (Nozhevnikova et al. 2019).

## 5.2 Carbon to Nitrogen Ratio (C/N)

The carbon-to-nitrogen (C/N) ratio plays a vital role in the composting process, significantly influencing microbial activity, degradation rates, and final product quality. Microorganisms require carbon as an energy source and nitrogen to synthesize amino acids, proteins, and nucleic acids. Ideally, they consume 15–30 times more carbon than nitrogen during aerobic fermentation.

Research indicates that optimal C/N ratios for composting range from 25 to 30. A ratio of approximately 30 ensures efficient carbon energy intake while promoting rapid microbial growth. Deviations from this range can impact composting duration; higher initial C/N ratios prolong the process. For instance, composting shrubs and wood, with recalcitrant carbon, takes around 18 months to mature, whereas household waste with a C/N ratio of 30 matures in just 7 months.

The C/N ratio's significance extends beyond process efficiency, as it also affects the characteristics of the finished compost. Maintaining an optimal ratio ensures balanced microbial activity, efficient nutrient cycling, and production of high-quality compost. Effective composting relies on striking this delicate balance between carbon and nitrogen availability.

## 5.3 Aeration

Adequate oxygen supply is crucial for aerobic composting, particularly in the initial stages. Insufficient oxygen hinders the growth of aerobic microorganisms, slowing down decomposition. Effective aeration addresses this need, providing oxygen while removing trapped gases, water vapor, and excess heat from the compost pile. This is especially vital in hot regions where overheating and fire risks are higher (Azim et al. 2018).

To ensure optimal aeration, compost managers regulate factors such as material quality (particle size and moisture content), pile size, ventilation, and turning frequency. By controlling these variables, composting facilities can maintain ideal conditions for microbial activity, efficient decomposition, and safe operating temperatures. Proper aeration is essential for successful composting.

## 5.4 Moisture

Microorganisms require water to facilitate nutrient and energy transport across cell membranes, making moisture a critical factor in composting. Maintaining optimal moisture levels can be challenging, especially in outdoor composting. Temperature monitoring helps resolve this issue by indicating when to turn, moisten, or ventilate the pile. Insufficient initial moisture (less than 30%) can cause dehydration, halting biological processes and resulting in physically stable but biologically unstable compost (Misra et al. 2003).

Conversely, high relative humidity (above 80%) leads to anaerobic conditions. Ideal moisture levels for composting materials range from 40% to 65%. Practically, starting with 50–60% moisture and finishing at 30% is advisable. During composting, microorganisms release metabolic water, increasing water content. However, temperature increases and forced ventilation reduce water content, releasing it as water vapor (Azim et al. 2018).

Moisture and aeration are interlinked, and optimal moisture content varies depending on the compost's physical composition and particle size. Each material's unique physical, chemical, and biological properties influence moisture relationships, making a universally applicable optimal moisture level impractical. Additional analysis of trace elements and physical characteristics is necessary to determine compost quality from municipal solid wastes. In summary, moisture management is crucial for effective composting, requiring careful balance to prevent dehydration or anaerobic conditions (Hassan et al. 2023). Monitoring temperature and adjusting moisture levels ensures optimal conditions for microbial activity and quality compost production.

## 5.5 Microorganism's Activity

Microbial activity in composting is significantly influenced by moisture content. Under dry conditions, activity decreases, while waterlogged conditions reduce aerobic activity due to limited air supply. Optimal moisture levels range from 40% to 60% on a bulk basis.

The composting process unfolds in distinct stages, driven by microbial development in response to temperature fluctuations. Initially, bacteria dominate, while fungi persist throughout, becoming predominant under specific conditions—water levels below 35% and temperatures above 60 °C. Actinomycetes emerge as key players during stabilization and curing, collaborating with fungi to break down resilient polymers.

## 5.6 Potential of Hydrogen (pH)

The pH level serves as a valuable indicator of composting progression, fluctuating throughout the various stages. Initially, pH decreases due to organic acid production, then increases during cooling and maturation as bacteria utilize these acids, eventually reaching neutral levels. This pH trend tracks compost maturation and stabilization. During the initial stage, pH may drop due to organic acid release and ammonia volatilization from simple organic substrate breakdown. Subsequent mineralization and degradation of easily degradable organics cause pH to rise. Nitrification can lead to an acidic pH finish. Raw materials, additions, and aeration conditions significantly impact pH variations (Antil et al. 2014).

Temperature influences pH evolution, promoting ammonia volatilization. In the thermophile phase, flora produces CO2 and organic acids (acid-genesis phase). Bacterial hydrolysis of protein and organic nitrogen generates ammonia, stabilizing pH (alkalization phase). Nitrogen is utilized to form humic compounds, while ammonia is lost through volatilization (Azim et al. 2018).

The composting process culminates in a stable phase, characterized by near-neutral pH. This stability results from the humus buffer's influence and sluggish reactions. Studies examining pH evolution in composting high-lignocellulosic materials reveal initial high pH levels, particularly during thermophilic phases, followed by dramatic decreases after 14 days, attributed to phenolic compound and organic acid synthesis (Wu et al. 2017). Composting's dynamic pH landscape underscores the importance of monitoring and managing factors like raw materials, aeration, and temperature to achieve optimal compost quality.

## 5.7 Lignin Content

Lignin is a key plant cell wall component; it poses significant resistance to microbial degradation due to its complex chemical composition. This structure has dual implications. On one hand, lignin reduces the bioavailability of other cell wall components, effectively lowering the actual carbon-to-nitrogen (C: N) ratio. On the other hand, lignin enhances porosity, creating an environment conducive to aerobic composting (Feng et al. 2011).

Interestingly, introducing lignin-decomposing fungi can have varying effects. In some cases, it may increase accessible carbon, accelerate composting, and minimize nitrogen loss. However, in other situations, it can lead to an elevated actual C: N ratio and compromised porosity, ultimately prolonging the composting process (Yang et al. 2021). This intricate relationship highlights the importance of carefully managing lignin's role in composting to optimize outcomes.

### 5.8 *Conductivity*

Compost conductivity measures the concentration of dissolved salts, providing valuable insight into its potential impact on plant health. Elevated salt levels can be detrimental to plants, making this assessment crucial. During the composting process, the breakdown of organic matter releases organic acids and soluble salts, increasing their concentration in water extracts. This rise in soluble salts serves as an indicator of compost stability and maturity (Antil et al. 2014; Parsa et al. 2018).

## 6 Agricultural Application of Compositing

Intensive agricultural practices, driven by growing demand for agricultural products, can degrade soil health and generate environmental issues. However, recycling organic waste through composting offers a sustainable solution for waste management (Li et al. 2018). Composting provides valuable organic matter to replenish depleted soil organic content, mitigating the impact of human activities. This approach improves various soil properties, enhancing soil quality (Bouajila and Sanaa 2011; Ye et al. 2019).

Effective fertilizer management, emphasizing organic fertilizers, is crucial for optimizing soil benefits. Compost application outcomes depend on soil- and compost- intrinsic properties, as well as application rates (Huang et al. 2011; Duong et al. 2012). Notably, the benefits of compost fertilizers may not be immediately apparent due to the gradual release of nutrients (Tittarelli et al. 2007). Subsequent sections will explore the diverse effects of compost application on soil properties.

### 6.1 *Nutrients Form in Compost*

Composting enhances soil fertility and mineralization, particularly in arid and semi-arid regions, where low precipitation and alkalinity prevail. Soil organic matter plays a vital role in maintaining fertility, reducing nutrient losses, and influencing characteristics such as water-holding capacity, productivity, biological activity, aeration, and structure (Rady et al. 2016). Nitrogen, primarily available in organic matter, is essential for plant growth. Compost, rich in decomposed organic waste, supplies crucial elements like nitrogen, carbon, sulfur, and phosphorus. Composting conditions and raw materials impact nutrient availability and humus stability.

Optimal compost composition requires balancing carbon and nitrogen levels. The C/N ratio should range from 25 to 30, with organic matter comprising 20–33.3% of the bio-compost. Moisture content should not exceed 40%, with 35% recommended for market compost. Ideal compost pH for agricultural applications is around 7.5. Compost quality standards also dictate salt content (approximately 3.89 g/L) and nutrient levels: phosphate (0.62% DM as $P_2O_5$) and potassium (1.01% DM). These guidelines ensure effective compost utilization in agriculture (Adugna 2016).

## 6.2 *Quality of Compost in Agriculture*

Compost quality is inherently linked to the characteristics of the raw materials, their nutritional composition, age, and production process. The nutritional balance and sufficiency of compost determine its suitability for various applications. Assessing compost quality involves evaluating physicochemical and agronomic parameters, as well as toxicity endpoints, such as pathogen content and seed germination (Chandna et al. 2013; Confesor et al. 2009; Neher et al. 2013).

Achieving advanced compost maturity and stability fosters the formation of humic substances, enhancing soil fertility and organic matter content without environmental risks (Cesaro et al. 2015; Jorge-Mardomingo et al. 2015). Stable nitrogenous compounds are less prone to volatilization, leaching, and denitrification, minimizing groundwater contamination. Nitrate, the preferred nitrogen form for plant growth, is more mobile but less likely to contaminate groundwater when compost is stable (Said-Pullicino et al. 2007; Ko et al. 2008; Muscolo et al. 2018).

However, immature compost can exhibit phytotoxicity due to organic acids, ammonia, and ethylene oxide production, hindering plant growth and seed germination (Ling et al. 2010). Research has shown that unstable organic matter correlates with lower germination indices. Moreover, applying immature compost can decrease soil oxygen levels, increase mineralization rates of organic carbon, and alter redox potential (Bertoncini et al. 2008; Perego et al. 2012). Optimizing compost maturity and stability is crucial for maximizing its agricultural benefits while minimizing environmental impacts (Alvarenga et al. 2015).

## 6.3 *Effects of Enzymatic and Biological Activities on Compost*

Compost serves as an ideal alternative fertilizer, profoundly impacting soil organic matter and microorganisms. Soil microbial diversity is crucial for nutrient cycling and biological processes, while enzymatic activities indicate the soil's biochemical functionality (Garcia et al. 2017). Organic fertilizers enhance these activities, with studies showing a 200% increase in β-glucosidase activity in amended soils. Manure application also boosts enzymatic activities and soil organic carbon (Bastida et al. 2007; Lupwayi et al. 2019).

Compost application effectively modifies soil microbial properties, augmenting biomass, respiration rates, and plant growth stages. Compost provides beneficial microorganisms, transforming insoluble matter into nutrients and degrading harmful substances. This improves soil conditions, supports micro- and macro-fauna biodiversity, and promotes rhizospheric microorganisms. Arbuscular mycorrhizal fungi

(AMF) are stimulated, increasing phosphorus absorption and the availability of limited elements. Organic amendment origin influences microbial activities, with compost exhibiting higher CO2 release rates. Composting offers additional benefits, including odor reduction and vector elimination (Cornejo et al. 2008; Li et al. 2015; Bünemann et al. 2018).

The long-term effects of organic fertilizer application are significant, enhancing soil biological properties like basal respiration, microbial biomass carbon, and enzymatic activities. The slow release of nutrients during organic matter decomposition sustains microbial biomass over extended periods (Clark and Zeto 2000; Oehl et al. 2004; Leon et al. 2006; Gosling et al. 2006; Dukare et al. 2011). Studies have shown that compost and manure-amended soils exhibit 20–40% higher microbial biomass carbon four years after application (Ginting et al. 2003; Medina et al. 2004; Zhen et al. 2014; Nannipieri et al. 2018). Overall, composting fosters soil health, supports biodiversity, and minimizes environmental impacts, making it a valuable strategy for sustainable agriculture (Paulin and Peter 2008).

## 6.4 Compost Effects on the Physical Properties of Soil

Soil structure plays a vital role in various soil processes, and compost addition significantly enhances soil health. Compost improves soil structure, facilitating gas and water transfer, seed germination, and root growth, while reducing erosion and water evaporation, and improving drainage (Osman 2012; Duong 2014; Novotná and Badalíková 2018). The formation of stable aggregates depends on organic matter content and dynamics, particularly in loam or clay soils. Compost application promotes aggregate stability, increasing soil fertility and structure. Studies show that compost enhances aggregate stability by 29–63% and improves soil structural stability (Annabi et al. 2007; Scotti et al. 2015). The degree of soil aggregate stability is influenced by compost application rate, maturity, intervals, feedstock, and soil type (Schmid and Murphy 2016).

Compost also reduces bulk density, providing a healthy soil environment. Increased soil porosity due to organic and inorganic fraction interactions contributes to this positive effect (Babalola et al. 2011). Research indicates that compost application decreases bulk density by 0.7–23% and reduces soil loss by 5–36%. Moreover, compost improves soil water retention, a critical factor for water use efficiency in agriculture (Kroulík et al. 2010; Vengadaramana and Justin 2012; Ozlu and Kumar 2018). Compost acts as a sponge in sandy soils, saving water and adds porosity to clay soils, preventing waterlogging. Compost application increases infiltration rates due to increased porosity, reduced bulk density, and microbiological activity (Martínez-Blanco et al. 2013; Logsdon et al. 2017; Zou et al. 2018).

Studies demonstrate significant increases in infiltration rates and water-holding capacity following compost application. For instance, compost addition increased infiltration rates by 18–25% and water-holding capacity by 50% and 34%, respectively (Carrizo et al. 2015). The benefits of compost in enhancing water-holding capacity are attributed to the amendment's water retention characteristics (Somerville et al. 2018). Overall, compost application is a promising natural fertilizer for addressing climate change and water scarcity challenges, particularly in arid and semi-arid regions (Brown and Cotton 2011; Skuras and Psaltopoulos 2012). By improving soil structure, fertility, and water retention, compost supports sustainable agriculture practices (Annabi et al. 2011; Kranz et al. 2020).

## 6.5 Compost Effects on the Chemical Properties of Soil

Soil physical structure and biological activity are significantly influenced by organic matter content, which determines soil suitability for agricultural activities. Compost addition enriches organic matter, enhancing soil fertility and nutrient levels (Choi 1999; Edwards and Hailu 2011). Compost contains essential macronutrients (N, P, K, Ca, S, Mg, C, O, and H) and trace elements (Fe, B, Cl, Mn, Zn, Cu, Mo, and Ni) vital for plant metabolism. The ideal compost C/N ratio ranges from 25 to 30, with carbon serving as an energy source and nitrogen essential for amino acid and protein synthesis. Research shows that compost application enhances nutrient cycling, improving sugarcane growth and increasing soil organic carbon and nitrogen levels (Johns 2015). Compost also increases microbial activity, with benefits lasting over time (Masmoudi et al. 2018).

Determining the optimal compost application rate depends on organic matter mineralization and nutrient release rates, influenced by environmental conditions and compost composition. Studies indicate that applying 7–10 Mg (dry matter) compost per hectare fulfills average soil organic matter demands, while exceeding 10 Mg promotes long-term increases. Replacing inorganic fertilizers with organic amendments benefits environmental protection by reducing nitrate leaching risks and greenhouse gas emissions. Compost application saves around 2282 kg $CO^{2\text{-}e}$ emissions per hectare over 20 years and sequesters carbon, accounting for 39% of greenhouse gas savings (Bhogal et al. 2007). Compost enhances cation exchange capacity (CEC), critical for nutrient retention, by accumulating compounds with negative charges. Increased CEC is observed with high compost application rates, improving soil fertility and reducing heavy metal immobilization (Gonzalez et al. 2010). Compost pH affects soil pH, with near-neutral or slightly alkaline composts elevating pH in acidic soils. However, stable composts rarely substantially increase soil pH due to low buffering capacity (Bendfeld et al. 2008; Taylor et al. 2016).

Long-term field trials demonstrate compost's equalizing effect on seasonal fluctuations in soil water, air, heat balance, and nutrient availability, enhancing crop yields. Compost tea positively impacts plant growth, essential oil content, and protein levels. Replacing 30% of N fertilizer with compost maintains maize yield and nitrogen uptake. Suitable compost application enhances maize yield parameters by 60% and improves soil fertility. Radish yield increases by 10% without phytotoxic effects. These findings underscore compost's significance in sustainable agriculture practices (Diacono and Montemurro 2010).

### 6.6 Compost Effects on Soil-Borne Pathogens and Diseases

Soil-borne pathogens pose a significant threat to agriculture, and conventional management methods are often expensive, environmentally harmful, and prone to pesticide resistance. Fortunately, composting offers a valuable solution (Moral et al. 2009). Mature composts have been shown to exhibit suppressive effects against phytopathogenic microorganisms, primarily due to the microbial activities within the compost. Compost can be utilized in various forms to control soil-borne pathogens, including raw compost, vermicompost, compost tea, and compost enriched with biopesticides (Li et al. 2023). The physical and chemical properties of compost play a crucial role in its suppressive effects, influencing the type and quantity of microorganisms present, as well as impacting pathogen growth, plant root health, and leaf nutrient status (Aviles et al. 2011; Singh et al. 2013).

Studies have demonstrated the efficacy of compost in inhibiting plant pathogens. For instance, humic substances extracted from compost suppressed the growth of Sclerotinia sclerotiorum, while stimulating the growth of beneficial fungi Trichoderma viride and Trichoderma harzianum (Ballardo et al. 2020). Compost's biotic and abiotic characteristics, including its water-soluble and humic fractions, have also been shown to reduce Pythium ultimum incidence in pea plants, minimizing root lesions and pathogen populations (Pascual et al. 2002). Compost teas have been effective in controlling various pathogens, including Rhizoctonia solani, Fusarium oxysporum, and Verticillium dahliae (Dionne et al. 2012; DeCorato 2020). The suppressive effects of compost are influenced by factors such as nutrient levels, particularly organic matter, and environmental conditions during maturation, which impact the recolonization of mesophilic microorganisms. By incorporating different types of compost, soil properties can be improved, leading to increased crop yields (Litterick et al. 2004; Alfano et al. 2011). Compost's multifaceted benefits make it a valuable tool in sustainable agriculture, offering a natural and effective solution to manage soil-borne pathogens and promote plant health.

## 7 Management Strategies of Compost

### 7.1 Tillage Managemont (TP)

Tillage practices (TP) significantly impact soil chemical, biological, and physical properties, influencing soil health and crop productivity. Various tillage methods, such as ploughing, strip tillage, and hand hoe tillage, have been evaluated for their effects on soil properties (Badagliacca et al. 2021; Barbosa 2020; Mihretie et al. 2022b; Nafi et al. 2020; Paye et al. 2023). Research shows that integrating ploughing with lime and phosphorus application can reduce soil pH (Njiru et al. 2023; Pan et al. 2023), extractable phosphorus, and exchangeable calcium, while conservation tillage practices like no-tillage, reduced tillage, mulching tillage, and strip-tillage enhance soil health and productivity (Liu et al. 2023c; Liu et al. 2022b; Sharma and Singh 2023).

Conservation tillage promotes soil accumulation, stability of soil aggregates, water infiltration, and mitigates erosion, leading to improved soil quality (Alam et al. 2024; Pramanik et al. 2024; Verma et al. 2024). No-tillage, in particular, enhances soil organic carbon, overall soil quality, soil temperature, organic matter content, nutrient availability, and microbial biomass (Toth et al. 2024; Zhang et al. 2024; Pi et al. 2024). Long-term no-tillage conserves soil moisture, maintains or increases organic matter, and improves crop production (Al-Shammary et al. 2023; Rocco et al. 2024). However, certain tillage practices, such as subsurface-sweeping and moldboard ploughing, can negatively impact soil chemical properties and reduce wheat yields (Madejón et al. 2023). Conventional tillage tends to accelerate organic matter decomposition and disrupt soil structure, negatively impacting biodiversity (Palsaniya et al. 2023; Hashimi et al. 2023). In contrast, conservation tillage promotes microbial abundance, diversity, and biodiversity by providing suitable habitats and food sources for soil organisms (Barbosa 2020; Liu et al. 2023a; Sharma and Singh 2023; Wang et al. 2020).

Tillage practices significantly influence biological properties, including soil microorganisms, organic matter decomposition, soil fauna diversity, enzyme activities, and overall biodiversity (Coulibaly et al. 2022; Denier et al. 2022). While conventional tillage exhibits higher enzymatic activity, conservation tillage promotes higher biodiversity (Malobane et al. 2020; Wen et al. 2023). Despite valuable insights from tillage practice studies, limitations such as variability, lack of control, time scale, observation scale, interactions with other management practices, generalizability, and absence of standardized protocols must be considered (Wang et al. 2020). Further research is necessary to fully understand the impacts of tillage practices on soil health and crop productivity.

## 7.2 Nutrients Management (NM)

Nutrient management (NM) plays a vital role in enhancing soil fertility and promoting sustainable agriculture, particularly in soil-crop systems. By implementing effective NM practices, farmers and land managers can improve soil fertility, mitigate environmental impacts, and foster sustainable agricultural methods (Chang et al. 2023; Naher et al. 2021; Zhang et al. 2023c). NM strategies encompass managing nutrient availability, balance, soil pH, cation exchange capacity, and salinity, while minimizing nutrient losses through leaching and runoff. Integrated NM approaches, combining organic manure and inorganic fertilizers, have been shown to enhance soil fertility and protect soil quality from degradation (Tiemann and Douxchamps 2023; Kong and Lu 2023; Zhang et al. 2023c; Dey et al. 2023; Naher et al. 2021; Somenahally et al. 2023; Tan et al. 2023; Melland et al. 2022). These holistic methods improve the availability of essential nutrients, such as nitrogen, phosphorus, potassium, and organic carbon, in soil. High-quality organic fertilizers have been found to increase wheat yields and improve soil chemistry (Ejaz et al. 2022; Tomar and Saikia 2022; Wang et al. 2023d; Madhurya et al. 2022; Tampio et al. 2022).

NM practices significantly influence soil biological properties, including microbial activity, soil biodiversity, organic matter decomposition, nutrient mineralization, soil structure formation, and overall soil health and resilience. Effective NM strategies promote soil vitality by supporting beneficial microorganisms, enhancing soil structure, and facilitating nutrient cycling. Research highlights the importance of integrated NM in maintaining soil fertility and sustainability (Neelima et al. 2022; Su et al. 2023; Borase et al. 2020; Zhang et al. 2023c; Koninger et al. 2021; Prack McCormick et al. 2022). For instance, studies have demonstrated that combined applications of organic manure and inorganic fertilizers improve soil fertility and reduce environmental degradation. Similarly, holistic NM approaches have been shown to enhance soil biodiversity, microbial activity, and nutrient availability.

Overall, adopting effective NM practices is crucial for maintaining soil health, promoting sustainable agriculture, and ensuring long-term ecosystem resilience. By understanding the complex relationships between NM and soil properties, farmers and land managers can implement targeted strategies to optimize soil fertility and mitigate environmental impacts (Reilly et al. 2023; Xie et al. 2022; Joshi et al. 2023; Neelima et al. 2022).

## *7.3 Crop Rotation Management (CR)*

Crop rotation (CR) plays a vital role in maintaining soil health and productivity, significantly influencing soil chemical and biological properties. By optimizing nutrient availability, organic matter content, pH balance, and nutrient balance, CR enhances soil fertility and mitigates pests and diseases (Liu et al. 2023c; Niether et al. 2023; Zhang et al. 2023a; Sun et al. 2022; Wu et al. 2024; Hanrahan et al. 2023; Jahan et al. 2016; Mwenda et al. 2023; Orek 2024). Research highlights the positive effects of CR on soil chemical properties, including improved nutrient cycling, increased organic matter inputs, and reduced soil degradation. CR also fosters a diverse and resilient soil microbial community, stimulating beneficial microorganisms and symbiotic relationships between legumes and nitrogen-fixing bacteria. Compared to monoculture systems, CR increases plant yields, reduces disease incidence, and improves soil microbial diversity. Effective implementation of CR restores soil fertility and health, mitigating the unilateral influence of plants on soil (Ma et al. 2024; Liu et al. 2023b, c; Theron et al. 2023; Wu et al. 2024; Ansari et al. 2024; Ma et al. 2023; Wang et al. 2024; Zhang et al. 2024).

Long-term studies demonstrate the benefits of CR, particularly when combined with organic fertilizers, on soil fertility, nutrient cycling, crop yields, and biodiversity. CR contributes to increased earthworm populations, improved soil structure, and enhanced ecosystem services. As a strategic approach in sustainable land management, CR advances agricultural sustainability, improves food security, and ensures increased food availability, accessibility, and utilization. By integrating CR into agricultural practices, farmers can promote soil health, reduce environmental degradation, and enhance ecosystem resilience (Yang et al. 2022; Denier et al. 2022; Zhao et al. 2023b). The multifaceted benefits of CR underscore its importance in maintaining soil productivity, fostering sustainable agriculture, and supporting global food security.

## *7.4 Organic Matter Management (OM)*

Organic matter (OM) management plays a critical role in shaping soil-crop systems, profoundly impacting soil health and fertility. Effective OM management strategies can significantly elevate soil organic carbon (SOC) content, influencing various chemical properties such as nutrient availability, cation exchange capacity, soil pH, fertility, and water-holding capacity. Research highlights the importance of OM in enhancing SOC content, nutrient cycling, and soil fertility. Distinct tillage systems, stubble management practices, and OM amendments like manure application and straw retention can improve soil chemical composition, increasing organic carbon, phosphorus, potassium, and magnesium levels (Fontúrbel et al. 2023; Gilmullina et al. 2023; Liu et al. 2024; Malone et al. 2023; Sao et al. 2023; Urbanski et al. 2023; Mabagala and Mng'ong'o 2022; Prats et al. 2023; Jeon et al. 2023; de Santana et al. 2022; Duan et al. 2022; Inagaki et al. 2023; Vedere et al. 2022). Moreover, OM management influences soil biological parameters, including microbial and enzyme activities, diversity, soil respiration, earthworm and soil fauna activity, disease suppression, and mycorrhizal associations.

Agronomic management practices, encompassing both organic and conventional approaches, can impact soil microarthropods, which are crucial for nutrient dynamics and soil microbial communities. Effective OM management holds substantial potential for improving soil-crop systems, contributing to sustainable agriculture practices and bolstering soil quality. By adopting optimal OM management strategies, farmers can enhance soil fertility, nutrient availability, water management, and overall soil health, ultimately supporting sustainable agriculture and environmental stewardship (Fang et al. 2023; Vedere et al. 2022; Jin et al. 2022, 2023; Miao et al. 2023; Shaaban et al. 2023; Ansari et al. 2023; Verrone et al. 2024; Hagner et al. 2023; Wen et al. 2023; Dignam et al. 2019; Ogundeji et al. 2022; Chen et al. 2023a; Yin et al. 2021). The multifaceted benefits of OM management underscore its importance in maintaining soil productivity, mitigating soil degradation, and promoting ecosystem resilience.

## 7.5 *Irrigation Management (IM)*

Irrigation management (IM) significantly impacts soil-crop systems through effective irrigation scheduling and techniques, influencing soil chemical and biological properties. Drip irrigation and deficit irrigation, for instance, affect soil salinity, pH, nutrient leaching, availability, erosion, and waterlogging. Well-managed irrigation scheduling helps mitigate nutrient losses via leaching or surface runoff (Abad et al. 2023; Ball et al. 2023; Díaz et al. 2021; Wang et al. 2023b; Zhang et al. 2023d).

Prudent IM implementation, including efficient water application, adequate drainage, and timely scheduling, is crucial for sustaining optimal agricultural conditions. However, using unconventional water resources like treated wastewater can elevate soil salinity and inhibit enzyme activities, posing threats to soil health and degradation. IM practices also impact soil properties like pH, cation exchange capacity, and organic matter content. Wastewater use introduces complexities in nutrient levels, including nitrogen, phosphorus, sodium, chloride, copper, zinc, and cadmium, risking heavy metal accumulation and soil quality degradation (Weldewahid et al. 2023; Zhang et al. 2022, 2023b; Ewald et al. 2022; Kama et al. 2023; Rath et al. 2021; Song et al. 2022; Bao et al. 2023; Majumdar et al. 2023; Moulia et al. 2023; Chen et al. 2023b; Zhu et al. 2024; Noia Júnior et al. 2023).

Furthermore, IM significantly influences biological properties, affecting microbial activity and diversity, organic matter decomposition, fauna activity, mycorrhizal associations, and disease dynamics. Adopting prudent IM practices is vital to maintaining soil health, preventing deterioration, and ensuring sustainable agriculture. Effective IM strategies can optimize soil fertility, nutrient availability, water management, and overall soil health, supporting sustainable agriculture and environmental stewardship. By understanding the complex relationships between IM and soil properties, farmers and land managers can implement targeted strategies to mitigate soil degradation and promote ecosystem resilience (Majumdar et al. 2023; Núnez and Schipanski 2023; Weldewahid et al. 2023; Benjlil et al. 2024; Ewald et al. 2022; Radini et al. 2023; Aganchich et al. 2022; Das et al. 2022; Mihretie et al. 2022a; Romero et al. 2022; Schattman et al. 2023).

## 7.6 *Mulching System Management (MS)*

Mulching (MS) significantly influences soil-crop systems through various mechanisms, primarily categorized into organic and inorganic materials. Organic mulching, using plant residues, wood chips, or compost, enhances soil chemical properties by increasing organic matter, nutrient availability, and water retention, while slightly affecting soil pH (Biswas et al. 2022; Palsaniya et al. 2023; Cai et al. 2023; Yang et al. 2023a, b). This contributes to improved nutrient cycling, fertility, and plant growth, as well as temperature regulation, weed suppression, and microbial activity. In contrast, inorganic mulches like plastic or gravel effectively inhibit weed growth, mitigate soil erosion, and retain moisture but have limited impact on soil organic matter concentration and biological activity (Biswas et al. 2022; Yang et al. 2023a; Zhang et al. 2023b; Micallef et al. 2023; Zhao et al. 2023a). While they may not directly enhance soil-crop system properties, appropriate mulching remains crucial for bolstering soil fertility, water use efficiency, and overall soil health (Gatea Al-Shammary et al. 2023; Palsaniya et al. 2023; Tuure et al. 2021; Wang et al. 2023a; McAmis et al. 2024).

Organic mulching positively affects soil biological properties, including microbial activity, organic matter accumulation, improved water and temperature conditions, erosion control, and increased earthworm activity. Incorporating organic mulches enriches soils with organic matter, fostering robust nutrient cycling and water retention (Carpena-Istan et al. 2024; Liu et al. 2021; Fan et al. 2023; Thakur and Kumar 2021; Xing et al. 2022). However, inorganic mulches may pose challenges to water infiltration under specific conditions. The impact of mulching regimes on soil-crop system properties varies depending on factors like mulch type and amount, climate, soil characteristics, and management strategies (Al-Shammary et al. 2020; Huang et al. 2023; Tesfuhuney et al. 2022; Bo et al. 2024). Effective mulching practices can optimize soil fertility, nutrient availability, water management, and overall soil health, supporting sustainable agriculture and environmental stewardship. By understanding the complex relationships between mulching and soil properties, farmers and land managers can implement targeted strategies to promote ecosystem resilience and mitigate soil degradation (Han et al. 2023; Huang et al. 2023; Al-Shammary et al. 2023; Liu et al. 2022a; Fan et al. 2023; Wang et al. 2023a).

## 8 Challenges and Potential Risks of Compost Application

Compost application in agriculture offers numerous benefits, including improved soil properties and fertility, enhanced crop yields, and reduced environmental pollution. However, to ensure safe and optimal compost application, understanding the best application rates and timing is crucial. Compost quality must meet specific characteristics, such as adequate organic matter content, minimal heavy metal concentrations, and optimal nutrient levels (Parchomenko and Borsky 2018). Despite its benefits, compost application poses potential risks, including increased soil salinity, heavy metal accumulation, nutrient leaching, and pathogen contamination (Pérez-Gimeno et al. 2015). These risks can be mitigated by considering factors like compost rate, soil type, application timing, and compost stability (Page et al. 2014).

Municipal organic waste composting offers significant environmental benefits, including reduced greenhouse gas emissions, extended landfill life, and decreased environmental pollution. Composting diverts organic waste from landfills, minimizing methane emissions and promoting carbon sequestration (Fagnano et al. 2011; Adamicová et al. 2017). This process also recycles nutrients, enhancing soil properties and reducing synthetic fertilizer use. Proper composting management practices, such as aeration and source separation, are essential for minimizing air pollutants and greenhouse gas emissions (Mukherjee and Reddy 2020). Effective composting can also mitigate water pollution by reducing leachate production and promoting soil health.

The economic benefits of composting include extended landfill life, reduced waste management costs, revenue generation from compost sales, and carbon credits. Medium-scale composting plants are often the most economically viable option, ensuring better process control and high-quality end-products. Compost-amended soils exhibit improved microbial communities, enhancing their resilience and promoting sustainable agriculture practices (Parra-Orobio et al. 2023). Efficient recycling of organic resources through composting contributes to soil functionality and fertility.

However, composting municipal organic waste also presents challenges, including potential environmental problems, operational difficulties, and management constraints. The high variability in organic waste composition affects compost quality, and impurities can complicate the composting process (Sileshi et al. 2019). Efficient source separation, pathogen reduction strategies, and thorough monitoring are necessary to ensure compost safety and quality. Bioaugmentation with specific microbial inoculants can enhance compost degradation and maturation, reducing pathogen loads. Despite these challenges, composting remains a vital sustainable waste management strategy, offering numerous benefits for soil amendment, waste reduction, and environmental protection (Hussain et al. 2022).

## 9 Conclusion

This comprehensive review underscores the critical role of agronomic management practices, including tillage, nutrient management, crop rotation, organic matter, irrigation, and mulching, in advancing sustainable agriculture and ensuring soil health. Effective implementation of these practices enhances soil fertility, nutrient cycling, and overall soil health, mitigating environmental degradation and promoting ecosystem resilience.

Composting emerges as a vital strategy for waste management, recycling nutrients, and improving soil quality. Proper composting techniques, consideration of site-specific conditions, and integration with other agronomic practices are essential for maximizing benefits. Scaling up sustainable agronomic practices and composting globally can significantly contribute to food security, climate change mitigation, soil restoration, and a resilient future. Further research is necessary to address existing challenges, optimize composting systems, and develop policy frameworks supporting widespread adoption of these environmentally beneficial practices.

## References

Abad FJ, Marín D, Imbert B, Virto I, Garbisu C, Santesteban LG (2023) Under-vine cover crops: impact on physical and biological soil properties in an irrigated Mediterranean vineyard. Sci Hortic 311:111797. https://doi.org/10.1016/j.scienta.2022.111797

Adamicová D, Radziemska M, Ridošková A, Bartoň S, Pelcová P, Elbl J, Kynický J, Brtnický M, Vaverková MD (2017) Environmental assessment of the effects of a municipal landfill on the content and distribution of heavy metals in Tanacetum vulgare L. Chemosphere 185:1011–1018. https://doi.org/10.1016/j.chemosphere.2017.07.034

Adugna G (2016) A review on impact of compost on soil properties, water use and crop productivity. Acad Res J Agric Sci Res 4(2):93–104

Aganchich B, Wahbi S, Yaakoubi A, El-Aououad H, Bota J (2022) Effect of arbuscular mycorrhizal fungi inoculation on growth and physiology performance of olive trees under regulated deficit irrigation and partial rootzone drying. S Afr J Bot 148:1–10. https://doi.org/10.1016/j.sajb.2022.02.005

Ajaweed AN, Hassan FM, Hyder NH (2022) Evaluation of physio-chemical characteristics of biofertilizer produced from organic solid waste using composting bins. Sustainability 14(7):4738. https://doi.org/10.3390/su14074738

Alam MJ, Islam MS, Mondol ATMAI, Naser HM, Salahin N, Alam MK, Islam MM, Akter S, Alam Z (2024) Cropping system-based fertilizer strategies for crop productivity and soil health under minimum tillage in grey terrace soil. Heliyon 10(1):e24106. https://doi.org/10.1016/j.heliyon.2023.e24106

Alfano G, Lustrato G, Lima G, Vitullo D, Ranalli G (2011) Characterization of composted olive mill wastes to predict potential plant disease suppressiveness. Biol Control 58(2):199–207. https://doi.org/10.1016/j.biocontrol.2011.04.015

Al-Shammary A, Kouzani A, Gyasi-Agyei Y, Gates W, Rodrigo-Comino J (2020) Effects of solarisation on soil thermal-physical properties under different soil treatments: a review. Geoderma 363:114137. https://doi.org/10.1016/j.geoderma.2020.114137

Al-Shammary AAG, Al-Shihmani LSS, Caballero-Calvo A, Fernández-Gálvez J (2023) Impact of agronomic practices on physical surface crusts and some soil technical attributes of two winter wheat fields in southern Iraq. J Soils Sediments:1–20

Alvarenga P, Mourinha C, Farto M, Santos T, Palma P, Sengo J, Morais M-C, Cunha-Queda C (2015) Sewage sludge, compost and other representative organic wastes as agricultural soil amendments: benefits versus limiting factors. Waste Manag 40:44–52. https://doi.org/10.1016/j.wasman.2015.01.027

Annabi M, Houot S, Francou C, Poitrenaud M, Le Bissonnais Y (2007) Soil aggregate stability improvement with urban composts of different maturities. Soil Sci Soc Am J 71(2):413–423. https://doi.org/10.2136/sssaj2006.0181

Annabi M, Le Bissonnais Y, Le Villio-Poitrenaud M, Houot S (2011) Improvement of soil aggregate stability by repeated applications of organic amendments to a cultivated silty loam soil. Agric Ecosyst Environ 144(1):382–389. https://doi.org/10.1016/j.agee.2011.06.015

Ansari J, Bardhan S, Eivazi F, Anderson SH, Mendis SS (2023) Bacterial community diversity for three selected land use systems as affected by soil moisture regime. Appl Soil Ecol 192:105100. https://doi.org/10.1016/j.apsoil.2022.105100

Ansari WA, Kumar M, Krishna R, Singh A, Zeyad MT, Tiwari P, Kumar SC, Chakdar H, Srivastava AK (2024) Influence of rice-wheat and sugarcane-wheat rotations on microbial diversity and plant growth promoting bacteria: insights from high-throughput sequencing and soil analysis. Microbiol Res 278:127533. https://doi.org/10.1016/j.micres.2023.127533

Antil RS, Raj D, Abdalla N, Inubushi K (2014) Physical, chemical and biological parameters for compost maturity assessment: a review. In: Composting for sustainable agriculture, pp 83–101

Arrigoni JP, Paladino G, Garibaldi LA, Laos F (2018) Inside the small-scale composting of kitchen and garden wastes: thermal performance and stratification effect in vertical compost bins. Waste Manag 76:284–293. https://doi.org/10.1016/j.wasman.2018.03.011

Avidov R, Saadi I, Krassnovsky A, Hanan A, Medina S, Raviv M, Chen Y, Laor Y (2017) Composting municipal biosolids in polyethylene sleeves with forced aeration: process control, air emissions, sanitary and agronomic aspects. Waste Manag 67:32–42. https://doi.org/10.1016/j.wasman.2017.05.011

Avidov R, Saadi I, Krasnovsky A, Medina S, Raviv M, Chen Y, Laor Y (2018) Using polyethylene sleeves with forced aeration for composting olive mill wastewater pre-absorbed by vegetative waste. Waste Manag 78:969–979. https://doi.org/10.1016/j.wasman.2018.06.033

Aviles M, Borrero C, Trillas M (2011) Review on compost as an inducer of disease suppression in plants grown in soilless culture. Dyn Soil Plant 5(1):1–11

Ayilara MS, Olanrewaju OS, Babalola OO, Odeyemi O (2020) Waste management through composting: challenges and potentials. Sustainability 12:4456. https://doi.org/10.3390/su12114456

Azim K, Soudi B, Boukhari S, Perissol C, Roussos S (2018) Composting parameters and compost quality: a literature review. Org Agric 8:141–158. https://doi.org/10.1007/s13165-017-0191-1

Babalola O, Adesodun J, Olasantan F, Adekunle A (2011) Responses of some soil biological, chemical and physical properties to short-term compost amendment. Int J Soil Sci 7:28–38. https://doi.org/10.3923/ijss.2011.28.38

Badagliacca G, Laudicina VA, Amato G, Badalucco L, Frenda AS, Giambalvo D, Ingraffia R, Plaia A, Ruisi P (2021) Long-term effects of contrasting tillage systems on soil C and N pools and on main microbial groups differ by crop sequence. Soil Tillage Res 211:104995. https://doi.org/10.1016/j.still.2021.104995

Baeta-Hall L, Sààgua MC, Bartolomeu ML, Anselmo AM, Rosa MF (2005) Bio-degradation of olive oil husks in composting aerated piles. Bioresour Technol 96:69–78. https://doi.org/10.1016/j.biortech.2004.05.014

Bagagiolo G, Vigoroso L, Pampuro N, Cavallo E (2022) The role of social interaction and personal characteristics in affecting the adoption of compost from organic fraction of municipal solid waste in Italy. Agronomy 12:445. https://doi.org/10.3390/agronomy12020445

Ball KR, Malik AA, Muscarella C, Blankinship JC (2023) Irrigation alters biogeochemical processes to increase both inorganic and organic carbon in arid-calcic cropland soils. Soil Biol Biochem 187:109189. https://doi.org/10.1016/j.soilbio.2022.109189

Ballardo C, Vargas-García MDC, Sánchez A, Barrena R, Artola A (2020) Adding value to home compost: biopesticide properties through bacillus thuringiensis inoculation. Waste Manag 106:32–43. https://doi.org/10.1016/j.wasman.2020.02.033

Bao H, Wu M, Meng X, Han H, Zhang C, Sun W (2023) Application of electrochemical oxidation technology in treating high-salinity organic ammonia-nitrogen wastewater. J Environ Chem Eng 11(5):110608. https://doi.org/10.1016/j.jece.2023.110608

Barbosa LAP (2020) Modelling the aggregate structure of a bulk soil to quantify fragmentation properties and energy demand of soil tillage tools in the formation of seedbeds. Biosyst Eng 197:203–215. https://doi.org/10.1016/j.biosystemseng.2020.06.008

Bastida F, Kandeler E, Hernandez T, Garcia C (2007) Long-term effect of municipal solid waste amendment on microbial abundance and humus-associated enzyme activities under semiarid conditions. Microb Ecol 55(4):651–661. https://doi.org/10.1007/s00248-007-9273-4

Bendfeld J, Tigges M, Splett M, Voss J (2008) GHG savings from biological treatment and application of compost. In: Proceedings of the 6th International Conference: ORBIT 2008 - Moving Organic Waste Recycling towards Resource Management and for Biobased Economy, pp 626–631

Benjlil H, Filali Alaoui I, Aït Hamza M, Braimi A, Oubidari T, Idhmida A, Ihitassen A, Tazi H, El Kherrak H, Paulitz T, Fossati-Gaschignard O, Ferji Z, Cherifi K, Mayad EH (2024) Nematodes associated with saffron II: bioindication for soil health assessment and impact of agricultural practices. Appl Soil Ecol 193:105111. https://doi.org/10.1016/j.apsoil.2023.105111

Bertoncini E, D'Orazio V, Senesi N, Mattiazzo M (2008) Effects of sewage sludge amendment on the properties of two Brazilian oxisols and their humic acids. Bioresour Technol 99(11):4972–4979. https://doi.org/10.1016/j.biortech.2007.09.063

Bher A, Cho Y, Auras R (2023) Boosting degradation of biodegradable polymers. Macromol Rapid Commun 44:2200769. https://doi.org/10.1002/marc.202200769

Bhogal A, Chambers BJ, Whitmore AP, Powlson DS (2007) The effect of reduced tillage practices and organic matter additions on the carbon content of arable soils (Scientific Report SP0561). Department of Environment, Food and Rural Affairs

Biruntha M, Karmegam N, Archana J, Karunai Selvi B, John Paul JA, Balamuralikrishnan B, Chang SW, Ravindran B (2020) Vermiconversion of biowastes with low-to-high C/N ratio into value-added vermicompost. Bioresour Technol 297:122398. https://doi.org/10.1016/j.biortech.2020.122398

Biswas T, Bandyopadhyay PK, Nandi R, Mukherjee S, Kundu A, Reddy P, Mandal B, Kumar P (2022) Impact of mulching and nutrients on soil water balance and actual evapotranspiration of irrigated winter cabbage (Brassica oleracea var. capitata L.). Agric Water Manag 263:107456. https://doi.org/10.1016/j.agwat.2022.107456

Bo L, Shukla MK, Mao X (2024) Long-term plastic film mulching altered soil physicochemical properties and microbial community composition in Shiyang River Basin, Northwest China. Appl Soil Ecol 193:105108. https://doi.org/10.1016/j.apsoil.2023.105108

Borase DN, Nath CP, Hazra KK, Senthilkumar M, Singh SS, Praharaj CS, Singh U, Kumar N (2020) Long-term impact of diversified crop rotations and nutrient management practices on soil microbial functions and soil enzymes activity. Ecol Indic 114:106322. https://doi.org/10.1016/j.ecolind.2020.106322

Bouajila K, Sanaa M (2011) Effects of organic amendments on soil physico-chemical and biological properties. J Mater Environ Sci 2(4):485–490

Brodie HL, Carr LE, Condon P (2000) A comparison of static pile and turned windrow methods for poultry litter compost production. Compost Sci Util 8(2):178–189. https://doi.org/10.1080/1065657X.2000.10702033

Brown S, Cotton M (2011) Changes in soil properties and carbon content following compost application: results of on-farm sampling. Compost Sci Util 19(2):87–96. https://doi.org/10.1080/1065657X.2011.10737030

Bünemann EK, Bongiorno G, Bai Z, Creamer RE, De Deyn G, De Goede R, Fleskens L, Geissen V, Kuyper TW, Mäder P et al (2018) Soil quality—a critical review. Soil Biol Biochem 120:105–125. https://doi.org/10.1016/j.soilbio.2017.12.015

Busato JG, Lima LS, Aguiar NO, Canellas LP, Olivares FL (2012) Changes in labile phosphorus forms during maturation of vermicompost enriched with phosphorus-solubilizing and diazotrophic bacteria. Bioresour Technol 110:390–395. https://doi.org/10.1016/j.biortech.2012.01.144

Cai S-s, Sun L, Wang W, Li Y, Ding J-l, Jin L, Li Y-m, Zhang J-m, Wang J-k, Wei D (2023) Straw mulching alters the composition and loss of dissolved organic matter in farmland surface runoff by inhibiting the fragmentation of soil small macroaggregates. J Integr Agric 23(5):1703–1717. https://doi.org/10.1016/S2095-3119(22)63653-6

Carpena-Istan V, Jurado MM, Estrella-Gonzalez MJ, Salinas J, Martinez-Gallardo MR, Toribio AJ, Lopez-Gonzalez JA, Suarez-Estrella F, Saez JA, Moral R, Lopez MJ (2024) Enhancing earthworm (Lumbricus terrestris) tolerance to plastic contamination through gut microbiome fortification with plastic-degrading microorganisms. J Hazard Mater 463:132836. https://doi.org/10.1016/j.jhazmat.2023.132836

Carrizo ME, Alesso CA, Cosentino D, Imhoff S (2015) Aggregation agents and structural stability in soils with different texture and organic carbon contents. Sci Agric 72(1):75–82. https://doi.org/10.1590/0103-9016-2014-0156

Cayuela M, Sánchez-Monedero M, Roig A (2006) Evaluation of two different aeration systems for composting two-phase olive mill wastes. Process Biochem 41(3):616–623. https://doi.org/10.1016/j.procbio.2005.08.012

Cesaro A, Belgiorno V, Guida M (2015) Compost from organic solid waste: quality assessment and European regulations for its sustainable use. Resour Conserv Recycl 94:72–79. https://doi.org/10.1016/j.resconrec.2014.12.007

Chandna P, Nain L, Singh S, Kuhad RC (2013) Assessment of bacterial diversity during composting of agricultural byproducts. BMC Microbiol 13(1):1–14. https://doi.org/10.1186/1471-2180-13-99

Chang T, Feng G, Paul V, Adeli A, Brooks JP, Jenkins JN (2023) Soil health assessment for different tillage and cropping systems to determine sustainable management practices in a humid region. Soil Tillage Res 233:105796. https://doi.org/10.1016/j.still.202

Chen P, Xie Q, Addy M, Zhou W, Liu Y, Wang Y, Cheng Y, Yanling C, Ruan R (2016) Utilization of municipal solid and liquid wastes for bioenergy and bioproducts production. Bioresour Technol 215:163–172. https://doi.org/10.1016/j.biortech.2016.04.065

Chen H, Levavasseur F, Montenach D, Lollier M, Morel C, Houot S (2022) An 18-year field experiment to assess how various types of organic waste used at European regulatory rates sustain crop yields and C, N, P, and K dynamics in a French calcareous soil. Soil Tillage Res 221:105415. https://doi.org/10.1016/j.still.2022.105415

Chen H-M, Shi F-X, Xu J-W, Liu X-P, Mao R (2023a) Tree mycorrhizal type controls over soil water-extractable organic matter quantity and biodegradation in a subtropical forest of southern China. For Ecol Manag 535:120900. https://doi.org/10.1016/j.foreco.2023.120900

Chen R, Li H, Wang J, Song Z (2023b) Critical factors influencing soil runoff and erosion in sprinkler irrigation: water application rate and droplet kinetic energy. Agric Water Manag 283: 108299. https://doi.org/10.1016/j.agwat.2023.108299

Choi K (1999) Optimal operating parameters in the composting of swine manure with wastepaper. J Environ Sci Health B 34(6):975–987. https://doi.org/10.1080/03601239909373235

Choudhary M, Panday SC, Meena VS, Singh S, Yadav RP, Mahanta D, Mondal T, Mishra PK, Bisht JK, Pattanayak A (2018) Long-term effects of organic manure and inorganic fertilization on sustainability and chemical soil quality indicators of soybean-wheat cropping system in the Indian mid-Himalayas. Agric Ecosyst Environ 257:38–46. https://doi.org/10.1016/j.agee.2018.01.029

Clark RB, Zeto SK (2000) Mineral acquisition by arbuscular mycorrhizal plants. J Plant Nutr 23(6): 867–902. https://doi.org/10.1081/PLN-100102946

Clay DE, Alverson R, Johnson JM, Karlen DL, Clay S, Wang MQ, Bruggeman S, Westhoff S (2019) Crop residue management challenges: a special issue overview. Agron J 111(1):1–3. https://doi.org/10.2134/agronj2018.06.0360

Coelho L, Osório J, Beltrão J, Reis M (2019) Efeito da aplicação de compostos orgânicos no controlo de infestantes na cultura de Stevia rebaudiana e nas propriedades do solo na região do Mediterrâneo. Revista Ciências Agrárias 42:111–120

Confesor R, Hamlett J, Shannon R, Graves R (2009) Potential pollutants from farm, food, and yard waste composts at differing ages: leaching potential of nutrients under column experiments. Part II Compost Sci Util 17(1):6–17. https://doi.org/10.1080/1065657X.2009.10702034

Cornejo P, Meier S, Borie G, Rillig MC, Borie F (2008) Glomalin-related soil protein in a Mediterranean ecosystem affected by a copper smelter and its contribution to cu and Zn sequestration. Sci Total Environ 406(1–2):154–160. https://doi.org/10.1016/j.scitotenv.2008.06.058

Coulibaly SFM, Aubert M, Brunet N, Bureau F, Legras M, Chauvat M (2022) Short-term dynamic responses of soil properties and soil fauna under contrasting tillage systems. Soil Tillage Res 215:105191. https://doi.org/10.1016/j.still.2021.105191

Das D, Bhattacharyya P, Ghosh B, Banik P (2016) Bioconversion and biodynamics of Eisenia foetida in different organic wastes through microbially enriched vermiconversion technologies. Ecol Eng 86:154–161. https://doi.org/10.1016/j.ecoleng.2015.11.009

Das D, Ullah H, Himanshu SK, Tisarum R, Cha-um S, Datta A (2022) Arbuscular mycorrhizal fungi inoculation and phosphorus application improve growth, physiological traits, and grain yield of rice under alternate wetting and drying irrigation. J Plant Physiol 278:153829. https://doi.org/10.1016/j.jplph.2022.153829

de Santana FB, Grunsky EC, Fitzsimons MM, Gallagher V, Daly K (2022) Diffuse reflectance mid-infrared spectroscopy combined with machine learning algorithms can differentiate spectral signatures in shallow and deeper soils for the prediction of pH and organic matter content. Catena 218:106552. https://doi.org/10.1016/j.catena.2022.106552

DeCorato U (2020) Agricultural waste recycling in horticultural intensive farming systems by on-farm composting and compost-based tea application improves soil quality and plant health: a review under the perspective of a circular economy. Sci Total Environ 738:139840. https://doi.org/10.1016/j.scitotenv.2020.139840

Dede C, Ozer H, Dede OH, Celebi A, Ozdemir S (2023) Recycling nutrient-rich municipal wastes into ready-to-use potting soil: An approach for sustainable resource circularity with inorganic porous materials. Horticulturae 9(3):203. https://doi.org/10.3390/horticulturae9030203

Denier J, Faucon M-P, Dulaurent A-M, Guidet J, Kervroëdan L, Lamerre J, Houben D (2022) Earthworm communities and microbial metabolic activity and diversity under conventional, feed and biogas cropping systems as affected by tillage practices. Appl Soil Ecol 169:104232. https://doi.org/10.1016/j.apsoil.2022.104232

Dey S, Purakayastha TJ, Sarkar B, Rinklebe J, Kumar S, Chakraborty R, Datta A, Lal K, Shivay YS (2023) Enhancing cation and anion exchange capacity of rice straw biochar by chemical modification for increased plant nutrient retention. Sci Total Environ 886:163681. https://doi.org/10.1016/j.scitotenv.2022.163681

Diacono M, Montemurro F (2010) Long-term effects of organic amendments on soil fertility. A review. Agron Sustain Dev 30(2):401–422. https://doi.org/10.1051/agro/2009044

Díaz M, Madejón E, López F, López R, Cabrera F (2002) Composting of vinasse and cotton gin waste by using two different systems. Resour Conserv Recycl 34(3):235–248. https://doi.org/10.1016/S0921-3449(01)00093-4

Díaz FJ, Sanchez-Hernandez JC, Notario JS (2021) Effects of irrigation management on arid soils enzyme activities. J Arid Environ 185:104330. https://doi.org/10.1016/j.jaridenv.2020.104330

Dignam BEA, O'Callaghan M, Condron LM, Raaijmakers JM, Kowalchuk GA, Wakelin SA (2019) Impacts of long-term plant residue management on soil organic matter quality, Pseudomonas community structure and disease suppressiveness. Soil Biol Biochem 135:396–406. https://doi.org/10.1016/j.soilbio.2019.04.017

Dinesh MS, Geetha KS, Vaishmavi V, Kale RD, Krishna-Murthy V (2010) Ecofriendly treatment of biomedical wastes using epigeic earthworms. J Indian Soc Hosp Waste Manag 9(1):5–20

Dionne A, Tweddell RJ, Antoun H, Avis TJ (2012) Effect of non-aerated compost teas on damping of pathogens of tomato. Can J Plant Pathol 34(1):51–57. https://doi.org/10.1080/07060661.2012.656014

Drózdz D, Malińska K, Kacprzak M, Mrowiec M, Szczypiór A, Postawa P, Stachowiak T (2020) Potential of fish pond sediments composts as organic fertilizers. Waste Biomass Valoriz 11(9): 5151–5163. https://doi.org/10.1007/s12649-020-01031-6

Duan Z, Tan X, Ali I, Wu X, Cao J, Xu Y, Shi L, Gao W, Ruan Y, Chen C (2022) Comparison of organic matter pools in water, suspended particulate matter, and sediments in eutrophic Lake Taihu, China: implication for dissolved OM tracking, assessment, and management. Sci Total Environ 845:157257. https://doi.org/10.1016/j.scitotenv.2022.157257

Dukare AS, Prasanna R, Dubey SC, Nain L, Chaudhary V, Singh R, Saxena AK (2011) Evaluating novel microbe amended composts as biocontrol agents in tomato. Crop Prot 30(3):436–442. https://doi.org/10.1016/j.cropro.2011.01.022

Duong TT (2014) Compost effects on soil properties and plant growth. EGU General Assembly:2013–2797

Duong TTT, Penfold C, Marschner P (2012) Differential effects of composts on properties of soils with different textures. Biol Fertil Soils 48(6):699–707. https://doi.org/10.1007/s00374-012-0685-3

Edwards S, Hailu A (2011) How to make compost and use. In: Ching LL, Edwards S, Nadia HS (eds) Climate change and food systems resilience in sub-Saharan Africa. FAO, pp 379–436

Ejaz MK, Aurangzaib M, Iqbal R, Shahzaman M, Habib-ur-Rahman M, El-Sharnouby M, Datta R, Alzuaibr FM, Sakran MI, Ogbaga CC, El Sabagh A (2022) The use of soil conditioners to ensure a sustainable wheat yield under water deficit conditions by enhancing the physiological and antioxidant potentials. Land 11(3):368. https://doi.org/10.3390/land11030368

Ewald M, Rusch D, Rißmann C, Trost B, Theuerl S, Ruess L (2022) Effects of irrigation and fertilization practice on soil nematode communities in arable land. Appl Soil Ecol 177:104546. https://doi.org/10.1016/j.apsoil.2022.104546

Fagnano M, Adamo P, Zampella M, Fiorentino N (2011) Environmental and agronomic impact of fertilization with composted organic fraction from municipal solid waste: a case study in the region of Naples, Italy. Agric Ecosyst Environ 141(1–2):100–107. https://doi.org/10.1016/j.agee.2011.02.018

Fan D, Jia G, Wang Y, Yu X (2023) The effectiveness of mulching practices on water erosion control: a global meta-analysis. Geoderma 438:116643. https://doi.org/10.1016/j.geoderma.2022.116643

Fang L, Lakshmanan P, Su X, Shi Y, Chen Z, Zhang Y, Sun W, Wu J, Xiao R, Chen X (2023) Impact of residual antibiotics on microbial decomposition of livestock manures in Eutric Regosol: implications for sustainable nutrient recycling and soil carbon sequestration. J Environ Sci (China)

Feng C, Zeng G, Huang D, Hu S, Zhao M, Lai C et al (2011) Effect of ligninolytic enzymes on lignin degradation and carbon utilization during lignocellulosic waste composting. Process Biochem 46(7):1515–1520. https://doi.org/10.1016/j.procbio.2011.04.015

Fontúrbel MT, Jiménez E, Merino A, Vega JA (2023) Contrasting immediate impact of prescribed fires and experimental summer fires on soil organic matter quality and microbial properties in the forest floor and mineral soil in Mediterranean black pine forest. Sci Total Environ:167669. https://doi.org/10.1016/j.scitotenv.2023.167669

Food and Agriculture Organization (FAO) (2021) Composting process and techniques (Martínez MM, Pantoja A (eds)). Food and Agriculture Organization (FAO)

Gai X, Liu H, Liu J, Zhai L, Wang H, Yang B, Ren T, Wu S, Lei Q (2019) Contrasting impacts of long-term application of manure and crop straw on residual nitrate-N along the soil profile in the North China Plain. Sci Total Environ 650:2251–2259. https://doi.org/10.1016/j.scitotenv.2018.10.072

Gajalakshmi S, Abbasi SA (2008) Solid waste management by composting: state of the art. Crit Rev Environ Sci Technol 38(3):311–400. https://doi.org/10.1080/10643380701413665

Garcia C, Hernandez T, Coll MD, Ondoño S (2017) Organic amendments for soil restoration in arid and semiarid areas: a review. AIMS Environ Sci 4:640–676

Garg V, Kaushik P (2005) Vermistabilization of textile mill sludge spiked with poultry droppings by an epigeic earthworm Eisenia foetida. Bioresour Technol 96(10):1063–1071. https://doi.org/10.1016/j.biortech.2004.09.007

Gaspar SS, Assis LLR, Ramos do Prado MP, Pedroso Miguel MG, dos Reis Ferreira GM, Schwan RF, Pasqual M, Rigobelo EC, Peron Castro R, Buttrós VH (2023) Diversity and enzymatic activity of the microbiota isolated from compost based on restaurant waste and yard trimmings. Front Sustain Food Syst 7:1013361. https://doi.org/10.3389/fsufs.2023.1013361

Gatea Al-Shammary AA, Lahmod NR, Fernández-Gálvez J, Caballero-Calvo A (2023) Effect of tillage systems combined with plastic film mulches and fertilizers on soil physical properties in a wheat-agricultural site in southern Iraq. Cuadernos de Investigación Geográfica 49(2):51–63. https://doi.org/10.18172/cig.4936

Ghosh M, Chattopadhyay G, Baral K (1999) Transformation of phosphorus during vermicomposting. Bioresour Technol 69(2):149–154. https://doi.org/10.1016/S0960-8524(98)00148-8

Gilmullina A, Rumpel C, Blagodatskaya E, Klumpp K, Bertrand I, Dippold MA, Chabbi A (2023) Is plant biomass input driving soil organic matter formation processes in grassland soil under contrasting management? Sci Total Environ 893:164550. https://doi.org/10.1016/j.scitotenv.2022.164550

Ginting D, Kessavalou A, Eghball B, Doran JW (2003) Greenhouse gas emissions and soil indicators four years after manure and compost applications. J Environ Qual 32(1):23–32. https://doi.org/10.2134/jeq2003.2300
Głąb T, Żabiński A, Sadowska U, Gondek K, Kopeć M, Mierzwa-Hersztek M, Tabor S (2018) Effects of co-composted maize, sewage sludge, and biochar mixtures on hydrological and physical qualities of sandy soil. Geoderma 315:27–35. https://doi.org/10.1016/j.geoderma.2017.11.021
Glawe U, Visvanathan C, Alamgir M (2005) Solid waste management in least developed Asian countries–a comparative analysis. In: Proceedings of the International Conference on Integrated Solid Waste Management in Southeast Asian Cities, Siem Reap, Cambodia
Gonawala SS, Jardosh H (2018) Organic waste in composting: a brief review. Int J Curr Eng Technol 8(1):36–38
Gonzalez M, Gomez E, Comese R, Quesada M, Conti M (2010) Influence of organic amendments on soil quality potential indicators in an urban horticultural system. Bioresour Technol 101(20): 8897–8901. https://doi.org/10.1016/j.biortech.2010.06.093
Gosling P, Hodge A, Goodlass G, Bending G (2006) Arbuscular mycorrhizal fungi and organic farming. Agric Ecosyst Environ 113(1–4):17–35. https://doi.org/10.1016/j.agee.2005.09.009
Grosu AC, Aldea CA, Matei F, Băbeanu NE (2023) Advances in grape marc composting: a review. Sci Bull Ser F Biotechnol 27(2)
Gunaruwan TL, Gunasekara WN (2016) Management of municipal solid waste in Sri Lanka: a comparative appraisal of the economics of composting. NSBM J Manag 2:27. https://doi.org/10.4038/nsbmjm.v2i0.27
Gupta R, Garg V (2008) Stabilization of primary sewage sludge during vermicomposting. J Hazard Mater 153(1–2):1023–1030. https://doi.org/10.1016/j.jhazmat.2007.09.047
Hagner M, Pohjanlehto I, Nuutinen V, Setälä H, Velmala S, Vesterinen E, Pennanen T, Lemola R, Peltoniemi K (2023) Impacts of long-term organic production on soil fauna in boreal dairy and cereal farming. Appl Soil Ecol 189:104944. https://doi.org/10.1016/j.apsoil.2022.104944
Han J, Yan X, Tang H (2023) Method of controlling tillage depth for agricultural tractors considering engine load characteristics. Biosyst Eng 227:95–106. https://doi.org/10.1016/j.biosystemseng.2022.12.003
Hanrahan BR, King KW, Rumora KR, Stinner JH (2023) Contrasting the influence of crop rotation on phosphorus balances and losses in agricultural fields across a tile-drained landscape in Ohio, USA. J Great Lakes Res. https://doi.org/10.1016/j.jglr.2023.102232
Hashim S, Waqas M, Rudra RP, Khan AA, Mirani AA, Sultan T, Ehsan F, Abid M, Saifullah M (2022) On-farm composting of agricultural waste materials for sustainable agriculture in Pakistan. Scientifica 2022:5831832. https://doi.org/10.1155/2022/5831832
Hashimi R, Huang Q, Dewi RK, Nishiwaki J, Komatsuzaki M (2023) No-tillage and rye cover crop systems improve soil water retention by increasing soil organic carbon in andosols under humid subtropical climate. Soil Tillage Res 234:105861. https://doi.org/10.1016/j.still.2022.105861
Hassan NYI et al (2023) Compositing: An eco-friendly solution for organic waste management to mitigate the effects of climate change. Innovare Journal of Social Sciences 11(4):1–7
Hrebecková T, Wiesnerová L, Hanˇc A (2019) Changes of enzymatic activity during a large-scale vermicomposting process with continuous feeding. J Clean Prod 239:118127. https://doi.org/10.1016/j.jclepro.2019.118127
Hsu JH, Lo SL (1999) Chemical and spectroscopic analysis of organic matter transformations during composting of pig manure. Environ Pollut 104(2):189–196. https://doi.org/10.1016/S0269-7491(98)00171-6
Huang W, Hall SJ (2017) Elevated moisture stimulates carbon loss from mineral soils by releasing protected organic matter. Nat Commun 8(1):1–10. https://doi.org/10.1038/s41467-017-01395-6
Huang PM, Li Y, Sumner ME (2011) Land application of wastes. In: Handbook of soil sciences. CRC Press, pp 620–645
Huang K, Xia H, Zhang Y, Li J, Cui G, Li F, Bai W, Jiang Y, Wu N (2020) Elimination of antibiotic resistance genes and human pathogenic bacteria by earthworms during vermicomposting of

dewatered sludge by metagenomic analysis. Bioresour Technol 297:122451. https://doi.org/10.1016/j.biortech.2020.122451

Huang F, Zhang Q, Wang L, Zhang C, Zhang Y (2023) Are biodegradable mulch films a sustainable solution to microplastic mulch film pollution? A biogeochemical perspective. J Hazard Mater 459:132024. https://doi.org/10.1016/j.jhazmat.2022.132024

Hussain M, Butt AR, Uzma F, Ahmed R, Irshad S, Rehman A, Yousaf B (2020) A comprehensive review of climate change impacts, adaptation, and mitigation on environmental and natural calamities in Pakistan. Environ Monit Assess 192(10):1–20. https://doi.org/10.1007/s10661-020-08351-6

Hussain MI, Khan ZI, Akhter P, Al-Hemaid FM, Al-Hashimi A, Elshikh MS, Ahmad K, Yang H-H (2022) Potential of organic amendments for heavy metal contamination in soil–coriander system: Environmental fate and associated ecological risk. Sustainability 14(16):11374. https://doi.org/10.3390/su141611374

Inagaki TM, Possinger AR, Schweizer SA, Mueller CW, Hoeschen C, Zachman MJ, Kourkoutis LF, Kögel-Knabner I, Lehmann J (2023) Microscale spatial distribution and soil organic matter persistence in top and subsoil. Soil Biol Biochem 178:108921. https://doi.org/10.1016/j.soilbio.2022.108921

Jahan MAHS, Hossain A, Sarkar MAR, Teixeira da Silva JA, Ferdousi MNS (2016) Productivity impacts and nutrient balances of an intensive potato-mungbean-rice crop rotation in multiple environments of Bangladesh. Agric Ecosyst Environ 231:79–97. https://doi.org/10.1016/j.agee.2016.06.023

Jeon I, Chung H, Kim SH, Nam K (2023) Use of clay and organic matter contents to predict soil pH vulnerability in response to acid or alkali spills. Heliyon 9(6):e17044. https://doi.org/10.1016/j.heliyon.2023.e17044

Jin W, Ge J, Shao S, Peng L, Xing J, Liang C, Chen J, Xu Q, Qin H (2022) Intensive management enhances mycorrhizal respiration but decreases free-living microbial respiration through its effect on microbial abundance and community in Moso bamboo forests. Pedosphere

Jin X, Cai J, Yang S, Li S, Shao X, Fu C, Li C, Deng Y, Huang J, Ruan Y, Li C (2023) Partial substitution of chemical fertilizer with organic fertilizer and slow-release fertilizer benefits soil microbial diversity and pineapple fruit yield in the tropics. Appl Soil Ecol 189:104974. https://doi.org/10.1016/j.apsoil.2022.104974

Johns C (2015) The chemical fertility of soils: soil nutrients and plant nutrition. Future Direction International Pty Ltd

Jorge-Mardomingo I, Jiménez-Hernández ME, Moreno L, DeLaLosa A, DeLaCruz MT, Casermeiro MÁ (2015) Application of high doses of organic amendments in a Mediterranean agricultural soil: an approach for assessing the risk of groundwater contamination. Catena 131:74–83. https://doi.org/10.1016/j.catena.2015.04.010

Joshi R, Nepal B, Sharma S, Begho T (2023) Nepalese farmers' perceptions of nitrogen inputs and attitudes to soil management: implications for soil health and environmentally sustainable farming. Soil Secur 12:100102. https://doi.org/10.1016/j.soilse.2022.100102

Kalamdhad AS, Singh YK, Ali M, Khwairakpam M, Kazmi A (2009) Rotary drum composting of vegetable waste and tree leaves. Bioresour Technol 100(22):6442–6450. https://doi.org/10.1016/j.biortech.2009.07.048

Kalemelawa F, Nishihara E, Endo T, Ahmad Z, Yeasmin R, Tenywa MM, Yamamoto S (2012) An evaluation of aerobic and anaerobic composting of banana peels treated with different inoculums for soil nutrient replenishment. Bioresour Technol 126:375–382. https://doi.org/10.1016/j.biortech.2012.09.055

Kama R, Liu Y, Zhao S, Hamani AKM, Song J, Cui B, Aidara M, Liu C, Li Z (2023) Combination of intercropping maize and soybean with root exudate additions reduces metal mobility in soil-plant system under wastewater irrigation. Ecotoxicol Environ Saf 266:115549. https://doi.org/10.1016/j.ecoenv.2022.115549

Kapoor J, Sharma S, Rana N (2015) Vermicomposting for organic waste management. Int J Rec Sci Res 6(7):7956–7960

Keng ZX, Tan JJM, Phoon BL, Khoo CC, Khoiroh I, Chong S, Supramaniam C, Singh A, Pan GT (2023) Aerated static pile composting for industrial biowastes: from engineering to microbiology. Bioengineering 10(8):938. https://doi.org/10.3390/bioengineering10080938

Kiyokawa T, Katayama H, Tatsuta Y, Takamatsu J, Ogasawara T (2021) Robotic waste sorter with agile manipulation and quickly trainable detector. IEEE Access 9:124616–124631. https://doi.org/10.1109/ACCESS.2021.3107448

Ko HJ, Kim KY, Kim HT, Kim CN, Umeda M (2008) Evaluation of maturity parameters and heavy metal contents in composts made from animal manure. Waste Manag 28(5):813–820. https://doi.org/10.1016/j.wasman.2007.05.012

Kong F, Lu S (2023) Inorganic amendments improve acidic paddy soils: effects on soil properties, Al fractions, and microbial communities. Chemosphere 331:138758. https://doi.org/10.1016/j.chemosphere.2022.138758

Koninger J, Lugato E, Panagos P, Kochupillai M, Orgiazzi A, Briones MJI (2021) Manure management and soil biodiversity: towards more sustainable food systems in the EU. Agric Syst 194:103251. https://doi.org/10.1016/j.agsy.2021.103251

Kranz CN, McLaughlin RA, Johnson A, Miller G, Heitman JL (2020) The effects of compost incorporation on soil physical properties in urban soils—a concise review. J Environ Manag 261:110209. https://doi.org/10.1016/j.jenvman.2020.110209

Kroulík M, Brant V, Masek J, Kovaříček P (2010) Influence of soil tillage treatment and compost application on soil properties and water infiltration. In: Trends in agricultural engineering, pp 343–349

Külcü R, Yaldiz O (2014) The composting of agricultural wastes and the new parameter for the assessment of the process. Ecol Eng 69:220–225. https://doi.org/10.1016/j.ecoleng.2014.03.091

Kulikowska D, Bernat K (2023) Composting of municipal sewage sludge and lignocellulosic waste: nitrogen transformations and humic substances molecular weight. Energies 16(2):376. https://doi.org/10.3390/en16020376

Kulikowska D, Gusiatin ZM, Bułkowska K, Klik B (2015) Feasibility of using humic substances from compost to remove heavy metals (Cd, Cu, Ni, Pb, Zn) from contaminated soil aged for different periods of time. J Hazard Mater 300:882–891. https://doi.org/10.1016/j.jhazmat.2015.07.069

Lelicińska-Serafiń K, Manczarski P, Rolewicz-Kalińska A (2023) An insight into post-consumer food waste characteristics as the key to an organic recycling method selection in a circular economy. Energies 16(5):1735. https://doi.org/10.3390/en16051735

Leon MCC, Stone A, Dick RP (2006) Organic soil amendments: impacts on snap bean common root rot (Aphanomyes euteiches) and soil quality. Appl Soil Ecol 31(3):199–210. https://doi.org/10.1016/j.apsoil.2005.09.004

Li J, Cooper JM, Lin Z, Li Y, Yang X, Zhao B (2015) Soil microbial community structure and function are significantly affected by long-term organic and mineral fertilization regimes in the North China Plain. Appl Soil Ecol 96:75–87. https://doi.org/10.1016/j.apsoil.2015.07.005

Li Y, Manandhar A, Li G, Shah A (2018) Life cycle assessment of integrated solid state anaerobic digestion and composting for on-farm organic residues treatment. Waste Manag 76:294–305. https://doi.org/10.1016/j.wasman.2018.03.014

Li Z, Zhang Q, Li Z, Qiao Y, Du K, Yue Z, Tian C, Leng P, Cheng H, Chen G, Li F (2023) Responses of soil CO2 emissions to tillage practices in a wheat-maize cropping system: a 4-year field study. Field Crop Res 294:108832. https://doi.org/10.1016/j.fcr.2022.108832

Liang C, Das KC, McClendon RW (2003) The influence of temperature and moisture contents regimes on the aerobic microbial activity of a biosolids composting blend. Bioresour Technol 86(2):131–137. https://doi.org/10.1016/S0960-8524(02)00174-7

Lin L, Xu F, Ge X, Li Y (2018) Improving the sustainability of organic waste management practices in the food-energy-water nexus: a comparative review of anaerobic digestion and composting. Renew Sust Energ Rev 89:151–167. https://doi.org/10.1016/j.rser.2018.03.065

Lin X, Chen C, Li H, Hei L, Zeng L, Wei Z, Chen Y, Wu Q (2022) Comprehensive recycling of fresh municipal sewage sludge to fertilize garden plants and achieve low carbon emission: a pilot study. Front Environ Sci 10:1023356. https://doi.org/10.3389/fenvs.2022.1023356

Ling N, Xue C, Huang Q, Yang X, Xu Y, Shen Q (2010) Development of a mode of application of bioorganic fertilizer for improving the biocontrol efficacy to Fusarium wilt. BioControl 55(5): 673–683. https://doi.org/10.1007/s10526-010-9296-7

Litterick AM, Harrier L, Wallace P, Watson CA, Wood M (2004) The role of uncomposted materials, composts, manures and compost extracts in reducing pest and disease incidence and severity in sustainable temperate agricultural and horticultural crop production—a review. Crit Rev Plant Sci 23(5):453–479. https://doi.org/10.1080/07352680490452861

Liu N, Zhou J, Han L, Ma S, Sun X, Huang G (2017) Role and multi-scale characterization of bamboo biochar during poultry manure aerobic composting. Bioresour Technol 241:190–199. https://doi.org/10.1016/j.biortech.2017.05.085

Liu H, Huang X, Yu X, Pu C, Sun Y, Luo W (2019) Dissipation and persistence of sulfonamides, quinolones and tetracyclines in anaerobically digested biosolids and compost during short-term storage under natural conditions. Sci Total Environ 684:58–66. https://doi.org/10.1016/j.scitotenv.2019.05.157

Liu T, Cheng J, Li XD, Shao M, Jiang C, Huang B, Zhu XC, Huang SH, Huang YL (2021) Effects of earthworm (Amynthas aspergillum) activities and cast mulching on soil evaporation. Catena 200:105104. https://doi.org/10.1016/j.catena.2021.105104

Liu B, Cheng X, He X, Bei Q, Dai Y, Wang Y, Zhu B, Zhang K, Tian X, Duan M, Xie X, Wang L (2022a) Effects of bio-mulching on wheat soil microbial community and carbon utilization efficiency in southwest China. Catena 214:106260. https://doi.org/10.1016/j.catena.2022.106260

Liu X, Liu H, Ren D, Liu C, Zhang Y, Wang S, Li Z, Zhang M (2022b) Interlinkages between soil properties and keystone taxa under different tillage practices on the North China Plain. Appl Soil Ecol 178:104551. https://doi.org/10.1016/j.apsoil.2022.104551

Liu K, Sozzi M, Gasparini F, Marinello F, Sartori L (2023a) Combining simulations and field experiments: effects of subsoiling angle and tillage depth on soil structure and energy requirements. Comput Electron Agric 214:108323. https://doi.org/10.1016/j.compag.2023.108323

Liu Q, Zhao Y, Li T, Chen L, Chen Y, Sui P (2023b) Changes in soil microbial biomass, diversity, and activity with crop rotation in cropping systems: a global synthesis. Appl Soil Ecol 186: 104815. https://doi.org/10.1016/j.apsoil.2022.104815

Liu Y, Chen S, Yu Q, Cai Z, Zhou Q, Bellingrath-Kimura SD, Wu W (2023c) Improving digital mapping of soil organic matter in cropland by incorporating crop rotation. Geoderma 438: 116620. https://doi.org/10.1016/j.geoderma.2022.116620

Liu C, Wu Z, He C, Zhang Y, Dong F, Huang W (2024) Effect of soil microbial community structure on the chemical compositions of different soil organic matter fractions in land uses of the Pearl River Estuary. Appl Soil Ecol 193:105126. https://doi.org/10.1016/j.apsoil.2023.105126

Logsdon SD, Sauer PA, Shipitalo M (2017) Compost improves urban soil and water quality. J Water Resour Protect 9(3):345–357. https://doi.org/10.4236/jwarp.2017.93025

Loh TC, Lee YC, Liang JB, Tan D (2005) Vermicomposting of cattle and goat manures by Eisenia foetida and their growth and reproduction performance. Bioresour Technol 96(1):111–114. https://doi.org/10.1016/j.biortech.2004.05.007

Lupwayi NZ, Zhang Y, Hao X, Thomas BW, Eastman AH, Schwinghamer TD (2019) Linking soil microbial biomass and enzyme activities to long-term manure applications and their nonlinear legacy. Pedobiologia 74:34–42. https://doi.org/10.1016/j.pedobi.2019.02.003

Ma X, Zhang Y, Wei F, Zhao L, Zhou J, Qi G, Ma Z, Zhu H, Feng H, Feng Z (2023) Applications of Chaetomium globosum CEF-082 improve soil health and mitigate the continuous cropping obstacles for Gossypium hirsutum. Ind Crop Prod 197:116586. https://doi.org/10.1016/j.indcrop.2022.116586

Ma T, Yang K, Yang L, Zhu Y, Jiang B, Xiao Z, Shuai K, Fang M, Gong J, Gu Z, Xiang P, Liu Y, Li J (2024) Different rotation years change the structure and diversity of microorganisms in the nitrogen cycle, affecting crop yield. Appl Soil Ecol 193:105123. https://doi.org/10.1016/j.apsoil.2023.105123

Mabagala FS, Mng'ong'o ME (2022) On the tropical soils; the influence of organic matter (OM) on phosphate bioavailability. Saudi J Biol Sci 29(5):3635–3641. https://doi.org/10.1016/j.sjbs.2022.03.017

Madejón P, Fernández-Boy E, Morales-Salmerón L, Navarro-Fernández CM, Madejón E, Domínguez MT (2023) Could conservation tillage increase the resistance to drought in Mediterranean faba bean crops? Agric Ecosyst Environ 349:108449. https://doi.org/10.1016/j.agee.2022.108449

Madhurya P, Latha M, Rao CS, Sree SP (2022) Effect of integrated nutrient management on physical and physico-chemical properties of soil. Int J Phys Soc Sci 34(24):1028–1033

Maina S, Kachrimanidou V, Koutinas A (2017) A roadmap towards a circular and sustainable bioeconomy through waste valorization. Curr Opin Green Sustain Chem 8:18–23. https://doi.org/10.1016/j.cogsc.2017.07.005

Majumdar A, Dubey PK, Giri B, Moulick D, Srivastava AK, Roychowdhury T, Bose S, Jaiswal MK (2023) Combined effects of dry-wet irrigation, redox changes and microbial diversity on soil nutrient bioavailability in the rice field. Soil Tillage Res 232:105752. https://doi.org/10.1016/j.still.2022.105752

Makan A, Fadili A, Oubenali M (2020) Interaction of physicochemical parameters during pressurized in-vessel composting of food waste. Bioresour Technol Rep 10:100350. https://doi.org/10.1016/j.biteb.2020.100350

Malinska K, Golańska M, Cáceres R, Rorat A, Weisser P, Ślęzak E (2017) Biochar amendment for integrated composting and vermicomposting of sewage sludge—the effect of biochar on the activity of Eisenia fetida and the obtained vermicompost. Bioresour Technol 225:206–214. https://doi.org/10.1016/j.biortech.2016.11.041

Malobane ME, Nciizah AD, Nyambo P, Mudau FN, Wakindiki IIC (2020) Microbial biomass carbon and enzyme activities as influenced by tillage, crop rotation and residue management in a sweet sorghum cropping system in marginal soils of South Africa. Heliyon 6(11):e05513. https://doi.org/10.1016/j.heliyon.2020.e05513

Malone Z, Berhe AA, Ryals R (2023) Impacts of organic matter amendments on urban soil carbon and soil quality: a meta-analysis. J Clean Prod 419:138148. https://doi.org/10.1016/j.jclepro.2022.138148

Mandpe A, Paliya S, Kumar S, Kumar R (2019) Fly ash as an additive for enhancing microbial and enzymatic activities in invessel composting of organic wastes. Bioresour Technol 293:122047. https://doi.org/10.1016/j.biortech.2019.122047

Manirakiza N, Seker C (2020) Effects of compost and biochar amendments on soil fertility and crop growth in a calcareous soil. J Plant Nutr 43(12):3002–3019. https://doi.org/10.1080/01904167.2020.1792425

Manzoor J, Sharma M (2019) Impact of biomedical waste on environment and human health. Environ Claims J 31(3):311–334. https://doi.org/10.1080/10440092.2019.1644141

Martin CCS, Brathwaite RA (2012) Compost and compost tea: principles and prospects as substrates and soil-borne disease management strategies in soil-less vegetable production. Biol Agric Horticult 28(1):1–33. https://doi.org/10.1080/01448765.2012.647436

Martínez-Blanco J, Lazcano C, Christensen TH, Muñoz P, Rieradevall J, Møller J, Antón A (2013) Compost benefits for agriculture evaluated by life cycle assessment. A review. Agron Sustain Dev 33(3):721–732. https://doi.org/10.1007/s13593-013-0138-1

Masmoudi S, Magdich S, Rigane H, Medhioub K, Rebai A, Ammar E (2018) Effects of compost and manure application rate on the soil Physico-chemical layers properties and plant productivity. Waste Biomass Valoriz 11(5):1883–1894. https://doi.org/10.1007/s12649-018-0309-9

McAmis S, Bae H, Ogram A, Rathinasabapathi B, Richter BS (2024) Living mulches present tradeoffs between soil nutrient cycling and competition during establishment of tea in an organic production system. Appl Soil Ecol 198:105350. https://doi.org/10.1016/j.apsoil.2023.105350

Medina A, Vassilev N, Alguacil M, Roldán A, Azcón R (2004) Increased plant growth, nutrient uptake, and soil enzymatic activities in a desertified Mediterranean soil amended with treated residues and inoculated with native mycorrhizal fungi and a plant growth-promoting yeast. Soil Sci 169(4):260–270. https://doi.org/10.1097/01.ss.0000125323.35154.7f

Meena K, Renu B (2009) Vermitechnology for sewage sludge recycling. J Hazard Mater 161(2–3): 948–954. https://doi.org/10.1016/j.jhazmat.2008.04.052

Melland AR, Bosomworth B, Cook FJ, Silburn DM, Eyles M (2022) Impacts of sugarcane (Saccharum sp.) soil and fertiliser management practices on nutrients and sediment in plot-scale runoff from simulated rainfall. Soil Tillage Res 216:105259. https://doi.org/10.1016/j.still.2021.105259

Meng L, Li W, Zhang S, Wu C, Lv L (2017) Feasibility of co-composting of sewage sludge, spent mushroom substrate and wheat straw. Bioresour Technol 226:39–45. https://doi.org/10.1016/j.biortech.2016.12.014

Meng X, Yan J, Zuo B, Wang Y, Yuan X, Cui Z (2020) Full-scale of composting process of biogas residues from corn stover anaerobic digestion: physical-chemical, biology parameters and maturity indexes during whole process. Bioresour Technol 302:122742. https://doi.org/10.1016/j.biortech.2020.122742

Miao J, Ji M, Xiao L, Liu F, Wu M, Sang W (2023) Unraveling the fascinating connection between hydrochar feedstock and methane emissions in rice paddy soil: insights from microorganisms and organic matter. Chem Eng J 472:144957. https://doi.org/10.1016/j.cej.2023.144957

Micallef SA, Callahan MT, McEgan R, Martinez L (2023) Soil microclimate and persistence of foodborne pathogens Escherichia coli O157:H7, Listeria monocytogenes, and Salmonella enterica newport in soil affected by mulch type. J Food Prot 86(11):100159. https://doi.org/10.4315/JFP-22-046

Michel F, O'Neill T, Rynk R, Gilbert J, Smith M, Aber J, Keener H (2022) Forced aeration composting, aerated static pile, and similar methods. In: The composting handbook. Academic Press, pp 197–269. https://doi.org/10.1016/B978-0-12-819645-1.00011-4

Mihretie FA, Tesfaye K, Hoogenboom G, Tsunekawa A, Molla A, Ebabu K, Sato S, Masutomi Y (2022a) Identifying low risk and profitable crop management practices for irrigated Teff production in northwestern Ethiopia. Eur J Agron 139:126572. https://doi.org/10.1016/j.eja.2022.126572

Mihretie FA, Tsunekawa A, Haregeweyn N, Adgo E, Tsubo M, Ebabu K, Masunaga T, Kebede B, Meshesha DT, Tsuji W, Bayable M, Berihun ML (2022b) Tillage and crop management impacts on soil loss and crop yields in northwestern Ethiopia. Int Soil Water Conserv Res 10(1):75–85. https://doi.org/10.1016/j.iswcr.2021.12.003

Misra, R. V., Roy, R. N., & Hiraoka, H. (2003). On-farm composting methods. Food and Agriculture Organization of the United Nations

Moral R, Paredes C, Bustamante M, Marhuenda-Egea F, Bernal M (2009) Utilisation of manure composts by high-value crops: safety and environmental challenges. Bioresour Technol 100(12):5454–5460. https://doi.org/10.1016/j.biortech.2009.03.064

Moulia V, Ait-Mouheb N, Lesage G, Hamelin J, Wéry N, Bru-Adan V, Kechichian L, Heran M (2023) Short-term effect of reclaimed wastewater quality gradient on soil microbiome during irrigation. Sci Total Environ 901:166028. https://doi.org/10.1016/j.scitotenv.2022.166028

Mukherjee A, Reddy M (2020) Metatranscriptomics: An approach for retrieving novel eukaryotic genes from polluted and related environments. 3 Biotech 10(2):71. https://doi.org/10.1007/s13205-020-0205-8

Muscolo A, Papalia T, Settineri G, Mallamaci C, Jeske-Kaczanowska A (2018) Are raw materials or composting conditions and time that most influence the maturity and/or quality of composts? Comparison of obtained composts on soil properties. J Clean Prod 195:93–101. https://doi.org/10.1016/j.jclepro.2018.05.155

Mwenda E, Muange EN, Ngigi MW, Kosgei A (2023) Impact of ICT-based pest information services on tomato pest management practices in the central highlands of Kenya. Sustain Technol Entrep 2(2):100036. https://doi.org/10.1016/j.stie.2023.100036

Nafi E, Webber H, Danso I, Naab JB, Frei M, Gaiser T (2020) Interactive effects of conservation tillage, residue management, and nitrogen fertilizer application on soil properties under maize-cotton rotation system on highly weathered soils of West Africa. Soil Tillage Res 196:104473. https://doi.org/10.1016/j.still.2020.104473

Naher UA, Haque MM, Khan FH, Ullah Sarkar MI, Ansari TH, Hossain MB, Chandra Biswas J (2021) Effect of long-term nutrient management practices on soil health and paddy yield of rice-rice-fallow cropping system in tropic humid climate. Eur J Soil Biol 107:103362. https://doi.org/10.1016/j.ejsobi.2021.103362

Nannipieri P, Trasar-Cepeda C, Dick RP (2018) Soil enzyme activity: a brief history and biochemistry as a basis for appropriate interpretations and meta-analysis. Biol Fertil Soils 54(1):11–19. https://doi.org/10.1007/s00374-017-1245-8

Neelima S, Babhulakar VP, Bhoyar KD, Wankhade MG, Shende TS (2022) Impact of nutrient management on soil chemical properties and maize (Zea mays L.) yield. Int J Environ Clim Change 12(12):1863–1870

Neher DA, Weicht TR, Bates ST, Le JW, Fierer N (2013) Changes in bacterial and fungal communities across compost recipes, preparation methods, and composting times. PLoS One 8(10):e79512. https://doi.org/10.1371/journal.pone.0079512

Niether W, Macholdt J, Schulz F, Gattinger A (2023) Yield dynamics of crop rotations respond to farming type and tillage intensity in an organic agricultural long-term experiment over 24 years. Field Crop Res 303:109131. https://doi.org/10.1016/j.fcr.2023.109131

Niwagaba C, Nalubega M, Vinnerås B, Sundberg C, Jonsson H (2009) Bench-scale composting of source-separated human faeces for sanitation. Waste Manag 29(2):585–589. https://doi.org/10.1016/j.wasman.2008.07.016

Njiru LG, Yegon JR, Mwithiga G, Micheni A, Gitari NJ, Mairura FS (2023) Restoring soil nutrient stocks using local inputs, tillage and sorghum-green gram intercropping strategies for drylands in Eastern Kenya. Heliyon 9(10):e20926. https://doi.org/10.1016/j.heliyon.2023.e20926

Noia Júnior R d S, Asseng S, García-Vila M, Liu K, Stocca V, dos Santos Vianna M, Weber TKD, Zhao J, Palosuo T, Harrison MT (2023) A call to action for global research on the implications of waterlogging for wheat growth and yield. Agric Water Manag 284:108334. https://doi.org/10.1016/j.agwat.2023.108334

Nordahl S, Preble C, Kirchstetter T, Scown C (2023) Greenhouse gas and air pollutant emissions from composting. Environ Sci Technol 57(4):2235–2247. https://doi.org/10.1021/acs.est.2c06369

Novotná J, Badalíková B (2018) The soil structure changes under varying compost dosage. Agriculture 64(4):143–148. https://doi.org/10.3390/agriculture6040143

Nozhevnikova AN, Mironov VV, Botchkova EA, Litti YV, Russkova YI (2019) Composition of a microbial community at different stages of composting and the prospects for compost production from municipal organic waste. Appl Biochem Microbiol 55(2):199–208. https://doi.org/10.1134/S0003683819020126

Núnez A, Schipanski M (2023) Changes in soil organic matter after conversion from irrigated to dryland cropping systems. Agric Ecosyst Environ 347:108392. https://doi.org/10.1016/j.agee.2022.108392

O'Connor D, Peng T, Zhang J, Tsang DC, Alessi DS, Shen Z, Bolan N, Hou D (2018) Biochar application for the remediation of heavy metal polluted land: a review of in situ field trials. Sci Total Environ 619:815–826. https://doi.org/10.1016/j.scitotenv.2017.11.146

Obersky L, Rafiee R, Cabral AR, Golding SD, Clarke WP (2018) Methodology to determine the extent of anaerobic digestion, composting and CH4 oxidation in a landfill environment. Waste Manag 76:364–373. https://doi.org/10.1016/j.wasman.2018.03.034

Oehl F, Sieverding E, Dubois D, Ineichen K, Boller T, Wiemken A (2004) Impact of long-term conventional and organic farming on the diversity of arbuscular mycorrhizal fungi. Oecologia 138(4):574–583. https://doi.org/10.1007/s00442-003-1475-1

Ogundeji AO, Meng L, Cheng Z, Hou J, Yin T, Zhang S, Liu X, Liu X, Li S (2022) Integrated crop practices management stimulates soil microbiome for Verticillium wilt suppression. Eur J Agron 140:126594. https://doi.org/10.1016/j.eja.2022.126594

Oliveira LS, Oliveira DS, Bezerra BS, Pereira BS, Battistelle RAG (2017) Environmental analysis of organic waste treatment focusing on composting scenarios. J Clean Prod 155:229–237. https://doi.org/10.1016/j.jclepro.2017.04.088

Orek C (2024) A review of management of major arthropod pests affecting cassava production in Sub-Saharan Africa. Crop Prot 175:106465. https://doi.org/10.1016/j.cropro.2023.106465

Osman KT (2012) Physical properties of soil. In: Soils. Springer, pp 49–65

Ozlu E, Kumar S (2018) Response of soil organic carbon, pH, electrical conductivity, and water stable aggregates to Long-term annual manure and inorganic fertilizer. Soil Sci Soc Am J 82(5): 1243–1251. https://doi.org/10.2136/sssaj2018.02.0072

Padmavathiamma PK, Li LY, Kumari UR (2008) An experimental study of vermi-biowaste composting for agricultural soil improvement. Bioresour Technol 99(10):1672–1681. https://doi.org/10.1016/j.biortech.2007.04.037

Page K, Harbottle M, Cleall P, Hutchings T (2014) Heavy metal leaching and environmental risk from the use of compost-like output as an energy crop growth substrate. Sci Total Environ 487: 260–271. https://doi.org/10.1016/j.scitotenv.2014.04.015

Palsaniya DR, Kumar TK, Chaudhary M, Choudhary M, Prasad M, Kumar S (2023) Tillage practices and mulching affect system productivity, profitability and energy use in Sesbania alley based food-fodder systems under rainfed agro-ecosystems of semi-arid tropics. Field Crop Res 302:109104. https://doi.org/10.1016/j.fcr.2023.109104

Pan P, Qi Z, Zhang T, Ma L (2023) Modeling phosphorus losses to subsurface drainage under tillage and compost management. Soil Tillage Res 227:105587. https://doi.org/10.1016/j.still.2022.105587

Pane C, Palese AM, Celano G, Zaccardelli M (2014) Effects of compost tea treatments on productivity of lettuce and kohlrabi systems under organic cropping management. Ital J Agron 9(3):153. https://doi.org/10.4081/ija.2014.641

Pane C, Spaccini R, Piccolo A, Celano G, Zaccardelli M (2019) Disease suppressiveness of agricultural greenwaste composts as related to chemical and bio-based properties shaped by different on-farm composting methods. Biol Control 137:104026. https://doi.org/10.1016/j.biocontrol.2019.104026

Papafilippaki A, Paranychianakis NV, Nikolaidis NP (2015) Effects of soil type and municipal solid waste compost as soil amendment on Cichorium spinosum (spiny chicory) growth. Sci Hortic 195:195–205. https://doi.org/10.1016/j.scienta.2015.09.012

Parchomenko A, Borsky S (2018) Identifying phosphorus hot spots: a spatial analysis of the phosphorus balance as a result of manure application. J Environ Manag 214:137–148. https://doi.org/10.1016/j.jenvman.2018.02.074

Parra-Orobio BA, Soto-Paz J, Hernández-Cruz JA, Gómez-Herreño MC, Domínguez-Rivera IC, Oviedo-Ocaña ER (2023) Evaluation of biochar as an additive in the co-composting of green waste and food waste. Sustainability 15(11):7437. https://doi.org/10.3390/su15117437

Parsa N, Khajouei G, Masigol M, Hasheminejad H, Moheb A (2018) Application of electrodialysis process for reduction of electrical conductivity and COD of water contaminated by composting leachate. Civil Eng J 4(5):1034–1045

Pascual JA, Garcia C, Hernandez T, Lerma S, Lynch JM (2002) Effectiveness of municipal waste compost and its humic fraction in suppressing Pythium ultimum. Microb Ecol 44(1):59–68. https://doi.org/10.1007/s00248-001-1051-9

Patil PM, Mahamuni PP, Shadija PG, Bohara RA (2019) Conversion of organic biomedical waste into value added product using green approach. Environ Sci Pollut Res 26(7):6696–6705. https://doi.org/10.1007/s11356-019-04267-7

Paulin, B., & Peter, O. M. (2008). Compost production and use in horticulture. Western Australian Agricultural Authority

Paye WS, Thapa VR, Ghimire R (2023) Limited impacts of occasional tillage on dry aggregate size distribution and soil carbon and nitrogen fractions in semi-arid drylands. Int Soil Water Conserv Res

Perego A, Basile A, Bonfante A, De Mascellis R, Terribile F, Brenna S, Acutis M (2012) Nitrate leaching under maize cropping systems in Po Valley (Italy). Agric Ecosyst Environ 147:57–65. https://doi.org/10.1016/j.agee.2011.10.021

Pérez-Gimeno A, Navarro-Pedreño J, Almendro-Candel MB, Gómez I, Jordan M (2015) Environmental consequences of the use of sewage sludge compost and limestone outcrop residue for soil restoration: salinity and trace elements pollution. J Soils Sediments 16(5):1012–1021. https://doi.org/10.1007/s11368-014-1063-6

Pergola M, Persiani A, Palese AM, Di Meo V, Pastore V, D'Adamo C, Celano G (2018) Composting: the way for a sustainable agriculture. Appl Soil Ecol 123:744–750. https://doi.org/10.1016/j.apsoil.2017.10.011

Pi H, Zhang X, Li S, Webb NP (2024) Influence of crop rotation, irrigation, fertilization, and tillage on the aggregate property and soil wind erosion potential in the floodplain of the Yellow River. Aeolian Res 67–69:100925. https://doi.org/10.1016/j.aeolia.2023.100925

Plaza C, Nogales R, Senesi N, Benitez E, Polo A (2008) Organic matter humification by vermicomposting of cattle manure alone and mixed with two-phase olive pomace. Bioresour Technol 99(10):5085–5089. https://doi.org/10.1016/j.biortech.2007.09.044

Pose-Juan E, Igual JM, Sánchez-Martín MJ, Rodríguez-Cruz MS (2017) Influence of herbicide triasulfuron on soil microbial community in an unamended soil and a soil amended with organic residues. Front Microbiol 8:378. https://doi.org/10.3389/fmicb.2017.00378

Prack McCormick B, El Mujtar VA, Cardozo A, Alvarez VE, Rodríguez HA, Tittonell PA (2022) Nutrient source, management system and the age of the plantation affect soil biodiversity and chemical properties in raspberry production. Eur J Soil Biol 111:103420. https://doi.org/10.1016/j.ejsobi.2022.103420

Pramanik P, Ghosh G, Ghosal P, Banik P (2007) Changes in organic–C, N, P and K and enzyme activities in vermicompost of biodegradable organic wastes under liming and microbial inoculants. Bioresour Technol 98(10):2485–2494. https://doi.org/10.1016/j.biortech.2006.09.022

Pramanik B, Kumar M, Naik BM, Singh SK, Kumar M, Singh SV (2024) Soil carbon-nutrient cycling, energetics, and carbon footprint in calcareous soils with adoption of long-term conservation tillage practices and cropping systems diversification. Sci Total Environ 912:169421. https://doi.org/10.1016/j.scitotenv.2023.169421

Prats SA, Serpa D, Santos L, Keizer JJ (2023) Effects of forest residue mulching on organic matter and nutrient exports after wildfire in North-Central Portugal. Sci Total Environ 885:163825. https://doi.org/10.1016/j.scitotenv.2022.163825

Qasim W, Lee MH, Moon BE, Okyere FG, Khan F, Nafees M, Kim HT (2018) Composting of chicken manure with a mixture of sawdust and wood shavings under forced aeration in a closed reactor system. Int J Recycl Org Waste Agric 7(3):261–267. https://doi.org/10.1007/s40093-018-0215-3

Radini S, González-Camejo J, Andreola C, Eusebi AL, Fatone F (2023) Risk management and digitalisation to overcome barriers for safe reuse of urban wastewater for irrigation—a review based on European practice. J Water Process Eng 53:103690. https://doi.org/10.1016/j.jwpe.2022.103690

Rady MM, Semida WM, Hemida KA, Abdelhamid MT (2016) The effect of compost on growth and yield of Phaseolus vulgaris plants grown under saline soil. Int J Recycl Org Waste Agric 5(2):311–321. https://doi.org/10.1007/s40093-016-0135-6

Rahman MO, Hussain A, Basri HA (2014) A critical review on waste paper sorting techniques. Int J Environ Sci Technol 11(2):551–564. https://doi.org/10.1007/s13762-013-0274-1

Ramachandra T, Bharath H, Kulkarni G, Han SS (2018) Municipal solid waste: generation, composition and GHG emissions in Bangalore, India. Renew Sust Energ Rev 82:1122–1136. https://doi.org/10.1016/j.rser.2017.09.104

Rasapoor M, Nasrabadi T, Kamali M, Hoveidi H (2009) The effects of aeration rate on generated compost quality, using aerated static pile method. Waste Manag 29(2):570–573. https://doi.org/10.1016/j.wasman.2008.07.015

Rath S, Zamora-Re M, Graham W, Dukes M, Kaplan D (2021) Quantifying nitrate leaching to groundwater from a corn-peanut rotation under a variety of irrigation and nutrient management practices in the Suwannee River Basin, Florida. Agric Water Manag 246:106634. https://doi.org/10.1016/j.agwat.2020.106634

Reilly K, Cavigelli M, Szlavecz K (2023) Agricultural management practices impact soil properties more than soil microarthropods. Eur J Soil Biol 117:103516. https://doi.org/10.1016/j.ejsobi.2022.103516

Reimer M, Hartmann TE, Oelofse M, Magid J, Bünemann EK, Möller K (2020) Reliance on biological nitrogen fixation depletes soil phosphorus and potassium reserves. Nutr Cycl Agroecosyst 118(2):273–291. https://doi.org/10.1007/s10705-020-10074-6

Ren X, Zeng G, Tang L, Wang J, Wan J, Wang J, Deng Y, Liu Y, Peng B (2018) The potential impact on the biodegradation of organic pollutants from composting technology for soil remediation. Waste Manag 72:138–149. https://doi.org/10.1016/j.wasman.2017.10.034

Roberts P, Edwards-Jones G, Jones DL (2007) In-vessel cocomposting of green waste with biosolids and paper waste. Compost Sci Util 15(4):272–282

Rocco S, Munkholm LJ, Jensen JL (2024) Long-term soil quality and C stock effects of tillage and cover cropping in a conservation agriculture system. Soil Tillage Res 241:106129. https://doi.org/10.1016/j.still.2023.106129

Rolewicz-Kalińska A, Lelicińska-Serafin K, Manczarski P (2020) The circular economy and organic fraction of municipal solid waste recycling strategies. Energies 13(17):4366. https://doi.org/10.3390/en13174366

Romero P, Navarro JM, Ordaz PB (2022) Towards a sustainable viticulture: the combination of deficit irrigation strategies and agroecological practices in Mediterranean vineyards. A review and update. Agric Water Manag 259:107216. https://doi.org/10.1016/j.agwat.2021.107216

Said-Pullicino D, Kaiser K, Guggenberger G, Gigliotti G (2007) Changes in the chemical composition of water-extractable organic matter during composting: distribution between stable and labile organic matter pools. Chemosphere 66(11):2166–2176. https://doi.org/10.1016/j.chemosphere.2006.06.069

Sanasam SD, Talukdar NC (2017) Quality compost production from municipality biowaste in mix with rice straw, cow dung, and earthworm Eisenia fetida. Compost Sci Util 25(2):141–151. https://doi.org/10.1080/1065657X.2016.1193424

Sao S, Ann V, Nishiyama M, Praise S, Watanabe T (2023) Tracing the pathways by which flood duration impacts soil bacteria through soil properties and water-extractable dissolved organic matter: a soil column experiment. Sci Total Environ 902:166524. https://doi.org/10.1016/j.scitotenv.2023.166524

Sarkar P, Chourasia R (2017) Bioconversion of organic solid wastes into biofortified compost using a microbial consortium. Int J Recycl Org Waste Agric 6(3):321–334. https://doi.org/10.1007/s40093-017-0166-3

Sayara T, Basheer-Salimia R, Hawamde F, Sánchez A (2020) Recycling of organic wastes through composting: process performance and compost application in agriculture. Agronomy 10(11):1838. https://doi.org/10.3390/agronomy10111838

Schattman RE, Jean H, Faulkner JW, Maden R, McKeag L, Nelson KC, Grubinger V, Burnett S, Erich MS, Ohno T (2023) Effects of irrigation scheduling approaches on soil moisture and vegetable production in the Northeastern U.S.A. Agric Water Manag 287:108428. https://doi.org/10.1016/j.agwat.2023.108428

Schmid C, Murphy J (2016) Effect of tillage and compost amendment on turfgrass establishment on a compacted sandy loam. J Soil Water Conserv 72(1):55–64. https://doi.org/10.2489/jswc.72.1.55

Scotti R, Bonanomi G, Scelza R, Zoina A, Rao M (2015) Organic amendments as sustainable tools to recover fertility in intensive agricultural systems. J Soil Sci Plant Nutr 15(2):333–352. https://doi.org/10.4067/S0718-95162015005000036

Sen B, Chandra T (2007) Chemolytic and solid-state spectroscopic evaluation of organic matter transformation during vermicomposting of sugar industry wastes. Bioresour Technol 98(5): 1680–1683. https://doi.org/10.1016/j.biortech.2006.05.043

Shaaban M, Wu Y, Núñez-Delgado A, Kuzyakov Y, Peng Q-A, Lin S, Hu R (2023) Enzyme activities and organic matter mineralization in response to application of gypsum, manure and rice straw in saline and sodic soils. Environ Res 224:115393. https://doi.org/10.1016/j.envres.2022.115393

Sharholy M, Ahmad K, Mahmood G, Trivedi RC (2008) Municipal solid waste management in Indian cities—a review. Waste Manag 28(3):459–467. https://doi.org/10.1016/j.wasman.2007.02.008

Sharma S, Singh P (2023) Tillage intensity and straw retention impacts on soil organic carbon, phosphorus and biological pools in soil aggregates under rice-wheat cropping system in Punjab, north-western India. Eur J Agron 149:126913. https://doi.org/10.1016/j.eja.2022.126913

Sileshi GW, Jama B, Vanlauwe B, Negassa W, Harawa R, Kiwia A, Kimani D (2019) Nutrient use efficiency and crop yield response to the combined application of cattle manure and inorganic fertilizer in sub-Saharan Africa. Nutr Cycl Agroecosyst 113(2):181–199. https://doi.org/10.1007/s10705-018-9945-5

Singh R, Singh R, Soni SK, Singh SP, Chauhan U, Kalra A (2013) Vermicompost from biodegraded distillation waste improves soil properties and essential oil yield of Pogostemon cablin (patchouli) Benth. Appl Soil Ecol 70:48–56. https://doi.org/10.1016/j.apsoil.2013.04.005

Singh YP, Arora S, Mishra VK, Singh A (2023) Composting of municipal solid waste using earthworms and Ligno-cellulolytic microbial consortia for reclamation of the degraded sodic soils and harnessing their productivity potential. Sustainability 15(4):2317. https://doi.org/10.3390/su15042317

Sinha RK, Bharambe G, Chaudhari U (2008) Sewage treatment by vermifiltration with synchronous treatment of sludge by earthworms: a low-cost sustainable technology over conventional systems with potential for decentralization. Environmentalist 28(4):409–420. https://doi.org/10.1007/s10669-008-9186-4

Sivaramanan S (2015) Global warming and climate change, causes, impacts and mitigation. Cent Environ Auth 2:1–26

Skuras D, Psaltopoulos D (2012) A broad overview of the main problems derived from climate change that will affect agricultural production in the Mediterranean area. In: BBuilding Resilience for Adaptation to Climate Change in the Agriculture Sector: Proceedings of a Joint FAO/OECD Workshop. Food and Agriculture Organization of the United Nations, pp 217–260

Smet E, Van Langenhove H, De Bo I (1999) The emission of volatile compounds during the aerobic and the combined anaerobic/aerobic composting of biowaste. Atmos Environ 33(10): 1295–1303. https://doi.org/10.1016/S1352-2310(98)00241-4

Solano M, Iriarte F, Ciria P, Negro MJ (2001) SE—structure and environment: performance characteristics of three aeration systems in the composting of sheep manure and straw. J Agric Eng Res 79(3):317–329. https://doi.org/10.1006/jaer.2001.0733

Somenahally AC, McLawrence J, Chaganti VN, Ganjegunte GK, Obayomi O, Brady JA (2023) Response of soil microbial communities, inorganic and organic soil carbon pools in arid saline soils to alternative land use practices. Ecol Indic 150:110227. https://doi.org/10.1016/j.ecolind.2022.110227

Somerville PD, May PB, Livesley SJ (2018) Effects of deep tillage and municipal green waste compost amendments on soil properties and tree growth in compacted urban soils. J Environ Manag 227:365–374. https://doi.org/10.1016/j.jenvman.2018.08.064

Song J-H, Her Y, Yu X, Li Y, Smyth A, Martens-Habbena W (2022) Effect of information-driven irrigation scheduling on water use efficiency, nutrient leaching, greenhouse gas emission, and plant growth in South Florida. Agric Ecosyst Environ 333:107954. https://doi.org/10.1016/j.agee.2022.107954

Soobhany N (2019) Insight into the recovery of nutrients from organic solid waste through biochemical conversion processes for fertilizer production: a review. J Clean Prod 241:118413. https://doi.org/10.1016/j.jclepro.2019.118413

Steen E, Brooks M (2015) Medical waste management—a review. J Environ Manag 163:98–108. https://doi.org/10.1016/j.jenvman.2015.07.045

Su J, Ji W, Sun X, Wang H, Kang Y, Yao B (2023) Effects of different management practices on soil microbial community structure and function in alpine grassland. J Environ Manag 327:116859. https://doi.org/10.1016/j.jenvman.2022.116859

Subramanian S, Sivarajan M, Saravanapriya S (2010) Chemical changes during vermicomposting of sago industry solid wastes. J Hazard Mater 179(1–3):318–322. https://doi.org/10.1016/j.jhazmat.2010.03.039

Sultana M, Jahiruddin M, Islam MR, Rahman MM, Abedin MA, Solaiman ZM (2021) Nutrient enriched municipal solid waste compost increases yield, nutrient content and balance in rice. Sustainability 13(2):1047. https://doi.org/10.3390/su13021047

Sun W, Villamil MB, Behnke GD, Margenot AJ (2022) Long-term effects of crop rotation and nitrogen fertilization on phosphorus cycling and balances in loess-derived Mollisols. Geoderma 420:115829. https://doi.org/10.1016/j.geoderma.2022.115829

Tampio E, Pettersson F, Rasi S, Tuomaala M (2022) Application of mathematical optimization to exploit regional nutrient recycling potential of biogas plant digestate. Waste Manag 149:105–113. https://doi.org/10.1016/j.wasman.2022.09.013

Tan M, Zong R, Lin H, Dhital YP, Ayantobo OO, Chen P, Li H, Chen R, Wang Z (2023) Responses of soil nutrient and enzyme activities to long-term mulched drip irrigation (MDI) after the conversion of wasteland to cropland. Appl Soil Ecol 190:104976. https://doi.org/10.1016/j.apsoil.2022.104976

Taylor MD, Kreis R, Rejtő L (2016) Establishing growing substrate pH with compost and limestone and the impact on pH buffering capacity. HortScience 51(9):1153–1158. https://doi.org/10.21273/HORTSCI11115-16

Tesfuhuney WA, Swart W, Van Rensburg LD, Wolmarans K, Walker S, Chung Yu H (2022) Soil microbial activity as influenced by crusted runoff strip length and mulch cover under in-field rainwater harvesting (IRWH). Phys Chem Earth, Parts A/B/C 128:103258. https://doi.org/10.1016/j.pce.2022.103258

Thakur M, Kumar R (2021) Light conditions and mulch modulate the damask rose (Rosa damascena Mill.) yield, quality, and soil environment under mid-hill conditions of the western Himalaya. Ind Crop Prod 163:113317. https://doi.org/10.1016/j.indcrop.2021.113317

Theron JS, Johannes van Coller G, Rose LJ, Labuschagne J, Swanepoel PA (2023) The effect of crop rotation and tillage practice on fusarium crown rot and agronomic parameters of wheat in South Africa. Crop Prot 166:106175. https://doi.org/10.1016/j.cropro.2022.106175

Tiemann T, Douxchamps S (2023) Opportunities and challenges for integrated smallholder farming systems to improve soil nutrient management in Southeast Asia. World Dev Sustain 3:100080. https://doi.org/10.1016/j.wds.2023.100080

Tiquia S, Tam N (1998) Composting of spent pig litter in turned and forced-aerated piles. Environ Pollut 99(2):329–337. https://doi.org/10.1016/S0269-7491(97)00087-3

Tittarelli F, Petruzzelli G, Pezzarossa B, Civilini M, Benedetti A, Sequi P (2007) Quality and agronomic use of compost. In: Solid waste: assessment, monitoring and remediation. Elsevier, pp 119–157

Tomar A, Saikia J (2022) Effect of integrated nutrient management on soil chemical and biological properties under dolichos bean cultivation. Int J Phys Soc Sci 34(12):99–105

Toth M, Stumpp C, Klik A, Strauss P, Mehdi-Schulz B, Liebhard G, Strohmeier S (2024) Long-term effects of tillage systems on soil health of a silt loam in Lower Austria. Soil Tillage Res 241:106120. https://doi.org/10.1016/j.still.2023.106120

Tripathi G, Bhardwaj P (2004) Comparative studies on biomass production, life cycles, and composting efficiency of Eisenia fetida (Savigny) and Lampito mauritii (Kinberg). Bioresour Technol 92(3):275–283. https://doi.org/10.1016/j.biortech.2003.09.005

Troy SM, Nolan T, Leahy JJ, Lawlor PG, Healy MG, Kwapinski W (2013) Effect of sawdust addition and composting of feedstock on renewable energy and biochar production from pyrolysis of anaerobically digested pig manure. Biomass Bioenergy 49:1–9. https://doi.org/10.1016/j.biombioe.2012.12.011

Trupiano D, Cocozza C, Baronti S, Amendola C, Vaccari FP, Lustrato G, Di Lonardo S, Fantasma F, Tognetti R, Scippa GS (2017) The effects of biochar and its combination with compost on lettuce (Lactuca sativa L.) growth, soil properties, and soil microbial activity and abundance. Int J Agron 2017:3158207. https://doi.org/10.1155/2017/3158207

Turcanu M (2018) Review of the fertilizing products regulation. European Parliament

Tuure J, Räsänen M, Hautala M, Pellikka P, Mäkelä PSA, Alakukku L (2021) Plant residue mulch increases measured and modelled soil moisture content in the effective root zone of maize in semi-arid Kenya. Soil Tillage Res 209:104945. https://doi.org/10.1016/j.still.2021.104945

Urbanski L, Kalbitz K, Rethemeyer J, Schad P, Kögel-Knabner I (2023) Unexpected high alkyl carbon contents in organic matter-rich sandy agricultural soils of Northwest Central Europe. Geoderma 439:116695. https://doi.org/10.1016/j.geoderma.2022.116695

Uyizeye OC, Thiet RK, Knorr MA (2019) Effects of community-accessible biochar and compost on diesel-contaminated soil. Biorem J 23(2):107–117. https://doi.org/10.1080/10889868.2019.1574693

Vedere C, Lebrun M, Honvault N, Aubertin M-L, Girardin C, Garnier P, Dignac M-F, Houben D, Rumpel C (2022) How does soil water status influence the fate of soil organic matter? A review of processes across scales. Earth Sci Rev 234:104214. https://doi.org/10.1016/j.earscirev.2022.104214

Vengadaramana A, Justin PTJ (2012) Effect of organic fertilizers on the water holding capacity of soil in different terrains of Jaffna peninsula in Sri Lanka. J Nat Prod Plant Resour 2(4):500–503

Ventorino V, Pascale A, Fagnano M, Adamo P, Faraco V, Rocco C, Fiorentino N, Pepe O (2019) Soil tillage and compost amendment promote bioremediation and biofertility of polluted area. J Clean Prod 239:118087. https://doi.org/10.1016/j.jclepro.2019.118087

Verma G, Dhaka AK, Singh B, Kumar A, Choudhary AK, Kumar A, Kamboj NK, Hasanain M, Singh S, Bhupenchandra I, Shabnam, Sanwal P, Kumar S (2024) Productivity, soil health, and carbon management index of soybean-wheat cropping system under double zero-tillage and natural-farming based organic nutrient management in north-Indian plains. Sci Total Environ 917:170418. https://doi.org/10.1016/j.scitotenv.2023.170418

Verrone V, Gupta A, Laloo AE, Dubey RK, Hamid NAA, Swarup S (2024) Organic matter stability and lability in terrestrial and aquatic ecosystems: a chemical and microbial perspective. Sci Total Environ 906:167757. https://doi.org/10.1016/j.scitotenv.2023.167757

Visvanathan C, Trankler J (2003) Municipal solid waste management in Asia: a comparative analysis. In: Proceedings of the Workshop on Sustainable Landfill Management, Chennai, India

Wang H, Guo Q, Li X, Li X, Yu Z, Li X, Yang T, Su Z, Zhang H, Zhang C (2020) Effects of long-term no-tillage with different straw mulching frequencies on soil microbial community and the abundances of two soil-borne pathogens. Appl Soil Ecol 148:103488. https://doi.org/10.1016/j.apsoil.2019.103488

Wang D, Li B, Ma J, Wang J, Wang H, Li W (2023a) Pseudoplastic liquid mulch film incorporating waste lignin and starch to improve its sprayability and available soil nitrogen. Chem Eng J 475:146392. https://doi.org/10.1016/j.cej.2023.146392

Wang H, Zheng C, Ning S, Cao C, Li K, Dang H, Wu Y, Zhang J (2023b) Impacts of long-term saline water irrigation on soil properties and crop yields under maize-wheat crop rotation. Agric Water Manag 286:108383. https://doi.org/10.1016/j.agwat.2022.108383

Wang Z, Deng H, Li F, Sun Y, Hong S (2023d) Optimized soil bacterial structure following grazing exclusion promotes soil nutrient cycling and plant growth. J Arid Environ 213:104977. https://doi.org/10.1016/j.jaridenv.2022.104977

Wang L, Lu P, Feng S, Hamel C, Sun D, Siddique KHM, Gan GY (2024) Strategies to improve soil health by optimizing the plant-soil-microbe-anthropogenic activity nexus. Agric Ecosyst Environ 359:108750. https://doi.org/10.1016/j.agee.2023.108750

Waqas M, Hashim S, Humphries UW, Ahmad S, Noor R, Shoaib M, Naseem A, Hlaing PT, Lin HA (2023) Composting processes for agricultural waste management: a comprehensive review. PRO 11(3):731. https://doi.org/10.3390/pr11030731

Wei Y, Li J, Shi D, Liu G, Zhao Y, Shimaoka T (2017) Environmental challenges impeding the composting of biodegradable municipal solid waste: a critical review. Resour Conserv Recycl 122:51–65. https://doi.org/10.1016/j.resconrec.2017.02.005

Weldewahid Y, Habtu S, Taye G, Teka K, Gessesse TA (2023) Effects of long-term irrigation practice on soil quality, organic carbon and total nitrogen stocks in the drylands of Ethiopia. J Arid Environ 214:104982. https://doi.org/10.1016/j.jaridenv.2022.104982

Wen L, Peng Y, Zhou Y, Cai G, Lin Y, Li B (2023) Effects of conservation tillage on soil enzyme activities of global cultivated land: a meta-analysis. J Environ Manag 345:118904. https://doi.org/10.1016/j.jenvman.2023.118904

Wu J, Zhao Y, Qi H, Zhao X, Yang T, Du Y et al (2017) Identifying the key factors that affect the formation of humic substance during different materials composting. Bioresour Technol 244:1193–1196. https://doi.org/10.1016/j.biortech.2017.05.086

Wu Q, Lawley Y, Congreves KA (2024) Soil health indicator responses to three years of cover crop and crop rotation in a northern semi-arid region, the Canadian prairies. Agric Ecosyst Environ 359:108755. https://doi.org/10.1016/j.agee.2023.108755

Xie B, Chen Y, Cheng C, Ma R, Zhao D, Li Z, Li Y, An X, Yang X (2022) Long-term soil management practices influence the rhizosphere microbial community structure and bacterial function of hilly apple orchard soil. Appl Soil Ecol 180:104627. https://doi.org/10.1016/j.apsoil.2022.104627

Xing J, Wang X, Hu C, Wang L, Xu Z, He X, Wang Z, Zhao P, Liu Q (2022) Effects of residual mulching films with different mulching years on the diversity of soil microbial communities in typical regions. Heliyon 8(12):e12180. https://doi.org/10.1016/j.heliyon.2022.e12180

Xu Z, Li G, Huda N, Zhang B, Wang M, Luo W (2020) Effects of moisture and carbon/nitrogen ratio on gaseous emissions and maturity during direct composting of cornstalks used for filtration of anaerobically digested manure centrate. Bioresour Technol 298:122503. https://doi.org/10.1016/j.biortech.2020.122503

Yadav A, Garg V (2009) Feasibility of nutrient recovery from industrial sludge by vermicomposting technology. J Hazard Mater 168(2):262–268. https://doi.org/10.1016/j.jhazmat.2009.02.078

Yadav A, Garg V (2011) Recycling of organic wastes by employing Eisenia fetida. Bioresour Technol 102(3):2874–2880. https://doi.org/10.1016/j.biortech.2010.10.086

Yadav KD, Tare V, Ahammed M (2010) Vermicomposting of source-separated human faeces for nutrient recycling. Waste Manag 30(1):50–56. https://doi.org/10.1016/j.wasman.2009.09.023

Yang H, Zhang H, Qiu H, Anning DK, Li M, Wang Y, Zhang C (2021) Effects of C/N ratio on lignocellulose degradation and enzyme activities in aerobic composting. Horticulturae 7(11):482. https://doi.org/10.3390/horticulturae7110482

Yang R, Qi Y, Yang L, Chen T, Deng A, Zhang J, Song Z, Ge B (2022) Rotation regimes lead to significant differences in soil macrofaunal biodiversity and trophic structure with the changed soil properties in a rice-based double cropping system. Geoderma 405:115424. https://doi.org/10.1016/j.geoderma.2021.115424

Yang C, Zhao Y, Long B, Wang F, Li F, Xie D (2023a) Biodegradable mulch films improve yield of winter potatoes through effects on soil properties and nutrients. Ecotoxicol Environ Saf 264:115402. https://doi.org/10.1016/j.ecoenv.2022.115402

Yang J-l, Ren L-q, Zhang N-h, Liu E-k, Sun S-k, Ren X-l, Jia Z-k, Wei T, Zhang P (2023b) Can soil organic carbon sequestration and carbon management index be improved by changing the film-mulching methods in semiarid region? J Integr Agric

Ye G, Lin Y, Liu D, Chen Z, Luo J, Bolan N, Fa J, Ding W (2019) Long-term application of manure over plant residues mitigates acidification, builds soil organic carbon and shifts prokaryotic diversity in acidic Ultisols. Appl Soil Ecol 133:24–33. https://doi.org/10.1016/j.apsoil.2018.10.005

Yin L, Dijkstra FA, Phillips RP, Zhu B, Wang P, Cheng W (2021) Arbuscular mycorrhizal trees cause a higher carbon to nitrogen ratio of soil organic matter decomposition via rhizosphere priming than ectomycorrhizal trees. Soil Biol Biochem 157:108246. https://doi.org/10.1016/j.soilbio.2021.108246

Zhan Y, Chang Y, Tao Y, Zhang H, Lin Y, Deng J, Ma T, Ding G, Wei Y, Li J (2023) Insight into the dynamic microbial community and core bacteria in composting from different sources by advanced bioinformatics methods. Environ Sci Pollut Res Int 30(10):8956–8966. https://doi.org/10.1007/s11356-022-21523-8

Zhang H, McGill E, Gomez CO, Carson S, Neufeld K, Hawthorne I, Smukler S (2017) Disintegration of compostable foodware and packaging and its effect on microbial activity and community composition in municipal composting. Int Biodeterior Biodegradation 125:157–165. https://doi.org/10.1016/j.ibiod.2017.05.011

Zhang L, Li L, Sha G, Liu C, Wang Z, Wang L (2020) Aerobic composting as an effective cow manure management strategy for reducing the dissemination of antibiotic resistance genes: an integrated meta-omics study. J Hazard Mater 386:121895. https://doi.org/10.1016/j.jhazmat.2020.121895

Zhang Y, Yang P, Liu X, Adeloye AJ (2022) Simulation and optimization coupling model for soil salinization and waterlogging control in the Urad irrigation area, North China. J Hydrol 607:127408. https://doi.org/10.1016/j.jhydrol.2022.127408

Zhang G, Ming B, Xie R, Chen J, Hou P, Xue J, Shen D, Li R, Zhai J, Zhang Y, Wang K, Li S (2023a) Reducing plastic film mulching and optimizing agronomic management can ensure food security and reduce carbon emissions in irrigated maize areas. Sci Total Environ 883:163507. https://doi.org/10.1016/j.scitotenv.2022.163507

Zhang J, Du L, Xing Z, Zhang R, Li F, Zhong T, Ren F, Yin M, Ding L, Liu X (2023b) Effects of dual mulching with wheat straw and plastic film under three irrigation regimes on soil nutrients and growth of edible sunflower. Agric Water Manag 288:108453. https://doi.org/10.1016/j.agwat.2022.108453

Zhang X, Li Q, Zhong Z, Huang Z, Wen X, Bian F, Yang C (2023c) Determining changes in microbial nutrient limitations in bamboo soils under different management practices via enzyme stoichiometry. Catena 223:106939. https://doi.org/10.1016/j.catena.2022.106939

Zhang Y, Hou R, Fu Q, Li T, Li M, Cui S, Dong W (2023d) Drip irrigation impacts on the root zone soil environment and enrichment characteristics of heavy metals in soybean. Agric Water Manag 288:108483. https://doi.org/10.1016/j.agwat.2022.108483

Zhang M, Song X, Wu X, Zheng F, Li S, Zhuang Y, Man X, Degrè A (2024) Microbial regulation of aggregate stability and carbon sequestration under long-term conservation tillage and nitrogen application. Sustain Prod Consump 44:74–86. https://doi.org/10.1016/j.spc.2023.11.005

Zhao B, O'Connor D, Zhang J, Peng T, Shen Z, Tsang DC, Hou D (2018) Effect of pyrolysis temperature, heating rate, and residence time on rapeseed stem derived biochar. J Clean Prod 174:977–987. https://doi.org/10.1016/j.jclepro.2017.10.272

Zhao B, Wang Y, Sun H, Xu Z (2022) Analysis of humus formation and factors driving the humification process during composting of different agricultural wastes. Front Environ Sci 10:954158. https://doi.org/10.3389/fenvs.2022.954158

Zhao C, Wang Y, Qiu Y, Xie Z, Zhang Y (2023a) Response of soil respiration to hydrothermal effects of gravel-sand mulch in arid regions of the loess plateau, China. Soil Tillage Res 231:105733. https://doi.org/10.1016/j.still.2022.105733

Zhao J, Liu Z, Lai H, Zhao M, Zhu Q, Zhao C, Yang D, Li X (2023b) The impacts of soil tillage combined with plastic film management practices on soil quality, carbon footprint, and peanut yield. Eur J Agron 148:126881. https://doi.org/10.1016/j.eja.2022.126881

Zhen Z, Liu H, Wang N, Guo L, Meng J, Ding N, Wu G, Jiang GM (2014) Effects of manure compost application on soil microbial community diversity and soil microenvironments in a temperate cropland in China. PLoS One 9(10):e108555. https://doi.org/10.1371/journal.pone.0108555

Zhu Z, Li J, Zhu D, Gao Z (2024) The impact of maize canopy on splash erosion risk on soils with different textures under sprinkler irrigation. Catena 234:107608. https://doi.org/10.1016/j.catena.2023.107608

Zorpas AA, Loizidou M (2008) Sawdust and natural zeolite as a bulking agent for improving quality of a composting product from anaerobically stabilized sewage sludge. Bioresour Technol 99(18):7545–7552. https://doi.org/10.1016/j.biortech.2008.07.015

Zou C, Li Y, Huang W, Zhao GK, Pu G, Su J, Coyne MS, Chen Y, Wang L, Hu X (2018) Rotation and manure amendment increase soil macro-aggregates and associated carbon and nitrogen stocks in flue-cured tobacco production. Geoderma 325:49–58. https://doi.org/10.1016/j.geoderma.2018.03.025

Zukowska G, Mazurkiewicz J, Myszura M, Czekała W (2019) Heat energy and gas emissions during composting of sewage sludge. Energies 12(24):4782. https://doi.org/10.3390/en12244782

# Reducing the Chemical Fertilizer in Agriculture by Composting and Improving the Self-Reliance Perspective for Long-Term Impact

**Yenda Damodhara Rao and Shweta Meshram**

## 1 Introduction

Chemical fertilizers are compounds that contain large amounts of nutrients in the form of small pellets or granules, and this supply provides nutrients to plants for their growth and survival (Terman 2018). These chemical fertilizers are an unavoidable risk to soil and the environment by changing the soil composition and by causing soil and groundwater pollution (Craswell 2021). In India, farmers use chemical fertilizers on an average of 121.4 kg/ha (Pahalvi et al. 2021). Prolonged usage of these chemical fertilizers leads to a decline in soil structure and soil quality, with the reduction of soil organic matter (SOM) (Ajayi et al. 2012). Due to these harmful effects, there is a need to reduce chemical fertilizers in agriculture by incorporating organic soil amendments.

### *1.1 Methods to Reduce the Usage of Chemical Fertilizers*

The biomass waste contains all types of nutrients that are required by plants, which acts as a good source if managed well. They contain high amounts of organic matter (OM), and the OM must be treated to remove pathogens and pests and used to fertilize agricultural fields (Chew et al. 2019). Chemical fertilizers release nutrients

Y. D. Rao
Department of Plant Pathology, Chandra Shekhar Azad University of Agriculture and Technology, Kanpur, Uttar Pradesh, India

S. Meshram (✉)
Biologicals lab, Bigbiological Pioneer, regional Center of Biotechnology (BScBionest incubator), Faridabad, Haryana, India
e-mail: shwetavinay04@gmail.com



R. K. Singh et al. (eds.), *Composting of Farm Residues by Nature-Based Solutions*,
https://doi.org/10.1007/978-3-032-12255-1_9

within a short period after application, but organic amendments need some time to mineralize (Shaji et al. 2021). The time taken to mineralize depends on soil moisture, organic matter composition, surrounding temperature, and type of composting (Cannavo et al. 2022). The application of these organic amendments is very promising to increase food production in a sustainable way, to reduce pollution, and to reduce reliance on chemical amendments (Bamdad et al. 2022). Organic fertilizers reduce the harmful effects caused by chemical fertilizers in the long run (Wang et al. 2019a, b). Biofertilizers also help in bioremediation of soils contaminated with fertilizers, pesticides, and hydrocarbons (Lau et al. 2023). The combined use of organic and chemical fertilizers results in good results in the fields (Ye et al. 2020). But in the long run, chemical fertilizers leave hazardous effects. The addition of organic manure offers numerous benefits for nature and reduces the problem of soil infertility in modern agriculture (Bhatt et al. 2019a, b). In an organic agricultural system, manuring recycles nutrients through farmyard manure, compost, and different types of organic and green manures (Iderawumi and Kamal 2022).

## 2 Biowaste Utilization for Fertilizer Production and Enhanced Plant Growth

The chemical fertilizers mainly contain nitrogen, phosphorus, and potassium (NPK). Using these fertilizers over the past decades all over the world has caused significant pollution of the environment (Yahaya et al. 2023). The harmful effects of continuous use of chemical fertilizers call for the recycling of biomass produced in various activities like compost, manure, food waste, and sewage sludge waste into organic fertilizers (Chojnacka et al. 2020). Organic materials used in agriculture are classified into several categories, including animal-based organic waste (manure), plant source and food waste (compost), and sewage sludge and household waste (urban waste) (Uysal et al. 2014). These wastes must be processed to enhance the nutrient content and promote their agricultural value for contribution to a more bio-friendly economy and environment (Sharma et al. 2024). Some of the important organic manuring products that are used in agriculture are animal manure, sewage sludge waste, food waste composting, and vermicomposting, which are used in place of chemical fertilizers in agriculture by composting and improve the self-reliance perspective for long-term impact.

## 3 Vermicomposting

**History**

Earthworms have been recognized for their contributions to increasing soil fertility by several civilizations throughout history. The ancient civilizations, like Greece and

**Table 1** Nutrients in vermicompost

| Nutrients | Percentage |
|---|---|
| Nitrogen | 0.5–0.9 |
| Phosphorous | 0.1–0.26 |
| Potassium | 0.15–0.256 |
| Organic carbon | 9.15–17.88 |
| Calcium and magnesium (Meq/100gm) | 22.67–47.6 |
| Sodium | 0.055–0.3 |
| Copper mg/kg | 2.0–9.5 |
| Iron mg/kg | 2.0–9.3 |
| Zinc mg/kg | 5.7–9.3 |
| Sulphur mg/kg | 128–548 |

Garg and Gupta (2009)

Egypt, held great respect for the earth for maintaining soil health. Pharaoh Cleopatra (69–30 B.C.) and other ancient Egyptians were among the first to realize that earthworms improved the fertility of the croplands in the Nile Valley, especially after the yearly floods. The importance of earthworms in improving the condition of the soil was also acknowledged by the ancient Greeks. Earthworms are the "intestines of the earth," as the renowned philosopher Aristotle (384–322 B.C.) memorably put it, emphasizing their critical role in soil health and ecological functioning (Medany 2011). The ancient Indian man Surpala suggested incorporating earthworms into the soil in the tenth century A.D. to improve the yield and quality of pomegranate fruits. Dr. Anatoly Igonin, a Russian scientist, emphasized the importance of earthworms in maintaining the entire biosphere centuries later. He emphasized their capacity to produce soil and increase its fertility while carrying out necessary tasks including neutralization, disinfection, protection, and productivity enhancement (Sinha et al. 2014).

Vermicomposting is defined as the collaborative biodegradation of the organic matter with the help of the earthworms and microorganisms (Abad and Shafiqi 2024). The vermicompost is rich in nutrients and minerals and organic acids like macro and micronutrients, growth hormones, vitamins, and some enzymes like lipases, cellulase, chitinase, and protease that are required by plants for good growth and development (Olle 2019) (Table 1). Additionally, the vermicompost contains many beneficial microorganisms. Farmers must apply different chemical fertilizers for all these nutrients and other compounds that are provided by the vermicompost. This vermicomposting requires a smaller area and less time when compared to other composting techniques (Kauser and Khwairakpam 2022). Unlike traditional composting, which takes several weeks to months for turning organic waste into manure, vermicomposting takes a few days to a week to turn raw organic matter into vermicompost. Vermicomposting technique needs a low temperature to keep worms active and to prevent worms from drying (Ahmad et al. 2021). In vermicomposting, microorganisms play a vital role in the biochemical breakdown of organic matter (Usmani et al. 2019). Vermicomposting is a self-regulating, self-sustaining, and self-improving technology that requires no or minimal energy for the composting process

(Walia and Kaur 2024). When compared to other organic manures, vermicompost stands superior (Manzoor et al. 2024). There are many advantages to using vermicompost in agricultural operations. It minimizes termite and insect infestations, inhibits weed growth, and helps lower irrigation water requirements. Vermicompost also improves agricultural results by encouraging quicker seed germination, speeding up the growth and development of seedlings, and raising the annual seed production in cereal crops and the fruit yield per plant in vegetable crops (Enebe and Erasmus 2023).

## *3.1 Impact of Vermicompost on Agricultural Crop Performance and Ways to Reduce Chemical Fertilizer Application*

**Enhance Nutrient Supply to Plants**

Vermicompost is rich in nitrogen (N), phosphorus (P), potassium (K), other micronutrients and other important enzymes required by plants for good growth and development (Olle 2019). Additionally, vermicompost contains exchangeable calcium (Ca) and phosphorus (P) in significant amounts (Bekele et al. 2018). Numerous studies have stated that earthworms have a positive influence on plant growth. Application of vermicompost also enhances seed germination, improves the growth and development of the seedlings (Wako and Muleta 2022). The use of vermiwash (that is, the liquid produced from the vermicompost pit) in garden peas results in a higher percentage of protein content and carbohydrate content when compared to garden pea that is cultivated with chemical fertilizers (Adhikary 2012). Replacing the application of chemical fertilizer with vermicompost enhances fruit quality. The enhancements include total soluble solids in fruit juice, pH of the fruit, enhanced levels of micro and macronutrients, protein, and carbohydrate content (Joshi et al. 2015). The continuous supply of these nutrients by vermicompost can lead to a reduction in the usage of chemical fertilizers.

## *3.2 Plant Growth and Protection*

The application of vermicompost shows a positive effect on plant growth (Adhikary 2012). The addition of vermicompost shows remarkable growth outcomes in various stages of the crop (Mistry 2015). The incorporation of vermicompost has good effects on germination, flowering, good growth, and overall improvement in crop yield (Kamar Zaman and Yaacob 2022). The application of vermicompost increases seed germination and supports good growth in mung bean (*Vigna radiata*) (Karmegam and Daniel 2000). Vermicompost application has a wide range of positive effects on plant growth and development and yield (Joshi et al. 2015).

The application of vermicompost results in good seed germination, taller stem growth, higher leaf count and good leaf area, good growth, enhanced dry weight of the leaf, greater root density, good root length, and good fruit production (Mahmud et al. 2020).

Incorporation of vermicompost in soil gives protection to plants against various pests and diseases and gives protection against various biotic and abiotic stresses without the need for the application of chemical fertilizers and pesticides (Rehman et al. 2023). This protection against various stresses to plants was provided by various mechanisms, including inducing resistance to plants, repelling pests and insects, suppressing pest growth and development, and exerting pesticidal activity on pests without the application of any chemicals (Sinha et al. 2011). Vermicompost contains actinomyces and antibiotics, which play an important role in enhancing biological resistance in plants and provide resistance to crop plants against various disease-causing organisms and pests (Yatoo et al. 2021). The incorporation of vermicompost and earthworms in agricultural practices results in a reduction of chemical pesticides as a spray and other forms (Sulaiman and Mohamad 2020). This results in a reduction of the application of chemical fertilizers and pesticides; in the long run, it leads to self-reliance and sustainability. By incorporating vermicompost and earthworms, farmers can foster a more resilient agricultural ecosystem while minimizing the reliance on chemical fertilizers and pesticides.

### 3.3 *Enhances Yield*

Vermicompost significantly increases the yield of various vegetables, fruits, and other crops (El-Goud and Amal 2020). The application of vermicompost has a notable effect on crop yield improvement, primarily affecting nutrient uptake (Song et al. 2022).In comparison with chemical fertilizers, vermicompost and vermiwash give a good yield in garden pea as increased pod length and no. of grains (Samatha and Rai 2022). The beneficial mechanism of vermicompost has been presented in Fig. 1. The seed yield and fodder yield of groundnut increase by the application of vermicast; therefore, the application of chemical fertilizers can be minimized (Cagasan and Orbeta 2020).

## 4 Impact of Animal Manure on Crop Performance and Ways to Reduce Chemical Fertilizer Application

Animal manure, such as cattle, pigs, poultry, and horses, as well as other organic waste materials from other agricultural waste products, is decomposed and used as the source of nutrition for both horticultural and field crops for good productivity (Loss et al. 2019) (Table 2). This manure can improve the soil structure and improve

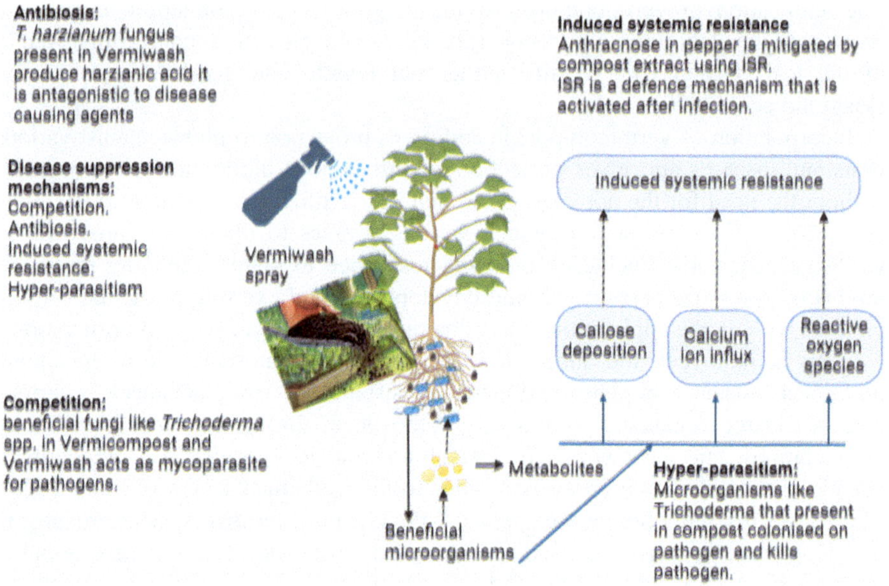


**Fig. 1** Representing the beneficial mechanisms of vermicompost and vermiwash to reduce the use of chemical fertilizers and pesticides

nutrient content in the soil without the use of chemical fertilizers. The use of organic animal manure in agriculture and horticulture dates to the Neolithic period about 12,000 years ago (Lorenz and Lal 2022). From the beginning of the 1950s, as the advancements increased, the productivity also increased by adding chemical fertilizers, which led to atmospheric pollution (Pimentel 1996). After the pollution hazards, the population moved toward organic fertilizers by rearing a greater number of animals and by providing more green manure. The animal manure provides optimal nutrition to crop plants and strengthens the soil. Application of animal manure to agricultural crops dates back several centuries (Zaib et al. 2023). The animal manure is organic in nature and contains all desirable nutrients like nitrogen (N), phosphorus (P), potassium (K), and some other micronutrients that are required by plants for good growth and development (Bhatt et al. 2019a, b). The management of poultry waste and animal waste is necessary as, by decomposition, these convert into manure and have good effects on soil aggregation in long-term application (Guo et al. 2019). The manure application results in good crop growth, better resistance against pests and diseases, and increased soil fertility (Bhunia et al. 2021).

The applications of animal manure led to an increase in total soil organic carbon content (TOC) over the years and helped in the reclamation of degraded soils (Abrar et al. 2023). The applications of animal manure increase nutrient content in soil and increase the cation exchange capacity of soil solely by avoiding the application of chemical fertilizers (Geng et al. 2019). Animal waste is also used in conservation farming, zero tillage, orchards, and perennial pastures (Carr et al. 2020). The increase of TOC by application of the animal manure increases nutrition adsorption,

**Table 2** Nutrient content in animal manure and their role in plant growth

| Nutrient | Percentage present in animal manure | Role of the nutrient in soil/plant growth | References |
|---|---|---|---|
| Nitrogen (N) | 0.5–5 | Essential for plant growth and development and for good chlorophyll production | Abubakar and Ismail (2012) |
| Phosphorous (P) | 0.2–2 | Promotes good flowering and root development | Smith and Van Dijk (1987) |
| Potassium (K) | 0.5–4 | Regulates enzyme activation and water movement in plants | Pushparaj and Gnanavelrajah (2023) |
| Calcium (Ca) | 0.5–3 | Improves plant's cell wall strength and soil structure | Nicholson et al. (1996) |
| Magnesium (Mg) | 0.1–0.6 | Central to photosynthesis and chlorophyll production | Strapazzon et al. (2021) |
| Copper (Cu) | 5–40 ppm | Necessary for enzymatic reaction in plants | Leeson and Caston (2008) |
| Sulphur (S) | 0.1–0.5 | Necessary for protein production and amino acid synthesis | Easha et al. (2015) |
| Manganese (Mn) | 50–300 ppm | Necessary for nitrogen metabolism and for photosynthesis | Jasinska et al. (2022) |
| Zinc (Zn) | 20–200 ppm | Supports the chlorophyll synthesis and enzyme production systems | Warncke et al. (2009) |
| Boron (Bo) | 1–50 ppm | Plays a vital role in cell wall formation and nutrient transport | Justice (2021) |
| Iron (Fe) | 100–2000 ppm | Crucial for enzyme function and chlorophyll synthesis | Zicari (2003) |
| Sodium (Na) | 0.1–1% | Influences foil salinity and pH | Rodrigues et al. (2021) |
| Moisture content | 50–90% | Indicates the water content in manure | Dadrasnia et al. (2021) |
| Organic matter | 20–60% | Improves the water retention and soil structure | Li et al. (2020) |

cation exchange capacity (Ferreira et al. 2022). Animal manure also contains magnesium (Mg) and calcium (Ca), which are fed to animals and not absorbed by the digestive tract (Pinotti et al. 2021). This results in the soils applied with the manures of pigs, poultry, cattle, and horses having high content of exchangeable calcium and magnesium in the soil. This is due to the movement of ions, including these sticks, to water-soluble organic substances, such as those derived from plant roots (Brunetto et al. 2012). The animal manure is rich in N, P, and K at the time of application. This manure contains N as $NH_4^+$. After it reaches the soil, it converts to nitrate $NO_3^-$, which is easily absorbed by plants (Vanotti et al. 2020). By applying animal manure, the pore size and porosity were increased with large-sized pores with better soil structure (Hartmann and Six 2023). The increased porosity and pore size, which are important for aeration and for soil-water relationships, help to maintain good soil structure (Ghosh et al. 2020). Water retention and infiltration are affected positively by the application of animal manure, which helps in increasing water availability to plants in water-deficient conditions (Rayne and Aula 2020; Roy et al. 2022). The increase in retention of water also depends on soil texture; fine-textured soils retain more water when compared to coarse-textured soils (Santos et al. 2022). The application of animal manure can stimulate plant root growth, thereby the plant can absorb more nutrients and water, so it can perform well and give good productivity (Asfaw 2022). In this way, the application of animal manure can reduce the use of chemical fertilizers.

## 5 Utilization of Food Waste Compost in Farming to Reduce the Use of Chemical Fertilizers

Over the last decade, food waste in this world has created environmental pollution and health hazards; it is a major issue globally (Palaniveloo et al. 2020). Rotting of the waste food releases methane gas, which causes environmental pollution, and the toxic leachates cause soil and water pollution (Ghosh et al. 2023). A solution to this food waste pollution is to turn that food waste into a useful product like manure, which is used to provide nutrients to plants instead of chemical fertilizers. By doing this, we can reduce the reliance on chemical fertilizers in farming. The food waste is generated from many sources, such as the food industry, houses, the baking industry, and hotels (Papargyropoulou et al. 2019). All this waste is accumulating in one place and causes pollution. By converting this waste into manure, we not only reduce pollution but also reduce the use of chemical fertilizers in farming, which can also reduce chemical pollution. Organic food waste composting is an aerobic process in which the food waste is broken down into simpler components that are used as manure (Palaniveloo et al. 2020). This food waste compost improves the soil texture and structure, and the water retention capacity of the soil (Dolit et al. 2022). At present, Germany is converting the highest amount (4280 kt) of food waste into manure, followed by Italy (2380 kt) and China (1350 kt) (Corden et al. 2019). The addition of earthworms to food waste can speed up the composting process (Manea et al. 2024). The food waste compost contains major macronutrients like N, P, and K (O'Connor et al. 2021). Also contains some trace elements and micronutrients, which are listed in Table 3. By adding this food waste compost to soil, we can avoid the usage of chemical fertilizers, and by doing so, the reliance on chemical fertilizers decreases, and the environmental pollution caused by the chemical fertilizers and food waste is reduced.

**Table 3** Representing the nutrients and other components present in food waste manure

| S. No | Nutrient/ mineral | Percentage/ range | Functions in plants |
|---|---|---|---|
| Macronutrients | | | |
| 1 | Nitrogen (N) | 0.5–2 | Essential for plant growth and development and for good chlorophyll synthesis |
| 2 | Phosphorous (P) | 0.1–1.0 | Promotes good root development and flowering |
| 3 | Potassium (K) | 0.5–1.5 | Provides disease resistance and supports overall plant health |
| 4 | Carbon (C) | 20–50 | Affects carbon-nitrogen ratio in soil |
| Micronutrients | | | |
| 5 | Calcium (Ca) | 1.0–5.0 | Improves soil structure and maintains pH |
| 6 | Magnesium (Mg) | 0.01–0.1 | Supports enzyme activity in plants |
| 7 | Sulphur (S) | 0.1–0.5 | Important for protein synthesis |
| Other components | | | |
| 8 | Moisture content | 25–60% | Impacts compost stability and microbial activity |
| 9 | Organic matter | 30–70 | Improves soil structure and fertility |
| 10 | pH | 6.0–8.0 | Indicates compost maturity, and this pH range is suitable for growth and development of most of the plants |
| Trace elements | | | |
| 11 | Iron (Fe) | 0.1–0.5 | Plays a crucial role in chlorophyll synthesis |
| 12 | Copper (Cu) | 0.002–0.02 | Enhances enzymatic activity |
| 13 | Zinc (Zn) | 0.005–0.05 | Plays a important role in plant's metabolism |
| 14 | Boron (B) | 0.001–0.02 | Plays a crucial role in cell wall strengthening and reproductive growth of the plant |
| 15 | Manganese (Mn) | 0.01–0.1 | Supports enzyme activity |

Wei et al. (2021)

## 6 Utilization of Sewage Sludge Manure in Farming to Reduce the Use of Chemical Fertilizers

Sewage sludge is the organic waste in the form of slurry that is obtained from municipal sewage, industrial sewage, and from pharmaceutical plants (Rorat et al. 2019), and can be used to remodulate the soil properties (Fig. 2). Sewage sludge is the only organic waste that contains a high concentration of phosphorus (Hušek et al. 2022) (Table 4). Phosphorus is the key element for animals and agriculture. Due to a deficiency of phosphorus, agricultural productivity gets affected severely (Khan et al. 2023). The source of phosphorus used in agriculture is nonrenewable phosphate rocks; continuous and non-judicious use of these rocks for phosphorus production leads to the extinction of these rocks. Sewage sludge has good properties

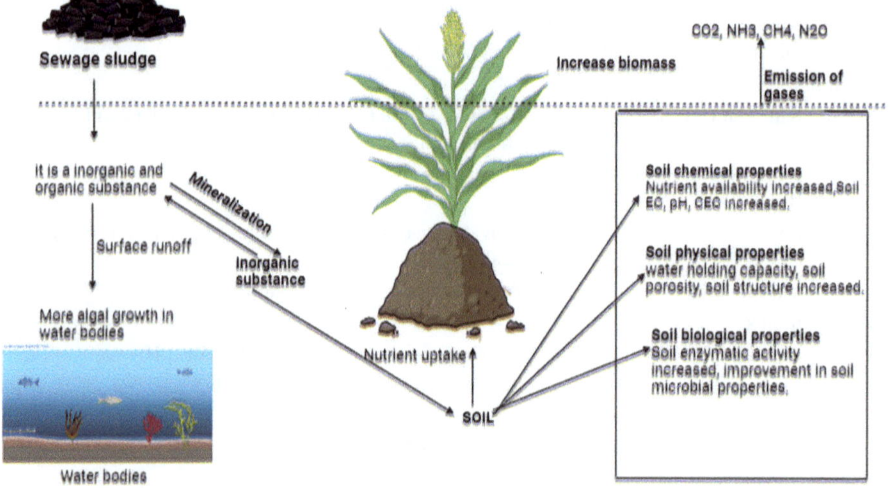


**Fig. 2** Representing the effects of sewage sludge on plant and soil properties

with nitrogen and phosphorus content, which can stimulate soil microbial activity, soil respiration, and soil enzyme activity because of the degradation of the organic matter (Picariello et al. 2020). The recovery of phosphorus from the sewage sludge is done by different methods like pyrolysis, composting, combustion, and hydrothermal carbonization (Liu et al. 2022). The combustion of sewage sludge produces products that contain a high proportion of phosphorus, which is used as fertilizer (Haggstrom et al. 2021). Sewage sludge contains N, P, K and other nutrients and minerals that help in the good growth and development of plants, and can replace the chemical fertilizers (Usman et al. 2012) (Table 5). Sewage sludge contains a good number of microorganisms that support plant growth and development; these microbial activities also play a vital role in maintaining soil fertility and soil quality (Nunes et al. 2021). Sewage sludge contains a high amount of organic matter and nutrients required by plants for good growth and development (Chen et al. 2022).

## 7 Conclusion

The continuous use of chemical fertilizers leads to environmental pollution and causes harmful effects on humans and animals, and also causes various diseases that may be irreversible in nature. To mitigate the harmful effects caused by chemical fertilizers, there is an urgent need to incorporate organic amendments in crop production, which leads to a reduction in reliance on chemical fertilizers. The organic fertilizers also contain many nutrients that are required by plants for good growth and development. On the other hand, chemical fertilizers supply only one or a few nutrients and minerals. To supply more nutrients, farmers must apply more

**Table 4** Representing the nutrients present in sewage sludge and their role in limiting the use of chemical fertilizers

| S. No | Nutrient/ mineral present | Percentage % by dry weight | Function of nutrients/description |
|---|---|---|---|
| Macronutrients | | | |
| 1 | Nitrogen (N) | 2–5 | Essential for plant growth and helps in chlorophyll synthesis |
| 2 | Phosphorus (P) | 0.5–3.5 | Key element for root growth and energy transfer in plants |
| 3 | Potassium (K) | 0.2–1.0 | Supports enzyme activity and water regulation in plants |
| Secondary nutrients | | | |
| 4 | Magnesium (Mg) | 0.5–1.5 | Central to chlorophyll and photosynthesis |
| 4 | Calcium (ca) | 1–5 | Plays a key role in soil pH. Regulation and cell wall stability |
| 6 | Sulphur (S) | 0.3–2.0 | Required for enzyme function and protein synthesis |
| Micronutrients | | | |
| 7 | Iron (Fe) | 1–6 | Essential for enzyme function and chlorophyll synthesis |
| 8 | Manganese (Mn) | 0.005–0.1 | Involves in nitrogen metabolism and photosynthesis |
| 9 | Zinc (Zn) | 0.01–0.2 | Plays a key role in plant hormone regulation, and enzyme function on plants |
| 10 | Boron (B) | 0.001–0.01 | Crucial for reproductive processes in plants and for cell wall strengthening |
| 11 | Copper (cu) | 0.005–0.1 | Important for photosynthesis and reproductive growth of plants |
| 12 | Molybdenum (Mb) | <0.001 | Essential for nitrogen fixation |
| Heavy metals | | | |
| 13 | Cadmium (cd) | 0.0001–0.001 | Processes potential toxicity if present excess levels |
| 14 | Chromium (Cr) | 0.001–0.1 | Impacts depend on its oxidation state |
| 15 | Lead (Pb) | 0.001–0.05 | High levels of Pb leads to toxicity |
| 16 | Mercury (Hg) | 0.00001–0.0005 | Highly toxic to plants if present in excess levels |
| 17 | Nickel (Ni) | 0.001–0.05 | Plants required Ni in small quantities, if present in high quantity leads to toxicity to plants |

Di Costanzo et al. (2021)

types of chemical fertilizers. The organic amendments, like vermicompost and animal manure, develop resistance in plants against diseases and pests that cause severe damage to crop plants. On the other hand, organic amendments also improve soil texture and structure, which gives the soil good water-holding capacity and good

**Table 5** Characteristics of different organic materials used for organic fertilizer production, which helps in the reduction of the use of chemical fertilizers

| S. No | Type material | Biomass source | Resulting effects | References |
|---|---|---|---|---|
| 1 | Organic manure and green manure | Green manure (*Trifolium alexandrianum*) and farmyard manure | By using these amendments in rice fields, the shoot biomass and good grain yield was observed | Javaid (2011) |
| 2 | Organic manure | Sheep manure | The uptake of heavy metals like Pb, Cd, and Zn by plants was controlled by application of sheep manure | Elouear et al. (2016) |
| 3 | Organic manure | Various livestock manure | It plays a significant role in promoting soil fertility. It also supports good support to coffee plants in water-deficient conditions by promoting good growth and yield | Chemura (2014) |
| 4 | Organic manure | Solid dairy manure and solid pig manure | Improves beneficial bacterial colonies and make nitrogen available to plants as compared to chemical fertilizers | Hamm et al. (2016) |
| 5 | Animal manure | Cattle and pig manure | Improves soil structure by reducing salt content in the soil | Guo et al. (2018) |
| 6 | Animal manure | Cattle or pig manure | Increases biological control organisms and make biocontrol organisms binds to plant roots | Guo et al. (2019) |
| 7 | Animal manure | Various livestock manure like pig, pigeon, and poultry. | Decreases soil pH and prevents the occurrence of secondary soil salinization | Li-Xian et al. (2007) |
| 8 | Sewage sludge | Waste water treatment plant | Make phosphorous available to crop plants for prolonged time bioavailability. Decreased Polycyclic Aromatic Hydrocarbon (PAH) concentrations and increased nutrient content | Franz (2008) |
| 9 | Sewage sludge ash | From wastewater treatment plant | It recovers phosphorous about 80% | Becker et al. (2019) |
| 10 | Industrial and municipal activated sewage sludge | Waste water treatment plants from municipal and industry water treatments | Makes bioavailability of phosphorous to plants for long time, and increased nutrient content with pyrolysis | Frišták et al. (2018) |
| 11 | Compost | Cattle manure and food waste | Increases nitrogen content and organic carbon level. | |

(continued)

**Table 5** (continued)

| S. No | Type material | Biomass source | Resulting effects | References |
|---|---|---|---|---|
| | | | Higher yield in maize crop with combination of cattle manure and food waste in the ratio of 50:50 | Wolka and Melaku (2015) |
| 12 | Dehydrated sewage sludge | Municipal wastewater treatment plant | Increases phosphate content in soil and makes bioavailability of phosphate to plants for long time, promotes temperature raise in soil this temperature raise can destroy soil inhabitant pathogens | Wang et al. (2019a, b) |
| 13 | Compost | Solid pig manure | It can reduce the use of commercial chemical and organic fertilizers and improves soil fertility | Romano et al. (2014) |
| 14 | Compost | Olive waste residues | The incorporation of olive waste residues can reduce the use of chemical fertilizers, the soil fertility increases by incorporation of this compost | Altieri and Esposito (2010) |
| 15 | Liquid compost | Liquid (Vermiwash) from vermicompost and liquid from compost | Used in soil reclamation and removes 21% of undissolved zinc from the topsoil | Chiang et al. (2016) |
| 16 | Compost | Wastage (Rice, vegetables, and meat) from food industry | Stabilizes soil EC after 96 h of application, provide good climate for beneficial microorganism reproduction | Jiang et al. (2015) |
| 17 | Vermicompost | Cattle manure and earthworms (*Eisenia fetida*) | Improves soil structure and makes more nutrient availability to plants. Improve growth and yield of the crop | Ayyobi et al. (2014) |
| 18 | Vermicompost | Cow's dung, sewage sludge, and earthworms (*Eisenia fetida*) | Reduces pathogen population in rhizosphere. Improves soil quality | Karimi et al. (2017) |
| 19 | Vermicompost | Degradable municipal solid waste and earthworms (*Eisenia fetida*) | Have high soil remediation capacity and reduces contaminants, reduces heavy metal toxicity | Soobhany (2018) |
| 20 | Vermicompost | Earthworms (*Eisenia fetida*), food waste from bakery industry, and cow dung | Enriched soil NPK content when compared to chemical fertilizers | Yadav and Garg (2019) |

(continued)

**Table 5** (continued)

| S. No | Type material | Biomass source | Resulting effects | References |
|---|---|---|---|---|
| 21 | Vermicompost | Earthworms (*Eisenia fetida*), wastewater sludge | Incorporation of this vermicompost increases nutrient content in soil when compared to chemical fertilizer application. It stimulates good growth and development of the crop plants with good weight, height, and productivity | Valdez-Perez et al. (2011) |

support to standing crop plants. By applying organic amendments, crop productivity increases and helps in soil remediation and removing heavy metals from contaminated soils. Organic amendments stabilize EC and Cation Exchange Capacity (CEC) of soil, which makes more nutrients available to plants. By the application of organic amendments, we can manage organic waste by reducing environmental pollution by producing valuable manure that is used in crop production, and self-reliance on chemical fertilizers and products is also reduced in the long run by the application of compost and other organic amendments to soil.

## References


Abad Q, Shafiqi S (2024) Vermicompost: significance and benefits for agriculture. J Res Appl Sci Biotechnol 3(2):202–207

Abrar MM, Shah SAA, Sun N, Mehmood K, Aziz T, Waqas MA et al (2023) Long-term manure application enhances organic carbon and nitrogen stocks in Mollisol subsoil. Land Degrad Dev 34(3):815–832

Abubakar BSUI, Ismail N (2012) Anaerobic digestion of cow dung for biogas production. ARPN J Eng Appl Sci 7(2):169–172

Adhikary S (2012) Vermicompost, the story of organic gold: a review. Agric Sci 03:905–917

Ahmad A, Aslam Z, Bellitürk K, Iqbal N, Naeem S, Idrees M et al (2021) Vermicomposting methods from different wastes: an environment friendly, economically viable and socially acceptable approach for crop nutrition: a review. Int J Food Sci Agric 5(1):58–68

Ajayi SO, Odesanya BO, Avwioroko AO, Adebambo GS, Okafor B (2012) Effects of long term fertilizer use on trace metal levels of soils in a farm settlement. J Agric Res Dev 2(2):44–51

Altieri R, Esposito A (2010) Evaluation of the fertilizing effect of olive mill waste compost in short-term crops. Int Biodeterior Biodegradation 64(2):124–128

Asfaw MD (2022) Effects of animal manures on growth and yield of maize (Zea mays L.). J Plant Sci Phytopathol 6(1):033–039

Ayyobi H, Olfati JA, Peyvast GA (2014) The effects of cow manure vermicompost and municipal solid waste compost on peppermint (Mentha piperita L.) in Torbat-e-jam and Rasht regions of Iran. Int J Recycl Org Waste Agric 3:147–153

Bamdad H, Papari S, Lazarovits G, Berruti F (2022) Soil amendments for sustainable agriculture: microbial organic fertilizers. Soil Use Manag 38(1):94–120

Becker GC, Wüst D, Köhler H, Lautenbach A, Kruse A (2019) Novel approach of phosphate-reclamation as struvite from sewage sludge by utilising hydrothermal carbonization. J Environ Manag 238:119–125

Bekele A, Kibret K, Bedadi B, Yli-Halla M, Balemi T (2018) Effects of lime, vermicompost, and chemical P fertilizer on selected properties of acid soils of Ebantu District, Western Highlands of Ethiopia. Appl Environ Soil Sci 2018(1):8178305

Bhatt MK, Labanya R, Joshi HC (2019a) Influence of long-term chemical fertilizers and organic manures on soil fertility-a review. Univ J Agric Res 7(5):177–188

Bhatt MK, Labanya R, Joshi HC (2019b) Influence of long-term chemical fertilizers and organic manures on soil fertility-a review. Univ J Agric Res 7(5):177–188

Bhunia S, Bhowmik A, Mallick R, Mukherjee J (2021) Agronomic efficiency of animal-derived organic fertilizers and their effects on biology and fertility of soil: a review. Agronomy 11(5): 823

Brunetto G, Comin JJ, Schmitt DE, Guardini R, Mezzari CP, Oliveira BS et al (2012) Changes in soil acidity and organic carbon in a sandy typic hapludalf after medium-term pig-slurry and deep-litter application. Rev Bras Ciênc Solo 36:1620–1628

Cagasan U, Orbeta H (2020) Enhancing yield and profitability of Peanut (Arachis hypogaea L.) to application of different organic foliar fertilizer. Eurasian J Agric Res 4(2):124–133

Cannavo P, Recous S, Valé M, Bresch S, Paillat L, Benbrahim M, Guénon R (2022) Organic fertilization of growing media: response of N mineralization to temperature and moisture. Horticulturae 8(2):152

Carr PM, Cavigelli MA, Darby H, Delate K, Eberly JO, Fryer HK et al (2020) Green and animal manure use in organic field crop systems. Agron J 112(2):648–674

Chemura A (2014) The growth response of coffee (Coffea arabica L) plants to organic manure, inorganic fertilizers and integrated soil fertility management under different irrigation water supply levels. Int J Recycl Org Waste Agric 3:1–9

Chen Z, Li Y, Peng Y, Mironov V, Chen J, Jin H, Zhang S (2022) Feasibility of sewage sludge and food waste aerobic co-composting: physicochemical properties, microbial community structures, and contradiction between microbial metabolic activity and safety risks. Sci Total Environ 825:154047

Chew KW, Chia SR, Yen HW, Nomanbhay S, Ho YC, Show PL (2019) Transformation of biomass waste into sustainable organic fertilizers. Sustainability 11(8):2266

Chiang PN, Tong OY, Chiou CS, Lin YA, Wang MK, Liu CC (2016) Reclamation of zinc-contaminated soil using a dissolved organic carbon solution prepared using liquid fertilizer from food-waste composting. J Hazard Mater 301:100–105

Chojnacka K, Moustakas K, Witek-Krowiak A (2020) Bio-based fertilizers: a practical approach towards circular economy. Bioresour Technol 295:122223

Corden C, Bougas K, Cunningham E, Tyrer D, Kreißig J, Zetti E et al (2019) Digestate and compost as fertilisers: risk assessment and risk management options. European Commission, London

Craswell E (2021) Fertilizers and nitrate pollution of surface and ground water: an increasingly pervasive global problem. SN Appl Sci 3(4):518

Dadrasnia A, de Bona Muñoz I, Yáñez EH, Lamkaddam IU, Mora M, Ponsá S et al (2021) Sustainable nutrient recovery from animal manure: a review of current best practice technology and the potential for freeze concentration. J Clean Prod 315:128106

Di Costanzo N, Cesaro A, Di Capua F, Esposito G (2021) Exploiting the nutrient potential of anaerobically digested sewage sludge: a review. Energies 14(23):8149

Dolit SAM, Asli UA, Khamis AK, Baharulrazi N, Azman NR, Yunus NA (2022) Improving the water retention characteristics of sandy soil using food waste compost amendment and indigenous microorganisms. Chem Eng Trans 97:211–216

Easha NJ, Rahman M, Uddin MK (2015) Complete characterization of ETP sludge from textile industry; focusing the reuse potentiality as Organic manure. Jahangirnagar Univ Environ Bull 4: 31–37

El-Goud A, Amal K (2020) Efficiency response of vermicompost and vermitea levels on growth and yield of eggplant (Solanum melongena, L.). Alexandria Sci Exch J 41(January–March):69–75

Elouear Z, Bouhamed F, Boujelben N, Bouzid J (2016) Application of sheep manure and potassium fertilizer to contaminated soil and its effect on zinc, cadmium and lead accumulation by alfalfa plants. Sustainable Environ Res 26(3):131–135

Enebe MC, Erasmus M (2023) Vermicomposting technology-A perspective on vermicompost production technologies, limitations and prospects. J Environ Manag 345:118585

Ferreira PAA, Ceretta CA, Lourenzi CR, De Conti L, Marchezan C, Girotto E et al (2022) Long-term effects of animal manures on nutrient recovery and soil quality in acid typic hapludalf under no-till conditions. Agronomy 12(2):243

Franz M (2008) Phosphate fertilizer from sewage sludge ash (SSA). Waste Manag 28(10): 1809–1818

Frišták V, Pipíška M, Soja G (2018) Pyrolysis treatment of sewage sludge: a promising way to produce phosphorus fertilizer. J Clean Prod 172:1772–1778

Garg VK, Gupta R (2009) Vermicomposting of agro-industrial processing waste. In: Biotechnology for agro-industrial residues utilisation: Utilisation of Agro-residues, pp 431–456

Geng Y, Cao G, Wang L, Wang S (2019) Effects of equal chemical fertilizer substitutions with organic manure on yield, dry matter, and nitrogen uptake of spring maize and soil nitrogen distribution. PLoS One 14(7):e0219512

Ghosh T, Maity PP, Das TK, Krishnan PRAMEELA, Bhatia ARTI, Bhattacharya P, Sharma DK (2020) Variation of porosity, pore size distribution and soil physical properties under conservation agriculture. Indian J Agric Sci 90(11):2051–2058

Ghosh A, Kumar S, Das J (2023) Impact of leachate and landfill gas on the ecosystem and health: research trends and the way forward towards sustainability. J Environ Manag 336:117708

Guo ZC, Zhang ZB, Zhou H, Rahman MT, Wang DZ, Guo XS et al (2018) Long-term animal manure application promoted biological binding agents but not soil aggregation in a vertisol. Soil Tillage Res 180:232–237

Guo Z, Zhang J, Fan J, Yang X, Yi Y, Han X et al (2019) Does animal manure application improve soil aggregation? Insights from nine long-term fertilization experiments. Sci Total Environ 660: 1029–1037

Haggstrom G, Hannl TK, Hedayati A, Kuba M, Skoglund N, Ohman M (2021) Single pellet combustion of sewage sludge and agricultural residues with a focus on phosphorus. Energy Fuel 35(12):10009–10022

Hamm AC, Tenuta M, Krause DO, Ominski KH, Tkachuk VL, Flaten DN (2016) Bacterial communities of an agricultural soil amended with solid pig and dairy manures, and urea fertilizer. Appl Soil Ecol 103:61–71

Hartmann M, Six J (2023) Soil structure and microbiome functions in agroecosystems. Nat Rev Earth Environ 4(1):4–18

Hušek M, Moško J, Pohořelý M (2022) Sewage sludge treatment methods and P-recovery possibilities: current state-of-the-art. J Environ Manag 315:115090

Iderawumi AM, Kamal TO (2022) Green manure for agricultural sustainability and improvement of soil fertility. Farming Manag 7(1):1–8

Jasinska A, Prasad R, Lisiecka J, Roszak M, Stoknes K, Mleczek M, Niedzielski P (2022) Combined dairy manure-food waste digestate as a medium for Pleurotus djamor—mineral composition in substrate and bioaccumulation of elements in fruiting bodies. Horticulturae 8(10):934

Javaid A (2011) Effects of biofertilizers combined with different soil amendments on potted rice plants. Chilean J Agric Res 71(1):157

Jiang Y, Ju M, Li W, Ren Q, Liu L, Chen Y et al (2015) Rapid production of organic fertilizer by dynamic high-temperature aerobic fermentation (DHAF) of food waste. Bioresour Technol 197: 7–14

Joshi R, Singh J, Vig AP (2015) Vermicompost as an effective organic fertilizer and biocontrol agent: effect on growth, yield and quality of plants. Rev Environ Sci Biotechnol 14:137–159
Justice MJN (2021) Effect of zinc and boron fertilization on the yield of bari tomato 17 (Solanum lycopersicum L.). MS Thesis. Department of Soil Science, Sher-e-Bangla Agricultural University, Dhaka 1207:1-2
Kamar Zaman AM, Yaacob JS (2022) Exploring the potential of vermicompost as a sustainable strategy in circular economy: improving plants' bioactive properties and boosting agricultural yield and quality. Environ Sci Pollut Res:1–17
Karimi H, Mokhtari M, Salehi F, Sojoudi S, Ebrahimi A (2017) Changes in microbial pathogen dynamics during vermicomposting mixture of cow manure–organic solid waste and cow manure–sewage sludge. Int J Recycl Org Waste Agric 6:57–61
Karmegam N, Daniel T (2000) Effect of biodigested slurry and vermicompost on the growth and yield of cowpea, Vigna unguiculata (L.) Walp. variety Cl.
Kauser H, Khwairakpam M (2022) Organic waste management by two-stage composting process to decrease the time required for vermicomposting. Environ Technol Innov 25:102193
Khan F, Siddique AB, Shabala S, Zhou M, Zhao C (2023) Phosphorus plays key roles in regulating plants' physiological responses to abiotic stresses. Plants 12(15):2861
Lau YY, Hernandes E, Kristanti RA, Wijayanti Y, Emre M (2023) Exploring the potential of composting for bioremediation of pesticides in agricultural sector. Ind Domest Waste Manag 3(1):47–66
Leeson S, Caston L (2008) Using minimal supplements of trace minerals as a method of reducing trace mineral content of poultry manure. Anim Feed Sci Technol 142(3–4):339–347
Li Z, Zhang X, Xu J, Cao K, Wang J, Xu C, Cao W (2020) Green manure incorporation with reductions in chemical fertilizer inputs improves rice yield and soil organic matter accumulation. J Soils Sediments 20:2784–2793
Liu Y, Li B, Guo D, Munir MT, Song L, Wu X, Huang Y (2022) Feasibility of using different hydrothermal processes for sewage sludge management in China. Sci Total Environ 838: 156154
Li-Xian Y, Guo-Liang L, Shi-Hua T, Gavin S, Zhao-Huan H (2007) Salinity of animal manure and potential risk of secondary soil salinization through successive manure application. Sci Total Environ 383(1–3):106–114
Lorenz K, Lal R (2022) Introduction to organic agriculture. In: Organic agriculture and climate change. Springer International Publishing, Cham, pp 1–38
Loss A, Couto RDR, Brunetto G, Veiga MD, Toselli M, Baldi E (2019) Animal manure as fertilizer: changes in soil attributes, productivity and food composition. Int J Res Granthaalayah 7(9):307
Mahmud M, Abdullah R, Yaacob JS (2020) Effect of vermicompost on growth, plant nutrient uptake and bioactivity of ex vitro pineapple (Ananas comosus var. MD2). Agronomy 10(9): 1333
Manea EE, Bumbac C, Dinu LR, Bumbac M, Nicolescu CM (2024) Composting as a sustainable solution for organic solid waste management: current practices and potential improvements. Sustainability 16(15):6329
Manzoor A, Naveed MS, Ali RMA, Naseer MA, Maqsood UH, Saqib M et al (2024) Vermicompost: a potential organic fertilizer for sustainable vegetable cultivation. Sci Hortic 336:113443
Medany, M. (2011). Vermiculture in Egypt: current development and future potential. Food and agriculture Organization of the United Nations Regional Office for the Near East Cairo, Egypt. Cairo, 99
Mistry JAYANTA (2015) Vermicompost, a best superlative for organic farming: a review. J Adv Stud Agric Biol Environ Sci 2(3):38–46
Nicholson FA, Chambers BJ, Smith KA (1996) Nutrient composition of poultry manures in England and Wales. Bioresour Technol 58(3):279–284

Nunes N, Ragonezi C, Gouveia CS, Pinheiro de Carvalho MÂ (2021) Review of sewage sludge as a soil amendment in relation to current international guidelines: a heavy metal perspective. Sustainability 13(4):2317

O'Connor J, Hoang SA, Bradney L, Dutta S, Xiong X, Tsang DC et al (2021) A review on the valorisation of food waste as a nutrient source and soil amendment. Environ Pollut 272:115985

Olle M (2019) Vermicompost, its importance and benefit in agriculture

Pahalvi HN, Rafiya L, Rashid S, Nisar B, Kamili AN (2021) Chemical fertilizers and their impact on soil health. In: Microbiota and Biofertilizers, Vol 2: Ecofriendly tools for reclamation of degraded soil environs, pp 1–20

Palaniveloo K, Amran MA, Norhashim NA, Mohamad-Fauzi N, Peng-Hui F, Hui-Wen L et al (2020) Food waste composting and microbial community structure profiling. Processes 8(6):723

Papargyropoulou E, Steinberger JK, Wright N, Lozano R, Padfield R, Ujang Z (2019) Patterns and causes of food waste in the hospitality and food service sector: food waste prevention insights from Malaysia. Sustainability 11(21):6016

Picariello E, Pucci L, Carotenuto M, Libralato G, Lofrano G, Baldantoni D (2020) Compost and sewage sludge for the improvement of soil chemical and biological quality of Mediterranean agroecosystems. Sustainability 13(1):26

Pimentel D (1996) Green revolution agriculture and chemical hazards. Sci Total Environ 188:S86–S98

Pinotti L, Manoni M, Ferrari L, Tretola M, Cazzola R, Givens I (2021) The contribution of dietary magnesium in farm animals and human nutrition. Nutrients 13(2):509

Pushparaj M, Gnanavelrajah N (2023) Assessment on the quality of compost produced by commercial producers and the potential for compost production in Kilinochchi District. In: Self-sustaining agriculture: way forward for food security and safety, p 181

Rayne N, Aula L (2020) Livestock manure and the impacts on soil health: a review. Soil Syst 4(4): 64

Rehman SU, De Castro F, Aprile A, Benedetti M, Fanizzi FP (2023) Vermicompost: enhancing plant growth and combating abiotic and biotic stress. Agronomy 13(4):1134

Rodrigues J, Alvarenga P, Silva AC, Brito L, Tavares J, Fangueiro D (2021) Animal slurry sanitization through pH adjustment: process optimization and impact on slurry characteristics. Agronomy 11(3):517

Romano E, Brambilla M, Bisaglia C, Pampuro N, Pedretti EF, Cavallo E (2014) Pelletization of composted swine manure solid fraction with different organic co-formulates: effect of pellet physical properties on rotating spreader distribution patterns. Int J Recycl Org Waste Agric 3: 101–111

Rorat A, Courtois P, Vandenbulcke F, Lemiere S (2019) Sanitary and environmental aspects of sewage sludge management. In: Industrial and municipal sludge. Butterworth-Heinemann, pp 155–180

Roy R, Núñez-Delgado A, Wang J, Kader MA, Sarker T, Hasan AK, Dindaroglu T (2022) Cattle manure compost and biochar supplementation improve growth of Onobrychis viciifolia in coal-mined spoils under water stress conditions. Environ Res 205:112440

Samatha B, Rai PK (2022) Standardization of different pre-sowing seed treatments with Jeevamrutha, Panchagavya, neem oil and Vermiwash on plant growth yield and yield attributing traits of cowpea (Vigna unguiculata L.) Var. SSC-06. Int J Plant Soil Sci 34(23):458–465

Santos JA, Gonzaga MIS, dos Santos WM, da Silva AJ (2022) Water retention and availability in tropical soils of different textures amended with biochar. Catena 219:106616

Shaji H, Chandran V, Mathew L (2021) Organic fertilizers as a route to controlled release of nutrients. In: Controlled release fertilizers for sustainable agriculture. Academic, pp 231–245

Sharma P, Bano A, Singh SP, Varjani S, Tong YW (2024) Sustainable organic waste management and future directions for environmental protection and techno-economic perspectives. Curr Pollut Rep 10(3):459–477

Sinha RK, Hahn G, Singh PK, Suhane RK, Anthonyreddy A (2011) Organic farming by vermiculture: producing safe, nutritive and protective foods by earthworms (Charles Darwin's friends of farmers). Am J Exp Agric 1(4):363

Sinha RK, Hahn G, Soni BK, Agarwal S (2014) Sustainable agriculture by vermiculture: earthworms and vermicompost can ameliorate soils damaged by agrochemicals, restore soil fertility, boost farm productivity and sequester soil organic carbon to mitigate global warming. Int J Agric Res Rev 2(8):99–114

Smith KA, Van Dijk TA (1987) Utilisation of phosphorus and potassium from animal manures on grassland and forage crops. In: Animal manure on grassland and fodder crops. Fertilizer or waste? Proceedings of an international symposium of the European Grassland Federation, Wageningen, The Netherlands, 31 August–3 September 1987. Springer Netherlands, pp 87–102

Song X, Li H, Song J, Chen W, Shi L (2022) Biochar/vermicompost promotes hybrid Pennisetum plant growth and soil enzyme activity in saline soils. Plant Physiol Biochem 183:96–110

Soobhany N (2018) Remediation potential of metalliferous soil by using extracts of composts and vermicomposts from municipal solid waste. Process Saf Environ Prot 118:285–295

Strapazzon AJ, Theves FAB, Wiesel PG, Lobo EA (2021) Efficiency in the treatment of three livestock manures through a composting process with mechanized turning. J Agric Nat Resour 4(1):140–153

Sulaiman ISC, Mohamad A (2020) The use of vermiwash and vermicompost extract in plant disease and pest control. In: Natural remedies for pest, disease and weed control. Academic, pp 187–201

Terman GL (2018) Fertilizer sources and composition. In: Handbook of soils and climate in agriculture. CRC Press, pp 295–330

Usman K, Khan S, Ghulam S, Khan MU, Khan N, Khan MA, Khalil SK (2012) Sewage sludge: an important biological resource for sustainable agriculture and its environmental implications. Am J Plant Sci 03:1708

Usmani Z, Kumar V, Rani R, Gupta P, Chandra A (2019) Changes in physico-chemical, microbiological and biochemical parameters during composting and vermicomposting of coal fly ash: a comparative study. Int J Environ Sci Technol 16:4647–4664

Uysal A, Demir S, Sayilgan E, Eraslan F, Kucukyumuk Z (2014) Optimization of struvite fertilizer formation from baker's yeast wastewater: growth and nutrition of maize and tomato plants. Environ Sci Pollut Res 21:3264–3274

Valdez-Perez MA, Fernández-Luqueño F, Franco-Hernandez O, Cotera LF, Dendooven L (2011) Cultivation of beans (Phaseolus vulgaris L.) in limed or unlimed wastewater sludge, vermicompost or inorganic amended soil. Sci Hortic 128(4):380–387

Vanotti MB, García-González MC, Szögi AA, Harrison JH, Smith WB, Moral R (2020) Removing and recovering nitrogen and phosphorus from animal manure. In: Animal manure: Production, characteristics, environmental concerns, and management, vol 67, pp 275–321

Wako FL, Muleta HD (2022) The role of vermicompost application for tomato production: a review. J Plant Nutr 46(1):129–144

Walia SS, Kaur T (2024) Vermitechnology: history and its applications. In: Earthworms and vermicomposting: species, procedures and crop application. Springer Nature Singapore, Singapore, pp 37–53

Wang H, Xu J, Liu X, Zhang D, Li L, Li W, Sheng L (2019a) Effects of long-term application of organic fertilizer on improving organic matter content and retarding acidity in red soil from China. Soil Tillage Res 195:104382

Wang X, Zheng G, Chen T, Shi X, Wang Y, Nie E, Liu J (2019b) Effect of phosphate amendments on improving the fertilizer efficiency and reducing the mobility of heavy metals during sewage sludge composting. J Environ Manag 235:124–132

Warncke DD, Dahl J, Jacobs L, Laboski C (2009) Nutrient recommendations for field crops in Michigan. Michigan State University Extension, East Lansing

Wei Y, Wang N, Lin Y, Zhan Y, Ding X, Liu Y et al (2021) Recycling of nutrients from organic waste by advanced compost technology-a case study. Bioresour Technol 337:125411

Wolka K, Melaku B (2015) Exploring selected plant nutrient in compost prepared from food waste and cattle manure and its effect on soil properties and maize yield at Wondo genet, Ethiopia. Environ Syst Res 4:1–7

Yadav A, Garg V (2019) Biotransformation of bakery industry sludge into valuable product using vermicomposting. Bioresour Technol 274:512–517

Yahaya SM, Mahmud AA, Abdullahi M, Haruna A (2023) Recent advances in the chemistry of nitrogen, phosphorus and potassium as fertilizers in soil: a review. Pedosphere 33(3):385–406

Yatoo AM, Ali MN, Baba ZA, Hassan B (2021) Sustainable management of diseases and pests in crops by vermicompost and vermicompost tea. A review. Agron Sustain Dev 41(1):7

Ye L, Zhao X, Bao E, Li J, Zou Z, Cao K (2020) Bio-organic fertilizer with reduced rates of chemical fertilization improves soil fertility and enhances tomato yield and quality. Sci Rep 10(1):177

Zaib M, Zubair M, Aryan M, Abdullah M, Manzoor S, Masood F, Saeed S (2023) A review on challenges and opportunities of fertilizer use efficiency and their role in sustainable agriculture with future prospects and recommendations. Curr Res Agric Farm 4(4):1–14

Zicari SM (2003) Removal of hydrogen sulfide from biogas using cow-manure compost. Doctoral dissertation, Cornell University

# Compost Application for Soil Health and Nutrient Restoration: Benefits and Limitations

**Mithlesh Kumar, S. N. Pandey, Sayyada Shakeel, and Jitendra Kumar**

## 1 Introduction

Compost is a nutrient-rich organic material produced by decomposition (aerobic, solid-state fermentation) of a variety of organic raw materials like crop residues, dung of dairy animals, poultry wastes, slaughter wastes, wastes produced by agricultural crops, and agro-industrial organic wastes. Compost improves soil structure, increases nutrient availability to plant roots, helps in soil aeration, and mitigates the effects of climate change. Compost is extensively used in conventional and organic agricultural systems and is a valuable source of different nutrient elements to crop plants. Continuous use of compost is helpful in increasing organic carbon, fertility, and the health of the soil. The increased soil organic carbon (SOC) contributes toward improvement and maintenance of soil health (Ladha et al. 2011). Application of compost in soil has many positive effects, which have promoted the use of compost in modern agricultural practices (Norris and Congreves 2018) after a lag of time in the era of the Industrial Revolution. Composting is a microorganism-mediated, thermophilic, aerobic and solid-state fermentation procedure carried out to transform a variety of organic materials into relatively more stable products, by which humic acid and other related compounds are formed. The feedstock commonly used for composting is agro-industrial, agricultural wastes, manure, sewage, food scraps, and other degradable organic matter, which are transformed into a stable form that persists in soil for a longer period as compared to the untreated raw organic wastes (Niamat et al. 2019). Due to modern agricultural practices, urbanization and industrialization, inoculation of pollutants, excessive use of inorganic fertilizers and lack of crop rotation cycle, the valuable soil resource is heavily depleted in SOC

M. Kumar
Department of Botany, Government Degree College, Todarpur, Hardoi, Uttar Pradesh, India

S. N. Pandey (✉) · S. Shakeel · J. Kumar
Department of Botany, University of Luckow, Lucknow, Uttar Pradesh, India



R. K. Singh et al. (eds.), *Composting of Farm Residues by Nature-Based Solutions*,
https://doi.org/10.1007/978-3-032-12255-1_10

content, available nutrient elements, increased bulk density, high pH, electrical conductivity (EC), lower cation exchange capacity (CEC) and poor soil microbial diversity (SMD) which leads to degradation of soil health. Excessive use of inorganic fertilizers in soil is a global problem because sole reliance on chemical fertilizers in soil ultimately leads to eutrophication of water bodies by runoff. The enrichment of nutrients in water bodies causes algal blooms and anoxia. For the improvement of physico-chemical properties of agricultural lands, application of compost is a reliable way, mainly for the improvement of soil structure, SOC, and microbial diversity (Stratton et al. 1995).

Compost application helps in the restoration of health and productivity, structural stability, optimal porosity and water-holding capacity (WHC), nutrient restoration and supply to plants, degradation of toxic pollutants, nutrient recycling, maintenance of optimal SMD, ample SOC and other ecosystem services and functions without harming the environment. These improvements in soil lead to better soil health and productivity. To produce healthy crops and better human health, the health of the soil is a crucial factor, and healthy soil plays a crucial role in ecosystem services and crop production.

## 2 Soil Health and Ecosystem Services

Soil health is a relatively new concept that came into existence a few decades ago and mainly relies on the biological properties of soil, rather than relying only on chemical and physical properties of soil, as in the previous concept of soil quality and fertility. Generally, soil health or soil quality is defined as the continuous capacity of soil to work as a dynamic living system within the boundaries of land use and ecosystem to promote the quality of air and water, to support biological productivity and to sustain plants, human and other animals' health. However, some researchers like Pankhurst et al. (1997) suggested that soil quality refers to the capacity of soil to support plant growth. The important character of healthy soil is that such soils keep a balance between environmental quality, productivity and health of plants, animals and human beings and these traits are significantly influenced by land use and management practices and decisions.

In recent years, there has been intense research on the effects of long-term agricultural management practices on soil quality and soil health. When we focus on soil health, it is necessary to decrease dependency on inorganic fertilizers and to rely on the reduction of mechanical impact by tillage, increased and frequent application of organic fertilizers, crop residue returns and crop rotation to enhance the diversity of plants both temporally and spatially and the reduction of erosion with contour plowing or grass strips (Karlen et al. 2019). Earlier, the concept of soil health was centered on specific properties of soil and its ability to sustain a wide range of ecosystem services, which can support sustainable, long-term cropping systems. Generally, the terms soil quality and soil health have been used commonly for crop production as well as environmental sustainability (Doran et al. 1996). In earlier data of agricultural research, soil health was only considered to be related to soil structure and nutrient content, but the biological component was not incorporated into it. The present-day research of soil must take into consideration the biological properties of soil as an essential component of soil health. Soil health takes into consideration the interaction between microbial communities of soil and plants, which affects plant growth (Berendsen et al. 2012). High plant growth and crop production require management of soil microbial richness, activity, specific function and community composition (Bonanomi et al. 2018). The concept of soil health is now connected to the emerging concept of one health, in which environment, human beings and animals are included. The scope of soil health ranges not only to human health but also includes the health of whole plants, while that of soil quality is centered on ecosystem services with reference to human beings (Bünemann et al. 2018).

Soil fertility refers to the natural, sustainable ability of soil for crop production or the capacity of soil to supply essential nutrient elements to plant roots (Agegnehu and Amede 2017) or soil fertility refers to function of soil in terms of crop production (Bünemann et al. 2018) and is managed at field scale by farmers for the efficient crop production and entirely focuses on better production of food material, fiber and fuel for human use (Patzel et al. 2000). The term soil quality refers to the ability of soil in terms of agricultural functions as well as in the context of immediate environment. Thus, soil quality incorporates the effects of soil on water quality and the health of plants and animals in whole ecosystems (Doran and Parkin 1994). Soil health includes scales, parties, functions, and valuation tools appropriate to soil fertility and soil quality, whose focus is beyond crop production and/or the benefits of human beings. The significant accomplishment of the concept of soil health is the incorporation of an immediate biological outlook to manage soil to cope with long-term sustainability challenges regarding crop production. In a biological perspective of soil health, there is a description of the association between soil health management and sustainable crop production, human health, water quality and mitigation of climate change. Soil provides many ecosystem services like water quality control, sustainable crop production, improvement of soil health and mitigation of climate change. These ecosystem services are taken into consideration during the management of soil health because they affect water quality and water use, the health of human beings and other animals, biodiversity, and climate (Jian et al. 2020). However, the basis of soil health is that the only management of nutrient bioavailability is not sufficient for optimum growth of plants (Bünemann et al. 2018).

## 3 Soil Health and Its Degradation

Soil is a nonrenewable natural resource associated with vital ecosystem services and provides a medium for the growth of plants, provides anchorage to roots, habitat for soil microbes, supply essential nutrient elements, supply oxygen and water to plant roots and act as an environmental natural buffer system that facilitates degradation and simplification of toxic xenobiotic and other harmful substances (Amoakwah et al. 2020). Soil is a complex mixture of soil separates (physical), chemical and biological factors. A fine balance between organic, inorganic, and biological components of soil is very crucial for ecosystem services and crop production but this equilibrium between soil physical, chemical, and biological component is heavily disturbed due to the introduction of many toxic substances in soil, extensive cultivation of crops without following the crop rotation cycle, irrigation with poor quality water, excess application of chemical fertilizers and continuous depletion of SOC lead to degradation of soil health. The alteration in optimal properties of soil leads to a decline in soil ecosystem services and crop production. Several land degradation forces are prevailing which are greatly responsible for alteration in physico-chemical and biological properties like increase in bulk density, deposition of high salt content (salinity and sodicity), low total and available nutrient content, high soil pH and EC, excess of heavy metal concentration like chromium (Cr), cadmium (Cd), mercury (Hg), arsenic (As), lead (Pb), and nickel (Ni) etc., land degradation due to erosion and inoculation of aromatic compounds like benzene, toluene, polyaromatic hydrocarbons (PAHs), and other persistent organic pollutants (POPs), polyfluoroalkyl, perfluoroalkyl, pesticides, herbicides, antibiotics, and microplastics etc. When incorporated in soil, these substances cause deterioration in physico-chemical and biological properties of soil (Evans et al. 2019). These degraded lands pose a severe threat to the survival of soil microflora, growth and yield of plants, and thus negatively affect soil health. Many of these pollutants can be taken up by plant roots and accumulated in different plant parts such as roots, stems, leaves, flowers and fruits. These toxic substances accumulated in plant tissues are supplied to organisms of higher trophic levels of the food chain and are injurious to the health of these organisms (Oliver and Gregory 2015).

According to FAO, integrated and interconnected tasks are required to accomplish sustainable agriculture, mainly to restore degraded lands and improve soil health. Thus, the necessity of the time is to protect, preserve, and improve the health and biological diversity of soil (FAO 2021). If mitigation of these degraded lands is not taken into consideration, it will cause irreversible and irreparable losses. So, there is an urgent need to introduce feasible mitigation strategies to maintain soil health. Some important strategies include better and efficient land use management practices, introduction of new amendments and suitable remedial machinery, etc.

## 4 Assessment of Soil Health

Assessment of soil health takes into consideration the physical, chemical, and biological properties of soil, instead of focusing on any particular property of soil (Moebius-Clune 2016) and includes soil structure, WHC, bulk density, pH, EC, extractable potassium, phosphorus, nitrogen, SMD, respiration, soil enzymes and SOC, etc. The selection of soil health indicators (SHIs) and the connection between assessment of soil health, ecological services, and productivity is a matter of intense discussion (Bagnall et al. 2023). The SHIs are the measurements of properties of soil that are helpful in the assessment of soil health. Several soil health indicators are taken as reference for measuring the response for nutrient cycling (mainly carbon and nitrogen dynamics), which influence other parameters of soil like soil pH, EC and nutrient cycling. At present, quantification and management of SMD are important components of soil health assessment and management practices. The SOC and microbial diversity are depleting rapidly, so there is an urgent need to conserve and improve SMD to restore and sustain soil health (Veresoglou et al. 2015). Haney soil health test (HSHT) (Chu et al. 2019) and Cornell comprehensive assessment of soil health-CASH (Gholoubi et al. 2018) are the most recent models of soil health assessment, which are used to quantify soil health under diverse cropping systems and these focus on soil biology like soil respiration, available nutrients to plants, SMD and bioavailable carbon and nitrogen. A parameter should fulfill several measures to be used as an indicator of soil health, which are appropriate to soil health, like ecosystem functions and services, sensitive, practical, and informative from a management point of view (Rinot et al. 2019). Currently, about half of the soil health indicators are being used in more than 20% of soil health analysis schemes which fulfill all the four measures; but there are also some parameters used as SHIs which do not fulfill these parameters as for example total organic carbon (TOC) fulfill three measures but it is not sensitive which means this is not changeable quickly while organic carbon fraction can be used as SHI as it is more sensitive (Van Wesemael et al. 2019). To date, many SHIs which have been developed and discussed include enzyme activity and microbial diversity (VeVerka et al. 2019), permanganate oxidizable carbon and mineralization (Thomas et al. 2019; Van Ex and Karlen 2019), carbon and nitrogen (C: N) ratio, physical properties of soil such as porosity, WHC, aggregate formation of soil particles, surface and subsurface penetration resistance (Van Ex and Karlen 2019), chemical properties of soil available fraction of micronutrients such as zinc (Zn), iron (Fe), copper (Cu), manganese (Mn) and potassium (K) and activity of alkaline phosphatase (involved in phosphorus cycling) (Thomas et al. 2019).

# 5 Soil Microflora and Soil Health

The microbial community is an important component of soil composition and greatly affects soil health, so the soil with an adequate population of plant growth-promoting microbes is good for soil health. Soils with a surface area of $10^1$ $m^2$ $g^{-1}$ are estimated to have bacterial density of $10^{10}$ cells $g^{-1}$ and 1000 cells per colony (Raynaud and Nunan 2014). At present, the role of SMD in sustaining soil fertility and the relationship between biological activity and physico-chemical properties of soil have been studied and well established. SMD is regarded as a very prominent parameter for the measurement of soil health. Biological activities of microbes play an important role in soil formation by influencing the processes of weathering of rock material and pedogenesis. The amount of SOM (a physical property) affects the dynamics of SMD and thus indicates soil health and productivity (Cynthia et al. 2016). SOM improves soil fertility by improving soil microbial population and by increasing WHC, which contributes to the growth and development of plants. Degradation in soil health causes microbial habitat destruction and a reduced food network of soil, which diminishes microbial population in soil. Decline in microflora in soil reduces the viability of the cropping system and increases disease and pest risks, which ultimately reduces crop productivity. Thus, the management of soil health is necessary to increase profit, and microbial diversity, which improves soil health and function, reduces input costs and increases the profitability of the crop. Soil microbes are actively involved in chemical transformation reactions, for example, microbes help in maintaining soil fertility by enhancing nutrient biogeochemical cycles like nitrogen, phosphorus and sulphur cycles, etc. Microbes enhance decomposition of introduced SOM as well as nutrient recycling, and this is a typical example of chemical transformation by soil microbes (Khatoon et al. 2017). Several groups of soil microorganisms intensify the quantity of nutrients in soil, for example, biological nitrogen-fixing bacteria/microbes fix atmospheric molecular nitrogen ($N_2$) into plant-absorbable forms (nitrate and ammonium ions). The soil microbial community is significantly changed when compost is inoculated with ammonifying and nitrifying microbes. Similarly, the mixing of nitrogen, combining mineral phosphorus with compost, resulted in stimulation of the growth of phosphate-solubilizing microbes.

## 6 Role of Compost in Restoration of Soil Health

In the last few decades, the consumption of chemical fertilizers has increased greatly due to the high demand for food grains for a rapidly expanding population, and this has also propelled agricultural market inputs like high demand for fertilizers and pesticides, etc. (FAO 2017). In the year 2020, the annual global market demand for fertilizers was expected to reach about 201.7 metric tons (Mt) (186.6 and 194.4 Mt. in the years 2016 and 2018, respectively), and this is still expected to grow by 25 metric tons annually (FAO 2019). Continuous excess use of chemical fertilizers leads to many environmental complications, such as soil degradation, groundwater pollution, nutrient loss, and a decline in the soil microbial population. Due to these drawbacks of chemical fertilizers, livestock-based biofertilizers have received significant attention and are currently defined as the green revolution. Biofertilizers are rich in organic carbon, nutrient elements, and supply many productive microbes to soils, which are useful in the maintenance and restoration of soil health, which is being degraded by various degradation forces (Chojnacka et al. 2020). Manure-based fertilizers provide a sustainable alternative to chemical fertilizers and are helpful in high crop production, improved soil health, sustainability and long-term ecosystem services. Organic fertilizers application is a very important resource to replace and/or recycle the organic matter produced due to crop cultivation and supply of essential nutrients to the soil (Song et al. 2017). The application of organic fertilizer is necessary to sustain and restore fertility and enhance the productivity of agricultural lands. Due to several adverse effects and increasing prices of synthetic chemical fertilizers, application of organic amendments like manure, compost, decomposed crop residue, and other organic matter is highly recommended. Composted material has highly degraded organic matter as compared to manure or crop residue, so it has more persistent and long-lasting effects than quickly decomposing manure and crop residue. The outcomes of compost application on soil properties depend on the various soil properties like particle size and concentration of nutrient element, which in turn depends on composting conditions, duration and feedstock composition (Duong et al. 2012).

Organic matter content and soil microbial activity are the main properties important for soil health and improving the quality of draining water. SOC content can be increased by increasing the use of organic amendments in soil by reducing tillage (Karlen et al. 2019). The increased SOC also increases aggregation and regulates microbial mineralization to carbon dioxide. Application of nitrogenous fertilizers in soil leads to the emission of a powerful greenhouse gas, nitrous oxide (Paustian et al. 2016). So the judicious use of nitrogenous fertilizers is crucial for the management of soil health. There are many types of organic fertilizers which are environmentally safe, sustainable, cost-effective, eco-friendly and readily available to farmers. There are many ways to produce compost, such as the decomposition of crop and plant residue, the generation of biochar by the process of pyrolysis, cattle manure, municipal solid waste and waste of sugarcane industry like bagasse, press mud and molasses. Some important organic fertilizers, based on their raw organic material, are as follows:

## 6.1 Cattle Manure

Cattle manure is a good source of organic fertilizers with a high content of essential macro- and micronutrients, and it also improves the physical, chemical, and biological properties of soil. Application of cattle manure in soil improves labile nitrogen fraction, SOC and microbial activity. An increase in microbial activity due to the addition of cattle manure, in turn, increases the disintegration and the maintenance of SOC as well as nutrient cycling. The improved microbial activity in composted soil promotes rapid decomposition and stabilization of SOC as well as biogeochemical cycling of nutrients. There are reports that the long-term application of manure improves the labile nitrogen and carbon fraction, SOM, SOC, and WHC (Maharjan et al. 2021). SOM content is positively correlated with soil WHC. Amendment of compost in soil affects physico-chemical properties of soil like infiltration rate, bulk density, hydraulic conductivity, aggregate stability, water content and porosity (Cogger 2005). Compact soils usually have a high bulk density, while less compact soils with low clay and high organic carbon content (OCC) have low bulk density.

## 6.2 Vermicompost

Vermicompost is also called worm castings and this is the most prominent product of vermiculture. Vermicompost is produced by earthworms as they consume and digest organic material, and it is rich in different essential nutrient elements (Zhou et al. 2022). Vermicompost is dark and odorless matter enriched with advantageous microbes, humic substances, and essential nutrients. Vermicompost improves soil structure, endorses beneficial microbes, enhances nutrient bioavailability and soil fertility (Mupambwa and Mnkeni 2018). The capacity of vermicompost to release nutrients slowly is favorable and makes it a model natural fertilizer for organic-based agriculture. Due to many physical and chemical properties of vermicompost, it is useful in improving soil properties. Vermicompost has considerably higher nutrient concentration than conventional garden compost. Generally, vermicompost contains twice as much concentration of both micro- and macronutrients. Many macronutrients like nitrogen, calcium, potassium, phosphorus, and magnesium are present in relatively higher concentrations and provide essential nutrients to plants. Along with macronutrients, vermicompost also provides micronutrients like zinc, iron, copper, magnesium, and boron, which are vital for different physiological processes in plants (Lim et al. 2015). Such ample availability of micro and macronutrients in vermicompost contributes to improving soil structure, fertility, and nutrient supply to plant roots and microbes. There is significant variation in the range of total nitrogen content in vermicompost depending on the way of organic input processing. Generally, the total nitrogen content ranges from 0.1% to 4% which can be further increased through nitrogen-rich raw material (Edwards and Bohlen 1996).

## 6.3 Poultry Manure

The expanding area of poultry farming produces a very high amount of waste material due to chicken droppings, which, after decomposition, produce a valuable manure loaded with organic carbon and minerals (Adekiya et al. 2019). The high nutrient and OCC of poultry manure provide essential nutrient elements to crop plants. Agricultural application of poultry manure for the improvement in soil has a significant effect and is used for such purposes globally. Poultry manure application in soil causes an increase in SOC, essential nutrient elements and physico-chemical properties of soil, which in turn improves crop yield (Adekiya et al. 2019). Poultry manure application results in an increase in nutrient element content in soil, which is due to the release of nutrients from decomposed organic matter. A significant increase in concentration of organic carbon, nitrogen, potassium, phosphorus, magnesium, and calcium (Adekiya et al. 2019).

## 6.4 *Municipal Solid Waste*

Municipal solid waste (MSW) is a random mixture of solid discards produced by rural and urban societies. The common constituents of this waste include garden litter, kitchen scraps and packaging. Kaza et al. (2018) reported that in 2016, 2.01 billion Mt. of MSW was produced globally, which is expected to reach 2.59 billion metric tons by 2030. Organic waste produced from households contains many essential nutrient elements and a significant amount of carbon (C). Of the total global waste produced, only about 19% is recovered through composting and recycling, 40% is disposed of in landfills, 11% is ignited through modern combustion processes, while the remaining portion is managed by burning or open dumping practices (Kaza et al. 2018). The landfill management of MSW requires huge space and has many environmental risks like pollution, hostile smell, fire risk and habitat destruction (Vaverkova et al. 2019). By composting, about 40–50% of the volume of MSW can be decreased. Roy et al. (2021) reported that composting or recycling of organic waste returns 191 million USD due to MSW management and savings from fertilizer costs. Rajaeifar et al. (2017) reported that in the year 1995, the European Union annually produced only 14 million tons (MT) of compost (about 6.2% of MSW produced), and this was increased to 36 MT by the year 2013 (about 15% of MSW produced). There are many countries that are converting a larger amount of MSW into compost, like Italy (18%), Luxembourg (20%), Switzerland (21%), Liechtenstein (23%), the Netherlands (27%), and Austria (31%) (Kaza et al. 2018). These countries are also securing high ranks in terms of wealth generated and per capita gross national income (World Bank 2023).

In 2017, India accounted for only 10% of MSW in compost, while 60% was dumped openly (Pujara et al. 2019), so the huge quantity of MSW produced can be converted into compost, which leads to income generation, environmental safety, sustainability and nutrients recycling. MSW-derived compost is applied in soil as a soil amendment or nutrient additive and mainly focuses on nitrogen supply. Besides nitrogen, MSW compost also supplies sufficient levels of phosphorus, similar to some mineral fertilizers, when supplied in comparatively high quantities (Turrion et al. 2018) and improves magnesium (Mg) and potassium (K) (Srivastava et al. 2016). In horticultural applications, MSW compost maintains WHC, reduces bulk density and supplies essential nutrient elements (Pujara et al. 2019).

## 6.5 *Biochar*

The global population is increasing rapidly at an alarming rate, and this expanding population also imposes pressure on the production and consumption of livestock to fulfill dietary protein requirements. The large production of livestock also produces a large amount of livestock manure; the disposal of this huge waste is a serious environmental problem across the world. The livestock manure produces a huge quantity of organic materials, gases, and many other substances which pose a substantial threat to natural resources (like soil and water) leading to their degradation and its mitigation cost is very high; for example, in Europe, the pollution cost due to the management of manure is estimated at about 12,300 million euros per year (Bernal et al. 2015). Anaerobic digestion of livestock manure has unlocked the prospects to remediate environmental load due to livestock manure waste production and to increase the production of biobased products. This anaerobic digestion of livestock is a powerful substitute for the production of bioenergy and an effective tool to abate emissions of greenhouse gases (Fagerström et al. 2018). The efficient management of livestock manure is important for the recovery of nutrients to generate novel, superior fertilizers, which greatly rely on the setup and supervision of the process. The unprocessed livestock manure has certain limitations, so it must be processed to remove harmful substances before application to the soil (Bernal et al. 2015). There exists a range of technologies for the processing (nutrients concentration, volume reduction of livestock manure), but the nutrient recovery approaches still do not fulfill environmental and economic requirements.

Biochar is the carbon-rich product formed by pyrolysis of feedstock in the absence of oxygen. Biochar amendment in soil improves physico-chemical properties like bulk density, porosity, chemical properties like soil pH, CEC and biological properties like microbial activity and available oxygen in the rhizosphere, which ultimately improve soil fertility (Zhang et al. 2019). To produce biochar, a range of feedstock is used, like tree leaves, straw, husk, corn stock, bagasse, and other agricultural wastes, industrial wastes, and waste of animals and birds have been extensively used. Over the last few decades, many scientists across the world have proposed the use of biochar to overcome poor soil fertility, to increase productivity and combat climate change due to its potential to enhance carbon sequestration, improve soil health, generate bioenergy, and immobilize different organic and inorganic pollutants. Application of biochar in soil improves soil nutrients, water retention and availability, soil porosity, soil structure and soil aeration, ultimately increasing growth and yield of plants (Singh et al. 2019). Hui (2021) reported that the addition of biochar in soil improves carbon sequestration and reduces the emission of greenhouse gases, and increases the soil carbon pool by changing the composition of the microbial community. Due to its large surface area and porous nature, biochar provides habitat to the microbial community of soil experiencing stress and its ash content supplies nutrients to plant roots and soil microbes (Kocsis et al. 2022). The increased surface area of biochar helps in the retention of nutrients and their release. The long-term stability and resistant nature are the main advantages of biochar over other organic amendments. In the recent decade, the application of biochar has gained significant notice for its potential to raise the soil carbon pool (Smith 2016). Biochar with a labile carbon pool has been reported to increase microbial activity in biochar-applied soil. Application of biochar and compost helps in a sufficient supply of available micro and macronutrients to soil microbes and plant roots. Nutrient availability and composition of biochar rely on conditions of pyrolysis and feedstock (Purakayastha et al. 2019).

## 7 Role of Compost in Nutrient Restoration

To restore the OCC of soil, application of organic fertilizers like animal manure, compost and biochar is the need of present-day agriculture to sustain soil health. There are some mechanisms through which organic amendments enhance carbon storage of soil, such as an increase in soil microbial biomass and activity, promoting soil water-stable aggregation and incorporation of recalcitrant carbon (Smith 2016). Application of compost in eroded soil is beneficial in many ways as it reduces crust formation, which in turn promotes percolation, reduces surface flow rates and enhances water storage in such soil. When compost is applied to the soil surface, it diminishes the evaporation rate, releases nutrients in the soil and makes a favorable environment in the rhizosphere, which ultimately promotes vegetative cover in such lands. The application of compost greatly helps in the restoration of degraded lands, for example, in eroded soils. Compost promotes forest cover as well as reduces the loads of dissolved solids in water flowing through such soil and thus ultimately reduces dumping of solid particulates in water bodies and lowland areas (Cogger 2005).

The effects of compost are not as quick as in the case of chemical fertilizers, but are visible after several months and persist for years. Compost releases essential nutrients like nitrogen, potassium, phosphorus and SOC in soil for the growth and development of plants. Compost application in soil not only enhances nutrient bioavailability but also improves soil structure. The application of compost in soil results in the mixing of humic acid precursors. The transformation of these humic acid precursors in humic acid, fulvic acid, humins and other nutritionally critical compounds causes better soil health and plant growth by virtue of the regulated supply of nutrient elements and thus checks nutrient loss by runoff with stormwaters and leaching. To prevent depletion in SOC and maintain an optimum level of SOC, regular amendment of organic material in the soil is necessary. It has been reported that compost inoculated with nutrient-transforming microbes and/or augmented with nutrients provides supplementary long-term benefits. In composted organic materials, the rate of release of nitrogen has been found to be slow, which leads to extension of the duration of the leaching process, as per the span of decomposition of the organic material applied. It is important to note that the persistence of chemical fertilizers is very short, and thus the rate of release of nitrogen is very high compared to that of compost. This is why the soils solely receiving chemical fertilizers are rapidly nitrogen-depleted (Claassen and Carey 2007). The Clean Water Act (CWA) amendments of 1987 started a major initiative for the improvement of the quality of stormwater and found that the quality of water in these stormwaters carries a heavy nutrient load, which promotes eutrophication. Use of compost in soil is also beneficial because it checks excessive nutrients leaching and thus reduces nutrient enrichment and algal blooming of water bodies (Glanville et al. 2004). The compost has uncertainty of its nutrient content (mainly nitrogen unavailability) and depends on the type and quality of material used and the conditions of composting. To overcome such uncertainty of nutrients, several mineral fertilizers are needed to mix with compost, which fulfill the nutrient requirements of crops and prevent environmental losses (Wortman et al. 2017). The uncertainty of nitrogen content in compost is due to the fact that most of the nitrogen in organic fertilizers is found in organic form, and this is not directly available to plant roots. The organic nitrogen is transformed into mineral nitrogen by different microbes through the process of mineralization. The process of mineralization is affected by a range of factors like water content, temperature, properties of soil and organic material (Robertson and Groffman 2007).

## 7.1 *Nitrogen in Manure*

In organic fertilizers (compost), nitrogen exists in both organic and inorganic forms. The primary inorganic form of nitrogen in manure is ammonium nitrogen ($NH_4$–N), which is released quickly and is immediately available for plant roots, while organic nitrogen is released very slowly and is available to plant roots after mineralization (Lpelc 2019). This inorganic form of nitrogen ($NH_4$–N) is prone to volatilization in the form of ammonia. Amalgamation of manure in soil is useful to prevent such nitrogen losses through volatilization. During the decomposition of organic matter, nitrogen undergoes two different stages: mineralization and immobilization. In mineralization, organic nitrogen is decomposed into nitrate ($NO_3^-$) and ammonium ions ($NH_4^+$), which can be absorbed by plant roots. Immobilization is the process by which microorganisms absorb nitrogen, or sometimes, when mineral nitrogen is associated with SOM and prevents its access to plant roots. This immobilized nitrogen will be available for plant roots once the microorganisms die and make this immobilized nitrogen free (Chaves et al. 2007). The degree and rate of immobilization of nitrogen depend on the composition of compost material; for example, compost with a high C: N ratio leads to high immobilization of nitrogen. Due to its immobilization, the rate of release of nitrogen is slowed when compost is supplemented with nitrogenous compounds.

Generally, ammonium is quickly transformed into nitrite by nitrifying bacteria in well-aerated soil, and this nitrite can move easily in deeper soil layers with irrigation or rainwater and increases the risk of leaching in groundwater. The rate of release of nitrogen in composted organic materials becomes slower than usual, due to which the leaching process may take years for the decomposition of the composting material. In the case of chemical fertilizers, the rate of release of nitrogen is higher, and the released nitrogen persists for a shorter period, due to which nitrogen is depleted quickly (Claassen and Carey 2007). The main factor that regulates nitrogen availability is the carbon (C) and nitrogen (N) ratio of the compost material. C/N ratio $<$ 25:1 promotes net mineralization, while that $>$25:1 promotes nitrogen immobilization, which is responsible for incorporation of mineral nitrogen from soil solution into microbial biomass (Myrold and Bottomley 2008). In some studies, this ratio of C/N has been reported to range from 14:1 to 19:1 in many organic amendments (Lazicki et al. 2020).

To obtain better results about persistence and release of nitrogen, nitrogen is recommended to be applied with compost fertilizers, which lead to better plant growth and crop production (Ahmad et al. 2006). The frequent application of stabilized and mineral-supplemented compost material in soil enhances SOC and soil organic nitrogen (SON) because of nitrogen mineralization, as compared to non-supplemented, less stabilized compost materials having a high C/N ratio. Application of organic amendments in soil adjusts the dynamics of nitrogen availability to plant roots by the mineralization of ample stock of SON produced due to frequent application of more stabilized organic amendments.

## 7.2 *Macronutrients in Compost*

Organic manure contains a significant amount of macronutrients (Table 1) like potassium (K) and phosphorus (P), which are easily available to plant roots. The required concentration of K and P in a fertilizer depends on soil concentration and the yield goals of crops (Szogi et al. 2015). Nutrient content varies greatly in animal manure as per the place of breeding. In Japanese animal manure, the concentration of nutrients (calcium, magnesium, potassium and nitrogen) is as follows: Poultry > swine > cattle (Harada et al. 1993). Besides these nutrients, animal manure also contains phosphorus and sulphur. Thus, when compost is applied to soil, it leads to an increase in the availability of macronutrients, which are essential for the growth of plants.

**Table 1** Comparative nutrient composition of different compost sources[a]

| Property | Nutrients composition | | |
|---|---|---|---|
| | Chicken | Cow | Swine |
| pH | 7.09 | 7.97 | 7.65 |
| EC (dS $m^{-1}$) | 3.76 | 4.45 | 4.18 |
| Total carbon (%) | 30.61 | 35.32 | 29.12 |
| Total nitrogen (%) | 5.81 | 2.54 | 2.88 |
| C/N | 5.28 | 13.89 | 10.11 |
| Total phosphorus (%) | 16.89 | 8.66 | 17.56 |
| K (mg $kg^{-1}$) | 20.61 | 23.43 | 22.80 |
| Ca (mg $kg^{-1}$) | 41.55 | 10.14 | 34.83 |
| Mg (mg $kg^{-1}$) | 8.88 | 7.23 | 12.87 |
| Na (mg $kg^{-1}$) | 3.73 | 5.22 | 4.81 |

[a]Adopted from Hwang et al. (2020)

### 7.3 Micronutrients in Compost

Animal manure also contains a significant quantity of micronutrients (Table 1) like zinc, copper, iron, boron and manganese, etc., these are required in very small amounts, which may vary with the type of crop and land. Zinc and boron cannot be fully supplied by manure application (Shahid et al. 2016). In calcareous soils of arid and semi-arid zones, the lower availability of micronutrients like zinc (Zn), manganese (Mn), iron (Fe), copper (Cu) and phosphorus and low organic matter content are common problems (Kumari et al. 2018). In spite of this, the calcareous soils have a relatively greater amount of total micronutrient content, but due to high soil pH, high calcium carbonate content and low organic matter content, the availability of most of the nutrient elements is very low (Rengel 2015).

## 8 Limitations of Compost Application in Soil

1. In the application of compost in agriculture, the main problem is that the total and available nutrient content is not well known, as, for example, nitrogen unavailability is especially uncertain in several types of compost material (Lazicki et al. 2020). The uncertainty of nitrogen content in compost is due to the fact that most of the nitrogen in organic fertilizers is found in organic form, and this is not directly available to plant roots.
2. The raw material, like biowastes, animal wastes and biosolids used for composting, may contain several metals and antibiotics (mainly in poultry, swine and dairy animal manure). The long-term and high amount of compost from these materials may lead to the accumulation of a high number of heavy metals and antibiotics in soil. The high buildup of antibiotics may cause antibiotic resistance in the environment (Arya et al. 2021).
3. High content of soil heavy metals may contaminate soil and water bodies through leaching and runoff, which limits the long-term use of compost in soil (Lillenbeg et al. 2010). To overcome these challenges, the raw organic materials must be processed before application to soil or conversion of organic matter into biochar (Rahman et al. 2020).
4. Bernal et al. (2015) reported some limitations of compost application, such as nutrients leaching, ammonia evaporation, and pathogen load, etc., are major limitations of direct application of manure in soil.
5. In many reports, it has been reported that the poultry manure has a high content of antibiotics, which is from the feeding of antibiotic-rich feed. This high antibiotic content in poultry manure is responsible for the dangerous transformation of normal bacteria to antibiotic-resistant bacteria.
6. A significant fraction of livestock manure is contributed by organic matter and many essential nutrients elements but it is important to estimate the specific nutrient content of organic manure (Lpelc 2019) because significant variation in

nutrients content of manure due to difference in manure composition at different locations but main emphasis should be given to locally derived manure (Malomo et al. 2018).

7. MSW has high organic and inorganic contamination, so the crop plant, especially salad and root crops, are vulnerable to contamination if MSW compost quality is not regulated carefully, because roots are in direct contact with soil and thus prone to high accumulation of toxic substances, and salad crops are consumed without cooking. Thus, the safety concern and quality of MSW compost are the main limiting factors to be used extensively (Farrell and Jones 2009).
8. In MSW, high concentrations of plastic are also problematic and reach above permissible limits, ultimately limiting plant growth.
9. The organic nitrogen is released very slowly, so it is not available to plants during the year of application and thus has no value for plants as a fertilizer.

## 9 Conclusion and Outlook

Due to modern agricultural practices and increasing load of food demand due to expanding global population, the health of agricultural land is depleting heavily in terms of soil organic matter, nutrient bioavailability and microbial population. After the second half of the twentieth century, the use of chemical fertilizers suppressed the use of compost-based organic fertilizers, leading to the degradation of soil health, which in turn affects the health of human beings and other animals. In present-day agricultural research, emphasis is given mainly to soil health, not only to soil fertility as in previous concepts. Soil microbial activity greatly affects soil processes like soil organic carbon, soil respiration, bioavailable phosphorus and potassium, which in turn affect soil health. The application of organic-based fertilizers is very critical in the restoration of soil health and productivity. Though extensive research has been carried out in understanding the role of compost in maintaining and restoring soil health, restoration and bioavailability of nutrients, still a huge amount of work is needed to carry out regarding compost application and soil health, in order to improving supply of nitrogen, phosphorus, potassium and improve microbial diversity in soil which ultimately improve soil health and crop production.

## References

Adekiya AO, Agbede TM, Aboyeji CM, Dunsin O, Simeon VT (2019) Effects of biochar and poultry manure on soil characteristics and the yield of radish. Sci Hortic 243:457–463

Agegnehu G, Amede T (2017) Integrated soil fertility and plant nutrient management in tropical agro-ecosystems: a review. Pedosphere 27(4):662–680

Ahmad R, Khalid A, Arshad M, Zahir ZA, Naveed M (2006) Effect of raw (un-composted) and composted organic waste material on growth and yield of maize (*Zea mays* L.). Soil Environ 25(2):135–142

Amoakwah E, Arthur E, Frimpong KA, Parikh SJ, Islam R (2020) Soil organic carbon storage and quality is impacted by corn cob biochar application on a tropical sandy loam. J Soils Sediments 20:1960–1969

Arya S, Williams A, Reina SV, Knapp CW, Kreft JU, Hobman JL, Stekel DJ (2021) Towards a general model for predicting minimal metal concentrations co-selecting for antibiotic resistance plasmids. Environ Pollut 275:116602

Bagnall DK, Rieke EL, Morgan CL, Liptzin DL, Cappellazzi SB, Honeycutt CW (2023) A minimum suite of soil health indicators for North American agriculture. Soil Secur 10:100084

Berendsen RL, Pieterse CM, Bakker PA (2012) The rhizosphere microbiome and plant health. Trends Plant Sci 17(8):478–486

Bernal MP, Bescós B, Burgos L, Bustamante M, Clemente R, Fabbri C, Bonmatí Blasi A (2015) Evaluation of manure management systems in Europe. Aragonese Society of Agroenvironmental Management, pp 1–180

Bonanomi G, Lorito M, Vinale F, Woo SL (2018) Organic amendments, beneficial microbes, and soil microbiota: toward a unified framework for disease suppression. Annu Rev Phytopathol 56(1):1–20

Bünemann EK, Bongiorno G, Bai Z, Creamer RE, de Deyn G, de Goede R, Fleskens L, Geissen V, Kuyper TW, M€ader P, Pulleman M, Sukkel W, van Groenigen JW, Brussaard L (2018) Soil quality e a critical review. Soil Biol Biochem 120:105–125

Chaves B, De Neve S, Boeckx P, Van Cleemput O, Hofman G (2007) Manipulating nitrogen release from nitrogen-rich crop residues using organic wastes under field conditions. Soil Sci Soc Am J 71(4):1240–1250

Chojnacka K, Moustakas K, Witek-Krowiak A (2020) Bio-based fertilizers: a practical approach towards circular economy. Bioresour Technol 295:122223

Chu M, Singh S, Walker FR, Eash NS, Buschermohle MJ, Duncan LA, Jagadamma S (2019) Soil health and soil fertility assessment by the haney soil health test in an agricultural soil in West Tennessee. Commun Soil Sci Plant Anal 50(9):1123–1131

Claassen VP, Carey JL (2007) Comparison of slow-release nitrogen yield from organic soil amendments and chemical fertilizers and implications for regeneration of disturbed sites. Land Degrad Dev 18:119–132

Cogger CG (2005) Potential compost benefits for restoration of soils disturbed by urban development. Compost Sci Util 13(4):243–251

Cynthia MK, Serita DF, Grandy AS (2016) Direct evidence for microbial- derived soil organic matter formation and its ecophysiological controls. Nat Commun 7:13630. https://doi.org/10.1038/ncomms13630

Doran JW, Parkin TB (1994) In: Doran JW, Coleman DC, Bezdicek DF, Stewart BA (eds) In defining soil quality for a sustainable environment, vol 32. Soil Science Society of America, pp 1–21

Doran JW, Sarrantonio M, Liebig MA (1996) Soil health and sustainability. Adv Agron:1–54

Duong TT, Penfold C, Marschner P (2012) Amending soils of different texture with six composts types: impact on soil nutrient availability, plant growth and nutrient uptake. Plant Soil 354:197–209

Edwards CA, Bohlen PJ (1996) Biology and ecology of earthworms, vol 3. Springer Science & Business Media

Evans AE, Mateo-Sagasta J, Qadir M, Boelee E, Ippolito A (2019) Agricultural water pollution: key knowledge gaps and research needs. Curr Opin Environ Sustain 36:20–27

FAO (2017). The future of food and agriculture trends and challenges. Food and Agriculture Organization of the United Nations, Rome, pp. 1–180

FAO (2019) World fertilizer trends and outlook to 2019. Food and Agriculture Organization of the United Nations, Rome, pp 1–38

FAO (2021) Sustainable food and agriculture. Available at: https://www.fao.org/sustainability/en/

Farrell M, Jones DL (2009) Critical evaluation of municipal solid waste composting and potential compost markets. Bioresour Technol 100(19):4301–4310

Gholoubi A, Emami H, Alizadeh A (2018) Soil quality change 50 years after forestland conversion to tea farming. Soil Res 56(5):509–517

Glanville TD, Persyn RA, Richard TL, Laflen JM, Dixon PM (2004) Environmental effects of applying composted organics to new highway embankments: Part 2. Water quality. Trans ASAE 47(2):471–478

Harada Y, Haga K, Osada T, Koshino M (1993) Quality of compost produced from animal wastes. Jpn Agric Res Q 26(4):238–246

Hui D (2021) Effects of biochar application on soil properties, plant biomass production, and soil greenhouse gas emissions: a mini-review. Agric Sci 12(03):213

Hwang HY, Kim H, Kim MS, Park SJ, Lee CH (2020) Co-composting of chicken manure with organic wastes: characterization of gases emissions and compost quality. Appl Biol Chem 63:1–10

Jian J, Du X, Stewart RD (2020) A database for global soil health assessment. Sci Data 7:16

Karlen DL, Veum KS, Sudduth KA, Obrycki JF, Nunes MR (2019) Soil health assessment: past accomplishments, current activities, and future opportunities. Soil Tillage Res 195:104365

Kaza S, Yao L, Bhada-Tata P, Van Woerden F (2018). What a waste 2.0: a global snapshot of solid waste management to 2050. World Bank Publications

Khatoon H, Solanki P, Narayan M, Tewari L, Rai J, Hina Khatoon C (2017) Role of microbes in organic carbon decomposition and maintenance of soil ecosystem. Int J Chem Stud 5(6): 1648–1656

Kocsis T, Ringer M, Bir B (2022) Characteristics and applications of biochar in soil–plant systems: a short review of benefits and potential drawbacks. Appl Sci 12(8):4051

Kumari K, Prasad J, Solanki IS, Chaudhary R (2018) Long-term effect of crop residues incorporation on yield and soil physical properties under rice-wheat cropping system in calcareous soil. J Soil Sci Plant Nutr 18(1):27–40

Ladha JK, Reddy CK, Padre AT, van Kessel C (2011) Role of nitrogen fertilization in sustaining organic matter in cultivated soils. J Environ Qual 40(6):1756–1766

Lazicki P, Geisseler D, Lloyd M (2020) Nitrogen mineralization from organic amendments is variable but predictable. J Environ Qual 49(2):483–495. https://doi.org/10.1002/jeq2.20030

Lillenbeg M, Herodes K, Kipper K, Nei L (2010) Plants uptake of some pharmaceutocals from fertilized soils. In: Proceedings of the 2010 international conference on environmental science and technology

Lim SL, Wu TY, Lim PN, Shak KPY (2015) The use of vermicompost in organic farming: overview, effects on soil and economics. J Sci Food Agric 95(6):1143–1156

Lpelc A (2019) Environmental benefits of manure application. https://lpelc.org/environmental-benefits-of-manure-application/. Accessed 15 Jan 2021

Maharjan B, Das S, Nielsen R, Hergert GW (2021) Maize yields from manure and mineral fertilizers in the 100-year-old Knorr–Holden Plot. Agron J 113(6):5383–5397

Malomo GA, Madugu AS, Bolu SA (2018) Sustainable animal manure management strategies and practices. In: Agricultural waste and residues, vol 119

Moebius-Clune BN (2016). Comprehensive assessment of soil health: the Cornell framework manual. Cornell University

Mupambwa HA, Mnkeni PNS (2018) Optimizing the vermicomposting of organic wastes amended with inorganic materials for production of nutrient-rich organic fertilizers: a review. Environ Sci Pollut Res 25:10577–10595

Myrold DD, Bottomley PJ (2008) Nitrogen mineralization and immobilization. Nitrogen Agric Syst 49:157–172

Niamat B, Naveed M, Ahmad Z, Yaseen M, Ditta A, Mustafa A, Xu M (2019) Calcium-enriched animal manure alleviates the adverse effects of salt stress on growth, physiology and nutrients homeostasis of *Zea mays* L. Plants 8(11):480

Norris CE, Congreves KA (2018) Alternative management practices improve soil health indices in intensive vegetable cropping systems: a review. Front Environ Sci 6(50). https://doi.org/10.3389/fenvs.2018.00050

Oliver MA, Gregory PJ (2015) Soil, food security and human health: a review. Eur J Soil Sci 66(2): 257–276

Pankhurst CE, Doube BM, Gupta VVSR (1997) In: Pankhurst C, Doube B, Gupta V (eds) Biological indicators of soil health. CAB International, pp 419–435

Patzel N, Sticher H, Karlen DL (2000) Soil fertility—phenomenon and concept. J Plant Nutr Soil Sci 163(2):129–142

Paustian K, Lehmann J, Ogle S et al (2016) Climate-smart soils. Nature 532:49–57. https://doi.org/10.1038/nature17174

Pujara Y, Pathak P, Sharma A, Govani J (2019) Review on Indian Municipal Solid Waste Management practices for reduction of environmental impacts to achieve sustainable development goals. J Environ Manag 248:109238

Purakayastha TJ, Bera T, Bhaduri D, Sarkar B, Mandal S, Wade P, Tsang DC (2019) A review on biochar modulated soil condition improvements and nutrient dynamics concerning crop yields: pathways to climate change mitigation and global food security. Chemosphere 227:345–365

Rahman GM, Rahman MM, Alam MS, Kamal MZ, Mashuk HA, Datta R, Meena RS (2020) Biochar and organic amendments for sustainable soil carbon and soil health. In: Carbon and nitrogen cycling in soil, pp 45–85

Rajaeifar MA, Ghanavati H, Dashti BB, Heijungs R, Aghbashlo M, Tabatabaei M (2017) Electricity generation and GHG emission reduction potentials through different municipal solid waste management technologies: a comparative review. Renew Sust Energ Rev 79:414–439

Raynaud X, Nunan N (2014) Spatial ecology of bacteria at the microscale in soil. PLoS One 9(1): 87217

Rengel Z (2015) Availability of Mn, Zn and Fe in the Rhizosphere. J Soil Sci Plant Nutr 15:397–409. https://doi.org/10.4067/S0718-95162015005000036

Rinot O, Levy GJ, Steinberger Y, Svoray T, Eshel G (2019) Soil health assessment: a critical review of current methodologies and a proposed new approach. Sci Total Environ 648:1484–1491

Robertson GP, Groffman PM (2007) Nitrogen transformations. In: Paul EA (ed) Soil microbiology, ecology, and biochemistry, 3rd edn. Academic, pp 341–364. https://doi.org/10.1016/B978-0-08-047514-1.50017-2

Roy ED, Esham M, Jayathilake N, Otoo M, Koliba C, Wijethunga IB, Fein-Cole MJ (2021) Compost quality and markets are pivotal for sustainability in circular food-nutrient systems: a case study of Sri Lanka. Front Sustain Food Syst 5:748391

Shahid M, Shukla AK, Bhattacharyya P, Tripathi R, Mohanty S, Kumar A, Nayak AK (2016) Micronutrients (Fe, Mn, Zn and Cu) balance under long-term application of fertilizer and manure in a tropical rice-rice system. J Soils Sediments 16:737–747

Singh R, Singh P, Singh H, Raghubanshi AS (2019) Impact of sole and combined application of biochar, organic and chemical fertilizers on wheat crop yield and water productivity in a dry tropical agro-ecosystem. Biochar 1:229–235

Smith P (2016) Soil carbon sequestration and biochar as negative emission technologies. Glob Chang Biol 22(3):1315–1324

Song K, Xue Y, Zheng X, Lv W, Qiao H, Qin Q, Yang J (2017) Effects of the continuous use of organic manure and chemical fertilizer on soil inorganic phosphorus fractions in calcareous soil. Sci Rep 7(1):1164

Srivastava V, De Araujo ASF, Vaish B, Bartelt-Hunt S, Singh P, Singh RP (2016) Biological response of using municipal solid waste compost in agriculture as fertilizer supplement. Rev Environ Sci Biotechnol 15:677–696

Stratton ML, Barker AV, Rechcigl JE (1995) Compost. In: Rechcigl JE (ed) Soil amendments and environmental quality. Lewis Publ., Boca Raton, pp 249–310

Szogi AA, Vanotti MB, Ro KS (2015) Methods for treatment of animal manures to reduce nutrient pollution prior to soil application. Curr Pollut Rep 1:47–56

Thomas BW, Hunt D, Bittman S, Hannam KD, Messiga AJ, Haak D, Hao X (2019) Soil health indicators after 21 year of no-tillage in south coastal British Columbia. Can J Soil Sci 99(2): 222–225

Turrión MB, Bueis T, Lafuente F, López O, San José E, Eleftheriadis A, Mulas R (2018) Effects on soil phosphorus dynamics of municipal solid waste compost addition to a burnt and unburnt forest soil. Sci Total Environ 642:374–382

Van Ex HM, Karlen DL (2019) Reanalysis validates 829 soil health indicator sensitivity and correlation with long-term crop yields. Soil Sci Soc Am J 83:721–732. https://doi.org/10.2136/sssaj2018.09.0338

Van Wesemael D, Vandaele L, Ampe B, Cattrysse H, Duval S, Kindermann M, Fievez V, De Campeneere S, Peiren N (2019) Reducing enteric methane emissions from dairy cattle: Two ways to supplement 3-nitrooxypropanol. J Dairy Sci 102:1780–1787. https://doi.org/10.3168/jds.2018-14534

Vaverková MD, Winkler J, Adamcová D, Radziemska M, Uldrijan D, Zloch J (2019) Municipal solid waste landfill–Vegetation succession in an area transformed by human impact. Ecol Eng 129:109–114

Veresoglou S, Halley J, Rillig M (2015) Extinction risk of soil biota. Nat Commun 6:8862. https://doi.org/10.1038/ncomms9862

World Bank (2023) World bank climate-smart agriculture. https://www.worldbankorg/en/topic/climate-smartagriculture

Wortman SE, Holmes AA, Miernicki E, Knoche K, Pittelkow CM (2017) First-season crop yield response to organic soil amendments: a meta-analysis. Agron J 109(4):1210–1217

Zhang P, Li Y, Cao Y, Han L (2019) Characteristics of tetracycline adsorption by cow manure biochar prepared at different pyrolysis temperatures. Bioresour Technol 285:121348

Zhou Y, Xiao R, Klammsteiner T, Kong X, Yan B, Mihai FC, Awasthi MK (2022) Recent trends and advances in composting and vermicomposting technologies: a review. Bioresour Technol 360:127591

# Improving the Nutrient Uptake in Crops by Compost Application in Diverse Agricultural Contexts

**Kshirod Chandra Sahoo, Dewali Roy, Pooja Tamuk, Aditya Pratap Singh, and Arkaprava Roy**

## 1 Introduction

### *1.1 Overview of Nutrient Uptake Challenges in Modern Agriculture*

Nutrient uptake challenges in modern agriculture are significant and multifaceted, requiring innovative solutions to ensure sustainable and efficient farming practices. Conventional farming methods often face issues with nutrient inefficiency, leading to environmental degradation and reduced crop yields. Hydroponics offers a potential solution by providing precise nutrient control and eliminating soilborne diseases and pests (Thapa et al. 2024). This method allows for optimized nutrient delivery and improved overall performance and yield. Similarly, nanotechnology has been recognized as a promising approach to increase nutrient use efficiency in agriculture

K. C. Sahoo
Department of Soil Science, Dr. Rajendra Prasad Central Agricultural University, Pusa, Bihar, India

D. Roy
Division of Soil Science and Agricultural Chemistry, Indian Agricultural Research Institute, Pusa, New Delhi, India

P. Tamuk
Department of Agricultural Sciences, Rajiv Gandhi University, Rono Hills, Doimukh, Arunachal Pradesh, India

A. P. Singh (✉)
Department of Genetics and Plant Breeding, Bidhan Chandra Krishi Viswavidyalaya, Mohanpur, Nadia, West Bengal, India

A. Roy
School of Crop Health Biology Research, ICAR-National Institute of Biotic Stress Management, Raipur, Chhattisgarh, India

 
R. K. Singh et al. (eds.), *Composting of Farm Residues by Nature-Based Solutions*,
https://doi.org/10.1007/978-3-032-12255-1_11

(Sen et al. 2022). For instance, nanofertilizers utilize nano-sized carriers for nutrient uptake by crops effectively, improving nutrient utilization in rice farming (Kumar et al. 2023). Interestingly, while nanofertilizers show promise in increasing crop yields and farmer profitability, organic farming faces challenges related to the release rate of nutrients and their availability, which may not always meet crop demands (Panday et al. 2024). This contradiction highlights the complexity of addressing nutrient uptake challenges across different agricultural approaches. Addressing nutrient uptake challenges requires a multifaceted approach. Precision agriculture techniques, such as site-specific nutrient management and the integration of AI-driven technologies for resource management, offer potential solutions (Sen et al. 2025). Additionally, the development of smart delivery systems and nano-enabled technologies (Lallawmkimi et al. 2025) can optimize fertilizer use and enhance nutrient uptake efficiency. However, successful implementation of these technologies requires careful consideration of environmental impacts, economic feasibility, and farmer education to ensure sustainable and effective nutrient management in modern agriculture.

## 1.2 Role of Compost as a Sustainable Soil Amendment for Improving Nutrient Availability

Compost serves as a sustainable soil amendment that significantly improves nutrient availability and overall soil health. It is an eco-friendly process of converting organic waste into inorganic nutrient-rich soil additive, which helps to increase soil fertility and health by improving soil quality, reducing greenhouse gas emissions, and efficiently conserving water (Hassan et al. 2023). The application of compost enhances soil structure and reduces erosion while suppressing plant diseases and decreasing dependence on synthetic fertilizers (Goldan et al. 2023). Microbial compost, in particular, plays a crucial role in improving soil health and fertility. It enhances nutrient acquisition (N, P, and K), promotes the production of phytohormones, siderophores, vitamins, and antibiotics, and controls pathogens through induced systemic resistance (Ahmed et al. 2023). The integration of compost with growth promoters like plant growth-promoting rhizobacteria (PGPR) has shown significant improvements in soil chemical properties, nutrient content, and microbial activity, leading to enhanced plant productivity (El-Akhdar et al. 2024). Interestingly, the combination of biochar–mineral complex (BMC) and compost has demonstrated synergistic effects on soil properties, particularly by increasing nitrate content and enriching bacterial nitrifiers (Ye et al. 2016). However, it is important to note that while compost generally improves soil quality, its effectiveness can vary depending on soil texture, application rates, and cropping systems (Matisic et al. 2024). Additionally, poor-quality compost or improper use can lead to negative effects such as soil acidity or alkalinity and water contamination (Goldan et al. 2023).

## 1.3 Significance of Addressing Diverse Agricultural Contexts

Addressing diverse agricultural contexts is crucial for developing effective and sustainable farming practices. Different soil types, climates, and farming systems require tailored approaches to maximize productivity and minimize environmental issues. In smallholder farming systems in Ethiopia, for example, climate-smart agriculture (CSA) practices like integrated nutrient management, water harvesting, and afforestation can address constraints such as small land size, resource scarcity, and soil degradation (Zerssa et al. 2021). However, the adoption of these practices is often limited due to various socio-economic factors, highlighting the need for context-specific solutions. The diversity of farming systems is further emphasized by the analytical framework presented in Therond et al. (2017), which describes three biotechnical types of farming systems (chemical-based, biological-based, and biodiversity-based) and four socio-economic contexts. This framework demonstrates the complexity of agricultural models and the importance of considering both biotechnical and socio-economic factors in developing sustainable farming practices. Addressing diverse agricultural contexts is essential for developing effective and sustainable farming practices. This approach allows for the integration of local knowledge, adaptation to specific environmental conditions, and consideration of socio-economic factors, ultimately leading to more resilient agricultural systems maintaining productivity (Bhati et al. 2024) Table 1.

## 1.4 Mechanistic Insights, Research Findings, and On-Farm Applications of Compost in Crop Nutrient Uptake

Compost derived from urban waste and agricultural byproducts has shown significant potential for enhancing nutrient uptake in crops. Studies have demonstrated that compost can effectively deliver plant-available calcium, magnesium, phosphorus, and potassium (Schröder et al. 2021). The mechanisms behind this improved nutrient uptake involve the stabilization of nutrients during the composting process, which controls their release rate once applied to the soil (Larney et al. 2006). Evidence supporting the efficacy of compost for nutrient uptake comes from various studies. A pot trial conducted with lettuce plants in which vermicompost from urban waste led to the highest yield and total uptake of macronutrients (P, K, Ca, and Mg) (Schröder et al. 2021). Similarly, a field experiment with barley showed that compost application increased nutrient uptake in plants, though not in grains (Ghouili et al. 2022). The addition of charred biomass to pig manure compost was found to facilitate organic matter decomposition and reduce DTPA-extractable Zn and Cu contents, potentially improving nutrient availability (Li et al. 2014). However, some contradictions and challenges exist. While compost application can increase nutrient recycling in urban ecosystems, it may lead to excessive phosphorus application, creating the potential for P loss through leaching and runoff (Shrestha et al. 2020).

**Table 1** Effects of compost application on yield, nutrient uptake, and stress mitigation in crops

| Crop | Compost type | Soil/climatic context | Application rate of compost | Yield effect over conventional chemical fertilization (CCF) | Improvement in nutrient availability and uptake over conventional chemical fertilization | Abiotic/biotic stress mitigation | Reference |
|---|---|---|---|---|---|---|---|
| Maize-Potato | Compost + Mineral Fertilizer (MF) | Southern Ethiopia | 6–16 t $ha^{-1}$ compost | Maize and potato yields increased by 22–54% and 8–18%, respectively, over MF alone | Compared to MF alone, N use efficiency (NUE) increased by upto 93%. Increase in available N, P, K, Ca, Mg, S, soil pH. Upto 11% decrease in bulk density (BD) | Can alleviate potential nutrient deficiency stress | Asaye et al. (2022) |
| Cowpea | Compost + Mineral Fertilizer (MF) | Saint-Louis, Senegal | 2.5–10 t $ha^{-1}$ compost | Upto ↑ 30% seed yield, ↑ 38% plant height, ↑ 53% stem diameter, ↑ 17% pod, ↑ 85% nodulation | N/A | N/A | Diatta et al. (2024) |
| Wheat | Vermicompost + PGPR | Cairo, Egypt | 3.75 kg vermicompost in 11.25 kg soil | Increased grain number, leaf area, plant height | N/A | Alleviated drought stress by reducing superoxide and hydrogen peroxide levels | Talaat and Abdel-Salam (2024) |
| Wheat seedlings | Cellulolytic microbe-enriched cow dung vermicompost | Faisalabad, Pakistan | 4–8 t $ha^{-1}$ | Improvement in physiological traits | Uptake of N ↑ 18.18%, P ↑ 27.53%, K ↑ 12.89%, Ca ↑ 8.89%, Mg ↑ 11.73%, S ↑11.80%; ↑ in micronutrient content | Alleviated drought stress | Ahmad et al. (2024) |

| | | | | | | | |
|---|---|---|---|---|---|---|---|
| Maize | Composted agricultural wastes | Arid agroecosystem of Riyadh, Saudi Arabia | 5–20 t $ha^{-1}$ | Growth, yield, and quality parameters improved upto 27.3, 31.8, and 17.2%, respectively. | ↑ organic matter (OM), and total N, P, and K content in soil | Can alleviate potential nutrient deficiency stress | Al-Suhaibani et al. (2021) |
| Maize | Soybean Compost + Biochar | Tropical Thailand | 0.56 kg $m^{-2}$ | Joint application of compost and biochar improved yield parameters, residual effect persisted for 5 years | Joint application of compost and biochar increased nutrient availability, with significant residual effect | Can alleviate potential nutrient deficiency stress | Wijitkosum et al. (2025) |
| Silage Corn | Date palm waste compost | Arid oasis agroecosystem | 10–40 t $ha^{-1}$ | Slightly lesser to at per yield. ↑ the compost application rate, ↑ yield | ↑ OM, total N, and available P and K. ↑ P uptake. | Can alleviate potential nutrient deficiency stress | El Janati et al. (2022) |
| Rice | Poultry manure and elephant grass compost | Sub-Saharan Ghana | 3.4 t $ha^{-1}$ | 20.4% ↓ yield than CCF, but 165% ↑ yield over control | ↓ Bulk density; ↑ porosity, moisture retention, organic matter, %N, available P, CEC, and microbial activity. | Can alleviate potential nutrient deficiency stress | Phares and Akaba (2022) |
| Rice | Municipal solid waste compost, rice straw–cow manure vermicompost, and *phumdi*-based compost | Manipur, North-east India | 3–3.75 t $ha^{-1}$ | Upto 33.6% ↑ grain yield and 26.21% ↑ straw yield, ↑ other growth parameters | N/A | N/A | Devi et al. (2025) |
| Tomatoes | *Streptomyces microflavus* G33 enriched compost | Sichuan Province, China | 0.2 kg $m^{-2}$ | Compost application increased yield by 87.1%, and by 119% | Increase in beneficial soil microbes | Compost application reduced bacterial wilt incidence by | Shen et al. (2021) |

(continued)

**Table 1** (continued)

| Crop | Compost type | Soil/climatic context | Application rate of compost | Yield effect over conventional chemical fertilization (CCF) | Improvement in nutrient availability and uptake over conventional chemical fertilization | Abiotic/biotic stress mitigation | Reference |
|---|---|---|---|---|---|---|---|
| | made from biogas residues, cattle manure and rice straw | | | when combined with *Streptomyces microflavus* G33. | | 19%, and by 56% when combined with *Streptomyces microflavus* G33. | |
| Organic barley | Date palm waste compost | Saving Oasis of Chenini, Gabes, Tunisia | 30 t $ha^{-1}$ | Upto 75% ↑ in grain yield, and 52% ↑ in straw yield | Improved plant nutrient uptake, expression of genes involved in N and P metabolism | Can alleviate potential nutrient deficiency stress | Ghouili et al. (2022) |
| Tomato | Fern manure compost extract + mycorrhiza | Center Region of Cameroon, equatorial climate | 0–10% of compost extract | Significant enhancement in floral abundance, fruit volume, fruit diameter, and overall yield | ↑ in available P and Mg | Upto 71.3% and 94.7% total reduction in incidence and severity of common tomato diseases, respectively. | Mbogning et al. (2024) |
| *Beta vulgaris* L. var. *cicla* | Compost from urban waste and sewage sludge | Viterbo, Italy; volcanic soil | 35 t $ha^{-1}$ | Similar yield to CCF, but ↑ than control | ↑ in SOC, total N, and CEC. ↓ the total As by 23%. ↓ in root and leaf uptake of As by 34 and 36%, respectively. | Alleviation of heavy metal toxicity stress | Stazi et al. (2022) |

Additionally, carbon and nitrogen losses during the composting process can result in greenhouse gas emissions (Larney et al. 2006). In conclusion, compost has demonstrated significant potential for enhancing nutrient uptake in various crops. Its applications range from urban horticulture to large-scale agriculture, with benefits including improved soil fertility, plant growth, and crop productivity (Ghouili et al. 2022; Yu et al. 2019). However, careful management of compost application rates is crucial to maximize benefits while minimizing potential environmental impacts (Shrestha et al. 2020). Future research should focus on optimizing nutrient availability, exploring alternative nitrogen sources, and assessing long-term effects of compost application on soil health and crop yields (Schröder et al. 2021).

### *1.5 Diversity of Agricultural Contexts (Conventional Versus Organic Farming)*

Organic farming generally supports greater biodiversity than conventional farming, with an average 34% increase observed (Smith and Halberg 2020), especially in intensive, homogeneous landscapes where it serves as a species refuge. Organic fields also show higher diversity in vegetation, seed banks, and seed rain (Roschewitz et al. 2005). However, in complex landscapes, conventional farms may host similar weed diversity. Despite biodiversity benefits, organic systems often yield 18% less than conventional ones (Gong et al. 2022), a trend also seen in natural farming (Das et al. 2024). The trade-off between yield and biodiversity varies by taxa and crop type, but overall, gains in biodiversity often mirror proportional yield losses.

### *1.6 Diversity of Agricultural Contexts (Tropical Versus Temperate Climate)*

Tropical regions generally support higher species richness than temperate ones due to higher speciation and lower extinction rates (Pyron and Wiens 2013), driven by climate factors like temperature and rainfall. While temperate agriculture faces climate change threats requiring adaptive strategies (e.g., soil cooling; Sabri et al. 2018), tropical Indigenous practices like swidden farming may enhance forest biodiversity (Downey et al. 2023). Climate change may further shift biodiversity patterns, such as increasing copepod diversity in extra-tropical zones (Rombouts et al. 2009), and influence the effectiveness of conservation agriculture across climates (Su et al. 2021). Understanding these contrasts is vital for sustainable agriculture and biodiversity conservation.

# 2 Nutrient Uptake in Crops: Mechanisms, Variability and Challenges

## *2.1 Fundamentals of Nutrient Uptake*

Nitrogen, phosphorus, and potassium are key macronutrients required for plant growth. Their availability directly influences root development, metabolism, and yield. Plants adapt to nutrient deficiencies via root architecture changes, secretion of root exudates, and symbiosis with microbes such as arbuscular mycorrhizae (Zhao et al. 2023). Uptake mechanisms are interconnected—e.g., the *Arabidopsis* nitrate transporter NPF7.3 also aids $K^+$ translocation under low nitrate (Zheng et al. 2016). Small signaling peptides like CEP1 and regulatory molecules such as microRNAs modulate uptake and nutrient homeostasis (Roy et al. 2022; Kulcheski et al. 2015). Enhancing nutrient use efficiency through molecular breeding or microbial bio-inoculants (e.g., PGPR) is critical for sustainable crop production (Rani et al. 2022). Micronutrients, though required in trace amounts, are essential for plant cellular functions. Plants absorb micronutrients using specific root transporters, chelators, and signaling pathways (Podar and Maathuis 2022). Soil texture, pH, redox potential, organic matter, and microbial activity govern their availability (Kumar et al. 2016). Chelation by ligands like nicotianamine and phytosiderophores facilitates transport and reduces toxicity (Seregin and Kozhevnikova 2020). For example, iron uptake is mediated through chelation and reduction systems. Understanding these processes is vital for addressing micronutrient deficiencies in soils and crops. Soil pH, organic matter, and microbial activity critically influence nutrient availability. pH affects nutrient solubility and microbial function, with optimal nutrient cycling occurring between pH 5.5 and 7.5 (Wang and Kuzyakov 2024). Organic matter improves cation exchange capacity, water retention, and microbial activity, enhancing nutrient release (Roy et al. 2024). However, high SOM does not always translate to high nutrient availability, especially in acidic soils with limited microbial activity (Balota et al. 2013). Microbial communities mediate extracellular enzyme activity, nutrient mineralization, and solubility near roots. Thus, soil health is key to efficient nutrient uptake.

Nutrient uptake is influenced by soil moisture, root traits, transporter expression, and environmental stress (Chtouki et al. 2022; Fig. 1). Drought limits uptake by reducing transpiration and nutrient flow. Modifying stomatal density affects water use and nutrient efficiency—lower densities help conserve water, while higher densities enhance uptake under well-watered conditions (Hepworth et al. 2015). Root hairs are essential for phosphate uptake; increased hair length improves efficiency more than density (Zygalakis et al. 2011). Trace metals can affect hormone signaling and root development, indirectly altering nutrient uptake (Lešková et al. 2021). Overall, optimizing root traits, soil moisture, and microbial interactions is crucial for improving nutrient use under changing climatic conditions.

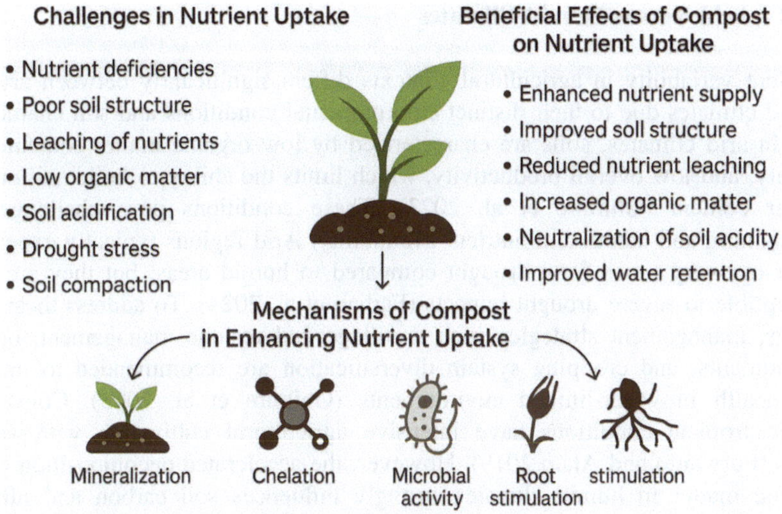


**Fig. 1** Compost as a solution for challenged nutrient uptake

## 2.2 *Nutrient Variability Across Agricultural Contexts*

### 2.2.1 Sandy Versus Clay Soils

Sandy and clay soils exhibit significant differences in nutrient dynamics and availability, impacting agricultural productivity and management practices. Sandy soils are low in nutrient- and water-holding capacities, making them prone to leaching and nutrient loss (Tahir and Marschner 2017). They have weak water-holding and temperature-buffering capacities, which can negatively impact plant growth, especially in the face of increased climate variability (Li et al. 2021). In contrast, clay soils have more organic carbon (OC) content and cation exchange capacity (CEC), contributing to better nutrient retention and availability (Riaz and Marschner 2020). Interestingly, the addition of clay to sandy soils can significantly improve their properties. Clay addition can increase water and nutrient retention, reduce leaching, and enhance organic carbon sequestration (Riaz and Marschner 2020). The effectiveness of clay addition depends on factors such as clay content, CEC, and the form in which clay is added (Riaz and Marschner 2020). For instance, finely ground clay soil is more effective in reducing nutrient leaching compared to larger clay peds (Tahir and Marschner 2017). In conclusion, while sandy soils present challenges for nutrient management, various amendment strategies can improve their agricultural potential. Integrating clay, organic matter, and biochar can enhance water retention, nutrient availability, and overall soil quality in sandy soils (Alghamdi et al. 2023). These approaches can help mitigate the inherent limitations of sandy soils and bridge the gap in nutrient variability between sandy and clay soils in agricultural contexts.

### 2.2.2 Arid Versus Humid Climates

Nutrient variability in agricultural contexts differs significantly between arid and humid climates due to their distinct environmental conditions and soil characteristics. In arid climates, soils are characterized by low organic matter content, poor fertility, and low overall productivity, which limits the ability to build soil organic matter content (Ghimire et al. 2023). These conditions pose challenges for maintaining soil health and nutrient availability. Arid regions typically experience faster recovery times from drought compared to humid areas, but they are more susceptible to severe drought impacts (Farhan et al. 2024). To address these challenges, management strategies such as tillage and residue management, organic amendments, and cropping system diversification are recommended to improve soil health in water-limited environments (Ghimire et al. 2023). Conversely, humid tropical conditions have intensive agricultural cultivation with various crops (Purwanto and Alam 2019). However, the accelerated decomposition of soil organic matter in humid climates strongly influences soil carbon and nitrogen dynamics. Implementation of inappropriate agricultural systems in humid regions can lead to decreased soil organic carbon and total soil nitrogen, causing soil degradation through erosion and leaching (Purwanto and Alam 2019). Interestingly, water use efficiency (WUE) tends to increase during drought periods in all climate regions, with a stronger effect observed in arid zones compared to humid regions (Farhan et al. 2024). In conclusion, nutrient variability across agricultural contexts in arid and humid climates requires tailored approaches to maintain soil health and productivity. For arid regions, focus should be on building SOM and improving water management, while humid areas need to balance intensive cultivation with proper carbon and nitrogen management. The benefit taken by both climate types from sustainable soil management practices, such as crop rotation, organic matter addition, and improved fertilization strategies, to neutralize the impacts of climate change and ensure long-term agricultural sustainability (Karri and Nalluri 2023).

## 2.3 *Compost as a Solution*

Compost is a valuable organic amendment that improves nutrient uptake, especially under stress conditions like salinity and drought. When combined with plant growth-promoting rhizobacteria (PGPR), compost enhances soil enzymes, ion selectivity, and plant water content, boosting the uptake of N, P, and K (Omara et al. 2022a, b). It also contributes to greenhouse gas mitigation by reducing emissions of $CO_2$, $CH_4$, and $N_2O$ (Brenzinger et al. 2018). However, the low N:P ratio in compost can lead to phosphorus overapplication, necessitating careful nutrient management (Shrestha et al. 2020). Integrating compost with mycorrhizae or Trichoderma has improved nutrient availability and crop productivity (Bibi et al. 2024). Compared to chemical fertilizers, organic fertilizers improve soil structure, microbial activity, and enzyme function. In chrysanthemum, liquid organic fertilizers increased root and shoot growth more than chemical options (Ji et al. 2017). However, chemical fertilizers often result in higher short-term yields, as seen in maize (Ikoyi et al. 2020). The most effective approach is integrated nutrient management—combining organic, inorganic, and biofertilizers—to enhance soil fertility and maximize yields (Susilowati et al. 2024; Bhatt et al. 2019). This strategy reduces dependence on chemical inputs while promoting sustainable agriculture.

# 3 Types and Composition of Compost

There are different types of compost as described below (Mladenov 2018):

## *3.1 Green Waste Compost*

Green waste compost refers to compost made largely from high-nitrogen, green plant materials. This typically includes grass clippings, garden trimmings, vegetable scraps, and other fresh yard waste. Green materials are rich in protein and nutrients, which fuel rapid microbial growth. Pure green waste can decompose quickly but may become anaerobic if not mixed with aerating agents (due to high moisture) (Mladenov 2018). Green waste compost usually has a relatively low carbon-to-nitrogen (C: N) ratio and provides a good amount of nitrogen to the soil. It is commonly produced by municipal programs that collect yard trimmings. For example, many cities in developed countries compost leaves and grass from parks and households, yielding a nutrient-rich soil conditioner for landscaping.

## *3.2 Brown Waste Compost*

Brown waste compost is derived mainly from dry, carbon-rich materials (often called *browns*). Browns include fallen leaves, straw, sawdust, wood chips, paper, and other fibrous or woody debris. Composting offers a sustainable management strategy for crop residues (Sen et al. 2024). These materials are high in cellulose and lignin but low in nitrogen. Composting pure brown waste is slow because microbes lack sufficient nitrogen; however, brown waste is excellent for bulking and aeration (Azim et al. 2018). In practice, brown waste compost is often amended with some a nitrogen source to aid decomposition. The finished compost from predominantly brown inputs tends to have a higher C: N ratio and a coarse texture (due to resistant lignin). It contributes stable carbon (humus) to soil. An example is leaf mold, a type of brown compost produced simply by piling and aging leaves over a long period (Mladenov 2018).

## 3.3 Manure-Based Compost

Manure compost is produced by composting animal manure, often mixed with bedding materials (straw, wood shavings). Farmyard manures (from cattle, horses, poultry, etc.) are rich in nitrogen and beneficial microbes, but can contain pathogens and weed seeds if not treated. Composting manure under managed aerobic conditions stabilizes its nutrients and reduces health risks. Manure-based composts are typically nutrient-rich (especially in nitrogen and phosphorus) and are widely used in agriculture as an organic fertilizer (Azim et al. 2018).

## 3.4 Food Waste Compost

Food waste compost is derived from kitchen scraps, plate waste, and other food processing residues. This feedstock is usually high in moisture and nitrogen (especially if it includes vegetables, fruits, and food scraps). Municipal waste are common base materials for composting (Prakash et al. 2022). Composting food waste can significantly reduce the volume of garbage sent to landfills and mitigate methane emissions from anaerobic decay (Azim et al. 2018). However, food waste can be challenging to compost because it tends to decompose very quickly and can become odorous or attract pests if not managed (Wang et al. 2022). Food waste compost, when properly cured, is rich in nutrients and organic matter and suitable for enriching soils in urban agriculture or landscaping (Fuchs and Cuijpers 2016).

## 3.5 Vermicompost

Vermicompost is the nutrient-rich organic material produced through the decomposition of organic waste by various species of earthworms, most commonly red wigglers, white worms, and other composting worms. In this process, known as vermicomposting, earthworms consume biodegradable materials such as kitchen scraps, dried leaves, and farm waste, digest them, and excrete them as vermicast or worm castings, which are high in essential nutrients like nitrogen, phosphorus, potassium, calcium, and magnesium (Chiarelotto et al. 2021). Vermicompost serves as an excellent organic fertilizer and soil conditioner, improving soil structure, enhancing water retention, and increasing microbial activity in the soil (Fuchs and Cuijpers 2016).

## 3.6 *Properties of Different Feedstocks Or Raw Materials Used for Composting*

The composition of nutrients depends entirely on the types of feedstocks used (Table 2), which is described below. To produce a sound and good-quality compost, the physical and chemical properties of the compost should be determined by the end of the processing period; therefore, the main objective of this research was to study some physical and chemical properties of compost made of different raw materials. These properties include pH, EC, total organic carbon, total organic matter, total nitrogen, total phosphorus, total potassium, C/N ratio, bulk density, moisture content, water-holding capacity and porosity (Policastro and Cesaro 2022).

## 3.7 *Mechanisms of Compost in Enhancing Nutrient Mobilization and Uptake*

Compost improves nutrient uptake not only by supplying essential elements but also by enhancing the *chemical, physical, and biological* properties of soil, creating a more favorable environment for plant growth and root development.

### 3.7.1 Chemical Improvements

One of the primary mechanisms is the *gradual mineralization* of compost's organic nutrients. Microbial decomposition converts organic forms of nitrogen, phosphorus, and sulfur into plant-available inorganic ions (e.g., $NH_4^+$, $NO_3^-$, $PO_4^{3-}$, $SO_4^{2-}$). This slow-release process provides a synchronized nutrient supply over time, in

**Table 2** Physical properties of the raw materials used in compost making (Khater 2015)

| Properties | Cattle manure | Herbal plants residues | Sugar cane plants residues |
|---|---|---|---|
| Bulk density (kg $m^{-3}$) | 750 | 335 | 426 |
| Moisture content (%) | 58.3 | 16.2 | 36.2 |
| Water-holding capacity | 3 | 3.5 | 3.3 |
| Porosity (%) | 41.57 | 80.62 | 69.96 |
| pH | 8.1 | 4.3 | 7.1 |
| EC (ds $m^{-1}$) | 4.2 | 1.3 | 3.1 |
| Total organic carbon (%) | 18.16 | 9.4 | 20 |
| Total organic matter (%) | 31.3 | 43.1 | 61.3 |
| Total N (%) | 0.93 | 1.35 | 1.62 |
| Total P (%) | 0.21 | 0.36 | 1.12 |
| Total K (%) | 0.17 | 0.42 | 1.36 |
| C/N ratio | 19.53:1 | 6.97:1 | 12.35:1 |

contrast to quick-release synthetic fertilizers. Since much of the nitrogen is initially in organic form, farmers often supplement compost with small amounts of mineral N to support early plant growth. Compost also improves the *availability of micronutrients* such as Fe, Zn, Mn, and Cu through *chelation and complexation* (Hale et al. 2021). Humic and fulvic acids in compost bind these metal ions, keeping them in soluble forms for root uptake—especially beneficial in calcareous or weathered soils. Chelation can increase the bioavailability of iron dramatically, even in high-pH soils where it typically precipitates. These humic substances also reduce phosphorus fixation by binding with $Fe^{3+}$, $Al^{3+}$, and $Ca^{2+}$, thereby enhancing phosphorus availability. Moreover, compost increases the *cation exchange capacity (CEC)* of soil, enabling better retention of positively charged nutrients ($K^+$, $Ca^{2+}$, $NH_4^+$) and reducing nutrient leaching. Its slightly alkaline pH (typically 6.5–8.0) helps buffer acidic or alkaline soils, stabilizing nutrient availability (Fuchs et al. 2018).

### 3.7.2 Physical Improvements

By adding organic matter, compost improves *soil structure*—reducing bulk density, increasing porosity, and enhancing aggregation. These changes lead to better aeration, infiltration, and root penetration. Improved structure also results in higher *water-holding capacity* and hydraulic conductivity, which are particularly important under drought-prone conditions. Field studies have demonstrated increased crop yields due to improved moisture retention and nutrient access (Mladenov 2018). In acidic soils, compost often has a liming effect, and it can raise pH by binding acidic aluminum and iron or by providing basic cations like $Ca^{2+}$ and $Mg^{2+}$ present in the compost (Goldan et al. 2023).

### 3.7.3 Biological Enhancements

Biologically, compost stimulates *microbial activity* by introducing diverse microorganisms (bacteria, fungi, actinomycetes) that drive nutrient cycling and organic matter decomposition. These microbes mineralize organic nutrients and contribute to a more active soil food web. One significant biological interaction promoted by compost is the symbiosis with *arbuscular mycorrhizal fungi (AMF)*. These fungi extend the root system via hyphae, improving access to immobile nutrients like phosphorus and zinc. Compost provides substrates and humic substances that encourage AMF colonization and efficiency (Pergola et al. 2018). In addition to fungi, compost enhances *beneficial bacterial communities*, further supporting nutrient transformation and plant uptake (Omara et al. 2022a, b). Compost also encourages the presence of *beneficial soil fauna*, such as earthworms, which fragment organic matter, increase aeration, and help distribute nutrients throughout the soil profile (Gutarowska et al. 2015).

## 3.8 Impact of Compost on Plants

Table 1 presents the effect of compost on different aspects of the plant and soil. Some of its important aspects are discussed below:

### 3.8.1 Impact on Root Morphology and Architecture Due to Compost

Given the improved physical, chemical, and biological conditions described above, plants grown in compost-amended soils often exhibit marked changes in root morphology and architecture. Root systems become more extensive and efficient. Key observed changes include greater root length and biomass, increased lateral branching and root density, thicker root diameters in certain cases, higher root-to-shoot ratios under nutrient-limited conditions and Exploration of a greater soil volume. Increased root surface area for absorption. Fine roots and root hairs provide the interface through which nutrients are absorbed. Compost-induced increases in root hair density and length mean the plant has much more surface area in contact with the soil solution (Kadir et al. 2016). Root hairs are particularly important for absorbing phosphorus, zinc, and other nutrients present at low concentrations. More root hairs (as stimulated by humic substances in compost) equate to more absorption sites, improving the efficiency of nutrient uptake per unit root mass. With compost fostering mycorrhizal networks and beneficial microbes, plants effectively gain additional "root extensions." Because compost breaks down slowly, it provides a sustained release of nutrients throughout the growing season. This matches well with plant uptake patterns, ensuring that roots have access to nutrients when they need them.

### 3.8.2 Impact on Crop Health Against Biotic Stressors

Compost fosters microbial communities that suppress pathogens and can induce systemic resistance in plants. Healthier roots with a rich microbiome are less prone to infection. There are documented cases where compost amendments nearly eliminated certain soilborne diseases in crops that are otherwise devastating. For instance, compost is known to naturally suppress *Pythium* and *Rhizoctonia* species that cause root rot in many crops (Gutarowska et al. 2015). The beneficial microbes in compost compete with these pathogens for resources and space in the rhizosphere and can even produce antifungal compounds. Additionally, with better nutrition (especially adequate potassium and micronutrients), plants often have thicker cell walls and biochemical defenses that make it harder for pathogens to invade. The improvements in plant nutritional status from compost thus tie directly into disease resistance (Gutarowska et al. 2015). As one study summarized, compost use enhances the ability of a farm to manage pathogens by creating a soil environment that promotes beneficial microbes and robust plant immunity (Amuah et al. 2022).

### 3.8.3 Effects on Crop Yield and Productivity

The ultimate test of any soil amendment in agriculture is its impact on crop yield and quality. By enhancing root development and nutrient uptake, compost application often leads to increased crop productivity. In low-fertility or degraded soils, compost can have dramatic yield effects (Hassanein and Abul-Soud 2010). Compost can also improve crop quality (which is an aspect of yield in a broad sense). Fruits and vegetables grown with compost often have higher brix (sugar content), better flavor, and improved nutritional content, attributed to more balanced nutrition and soil microbial activity (Abdel-Haleem et al. 2022). While yield (quantity) might be the primary metric for staple crops, quality improvements are crucial for market value in horticultural crops, and compost has been linked to such enhancements (e.g., better tomato flavor and higher antioxidant levels in organically fertilized systems) (Hassanein and Abul-Soud 2010).

# 4 Compost Application in Diverse Agricultural Contexts

## *4.1 Temperate Agricultural Systems*

Compost application in temperate agricultural systems has shown various effects on soil fertility, crop production, and environmental impacts. A long-term simulation study in northern Germany examined the effects of compost applications on soil organic matter content, nitrate leaching and crop production in loamy and sandy soils (Gerke et al. 1999). The results suggested that differences between compost types (matured and non-matured) were less significant compared to site-specific properties and management practices over a 50-year application period. Interestingly, the study found that nitrate leaching was highly sensitive to different compost application scenarios in sandy soils, while the effect was relatively small in loamy soils. The research also revealed that relatively high crop yields can be achieved by a combination of 10 t compost/ha/yr and soil-N min adjusted mineral nitrogen additions of about 20 kg $ha^{-1}$ $year^{-1}$ to a winter-grain dominated crop rotation, which results in low nitrate concentrations in drainage water in sandy soil. In conclusion, compost application in temperate agricultural systems can be beneficial for soil fertility and crop production, but careful consideration of soil type, application rates, and complementary management practices is crucial. Further optimization may be possible by applying reduced rates of economically interesting non-matured compost (Gerke et al. 1999). However, it is important to note that the effectiveness and security of compost application in agricultural soils continue to generate concern, and improving compost quality or combining it with other remediation strategies may be necessary for safer and more effective soil management (Huang et al. 2016).

Compost application in temperate agricultural systems can have significant effects on soil properties and crop production, with outcomes varying based on soil types and crops. In loamy and sandy soils, application of compost on a long-term basis can increase soil organic matter content and supply plant nutrients, potentially improving crop yields (Gerke et al. 1999). However, the effects on nitrate leaching differ between soil types, with sandy soils being more sensitive to different compost application scenarios compared to loamy soils. Interestingly, while organic farming practices, including compost application, generally improve soil quality, they may result in lower crop productivity (5–58%) compared to conventional systems (Aulakh et al. 2022). This contradiction highlights the complexity of balancing soil health improvements with crop yield expectations in temperate agricultural systems. In conclusion, compost application in temperate agricultural systems can enhance soil fertility and crop resilience, particularly when combined with other conservation practices (Singh et al. 2024). For optimal results, farmers should consider soil type-specific application rates and timing, as well as crop rotations. While short-term transitions may result in yield reductions, long-term benefits to soil health and sustainability can be achieved through integrated organic management practices.

## 4.2 *Tropical and Subtropical Systems*

Compost application in tropical and subtropical systems with high rainfall can have both positive and negative effects on soil properties and nutrient dynamics. In a study on an Oldsmar sand soil in Florida, application of various composts at 100 Mg $ha^{-1}$ resulted in significant nutrient leaching, with nitrate, ammoniacal N and phosphate concentrations in leachate reaching as high as 246, 29, and 7 mg $L^{-1}$, respectively (Li et al. 1997). The amount of N and P leached accounted for 3.3–15.8% of total N and 0.2–2.8% of total P in the compost, with leaching peaks occurring after only 6.8–9.1 cm of rainfall equivalent. Interestingly, a study on tropical acid soils found that the combined use of urea and compost could maintain soil pH and reduce nitrogen losses through leaching (Latifah et al. 2017). This approach increased soil pH, buffering capacity, total nitrogen, organic nitrogen fractions, and inorganic nitrogen availability, suggesting that compost can be effectively used to regulate nitrogen availability in tropical soils when combined with mineral fertilizers. In conclusion, while compost application in tropical and subtropical systems can improve soil properties and nutrient availability, careful management is necessary to minimize nutrient leaching, especially in areas with high rainfall and sandy soils. The combined use of compost with mineral fertilizers or biochar may offer a more balanced approach to nutrient management in these systems, potentially reducing leaching losses while maintaining soil fertility and crop productivity (Cao et al. 2017).

## 4.3 Arid and Semiarid Systems

Compost application in arid and semiarid systems has shown promising results for improving soil moisture and nutrient retention, as evidenced by several case studies. In California, a field survey demonstrated that compost application significantly increased soil organic carbon to three times compared to control soils. This increase was accompanied by improvements in soil microbial activity (2.23 times higher), gravimetric water content (1.57 times higher), and reduced bulk density (0.87 times lower). Notably, water infiltration rates and water-holding capacity increased in coarser-textured soils (Brown and Cotton 2011). A study on date palm-based organic amendments in sandy loam and silty loam soils found that biochar was more effective than compost in improving soil water retention. However, when combined with sand supplementation, both compost and biochar showed significant improvements in available water capacity. For instance, in a sandy loam soil enriched with 2/3 sand, compost, and biochar amendments increased available water capacity by 26% and 80%, respectively, compared to unamended soil (Le Guyader et al. 2024). Interestingly, some studies have reported contradictory findings. In some areas of irrigated arid and semiarid soils, crop productivity was affected much less by organic matter content than the presumed amount. This suggests that other factors, such as soil texture, may play a more critical role in determining soil fertility in these regions (Hag Husein et al. 2021). So compost application generally improves soil moisture and nutrient retention in arid and semiarid systems; the effectiveness can vary depending on soil type, application rates, and local conditions. The integration of compost with other soil management practices, such as mulching and reduced tillage, may provide more comprehensive benefits for sustainable agriculture in water-scarce regions (Matisic et al. 2024).

## 4.4 Organic Versus Conventional Farming

Compost application in organic farming systems has shown significant benefits for plant growth, crop productivity and soil fertility compared to conventional farming methods. In a field experiment with organic barley production, the application of compost alone or in combination with compost water extract increased yield parameters, photosynthetic pigments, and nutrient uptake compared to control plants (Ghouili et al. 2022). This suggests that compost promotes barley plant growth, development, and productivity under organic farming conditions. Interestingly, while organic farming management generally results in higher soil organic carbon, nitrogen, and available nutrients like phosphorus, potassium, iron, and zinc compared to conventional systems, crop yields can be lower in organic systems, at least initially. A study in Spain found that organic crop yields were on average 23% lower than conventional yields, though only two crops showed statistically significant differences (Herencia et al. 2008). However, this yield gap may decrease over time as soil health improves with continued organic management. In conclusion, compost application in organic farming systems can significantly enhance soil fertility, nutrient availability, and crop performance compared to conventional methods relying on synthetic fertilizers. While initial yields may be lower in organic systems, the long-term benefits to soil health and sustainability make compost a valuable tool for organic farmers. Further research is needed to optimize compost formulations and application methods to maximize nutrient uptake and crop productivity in organic farming systems (Sani and Yong 2021).

### *4.5 Small Holder Versus Large-Scale Farming*

Compost application has shown positive effects on soil fertility and crop yields in both smallholder and large-scale farming systems. In smallholder farming in Ethiopia, applying 6 t compost $ha^{-1}$ with 34.5 kg N & 10 kg P $ha^{-1}$ resulted in the highest grain yield and was economically profitable with an 844% marginal rate of return (Demelash et al. 2014). Similarly, in large-scale farming in China, farmers with larger landholdings prefer to apply organic fertilizers and increase the intensity of application compared to small-scale farmers (Chen et al. 2022). Interestingly, the benefits of compost application extend beyond immediate yield increases. In Ethiopia, residual effects of compost application were observed in subsequent years, with 8 t compost $ha^{-1}$ combined with inorganic fertilizers leading to a 271% yield increase over the control in the following season (Demelash et al. 2014). Additionally, long-term compost use in greenhouse cultivation improved soil fertility and produced equal yields and nutrient composition in crops compared to mineral fertilization (Herencia et al. 2007). In conclusion, while both smallholder and large-scale farmers can benefit from compost application, the scale of farming operations may influence adoption rates and application intensity. Smallholder farmers may face challenges in accessing and applying compost, while large-scale farmers may have more resources to implement organic fertilization practices. Regardless of scale, compost application has demonstrated potential for improving soil health, crop yields, and long-term agricultural sustainability across different farming systems.

## 5 Empirical Evidence and Case Studies

Numerous studies have confirmed compost's positive effects on soil health, nutrient uptake, and crop yield. For instance, integrating compost with plant growth-promoting rhizobacteria (PGPR) improved soil chemical properties and microbial activity in sandy soils, enhancing wheat productivity (El-Akhdar et al. 2024). Combining compost with inorganic fertilizers increased rice grain and straw yields compared to inorganic inputs alone (Gosal et al. 2017), while enriched compost from agricultural waste and minerals boosted potato and soybean yields and soil fertility (Biswas 2010). However, potential challenges exist. In urban agriculture, compost applied based on nitrogen demand sometimes led to phosphorus overapplication and leaching risks (Shrestha et al. 2020). Compost also outperformed biochar in improving soil fertility and maize growth in alkaline soils (Manirakiza and Şeker 2020). Long-term use of compost improves soil fertility sustainably without exceeding safe heavy metal thresholds (Wang et al. 2016). In Kenya, combining 1 t $ha^{-1}$ each of manure and crop residue increased maize yields by 17–51% over manure alone, with conservation tillage further enhancing outcomes (Kihara et al. 2011). In India, green manure plus urea matched the rice yield of urea alone, while FYM with nitrogen initially reduced yields but proved beneficial after 5 years of use. FYM also increased wheat yield by 6% compared to other organics. Enriched compost applied at 4 t/ha with 50% inorganic fertilizer increased potato yield by 43.3% and boosted nutrient uptake, with residual benefits for soybean (Biswas 2010). Overall, optimizing compost type, rate, and integration with fertilizers under local conditions can substantially enhance soil health and productivity.

# 6 Practical Considerations for Compost Application

Composting is a sustainable and efficient method for treating organic waste, converting 40–60% of the original volume into nutrient-rich compost (Breitenbeck and Schellinger 2013). It involves the aerobic breakdown of organic matter by microorganisms, earthworms, and insects in the presence of air and moisture, yielding a final product as compost. Composting reduces greenhouse gas emissions, lowers reliance on chemical fertilizers, and enriches soil with essential nutrients like nitrogen, phosphorus, and potassium. These benefits enhance soil fertility, boost crop yields, and increase organic matter and total nitrogen in topsoil (Pane et al. 2014). Therefore, incorporating compost into agricultural soil can enhance crop yields and increase soil organic matter due to its rich nutrient content and the presence of beneficial plant growth-promoting organisms (Luo et al. 2017).

## *6.1 Best Practices for Compost Use*

As agricultural production rises to meet growing food demand, so does agricultural waste. Recycling organic waste into compost helps convert it into valuable nutrients for farming. Compost improves soil organic carbon, enhances structure through stable aggregates, and aids in water and nutrient retention (Misra et al. 2003). Effective composting requires proper site selection, material preparation, and regular turning. For optimal results, it is important to choose the right compost type, apply it correctly, and consider timing, rate, and crop-specific needs. Considering the environmental harm caused by intensive chemical fertilizer use, compost and other organic inputs offer a sustainable alternative. However, their effectiveness depends on proper application methods, timing, and optimal rates.

### 6.1.1 Application Method

Surface application involves spreading compost evenly across the soil without disturbing plant roots, which helps retain moisture, reduce erosion, and support microbial activity (Ho et al. 2022). Earthworms and rain play a significant role in mixing surface-applied compost with soil slowly and naturally. Surface application of compost supports key principles of soil conservation, such as reducing soil disturbance and maintaining consistent ground cover. This helps preserve soil structure, lowers the need for tillage, and functions as a protective mulch while reducing erosion, minimizes evaporation, and enhances microbial activity and bioturbation—processes vital for stabilizing soil aggregates and gradually incorporating organic matter into deeper layers (Reimer et al. 2023). It preserves soil structure by reducing soil compaction and promotes long-term productivity, though its effects may be limited by shallow integration. In contrast, incorporating compost into the soil leads to better soil aggregation and can enhance water infiltration, particularly in compacted soils lacking organic matter. Both methods have benefits, and the choice depends on soil conditions and specific goals. Data show that both surface and incorporated compost have lasting benefits, with effects persisting over 5 years. Both surface and incorporated compost improved soil carbon and nitrogen, and reduced bulk density compared to the control, with incorporation showing stronger effects due to greater organic matter input (Kranz et al. 2020). Incorporation also lowered the soil C: N ratio, reduced subsurface compaction, and had more apparent positive improvement in soil aggregation, though wet sieving showed no significant treatment differences (Curtis and Claassen 2005).

Surface compost application initially improved soil moisture as measured by tensiometers, but the effect lessened or declined as plants matured (Cogger et al. 2008). Chemical properties were similar between the surface-applied and incorporated compost, though incorporated compost led to slightly lower pH and exchangeable K. From the second year onward, incorporated compost showed higher residual nitrate-N due to increased nitrogen mineralization compared to surface-applied compost (Cox et al. 2001;Cogger et al. 2008). Data indicate that compost incorporation has a stronger impact on soil carbon, nitrogen, and bulk density than surface application. However, surface application remains beneficial when incorporation is not practical.

### 6.1.2 Time of Compost Application

The application of compost at the right time enhances nutrient use efficiency and minimizes environmental risks (Ho et al. 2022). The time of application depends on agronomic, economic, environmental, and social factors, including cropping plans, soil conditions, weather, regulations, and community considerations. The best results are obtained when compost is applied at the peak crop nutrient demand to align nutrient release with plant uptake. According to Viaenea et al. (2016) and Azim et al. (2018), the right time for application in a field depends on

1. *Operational considerations* include crop schedules, previous nutrient use, manure storage limits, and compliance with regulations.
2. *Site-specific conditions*, such as soil type and existing nutrient levels.
3. *Environmental factors* include seasonality, weather patterns, and the risk of runoff.
4. *Social aspects* such as proximity to neighbors and timing of local activities or events.

Compost is often applied in the fall when soils are drier and more time is available, but the risk of nutrient loss by runoff is higher due to the absence of crops to take up nutrients and the lack of ground cover. Spring application of compost is suitable for annual crops but challenged by a short growing season, limited time, and potential seeding delays due to wet soils or equipment constraints, while late summer or early fall application of composted manure allows nutrients to stabilize in the soil and results in greater yield benefits (Loecke et al. 2004). Nutrient loss risks from late applications can be minimized through incorporation, cover crop/ mulch, and adherence to setback guidelines like composting away from waterways, drainage ditches, or areas prone to runoff to reduce nutrient loss and environmental impact (Kranz et al. 2020). Post-seeding or in crop compost application provides additional/extended windows for compost application. Solid/liquid separation enables the liquid portion of manure to be applied via irrigation to growing crops without harming them. Thus, understanding application timing trade-offs is crucial to maximizing compost benefits while minimizing environmental impacts.

## 6.2 *Quality Control and Standardization*

### 6.2.1 Importance of Compost Quality

The addition of compost to soil enhances crop yield, organic matter, and humus content, and imparts disease resistance, improving overall soil fertility and plant health (Luo et al. 2017). However, these benefits depend on the availability and adoption of high-quality, mature, pathogen-free compost. Poor-quality compost can cause more harm than benefit, making compost quality an essential criterion for sustainable agriculture and environmental safety. The direct application of fresh or immature organic waste is not recommended due to risks like nutrient imbalances, phytotoxicity, pathogens, and heavy metals, which can harm plant growth and decrease overall crop productivity (Suarez-Estrella et al. 2013; Cerda et al. 2018). These issues are addressed by proper composting mechanisms, which convert waste into a nutrient-rich, pathogen-free product, mature compost. Mature, good-quality pathogen-free compost provides a stable, nutrient-rich amendment that promotes healthy plant growth, prevents disease, protects health, and reduces reliance on synthetic fertilizers and pesticides due to its cost-effectiveness (Papafilippaki et al. 2015; Sanasam and Talukdar 2017).

Despite the use of compost for several decades as a part of waste management and as a low-cost nutrient source, concerns remain about compost quality, particularly the presence of pathogens like bacteria, viruses, parasites and weed/foreign seed that can pose health risks to both crop and human if not properly managed (Scopetani et al. 2022). These risks may reduce compost application as a soil amendment. Compost can be sterilized to eliminate pathogens, as it may also destroy beneficial microbes. Therefore, it is important to be aware of composting standards and testing protocols.

### 6.2.2 Standards in Composting

Compost quality affects its effectiveness, with maturity being a key indicator. Mature compost is dark, earthy-smelling, moist, and free of foul odors or ammonia off-odors (Azim et al. 2018). While the composting process matters, feedstock choice is the main quality determinant. To ensure the production of high-quality compost, certain standards regarding nutrient content, safety and the environment are mentioned in Table 3.

### 6.2.3 Testing Protocols

A mature compost typically has characteristics like the absence of ammonia smell, maintains a low temperature even after turning while moist, has a dark, crumbly texture with a pleasant earthy scent, and its original materials are no longer easily identifiable. However, these quick visual and sensory checks should be confirmed with detailed laboratory testing (Szyłak-Szydłowski and Kos 2024; Sullivan and Miller 2001) for the safety in compost use for agriculture. Some of the commonly used parameters that are helpful in assessing the maturity of compost are mentioned in Table 4.

## *6.3 Challenges and Limitations*

Compost application has well established its role in the improvement of soil quality, fertility, and environmental health (Ho et al. 2022). Though composting is beneficial, it faces several challenges, including moisture and aeration control, odor, pests, contamination, and pathogen risks (Pereira and Fiore 2024). In India, despite being a major food producer, on-farm composting remains uncommon due to limited access

**Table 3** Parameters affecting compost quality

| Parameter | Description |
|---|---|
| Moisture level | The ideal moisture content for composting is 45–50%, enabling microbes to effectively break down organic matter (Razmjoo et al. 2015). Moisture below 30% limits bacterial activity and slow down decomposition, while levels above 65% reduce porosity, causing anaerobic conditions and unpleasant odor emissions (Roman et al. 2015). Monitoring temperature helps manage moisture by indicating when to turn, moisten, or ventilate the compost pile (Azim et al. 2018). |
| Oxygen level | Oxygen levels in compost depend on porosity, which is influenced by particle size, shape, and moisture content. At least a minimum rate of 5% oxygen is needed to maintain aerobic conditions, while levels below 1% cause anaerobic zones (Mustin 1987). Oxygen levels below 5% indicate poor aeration and slow decomposition due to excess moisture. Addition of organic materials with high moisture absorption and fine particles can enhance aeration and evaporation. The ideal range is 15–20% for optimal microbial activity. Oxygen demand is highest (15–20%) during initial decomposition due to intense microbial activity, then drops during stabilization of compost (5–10%) and further during maturation. $CO_2$ levels should stay below 15%, which corresponds to about 6% oxygen. The oxygen levels above 20% suggest over-aeration and reduced moisture, also slow down decomposition. Compaction of such soil will reduces porosity. |
| Temperature | Temperature is a key factor in composting, which drives microbial activity and ensure compost sanitation (Azim et al. 2018). The heat flux is generated due to the biological activity is an indicator of the successful composting which can be regulated by ventilation or turning. The temperature range of 35–40 °C (mesophilic phase) is considered ideal range for microbial growth and decomposition of compost, while temperature range of 50–65 °C (thermophilic phase) is ideal for pathogen inactivation and weed seed destruction. Temperature more than 65 °C indicate slow biological activity and undesirable chemical modifications of the organic matter caused due to poor ventilation and low moisture. Thus, a continuous temperature monitoring is essential for balancing microbial activity and compost quality. |
| pH | Organic material can be composted across a wide pH range (3–11), with an optimal range of 5.5–8 for microbial activity. Initially, pH drops during composting due to the acid-genesis phase caused by organic acid and $CO_2$ release (Poincelot 1972). As composting progresses, pH rises in the alkalization phase due to ammonia production from protein breakdown especially during the thermophilic phase. In the stabilization phase, pH levels stabilize as ammonia volatilizes and microbes use nitrogen for humus formation. The final maturation phase brings the pH close to neutral, indicating compost stability. The acidic or alkaline pH levels of compost are indication of immature compost. |
| C/N | The carbon-to-nitrogen (C:N) ratio is a key indicator of compost quality, with an optimal range of 10–20:1 as ideal since nutrient in compost within this are less likely to immobilize. The composts with C:N above 25:1 are unacceptable for use in cropping systems as it may immobilize nitrogen and hence are unsuitable for crops. C:N is useful for prescreening of compost quality. For mature compost, nitrate should be ≥100 ppm, with optimal plant growth observed at 600–2000 ppm, while ammonium should be <1000 ppm, with an ammonium-to-nitrate ratio below 1:1. |
| Particle size | Particle size affects compost aeration and compaction (Ebo et al. 2022). Particles <3.1 mm cause tight packing, and can lead to anaerobic conditions in composting. Compaction from machinery or pile weight also reduces porosity. The ideal |

(continued)

**Table 3** (continued)

| Parameter | Description |
|---|---|
| | particle size range is 3.1–12.7 mm, promoting adequate pore space and evaporation. Sizes >12.7 mm create large pores and excess aeration, which can slow decomposition by reducing temperature, and regards the composting process. Reduction in material size by chopping/shredding is recommended. |

**Table 4** Parameters indicative of compost maturity

| Parameters | Description |
|---|---|
| pH | The pH is one of the chemical indicators of compost maturity. Immature composts are typically acidic, while mature composts have a pH between 7 and 9 (Avnimelech et al. 1996; Albrecht 2007). |
| CEC | Cation exchange capacity (CEC) tends to increase during composting of organic materials due to humification and the formation of functional groups. A CEC above 60 meq/100 g of organic matter indicates mature compost (Wichuk and McCartney 2013). |
| C/N ratio | The C/N ratio decreases during composting and is a key indicator of maturity of compost. The C/N ratio below 20, ideally under 15, signifies mature compost (Albrecht 2007; Iglesias-Jimenez and Perez-Garcia 1991). |
| $NO_3^-/NH_4^+$ ratio | The presence of nitrates in compost indicates compost maturity, as nitrifying microbes convert ammonium ($NH_4^+$) to nitrate ($NO_3^-$). The $NO_3^-/NH_4^+$ ratio is often used as a maturity indicator, though results may vary (Francou 2003). |
| Humification ratio (HA/FA) | During composting, the humic acid to fulvic acid (HA/FA) ratio increases, with values <1 indicates immature compost and >1–3 indicates mature compost (Jouraiphy et al. 2005). |
| Measurements of respiration | Respiration of microorganisms is also an indication of compost maturity, with immature compost showing high oxygen demand and carbon dioxide production due to active microbial degradation, while mature compost is more stable and less biologically active (Bernal et al. 1998; Lasaridi and Stentiford 1998). |
| Enzymatic activity | High protease, lipase, and cellulase activity are observed during the active phase of composting (Mondini et al. 2004; Goyal et al. 2005). The enzyme activity stabilized as the compost matures. |
| Self-heating test | Heat generated by microbial activity in immature composts indicates ongoing decomposition. Thus, measuring temperature rise after moistening helps assess stage of compost decomposition (Leifeld and Siebert 2001). |
| Ammonia test | An *ammonia test* evaluates compost *maturity* and *stability* by detecting residual *ammonium* ($NH_4^+$), which is high (>*500 mg/kg* $NH_4^+$) in immature composts and decreases as compost matures (<*100 mg/kg* $NH_4^+$). Also the ratio of $NH_4^+/NO_3^- < 1$ indicate maturity and $> 3$ indicate immaturity of compost.The qualitative smell test indicate immature compost have strong smell of ammonia, while lack of ammonia smell indicate mature compost. |
| Microbial populations | The fungivorous/bacterivorous ratio increases as compost matures, ranging from 0 to 11.9, which makes it a useful potential indicator of compost maturity (Daramola et al. 2022). |

to suitable composting materials, transportation, storage, contamination concerns, and regulatory requirements.

### 6.3.1 Logistical Barriers

With growing emphasis on eco-friendly reuse of organic waste, composting offers a sustainable way to recycle it into nutrient-rich material for agriculture, reducing reliance on chemical fertilizers (Ho et al. 2022). Compost enhances soil structure, fertility, and productivity, and lowers input costs. However, barriers such as limited access to organic materials, logistics, and adoption challenges hinder on-farm composting (Viaenea et al. 2016; Ayilara et al. 2020).

A key barrier to composting is the shortage of organic materials due to competing uses such as biofuel production, livestock feed, and household fuel (Ben-Iwo et al. 2016). In rural India, for instance, 85% of households rely on biomass for cooking (UNIDO report), which is a health hazard as it produces extremely high levels of indoor air pollution. Organic waste is also diverted for energy production, and livestock manure is often underutilized. These competing demands limit the availability of materials for composting. To address this, alternatives like woody byproducts from pruning or landscape management, which are less suited for combustion, could be redirected for composting. Compost releases nutrients slowly compared to fast-acting fertilizers like mineral fertilizers and slurry. While it enriches soil organic carbon and can reduce reliance on synthetic fertilizers, its limited nitrogen and phosphorus availability may not meet immediate crop demands (Sayara et al. 2020; Viaenea et al. 2016). A balanced approach combining compost with quick-release fertilizers and practices like intercropping is required to optimize nutrient supply and minimize leaching. However, compost supply is often unreliable, and farmers face uncertainty about its availability when needed. Compost is required to be applied at very high quantities, and more so, often it is difficult to obtain disease-free/weed-free mature compost in large amounts. High transport costs of biomass, reliance on certified contractors, and free manure transport make compost less competitive. Limited viable transport range and inconsistent compost quality and supply further discourage agricultural use (Azim et al. 2018).

Composting requires significant time for monitoring and turning, often conflicting with primary farming tasks. Seasonal feedstock availability and fast decay add to the challenge. Combined with high costs and delayed benefits, this limits the short-term profitability of on-farm composting and makes composting less attractive compared to immediate returns from synthetic fertilizers. Maintaining and assuring the availability of compost quality is another important concern. High composting temperatures may reduce beneficial organisms, leading to biologically poor compost. Risks of weeds, diseases, and inconsistent quality deter use. Additionally, uncertainty about nitrogen release timing and the broad definition of compost further hinder farmer confidence (Gajalakshmi and Abbasi 2008; Pereira and Fiore 2024). In a country like India, free and abundant manure from livestock farms reduces the incentive to use compost, further limiting the quality of compost application. The production of quality compost requires costly equipment like compost turners, $CO_2$ sensors, and shredders—often unaffordable for individual farmers. Additionally, high fuel use, large feedstock volume-to-weight ratio, and transport costs further raise expenses, limiting on-farm composting adoption. Most important among all is the lack of knowledge about proper composting methods, application practices, economic benefits, and regulatory requirements, which prevents many farmers from producing or using compost effectively (Kranz et al. 2020; Azim et al. 2018; Pereira and Fiore 2024).

### 6.3.2 Potential Risks

While the benefits of composting are widely recognized, many remain unaware of its potential risks. These include health hazards from pathogens, chemical pollutants, allergens, and molds. Compost piles may also attract pests and emit unpleasant odors, affecting nearby communities. Compost quality depends on the absence of contaminants such as human pathogens, plastics, weed seeds, heavy metals, and pesticide residues (Scopetani et al. 2022).

#### Heavy Metal Contamination

Long-term compost use may lead to the accumulation of potentially toxic elements such as zinc (Zn), nickel (Ni), and copper (Cu) in soil. It can lead to environmental risks despite infrequent monitoring of plant uptake with compost fertilization (Pereira and Fiore 2024). The problem of heavy metal contamination is more severe when municipal solid waste (MSW) is used as raw material for compost production. Over half of MSW is putrescible organic matter, making composting an efficient and eco-friendly solution (Horrocks et al. 2016). However, MSW-derived compost often contains heavy metals like Cd, Cu, Pb, Ni, and Zn, which pose risks of soil and crop contamination. These metals are not degraded during composting and may become concentrated. While composting may reduce metal bioavailability, risks of leaching into groundwater remain, especially in acidic soils or when metal concentrations are high (Gonçalves et al. 2018).

#### Microplastic Contamination

Microplastic contamination in compost is a growing environmental concern, stemming from improper waste disposal and non-biodegradable plastics in bio-waste and sewage sludge. These plastics introduce hazardous substances into soils, impacting soil health and biota. Plastics can also adsorb and release toxic additives like plasticizers, flame retardants, and pigments, posing risks to ecosystems and human health (Hahladakis et al. 2018; Braun et al. 2021).

Nutrient Imbalances

Compost, as a multi-nutrient fertilizer, often causes nutrient imbalances when not properly integrated with other nutrient sources (Zikeli et al. 2017; Tittarelli et al. 2017; Reimer et al. 2023). Its nutrient composition rarely aligns with plant needs, leading to excesses—especially phosphorus and potassium—when applied based on nitrogen demand. Variability in nutrient availability, nitrogen loss during composting, and mismatched nutrient ratios can hinder plant growth. Long-term use requires additional fertilizers, such as phosphorus-free potassium sources, to maintain nutrient balance (Reimer et al. 2023).

### *6.4 Integration with Other Practices*

Combining compost with practices like precision agriculture, irrigation, and cover cropping enhances soil health and crop outcomes (Xing and Wang 2024). Precision Compost Strategy (PCS) tailors compost use to specific conditions—especially effective in warm, dry, acidic, or sandy/clay soils—boosting yields by up to 40% (Zhao et al. 2021). Compost improves water retention with irrigation and boosts fertility with cover crops, which also reduce erosion and external nutrient needs. However, results vary due to diverse compost types, shifting feedstocks (e.g., food waste, sewage sludge), intensive cropping systems, and altered application methods. Modern systems often pair compost with fertilizers in degraded soils under climate stress, requiring strategic, site-specific compost use. PCS aims to match compost inputs to crop and soil needs, minimizing nutrient excesses and environmental impacts while advancing sustainable, circular agriculture (Abdou et al. 2016; Jeong et al. 2018).

## 7 Knowledge Gaps

Compost application has been shown to have potential benefits for crop production by supplying plant nutrients as well as increasing soil fertility. However, there is a lack of comprehensive understanding of its long-term impacts across diverse agricultural settings. While (Gerke et al. 1999) provide insights into the effects of compost application on soil organic matter content, nitrogen availability and crop production over a 50-year period, it focuses primarily on loamy and sandy soils in a specific region. Interestingly, the research reveals some contradictions and nuances. He also suggests that the differences between compost types (matured and non-matured) are less compared to other properties and management practices when considering long-term application. This highlights the need for more research on how compost interacts with various soil types, climatic conditions, and agricultural practices across different regions. There is a broader research gap in understanding the long-term effects of compost application on crop productivity, soil health, and environmental impacts across diverse agricultural contexts. Future studies should focus on examining these effects in various soil types, climatic zones, and cropping systems, while also considering the interaction of compost with other agricultural practices such as tillage, crop rotation, and integrated disease-pest management. This research would be crucial for developing sustainable agricultural practices that balance productivity with environmental conservation (Chen et al. 2024).

## 8 Conclusion

Compost is a versatile and sustainable soil amendment that significantly enhances nutrient uptake and crop productivity across diverse agricultural contexts. By improving soil physical, chemical, and biological properties, compost addresses nutrient inefficiencies, reduces reliance on chemical fertilizers, and mitigates environmental impacts. Field studies show substantial yield improvements in crops like maize, wheat, and tomatoes, with benefits varying by soil type, climate, and compost quality. However, challenges such as heavy metal contamination, nutrient imbalances, and logistical constraints require careful management through practices like soil testing, source separation, and precision application. Integrating compost with techniques such as cover cropping and irrigation can further optimize outcomes. Future research should focus on long-term impacts of compost on soil microbiomes and standardized protocols for compost quality to maximize its role in sustainable agriculture. Farmers and policymakers are encouraged to adopt context-specific composting strategies to enhance soil health and food security.

## References

Abdel-Haleem ES, Farrag HM, Abeer BAKR, Abdelrasheed KG (2022) Combined use of compost, compost tea, and vermicompost tea improves soil properties, and growth, yield, and quality of (Allium cepa L.). Not Bot Horti Agrobot Cluj-Napoca 50(1):12565

Abdou G, Ewusi-Mensah N, Nouri M, Tetteh FM, Safo EY, Abaidoo RC (2016) Nutrient release patterns of compost and its implication on crop yield under Sahelian conditions of Niger. Nutr Cycl Agroecosyst 105:117–128

Ahmad A, Aslam Z, Abbas RN, Bellitürk K, Hussain S, Hussain S, Ahmad M, Zulfiqar U, Moussa IM, Elshikh MS (2024) Enhancing wheat crop resilience to drought stress through cellulolytic microbe-enriched cow dung vermicompost. ACS Omega 9(2):2123–2133. https://doi.org/10.1021/acsomega.3c04402

Ahmed T, Shahid M, Qi Y, Li B, Xu L, Noman M, Yao Y, Qi X, Ali HM, Hussain S, El-Kott AF, Masood HA, Negm S (2023) Fertilization of microbial composts: a technology for improving stress resilience in plants. Plants (Basel, Switz) 12(20):3550. https://doi.org/10.3390/plants12203550

Albrecht R (2007) Co-compostagedeboues destationd'épuration et de déchets verts: Nouvelle méthodologie du suivi des transformations de la matière organique. Dissertation, Faculté des Sciences et Technique St Jérôme, L'Université Paul Cézanne, Marseille, France, p 189

Alghamdi AG, Ibrahim HM, Majrashi MA (2023) Improving the physical properties and water retention of Sandy soils by the synergistic utilization of natural clay deposits and wheat straw. Sustainability 16(1):46. https://doi.org/10.3390/su16010046

Alromian FM (2020) Effect of type of compost and application rate on growth and quality of lettuce plant. J Plant Nutr 43(18):2797–2809

Al-Suhaibani N, Selim M, Alderfasi A, El-Hendawy S (2021) Integrated application of composted agricultural wastes, chemical fertilizers and biofertilizers as an avenue to promote growth, yield and quality of maize in an arid agro-ecosystem. Sustainability 13(13):7439. https://doi.org/10.3390/su13137439

Amuah EEY, Fei-Baffoe B, Sackey LNA, Douti NB, Kazapoe RW (2022) A review of the principles of composting: understanding the processes, methods, merits, and demerits. Org Agric 12(4):547–562

Asaye Z, Kim DG, Yimer F, Prost K, Obsa O, Tadesse M, Gebrehiwot M, Brüggemann N (2022) Effects of combined application of compost and mineral fertilizer on soil carbon and nutrient content, yield, and agronomic nitrogen use efficiency in maize-potato cropping systems in Southern Ethiopia. Land 11(6):784. https://doi.org/10.3390/land11060784

Aulakh CS, Sharma S, Thakur M, Kaur P (2022) A review of the influences of organic farming on soil quality, crop productivity and produce quality. J Plant Nutr 45(12):1884–1905. https://doi.org/10.1080/01904167.2022.2027976

Avnimelech Y, Bruner M, Ezrony I, Sela R, Kochba M (1996) Stability indexes for municipal solid waste compost. Compost Sci Util 4(2):13–20

Ayilara MS, Olanrewaju OS, Babalola OO, Odeyemi O (2020) Waste management through composting: challenges and potentials. Sustainability 12(11):4456. https://doi.org/10.3390/su12114456

Azim K, Soudi B, Boukhari S, Perissol C, Roussos S, Thami Alami I (2018) Composting parameters and compost quality: a literature review. Org Agric 8:141–158

Balota EL, Amaral H, Yada IF, Coyne MS, Nakatani AS, Dick RP (2013) Long-term land use influences soil microbial biomass p and s, phosphatase and arylsulfatase activities, and s mineralization in a brazilian oxisol. Land Degrad Dev 25(4):397–406. https://doi.org/10.1002/ldr.2242

Ben-Iwo J, Manovic V, Longhurst P (2016) Biomass resources and biofuels potential for the production of transportation fuels in Nigeria. Renew Sust Energ Rev 63:172–192. https://doi.org/10.1016/j.rser.2016.05.050

Bernal M, Navarro A, Sánchez-Monedero M, Roig A, Cegarra J (1998) Influence of sewage sludge compost stability and maturity on carbon and nitrogen mineralization in soil. Soil Biol Biochem 30(3):305–313

Bhatt KM, Joshi HC, Labanya R (2019) Influence of long-term chemical fertilizers and organic manures on soil fertility - A review. Universal Journal of Agricultural Research 7(5):177–188. https://doi.org/10.13189/ujar.2019.070502

Bhati P, Saikia AR, Nengparmoi T, Chaudhary S, Talukdar N, Hazarika S, Bahadur R (2024) Integrated farming systems for environment sustainability: a comprehensive review. J Sci Res Rep 30(1):143–155. https://doi.org/10.9734/jsrr/2024/v30i11834

Bibi H, Rahim HU, Khan AA, Haris M, Iqbal M, Ali R, El-Sheikh MA, Kaushik P (2024) Harmonized tripartite approach: Enhancing nutrient accessibility, uptake, and wheat productivity through Trichoderma harzianum, compost, and phosphorus synergy. Journal of King Saud University - Science 36(3):103106. https://doi.org/10.1016/j.jksus.2024.103106

Biswas DR (2010) Nutrient recycling potential of rock phosphate and waste mica enriched compost on crop productivity and changes in soil fertility under potato–soybean cropping sequence in an Inceptisol of Indo-Gangetic Plains of India. Nutr Cycl Agroecosyst 89(1):15–30. https://doi.org/10.1007/s10705-010-9372-6

Braun M, Mail M, Heyse R, Amelung W (2021) Plastic in compost: prevalence and potential input into agricultural and horticultural soils. Sci Total Environ 760:143335. https://doi.org/10.1016/j.scitotenv.2020.143335

Breitenbeck G, Schellinger D (2013) Calculating the reduction in material mass and volume during composting. Compost Sci Util 12:365–371. https://doi.org/10.1080/1065657X.2004.10702206

Brenzinger K, Drost SM, Korthals G, Bodelier PLE (2018) Organic residue amendments to modulate greenhouse gas emissions from agricultural soils. Front Microbiol 9. https://doi.org/10.3389/fmicb.2018.03035

Brown S, Cotton M (2011) Changes in soil properties and carbon content following compost application: results of on-farm sampling. Compost Sci Util 19(2):87–96. https://doi.org/10.1080/1065657x.2011.10736983

Cao Y, Li J, Qi Y, Gao Y (2017) Biochar-enhanced composts reduce the potential leaching of nutrients and heavy metals and suppress plant-parasitic nematodes in excessively fertilized cucumber soils. Environ Sci Pollut Res 25(8):7589–7599. https://doi.org/10.1007/s11356-017-1061-4

Cerda A, Artola A, Font X, Barrena R, Gea T, Sánchez A (2018) Composting of food wastes: status and challenges. Bioresour Technol 248:57–67

Chen Y, Liu Y, Fu X (2022) Effect of farmland scale on farmers' application behavior with organic fertilizer. Int J Environ Res Public Health 19(9):4967. https://doi.org/10.3390/ijerph19094967

Chen Q, Zhong G, Lu Y, Song Y, An Y (2024) Soil microorganisms: their role in enhancing crop nutrition and health. Diversity 16(12):734. https://doi.org/10.3390/d16120734

Chiarelotto M, Restrepo JCPS, Lorin HEF, Damaceno FM (2021) Composting organic waste from the broiler production chain: a perspective for the circular economy. J Clean Prod 329:129717

Chtouki M, Naciri R, Oukarroum A, Laaziz F, Nguyen F, Garré S (2022) Interactive effect of soil moisture content and phosphorus fertilizer form on chickpea growth, photosynthesis, and nutrient uptake. Sci Rep 12(1). https://doi.org/10.1038/s41598-022-10703-0

Cogger C, Hummel R, Hart J, Bary A (2008) Soil and Redosier dogwood response to incorporated and surface-applied compost. HortScience 43(7):2143–2150

Cox D, Bezdicek D, Fauci M (2001) Effects of compost, coal ash, andstraw amendmentson restoring the quality of eroded Palouse soil. Biol Fertil Sóils 33:365–372

Curtis MJ, Claassen VP (2005) Compost incorporation increases plant available water in a drastically disturbed serpentine soil. Soil Sci 170:939–953

Daramola FY, Orisajo SB, Osemwegie OO (2022) Nematode management prospects in composting. In: Chaudhary KK, Meghvansi MK (eds) Sustainable management of nematodes in agriculture, Vol.1: organic management, Sustainability in plant and crop protection, vol 18. Springer, Cham. https://doi.org/10.1007/978-3-031-09943-4_3

Das A, Singh S, Rani K, Gouda HS, Mahakuda B, Roy A, Thakur JK, Chakraborty R (2024) Natural farming: status, challenges and prospects. In: Mohapatra B, Singh AK, Gaikwad SB, Singh K, Ghosh A (eds) Futuristic trends in agriculture engineering & food sciences volume 3 book 23. IIP Iterative International Publishers, Chikkamagaluru, pp 225–236. https://doi.org/10.58532/V3BCAG23CH18

Demelash N, Ziadat F, Sommer R, Tesfaye S, Dayu W (2014) Current and residual effects of compost and inorganic fertilizer on wheat and soil chemical properties. Nutr Cycl Agroecosyst 100(3):357–367. https://doi.org/10.1007/s10705-014-9654-5

Devi NT, Devi AS, Singh KR (2025) Impact of chemical fertilizer and composts application on growth and yield of rice in Northeast India. Sci Rep 15(1):8575. https://doi.org/10.1038/s41598-025-93684-0

Diatta AA, Kanfany G, Camara B, Bassène C, Manga AG, Seleiman M, Mbow C, Schillaci C (2024) Compost as an alternative to inorganic fertilizers in cowpea [Vigna unguiculata (L.) Walp.] production. Legume Sci 6(3):e247. https://doi.org/10.1002/leg3.247

Downey SS, Walker M, Moschler J, Penados F, Peterman W, Pop J, Qin R, Scaggs SA, Song S (2023) An intermediate level of disturbance with customary agricultural practices increases species diversity in Maya community forests in Belize. Commun Earth Environ 4(1). https://doi.org/10.1038/s43247-023-01089-6

Ebo E, Amuah Y, Fei B, Kazapoe RW (2022) A review of the principles of composting: understanding the processes, methods, merits, and demerits. Org Agric 12:547–562

El-Akhdar I, Shabana MMA, El-Khateeb NMM, Elhawat N, Alshaal T (2024) Sustainable wheat cultivation in Sandy soils: impact of organic and biofertilizer use on soil health and crop yield. Plants (Basel, Switz) 13(22):3156. https://doi.org/10.3390/plants13223156

Farhan M, Zhao X, Tariq A, Wang S, Wu T, Yang J (2024) Impact assessment of agricultural droughts on water use efficiency in different climatic regions of Punjab Province Pakistan using MODIS time series imagery. Hydrol Process 38(7). https://doi.org/10.1002/hyp.15232

Francou C (2003) Stabilisation dela matière organique au cours du compostage de déchets urbains: Influence de la nature des déchets et du procédé de compostage– Recherche d'indicateurs pertinents. Dissertation, Institut national agronomique Paris- Grignon, p 289

Fuchs JG, Cuijpers WJ (2016) Compost types, feedstocks and composting methods. In: Handbook for composting and compost use in organic horticulture, vol 82(6). pp 29–43

Gajalakshmi S, Abbasi SA (2008) Solid waste management by composting: state of the art. Crit Rev Environ Sci Technol 38:311–400

Gerke HH, Arning M, Stöppler-Zimmer H (1999) Modeling long-term compost application effects on nitrate leaching. Plant Soil 213(1/2):75–92. https://doi.org/10.1023/a:1004452515423

Ghimire R, Noble Strohm T, Schipanski M, Acosta-Martinez V, Shukla MK, Angadi SV, Adebayo O, Fonte SJ, Bista P, Mikha MM, Thapa VR, Slaughter LC (2023) Soil health assessment and management framework for water-limited environments: examples from the Great Plains of the USA. Soil Syst 7(1):22. https://doi.org/10.3390/soilsystems7010022

Ghouili E, Hidri Y, M'Hamed HC, Somenahally A, Xue Q, Znaïdi IEA, Jebara M, Ouertani RN, Muhovski Y, Riahi J, Abid G, Sassi K (2022) Date palm waste compost promotes plant growth and nutrient transporter genes expression in barley (Hordeum vulgare L.). S Afr J Bot 149:247–257. https://doi.org/10.1016/j.sajb.2022.06.018

Goldan E, Nedeff V, Culea M, Chitimus D, Mosnegutu E, Irimia O, Panainte-Lehadus M, Tomozei C, Barsan N (2023) Assessment of manure compost used as soil amendment—A review. Processes 11(4):1167. https://doi.org/10.3390/pr11041167

Gonçalves ATT, Moraes FTF, Marques GL, Lima JP, Lima RDS (2018) Urban solid waste challenges in the BRICS countries: a systematic literature review. Rev Ambient Água 13:e2157

Gong S, Batáry P, Liu Y, Tscharntke T, Hodgson JA, Zou Y, Van Der Werf W, Knops JMH (2022) Biodiversity and yield trade-offs for organic farming. Ecol Lett 25(7):1699–1710. https://doi.org/10.1111/ele.14017

Gosal SK, Gill GK, Sharma S, Walia SS (2017) Soil nutrient status and yield of rice as affected by long-term integrated use of organic and inorganic fertilizers. J Plant Nutr 41(4):539–544. https://doi.org/10.1080/01904167.2017.1392570

Goyal S, Dhull S, Kapoor K (2005) Chemical and biological changes during composting of different organic wastes and assessment of compost maturity. Bioresour Technol 96(14): 1584–1591

Gutarowska B, Skóra J, Stępień Ł, Szponar B, Otlewska A, Pielech-Przybylska K (2015) Assessment of microbial contamination within working environments of different types of composting plants. J Air Waste Manage Assoc 65(4):466–478

Hag Husein H, Lucke B, Sahwan W, Bäumler R (2021) A contribution to soil fertility assessment for arid and semi-arid lands. Soil Syst 5(3):42. https://doi.org/10.3390/soilsystems5030042

Hahladakis JN, Velis CA, Weber R, Iacovidou E, Purnell P (2018) An overview of chemical additives present in plastics: migration, release, fate and environmental impact during their use, disposal and recycling. J Hazard Mater 344:179–199. https://doi.org/10.1016/j.jhazmat.2017.10.014

Hale L, Curtis D, Azeem M, Montgomery J, Crowley DE, McGiffen ME Jr (2021) Influence of compost and biochar on soil biological properties under turfgrass supplied deficit irrigation. Appl Soil Ecol 168:104134

Hassan NYI, Abdelhamid AN, El Wahed NHA, Ashraf M, Abdelfattah EA (2023) Composting: an eco-friendly solution for organic waste management to mitigate the effects of climate change. Innovare J Soc Sci:1–7. https://doi.org/10.22159/ijss.2023.v11i4.48529
Hassanein MK, Abul-Soud M (2010) Effect of different compost types and application methods on growth and yield of three maize hybrids. J Appl Sci Res 6(9):1387–1399
Hepworth C, Gray JE, Doheny-Adams T, Hunt L, Cameron DD (2015) Manipulating stomatal density enhances drought tolerance without deleterious effect on nutrient uptake. New Phytol 208(2):336–341. https://doi.org/10.1111/nph.13598
Herencia JF, Ruiz-Porras JC, Melero S, Garcia-Galavis PA, Morillo E, Maqueda C (2007) Comparison between organic and mineral fertilization for soil fertility levels, crop macronutrient concentrations, and yield. Agron J 99(4):973–983. https://doi.org/10.2134/agronj2006.0168
Herencia JF, Maqueda C, Melero S, Garcia Galavís PA, Ruiz JC (2008) A short-term comparison of organic*v.* conventional agriculture in a silty loam soil using two organic amendments. J Agric Sci 146(6):677–687. https://doi.org/10.1017/s0021859608008071
Ho TTK, Tra VT, Le TH, Nguyen NKQ, Tran CS, Nguyen PT, Vo TDH, Thai VN, Bui XT (2022) Compost to improve sustainable soil cultivation and crop productivity. Case Stud Chem Environ Eng 6:100211. ISSN 2666-0164. https://doi.org/10.1016/j.cscee.2022.100211
Horrocks A, Curtin D, Tregurtha C, Meenken E (2016) Municipal compost as a nutrient source for organic crop production in New Zealand. Agronomy 6:35. https://doi.org/10.3390/agronomy6020035
Huang M, Liu C, Li Z, Huang B, Luo N, Zeng G, Zhu Y (2016) Compost as a soil amendment to remediate heavy metal-contaminated agricultural soil: mechanisms, efficacy, problems, and strategies. Water Air Soil Pollut 227(10). https://doi.org/10.1007/s11270-016-3068-8
Iglesias-Jimenez E, Perez-Garcia V (1991) Composting of domestic refuse and sewage sludge. I. Evolution of temperature, pH, C/N ratio and cation exchange capacity. Resour Conserv Recy 6(1):45–60
Ikoyi I, Chaves C, Ahmed M, Egeter B, Schmalenberger A, Fowler A (2020) Responses of soil microbiota and nematodes to application of organic and inorganic fertilizers in grassland columns. Biol Fertil Soils 56(5):647–662. https://doi.org/10.1007/s00374-020-01440-5
Janati ME, Akkal-Corfini N, Robin P, Oukarroum A, Sabri A, Thomas Z, Chikhaoui M, Bouaziz A (2022) Compost from date palm residues increases soil nutrient availability and growth of silage corn (Zea mays L.) in an arid agroecosystem. J Soil Sci Plant Nutr 22(3):3727–3739. https://doi.org/10.1007/s42729-022-00922-9
Jeong ST, Kim GW, Hwang HY, Kim PJ, Kim SY (2018) Beneficial effect of compost utilization on reducing greenhouse gas emissions in a rice cultivation system through the overall management chain. Sci Total Environ 613-614:115–122
Ji R, Shi W, Dong G, Min J (2017) Effects of liquid organic fertilizers on plant growth and rhizosphere soil characteristics of chrysanthemum. Sustainability 9(5):841. https://doi.org/10.3390/su9050841
Jouraiphy A, Amir S, El Gharous M, Revel J, Hafidi M (2005) Chemical and spectroscopic analysis of organic matter transformation during composting of sewage sludge and green plant waste. Int Biodeterior Biodegradation 56(2):101–108
Kadir AA, Azhari NW, Jamaludin SN (2016) An overview of organic waste in composting. In: MATEC Web of Conferences, vol 47. EDP Sciences, p 05025
Karri V, Nalluri N (2023) Enhancing resilience to climate change through prospective strategies for climate-resilient agriculture to improve crop yield and food security. Plant Sci Today. https://doi.org/10.14719/pst.2140
Khater ESG (2015) Some physical and chemical properties of compost. Int J Waste Resour 5(1): 72–79
Kihara J, Bationo A, Mugendi DN, Martius C, Vlek PLG (2011) Conservation tillage local organic resources and nitrogen fertilizer combinations affect maize productivity soil structure and nutrient balances in semi-arid Kenya Nutrient Cycling in Agroecosystems 90(2):213–225. https://doi.org/10.1007/s10705-011-9423-7

Kranz CN, McLaughlin RA, Jackson A, Miller G, Heitman JL (2020) The effects of compost incorporation on soil physical properties in urban soils—A concise review. J Environ Manage 261:110209

Kulcheski FR, Côrrea R, Gomes IA, De Lima JC, Margis R (2015) NPK macronutrients and microRNA homeostasis. Front Plant Sci 6(242). https://doi.org/10.3389/fpls.2015.00451

Kumar A, Choudhary AK, Pooniya V, Suri VK, Singh U (2016) Soil factors associated with micronutrient acquisition in crops – Biofortification perspective. In Biofortification of Food Crops (pp. 159–176). Springer. ISBN: 978-81-322-2714-4

Kumar Bhatt M, Joshi HC, Labanya R (2019) Influence of long-term chemical fertilizers and organic manures on soil fertility - a review. Univers J Agric Res 7(5):177–188. https://doi.org/10.13189/ujar.2019.070502

Kumar M, Singh YK, Maurya SK, Sachan R, Maurya D, Gautam M, Tiwari A (2023) Efficient use of nano-fertilizer for increasing productivity and profitability along with maintain sustainability in rice crop: a review. Int J Environ Clim Change 13:1358–1368. https://doi.org/10.9734/IJECC/2023/v13i102788

Lallawmkimi C, Patil M, Upadhyay S, Dhanakrishna M, Niren B, Kumar A, Sadaya G, Panigrahi C (2025) Application of nanotechnology in agriculture: opportunities and challenges in the context of environmental sustainability. Arch Curr Res Int 25:37–53. https://doi.org/10.9734/acri/2025/v25i11035

Larney FJ, Buckley KE, Sullivan DM, Eghball B (2006) The role of composting in recycling manure nutrients. Can J Soil Sci 86(4):597–611. https://doi.org/10.4141/s05-116

Lasaridi K, Stentiford E (1998) A simple respirometric technique for assessing compost stability. Water Res 32(12):3717–3723

Latifah O, Ahmed OH, Majid NMA (2017) Soil pH buffering capacity and nitrogen availability following compost application in a tropical acid soil. Compost Sci Util 26(1):1–15. https://doi.org/10.1080/1065657x.2017.1329039

Le Guyader E, Morvan X, Girods P, Fontana S, Delgado-Iniesta MJ, Boumaraf B, Miconnet V, Kavvadias V, Moussa M, Tirichine A, Sbih M, Marin B, Gommeaux M, Intrigliolo DS (2024) Influence of date palm-based biochar and compost on water retention properties of soils with different sand contents. Forests 15(2):304. https://doi.org/10.3390/f15020304

Leifeld J, Siebert S, Kögel-Knabner I (2001) Stabilization of composted organic matter after application to a humus-free Sandy Mining soil. J Environ Qual 30(2):602

Lešková A, Javot H, Giehl RFH (2021) Metal crossroads in plants: modulation of nutrient acquisition and root development by essential trace metals. J Exp Bot 73(6):1751–1765. https://doi.org/10.1093/jxb/erab483

Li YC, Stoffella PJ, Alva AK, Calvert DV, Graetz DA (1997) Leaching of nitrate, ammonium, and phosphate from compost amended soil columns. Compost Sci Util 5(2):63–67. https://doi.org/10.1080/1065657x.1997.10701875

Li R, Wang Q, Zhang Z, Zhang G, Li Z, Wang L, Zheng J (2014) Nutrient transformation during aerobic composting of pig manure with biochar prepared at different temperatures. Environ Technol 36(7):815–826. https://doi.org/10.1080/09593330.2014.963692

Li L, Zhang YJ, Novak A, Yang Y, Wang J (2021) Role of biochar in improving sandy soil water retention and resilience to drought. Water 13(4):407

Loecke TD, Liebman M, Cambardella CA, Richard TL (2004) Corn response to composting and time of application of solid swine manure. Agron J 96(1). https://doi.org/10.2134/agronj2004.2140

Luo X, Liu G, Xia Y, Chen L, Jiang Z, Zheng H, Wang ZJ (2017) Use of biochar- compost to improve properties and productivity of the degraded coastal soil in the Yellow River Delta, China. J Soils Sediments 17:780–789

Manirakiza N, Şeker C (2020) Effects of compost and biochar amendments on soil fertility and crop growth in a calcareous soil. J Plant Nutr 43(20):3002–3019. https://doi.org/10.1080/01904167.2020.1806307

Matisic M, Dugan I, Bogunovic I (2024) Challenges in sustainable agriculture—the role of organic amendments. Agriculture-Basel 14(4):643. https://doi.org/10.3390/agriculture14040643

Mbogning S, Okiobe ST, Theuerl S, Nwaga D (2024) Synergistic interplay between arbuscular mycorrhizal fungi and fern manure compost tea suppresses common tomato phytopathogens and pest attacks on-farm. Front Hortic 3:1253616. https://doi.org/10.3389/fhort.2024.1253616

Misra RV, Roy RN, Hiraoka H (2003) On-farm composting methods. UN-FAO, Rome. (also available at: http://www.fao.org/docrep/007/y5104e/y5104e00.htm#Contents)

Mladenov M (2018) Chemical composition of different types of compost. J Chem Technol Metall 53(4):712–716

Mondini C, Fornasier F, Sinicco T (2004) Enzymatic activity as a parameter for the characterization of the composting process. Soil Biol Biochem 36(10):1587–1594

Mustin M (1987) Lecompost: Lagestiondela matière organique. Francois Dubusc, Paris. p 954

Omara AED, Hafez EM, Osman HS, Rashwan E, El-Said MA, Alharbi K et al (2022a) Collaborative impact of compost and beneficial rhizobacteria on soil properties, physiological attributes, and productivity of wheat subjected to deficit irrigation in salt affected soil. Plants 11(7):877

Omara AE-D, Alharbi K, Osman HS, Rashwan E, Hafez EM, El-Said MAA, Gowayed SM, Abd El-Moneim D (2022b) Collaborative impact of compost and beneficial Rhizobacteria on soil properties, physiological attributes, and productivity of wheat subjected to deficit irrigation in salt affected soil. Plants 11(7):877. https://doi.org/10.3390/plants11070877

Panday D, Bhusal N, Das S, Ghalehgolabbehbahani A (2024) Rooted in nature: the rise, challenges, and potential of organic farming and fertilizers in agroecosystems. Sustainability 16(4):1530. https://doi.org/10.3390/su16041530

Pane C, Palese AM, Celano G, Zaccardelli M (2014) Effects of compost tea treatments on productivity of lettuce and kohlrabi systems under organic cropping management. Ital J Agron 9:153–156

Papafilippaki A, Paranychianakis NV, Nikolaidis NP (2015) Effects of soil type and municipal solid waste compost as soil amendment on *Cichorium spinosum* (spiny chicory) growth. Sci Hortic 195:195–205

Pereira VR, Fiore FA (2024) Opportunities and barriers to composting in a municipal context: ACase study in São José dos Campos, Brazil. Sustainability 16:3359. https://doi.org/10.3390/su16083359

Pergola M, Persiani A, Palese AM, Di Meo V, Pastore V, D'Adamo C, Celano G (2018) Composting: the way for a sustainable agriculture. Appl Soil Ecol 123:744–750

Phares CA, Akaba S (2022) Co-application of compost or inorganic NPK fertilizer with biochar influences soil quality, grain yield and net income of rice. J Integr Agric 21(12):3600–3610. https://doi.org/10.1016/j.jia.2022.07.041

Podar D, Maathuis FJM (2022) Primary nutrient sensors in plants. IScience 25(4):104029. https://doi.org/10.1016/j.isci.2022.104029

Poincelot RP (1972) Biochemistry and methodology of composting. Conn Agric Exp Stn Bull 727: 1–38. https://archive.org/details/biochemistrymeth00poin

Policastro G, Cesaro A (2022) Composting of organic solid waste of municipal origin: the role of research in enhancing its sustainability. Int J Environ Res Public Health 20(1):312

Prakash C, Roy A, Gawdiya S (2022) Utilization of municipal wastes in agriculture. Food Sci Rep 3(6):13–16

Purwanto BH, Alam S (2019). Impact of intensive agricultural management on carbon and nitrogen dynamics in the humid tropics. Soil Science and Plant Nutrition 66(1):50–59. https://doi.org/10.1080/00380768.2019.1705182

Pyron RA, Wiens JJ (2013) Large-scale phylogenetic analyses reveal the causes of high tropical amphibian diversity. Proc R Soc Lond B Biol Sci 280(1770):20131622. https://doi.org/10.1098/rspb.2013.1622

Rani K, Tigga P, Roy A, Das A, Trivedi A (2022) Chapter-9 nutrient availability and plant productivity through PGPR: mechanisms, potential and constraints. Agric Biotechnol 75: 138. ISBN: 978-93-5570-234-0

Razmjoo P, Pourzamani H, Teiri H, Hajizadeh Y (2015) Determination of an empirical formula for organic composition of mature compost produced in Isfahan-Iran composting plant in 2013. Int J Environ Health Eng 4:3. http://www.ijehe.org/text.asp?2015/4/1/3/153988

Reimer M, Kopp C, Hartmann T, Zimmermann H, Ruser R, Schulz R, Müller T, Möller K (2023) Assessing long term effects of compost fertilization on soil fertility and nitrogen mineralization rate. J Plant Nutr Soil Sci 186:217–233

Riaz M, Marschner P (2020) Sandy soil amended with clay soil: effect of clay soil properties on soil respiration, microbial biomass, and water extractable organic C. J Soil Sci Plant Nutr 20(4): 2465–2470. https://doi.org/10.1007/s42729-020-00312-z

Roman P, Martinez MM, Pantoja A (2015) Farmer's compost handbook: experiences in Latin America. FAO, Rome. ISBN: 978–92–5-107845-7. http://www.fao.org/3/a-i3388e

Rombouts I, Beaugrand G, Ibañez F, Gasparini S, Chiba S, Legendre L (2009) Global latitudinal variations in marine copepod diversity and environmental factors. Proceedings of the Royal Society B: Biological Sciences 276(1670):3053–3062

Roschewitz I, Thies C, Tscharntke T, Gabriel D (2005) The effects of landscape complexity on arable weed species diversity in organic and conventional farming. J Appl Ecol 42(5):873–882. https://doi.org/10.1111/j.1365-2664.2005.01072.x

Roy S, Torres-Jerez I, Antonelli E, Zhang S, Udvardi M, Sanchez B, York LM, Griffiths M, Jain D, Krom N, Scheible W-R (2022) Application of synthetic peptide CEP1 increases nutrient uptake rates along plant roots. Front Plant Sci 12(9384). https://doi.org/10.3389/fpls.2021.793145

Roy A, Jadhav KP, Ash M, Dash AK, Raza MB, Reddy DD, Srinivasa Rao C (2024) Carbon farming strategies in Indian agriculture: perspective and challenges. In: Srinivasa Rao C, Dhandapani A, Kumar S (eds) Research and technology advancements in agriculture. ICAR-National Academy of Agricultural Research Management, Hyderabad, pp 379–410. ISBN: 978-93-340-3808-8

Sabri NSA, Hara H, Zakaria Z, Mohamad SE, Jaafar AB (2018) Importance of soil temperature for the growth of temperate crops under a tropical climate and functional role of soil microbial diversity. Microbes Environ 33(2):144–150. https://doi.org/10.1264/jsme2.me17181

Sanasam SD, Talukdar NC (2017) Quality compost production from municipality biowaste in mix with rice straw, cow dung, and earthworm Eisenia fetida. Compos Sci Util 25:141–151

Sani MNH, Yong JWH (2021) Harnessing synergistic biostimulatory processes: a plausible approach for enhanced crop growth and resilience in organic farming. Biology 11(1):41. https://doi.org/10.3390/biology11010041

Sayara T, Basheer-Salimia R, Hawamde F, Sánchez A (2020) Recycling of organic wastes through composting: process performance and compost application in agriculture. Agronomy 10:1838. https://doi.org/10.3390/agronomy10111838

Schröder C, Häfner F, Krause A, Larsen OC (2021) Urban organic waste for urban farming: growing lettuce using vermicompost and thermophilic compost. Agronomy 11(6):1175. https://doi.org/10.3390/agronomy11061175

Scopetani C, Chelazzi D, Cincinelli A, Martellini T, Leiniö V, Pellinen J (2022) Hazardous contaminants in plastics contained in compost and agricultural soil. Chemosphere 293: 133645. https://doi.org/10.1016/j.chemosphere.2022.133645

Sen M, Roy A, Rakshit D, Sarkar A, Dey S, Chanda S (2022) Principles of nanofertilizers and its potential in enhancing nutrient use efficiency. In: Meena S, Roy A, Meena AR, Kumar L (eds) Emerging research trends in soil science and horticulture. Blurose Publishers, Delhi, pp 114–136. ISBN: 978-93-5628-309-1

Sen M, Roy A, Rani K, Nalia A, Das T, Tigga P, Rakshit D, Atta K, Mandal. S, Vishwanath, Das A (2024) Crop residue: status, distribution, management, and agricultural sustainability. In: Meena VS, Rakshit A, Meena MD, Baslam M, Fattah IMR, Lam SS, Kaba JS (eds) Waste management for sustainable and restored agricultural soil. Academic Press, London, pp 167–201. https://doi.org/10.1016/B978-0-443-18486-4.00017-8

Sen M, Garai S, Dutta S, Rakshit D, Roy A, Das R, Das S (2025) Harnessing artificial intelligence for circular economy in agriculture: reducing waste and maximising efficiency. In: Integrating agriculture, green marketing strategies, and artificial intelligence. IGI Global Scientific Publishing, pp 317–362. https://doi.org/10.4018/979-8-3693-6468-0.ch011

Seregin IV, Kozhevnikova AD (2020) Low-molecular-weight ligands in plants: role in metal homeostasis and hyperaccumulation. Photosynth Res 150(1–3):51–96. https://doi.org/10.1007/s11120-020-00768-1

Shen T, Lei Y, Pu X, Zhang S, Du Y (2021) Identification and application of Streptomyces microflavus G33 in compost to suppress tomato bacterial wilt disease. Appl Soil Ecol 157: 103724. https://doi.org/10.1016/j.apsoil.2020.103724

Shrestha P, Kay A, Small GE (2020) Quantifying nutrient recovery efficiency and loss from compost-based urban agriculture. PLoS One 15(4):e0230996. https://doi.org/10.1371/journal.pone.0230996

Singh NK, Sachan K, Ranjitha G, Chandana S, Manoj P, Panotra N, Katiyar D (2024) Building soil health and fertility through organic amendments and practices: a review. Asian J Soil Sci Plant Nutr 10(1):175–197. https://doi.org/10.9734/ajsspn/2024/v10i1224

Smith A, Halberg N (2020) Assessment of the environmental sustainability of organic farming: Definitions, indicators and the major challenges. Canadian Journal of Plant Science 92(6):981–996. https://doi.org/10.4141/cjps2012-035

Stazi SR, Allevato E, Marabottini R, Digiesi L, Vannini A, Chilosi G (2022) Use of compost in the uptake mitigation of arsenic in Beta vulgaris L. var. cicla. J Sci Food Agric 102(14):6596–6602. https://doi.org/10.1002/jsfa.12026

Su Y, Makowski D, Gabrielle B (2021) The impact of climate change on the productivity of conservation agriculture. Nat Clim Chang 11(7):628–633. https://doi.org/10.1038/s41558-021-01075-w

Suarez-Estrella F, Arcos-Nievas MA, Lopez MJ, Vargas-García MC, Moreno J (2013) Biological control of plant pathogens by microorganisms isolated from agroindustrial composts. Biol Control 67:509–515. https://doi.org/10.1016/j.biocontrol.2013.10.008

Sullivan DM, Miller RO (2001) Compost quality attributes, measurements and variability. In: Stofella PJ, Kahn BA (eds) Compost utilization in horticultural cropping systems. CRC Press, Boca Raton, pp 95–120

Susilowati LE, Sukartono S, Akbar MF, Kusumo BH, Suriadi A, Leksono AS, Fahrudin F (2024) Assessing the synergistic effects of inorganic, organic, and biofertilizers on rhizosphere properties and yield of maize. SAINS TANAH - J Soil Sci Agroclimatol 21(1):104. https://doi.org/10.20961/stjssa.v21i1.85373

Szyłak-Szydłowski M, Kos W (2024) Application of sensory methods to evaluate the effectiveness of solutions to reduce the exposure to odour nuisance and ammonia emissions from the compost heaps. Sensors 24(13):4200. https://doi.org/10.3390/s24134200

Tahir S, Marschner P (2017) Clay addition to sandy soil reduces nutrient leaching—effect of clay concentration and ped size. Commun Soil Sci Plant Anal 48(15):1813–1821. https://doi.org/10.1080/00103624.2017.1395454

Talaat NB, Abdel-Salam SA (2024) An innovative, sustainable, and environmentally friendly approach for wheat drought tolerance using vermicompost and effective microorganisms: upregulating the antioxidant defense machinery, glyoxalase system, and osmotic regulatory substances. BMC Plant Biol 24(1):866. https://doi.org/10.1186/s12870-024-05550-2

Thapa U, Hansda N, Kundu S, Giri A, Tamang D, Rahaman AO (2024) Advancements in hydroponic systems: a comprehensive review. Arch Curr Res Int 24:317–328. https://doi.org/10.9734/acri/2024/v24i11973

Therond O, Roger-Estrade J, Duru M, Richard G (2017) A new analytical framework of farming system and agriculture model diversities. A review. Agron Sustain Dev 37(3). https://doi.org/10.1007/s13593-017-0429-7

Tittarelli F, Båth B, Ceglie FG, García MC, Möller K, Reents HJ, Védie H, Voogt W (2017) Soil fertility management in organic greenhouse: an analysis of the European context. Acta Hortic 1164:113–126

Viaenea J, Lanckerb JV, Vandecasteelea B, Willekensa K, Bijttebierb J, Ruysschaerta G, Nevec SD, Reubens B (2016) Opportunities and barriers to on-farm composting and compost application: a case study from northwestern Europe. Waste Manag 48:181–192

Wang C, Kuzyakov Y (2024) Soil organic matter priming: the pH effects. Glob Chang Biol 30(6). https://doi.org/10.1111/gcb.17349

Wang F, Wang Z, Zhao D, Ma Z, Kou C (2016) Responses of wheat yield, macro- and micronutrients, and heavy metals in soil and wheat following the application of manure compost on the North China plain. PLoS One 11(1):e0146453. https://doi.org/10.1371/journal.pone.0146453

Wang Y, Tang Y, Yuan Z (2022) Improving food waste composting efficiency with mature compost addition. Bioresour Technol 349:126830

Wichuk K, McCartney D (2013) Compost stability and maturity evaluation—a literature review. J Environ Eng Sci 8(5):601–620

Wijitkosum S, Sriburi T, Toonsiri P (2025) Effect of one-time application of biochar and compost on soil and maize during 5-time consecutive periods of crop cultivation. Emerg Sci J 9(1): 114–130. https://doi.org/10.28991/ESJ-2025-09-01-07

Xing Y, Wang X (2024) Precision agriculture and water conservation strategies for sustainable crop production in arid regions. Plants 13(22):3184. https://doi.org/10.3390/plants13223184

Ye J, Zhang R, Nielsen S, Joseph SD, Huang D, Thomas T (2016) A combination of biochar-mineral complexes and compost improves soil bacterial processes, soil quality, and plant properties. Front Microbiol 7. https://doi.org/10.3389/fmicb.2016.00372

Yu Y-Y, Luo Y-M, Li J-G, Guo J-H, Li S-M, Qiu J-P (2019) Combination of agricultural waste compost and biofertilizer improves yield and enhances the sustainability of a pepper field. J Plant Nutr Soil Sci 182(4):560–569. https://doi.org/10.1002/jpln.201800223

Zapałowska A, Jarecki W (2024) The impact of using different types of compost on the growth and yield of corn. Sustainability 16(2):511

Zerssa G, Eichler-Löbermann B, Feyssa D, Kim D-G (2021) Challenges of smallholder farming in Ethiopia and opportunities by adopting climate-smart agriculture. Agriculture (Basel) 11(3): 192. https://doi.org/10.3390/agriculture11030192

Zhao S, Schmidt S, Gao H, Li T, Chen X, Hou Y, Chadwick D, Tian J, Dou Z, Zhang W, Zhang F (2021) A prominent role for precision composting in sustainable agriculture. Research Square. https://doi.org/10.21203/rs.3.rs-745873/v1

Zhao B, Yi K, Murray JD, Wang E, Yu N, Jia X (2023) Microbe-dependent and independent nitrogen and phosphate acquisition and regulation in plants. New Phytol 242(4):1507–1522. https://doi.org/10.1111/nph.19263

Zheng Y, Drechsler N, Rausch C, Kunze R (2016) The Arabidopsis nitrate transporter NPF7.3/NRT1.5 is involved in lateral root development under potassium deprivation. Plant Signal Behav 11(5):e1176819. https://doi.org/10.1080/15592324.2016.1176819

Zikeli S, Deil L, Möller K (2017) The challenge of imbalanced nutrient flows in organic farming systems: a study of organic greenhouses in southern Germany. Agric Ecosyst Environ 244:1–13

Zygalakis KC, Kirk GJD, Jones DL, Wissuwa M, Roose T (2011) A dual porosity model of nutrient uptake by root hairs. New Phytologist 192(3):676–688

# Nanotechnology Advances in Composting

**Faria Fatima, Deepti Srivastava, Abdul Mazeed, and P. Smriti Rao**

## 1 Introduction

### *1.1 Compost: A Catalyst for Soil Vitality*

As the world's population grows, there is a greater need for food, medicine, fiber, and other resources. The world can no longer sustain this rapidly expanding population. The use of insecticides, pesticides, herbicides, fertilizers, compost, and other materials to increase yield is widespread. The results of biodegradation might be affected by the prevalence of environmental factors, including temperature, oxygen, pH, nutrients, and moisture. (Rajmohan et al. 2020). Efficient composting of organic waste has the potential (Fig. 1) to be a part of sustainable agriculture in the near future.

Scientists are currently very interested in nanoparticles (NPs) because of their superior qualities, which include deliverability, stability, reactivity, and recyclability (Shaik and Franchetti 2022). Higher soil reclamation rates are produced by this increased reactivity, and in situ applications benefit from easier delivery. Zeolite, graphene oxide, phosphate, iron sulfide, iron oxide ($Fe_3O_4$), nano-zero-valent iron (nZVI), carbon nanotubes, and other NPs are frequently utilized for soil reclamation (Mukhopadhyay et al. 2022). The kind, concentration, and enzyme activity of the soil all affect how the NPs and soil react. Table 1 lists the microorganisms that are involved at different phases of the composting process.

$Fe_3O_4$ NPs have reactive sites and the innate ability to use surface binding to absorb different pollutants from the soil. They have advanced in a variety of applications, including soil remediation, due to their environment friendly nature and adsorption performance. As a result, they can efficiently extract a variety of heavy metals from soils, including nickel, copper, zinc, arsenic (As), lead, and

F. Fatima · D. Srivastava · A. Mazeed · P. Smriti Rao (✉)
Department of Agriculture, Integral University, Lucknow, UP, India



R. K. Singh et al. (eds.), *Composting of Farm Residues by Nature-Based Solutions*,
https://doi.org/10.1007/978-3-032-12255-1_12

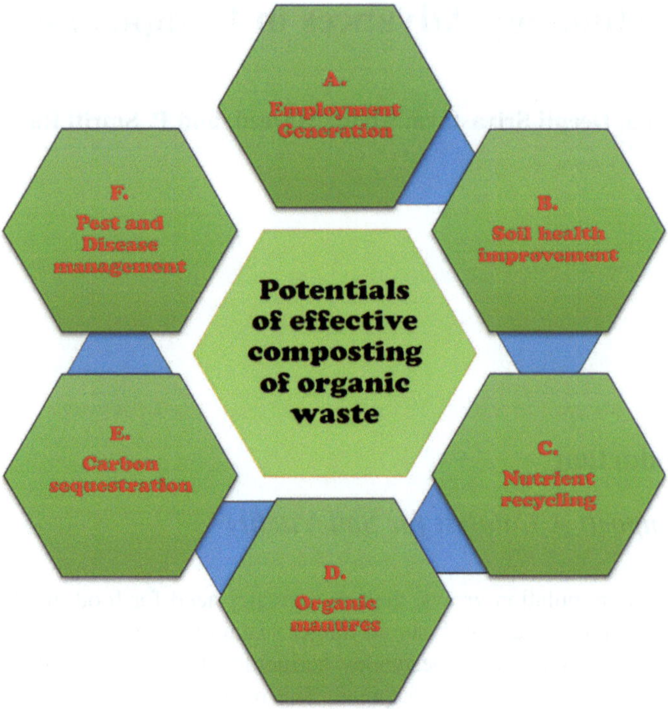


**Fig. 1** Potential of effective composting of organic waste

chromium (Latif et al. 2020). Additionally, the goal of nano-zero-valent iron (nZVI) was to break down a variety of organic pollutants, such as pesticides and petrochemicals (Sharma et al. 2023a). ZVI has demonstrated significant promise for the elimination of pollutants as a crucial, environment safe reducing agent. Heavy metal ion adsorption processes on ZVI-based materials from the environment are attributed to chemical reactions and monolayer molecular sorption, according to a number of studies, batch adsorption investigations, and theoretical studies (Sharma et al. 2023b). They have the ability to extract various heavy metals from the soil. Additionally, they have reducing action, which means that high-oxidation metal ions like uranium and chromium change into low-oxidation states and impair mobility, solubility, and toxicity. By immobilizing lead, cadmium, and copper, the application of graphene oxide nanoparticles (NPs) helps control the pollution caused by heavy metals in the soil (Mathur et al. 2022).

Carbon nanotubes and NPs based on iron sulfide and phosphate are also widely used to extract heavy metals and other organic pollutants from soils (Chaudhary et al. 2023). No matter what kind of soil is used, availability is greatly decreased when nZVI is applied and increased when graphene oxide nanoparticles (nGOx) are used. Flexible options for immobilizing, as in stabilization procedures, or mobilizing, as in (phyto)extraction tactics, are made possible by the unique actions of nZVI

**Table 1** Microbial communities and their roles across different composting phases

| Phase | Key microorganisms | Role/activity | Temperature range |
|---|---|---|---|
| Mesophilic phase | Bacteria (e.g., *Bacillus, Pseudomonas*) | Basic organic materials' initial breakdown (proteins, carbohydrates). Rapid mesophilic bacterial growth | 10–40 °C |
| | Fungi (e.g., *Aspergillus, Penicillium*) | | |
| Thermophilic phase | Thermophilic bacteria (e.g., *Thermus, Actinomyces*) | Decomposition of more complex compounds (cellulose, lignin, fats) | 40–70 °C |
| | Thermophilic fungi (e.g., *Thermomyces, Scytalidium*) | Pathogen destruction due to high heat | |
| Cooling phase | Actinomycetes | Decomposition of remaining resistant compounds (e.g., lignocellulose) | 40–25 °C |
| | Fungi (e.g., *Trichoderma*) | Reduced microbial activity. | |
| Maturation phase | Fungi (e.g., *Cladosporium, Penicillium*) | Stabilization of compost, humus formation and nutrient enrichment | Ambient temperature |
| | Actinomycetes | | |
| | Nitrogen-fixing bacteria (e.g., *Azotobacter*) | | |

and nGOx (Ibrahim and El-Sherbini 2024). By enhancing the nutritional qualities, water retention capacity, clay shift proportions, and hazardous material sequestration, zeolite-based NPs can improve the properties of the soil. They have the ability to absorb radionuclides like cesium and strontium. In addition to all of these benefits, adding them raises pH and gets rid of the different heavy metals in the soil (Liu et al. 2018).

Apart from reclaiming soil, NPs also have a vital part in composting. NP-mediated composting is a method that gives plants nourishment through metal nanoparticles as well as compost (Kaur et al. 2023). Nevertheless, nothing is known about how well the microbial bioprocess performs in NP-mediated composting. This bioprocess performance may be useful in gaining a detailed insight into NP behavior in the compost. With a thorough background of earlier research, this review article seeks to close the knowledge gap by achieving the following goals: (a) Being aware of the microbial bioprocessing that takes place during composting. (b) Evaluation of composting facilitated by NP, (c) Examining how microbial bioprocessing is affected by NP-mediated composting, (d) The LCA conceptual framework for nanocomposting, and (e) Understanding the concept of nanozymes in the process of composting.

## 2 Composting Method

In order to create microbial biomass for their uses, microbes in the presence of carbon (C), nitrogen (N), and oxygen (O) undergo a number of metabolic activities during the composting process. The composting process is divided into three stages: mesophilic, thermophilic, and maturation (Cao et al. 2020). Bacteria and fungi hydrolyze sugar, amino acids, and lipids during the mesophilic phase. Actinobacteria cause the hydrolysis of polymers, including cellulose, hemicellulose, lignin, and proteins, during the thermophilic phase. At the same time, infections are eradicated during this phase. Later on, this process leads to a cooling phase, as mesophilic bacteria recolonizes the area. Lastly, the maturation phase aids in the formation of the humic chemicals' precursors (Sharma et al. 2023c). Because of their special qualities, which include a greater surface-to-volume ratio and mechanical, electrical, and physicochemical characteristics, among others, nanomaterials are quickly becoming the answer for many industries (Gattupalli et al. 2023). Numerous investigations have been carried out regarding their effects on the target molecule, toxicity, industrial applicability, etc.

Nevertheless, there are not many studies on how NPs affect the microbial bioprocess during composting (Roshan et al. 2023). When using purposeful applications, the kind of NPs used in composting is determined by the soil's needs. Silver, zinc oxide (ZnO), copper oxide (CuO), $Fe_3O_4$, titanium dioxide ($TiO_2$), and other NPs have generally been employed with good results (Kaur et al. 2023). Figure 2

**Fig. 2** Steps involved in the organic waste composting process

**Table 2** Role of nanofertilizers in agriculture

| Type of nanofertilizer | Role in agriculture | Benefits | Examples | Reference |
|---|---|---|---|---|
| Nano-nitrogen fertilizers | Release nitrogen slowly to enhance nitrogen use efficiency | Increased nitrogen availability for plants | Nano-urea | Verma et al. (2023) |
| | Reduce nitrogen loss through volatilization | Improved crop yields and reduced environmental pollution | Nano-ammonium nitrate | Abhiram (2023) |
| Nano-phosphorus fertilizers | Enhance phosphorus uptake by plants | Improved root growth and early-stage plant development | Nano-hydroxyapatite | Jadav et al. (2024) |
| | Overcome phosphorus fixation in soil | Increased crop production | Nano-phosphate | Poudel et al. (2023) |
| Nano-potassium fertilizers | Improve potassium uptake and efficiency in crops | Increased drought tolerance | Nano-potassium nitrate | Wang et al. (2024a) |
| | Enhance stress tolerance and water use efficiency | Enhanced overall plant health and productivity | Nano-potassium sulfate | Rady et al. (2023) |
| Nano-zinc fertilizers | Provide bioavailable zinc to plants | Improved crop quality, especially in cereals | Nano-zinc oxide | Kumar et al. (2021) |
| | Correct zinc deficiency in soil, improving crop nutrition | Enhanced resistance to diseases | Nano-zinc sulfate | Singh et al. (2023) |
| Nano-iron fertilizers | Improve iron absorption by plants, especially in calcareous soils | Enhanced growth of leafy crops | Nano-iron oxide | Faisal (2025) |
| | Promote photosynthesis in crops | Reduced iron chlorosis, leading to healthier plants | Nano-iron chelate | Szuplewska et al. (2023) |

represents the process involved in composting, and Table 2 displays the various roles of NPs in agriculture.

## 3 Green Synthesis and Nanoparticle Production

Green synthesis focuses on developing eco-friendly chemical technologies to minimize hazardous substances in chemical processes. This approach aims to reduce pollution, conserve nonrenewable resources, and shorten synthesis time. Paul

J. Anastas, the father of green chemistry, described it as the use of alternative methods to prevent pollution in both synthesis and byproduct treatment (Andraos and Matlack 2022).

## 3.1 Nanoparticle Synthesis Approaches

1. Top-down approach—Physical methods like grinding or abrasion produce nanoparticles.
2. Bottom-up approach—Nanoparticles form from atomic or molecular building blocks through:
   - Chemical synthesis—Reaction-based formation of nanoparticles.
   - Self-assembly—Molecules arrange themselves through interactions.
   - Positional assembly—Atoms/molecules are deliberately placed (impractical for industry).

The bottom-up approach is preferred due to its lower equipment needs and faster synthesis (Kaur et al. 2020).

## 3.2 Green Synthesis of Nanoparticles

Biological resources like plants, algae, bacteria, or fungi serve as key agents, with natural compounds (terpenes, polyphenols, alkaloids, etc.) playing a crucial role. Factors such as metal ion concentration, pH, reaction time, and temperature affect nanoparticle characteristics (Vaseghi et al. 2018).

## 3.3 Phases of Green Synthesis

- *Initial Phase*—Aqueous plant extracts or microbial culture media are prepared with precursor metal salts.
- *Activation Phase*—Chemical reduction and nucleation occur, initiating nanoparticle formation.
- *Growth Phase*—Small nanoparticles fuse into larger aggregates, influenced by external factors.
- *Termination Phase*—Nanoparticles stabilize, and their final shape and properties are determined.

Green synthesis is increasingly recognized as a sustainable method within the bottom-up approach (Vaseghi et al. 2018).

### 3.4 Biological Synthesis of Nanoparticles

Nanoparticles' special uses have drawn a lot of interest in the fields of biology, medicine, and electronics. An environmentally acceptable substitute for traditional synthesis techniques is green synthesis, which uses bacteria, fungi, algae, and plants instead of high-energy procedures and hazardous chemicals. This biological method is sustainable, easy to use, and dependable. Both external (using bacterial secretions for quick synthesis) and intracellular (inside cells, requiring cell killing for extraction) methods are used by bacteria to generate nanoparticles. Proteins and enzymes help fungi increase the stability of nanoparticles, whereas natural bioaccumulation and reduction processes are used by algae. Among all the biological sources, plants are the most commonly used because they provide quick and effective synthesis through extracellular, intracellular, or phytochemical-mediated pathways (Kaur et al. 2020).

### 3.5 Important Elements in Green Synthesis

The shape, size, and yield of nanoparticles are greatly influenced by the reaction parameters, which include temperature, pH, and time. While pH levels define nucleation centers and affect the sorts of nanoparticles generated (e.g., alkaline pH for magnetic nanoparticles), higher temperatures speed up processes and increase particle size. Similarly, longer reaction times provide extended contact periods, promoting the formation of larger nanoparticles and improving overall yield. To maximize the characteristics of nanoparticles, these variables need to be carefully managed (Vaseghi et al. 2018).

### 3.6 The Synthesis Mechanism of Green Nanoparticles

Proteins, flavonoids, terpenoids, tannins, and sugars are examples of biological substances that actively contribute to the creation of nanoparticles by stabilizing the final structures and lowering metal ions. While flavonoids provide reducing protons to aid in metal ion conversion, polyphenols bind metal ions and promote nucleation. Saccharides, such as glucose, function as reducing agents and undergo oxidation processes to create nanoparticles. Furthermore, the concentration and location of amino acids within the synthesis medium affect the yield and appearance of nanoparticles. This environment friendly synthesis technique provides an effective and sustainable means of creating nanoparticles for a range of uses (Kaur et al. 2020).

## 4 Enhancing Composting Efficiency with Green Nanoparticles

The production of metallic nanoparticles by plant species and microbes is environmentally benign and advantageous (Das et al. 2017). A safe method that necessitates mild conditions and yields less-toxic or nontoxic end products is the biological production of NPs using microbial cell factories (Agrawal et al. 2022). Fungi such as endophytic fungus, Penicillium sp., *Neurospora crassa*, and others, as well as cyanobacteria like Limnothrix sp., Anabaena sp., Synechocystis sp., and others, can produce NPs. Certain bacterial species have the ability to use special defense mechanisms to combat the toxicity of heavy metal ions (Pal et al. 2022). Metal and metal oxide nanoparticles (NPs) are produced by actinobacteria and can be utilized in biological systems, including agriculture and the environment friendly manufacturing of nanomaterials. To limit the size of the metal and lessen negative effects, biologically generated nanomaterials are supplemented with plant defense capping molecules, such as flavonoids, proteins, sugars, polyphenols, alkaloids, esters, and organic acid group components (Godeto et al. 2023). Therefore, it would be preferable to employ green-synthesized NPs to reduce toxicity in the event that NPs are purposefully applied during the composting process. Table 3 shows a selection of the green NPs.

## 5 Nanoparticles and Their Interactions with Microorganisms

Solid waste, water discharge, industrial pollutants, and other sources allow NPs to infiltrate the ecosystem. In addition to their numerous structural and catalytic roles, the beneficial microbial composition of the soil has been devastated by the unchecked release of NPs and their persistence in soil ecosystems (Pathan et al. 2020). Some NPs impair normal cell activity by adhering to the surface of microbial cells or becoming trapped in the mycelium or cells of bacteria and fungi. Membrane disintegration, protein oxidation, oxidative stress, intracellular pH, mitochondrial malfunction, nucleic acid damage, cell wall adhesion, and gene expression are some of the ways that NPs interact with microorganisms. Some scholars have provided a comprehensive explanation of their intricate process (Pal et al. 2021; Wang et al. 2024b; Tee et al. 2016). Biologically produced NPs can help lessen negative impacts, and methods for the safe disposal of metal NPs in soil ecosystems are advised, despite the fact that NPs are poisonous to beneficial soil bacteria.

**Table 3** Different sources of green nanoparticles

| Source | Description | Types of green nanoparticles | Applications |
|---|---|---|---|
| Plant extracts | Plants contain bioactive compounds that can reduce metal ions into nanoparticles | Silver (Ag) NPs<br>Gold (Au) NPs<br>Copper (Cu) NPs | Antimicrobial agents<br>Drug delivery<br>Water purification |
| Fungi | Fungi, particularly mushrooms, are effective in synthesizing nanoparticles due to their metabolic processes | Silver (Ag) NPs<br>Zinc oxide (ZnO) NPs | Medicine<br>Environmental cleanup<br>Agricultural applications |
| Bacteria | Certain bacteria can bioreduce metal ions to produce nanoparticles through enzymatic reactions | Silver (Ag) NPs<br>Gold (Au) NPs<br>Iron (Fe) NPs | Biosensing<br>Wastewater treatment<br>Antibacterial coatings |
| Algae | Algae produce compounds that help in the formation of nanoparticles, often with eco-friendly processes | Silver (Ag) NPs<br>Iron oxide ($Fe_2O_3$) NPs<br>Titanium dioxide ($TiO_2$) NPs | Solar energy<br>Environmental remediation<br>Agriculture |
| Seaweeds | Rich in polysaccharides, seaweeds are used for nanoparticle synthesis with minimal toxic by-products | Gold (Au) NPs<br>Silver (Ag) NPs | Antioxidant properties<br>Bioremediation<br>Cosmetics |
| Citrus fruits (e.g., lemon, orange) | The peels of citrus fruits contain natural acids and compounds that can help in nanoparticle formation | Silver (Ag) NPs<br>Copper (Cu) NPs | Antibacterial<br>Anticancer<br>Food preservation |
| Tea leaves | Tea leaves contain polyphenols, which act as reducing agents for nanoparticle synthesis | Silver (Ag) NPs<br>Gold (Au) NPs | Antioxidant<br>Antimicrobial<br>Water purification |
| Herbs (e.g., rosemary, mint) | Herbal extracts are rich in flavonoids, alkaloids, and t erpenoids, promoting eco-friendly nanoparticle synthesis | Silver (Ag) NPs<br>Zinc oxide (ZnO) NPs | Antimicrobial<br>Drug delivery<br>Agriculture |
| Fruit extracts (e.g., banana, pomegranate) | Certain fruit extracts contain phytochemicals that reduce metal ions to form nanoparticles | Silver (Ag) NPs<br>Copper oxide (CuO) NPs | Antioxidant<br>Antibacterial<br>Cosmetics |

# 6 Nanoparticles as Modifiers of Microbial and Enzyme Activities

NPs' effects on microbial and enzyme activity are contingent upon the soil type, incubation duration, NP size, synthesis technique, and reagents (Gauba et al. 2023). The authors also describe how different nanomaterials' sizes affect their toxicity. According to all of these studies, because smaller particles have a bigger surface area and can cause more responses, they are more dangerous than larger ones. Furthermore, ionization is caused by the smaller size, more edges, and sharper corners, which leads to the production of reactive oxygen species (ROS) and toxicity (Wang et al. 2021). Another factor contributing to nanomaterial toxicity is their greater surface area, which helps them react with everything. If that reaction is improper or unfavorable, the NP will turn hazardous (Khan et al. 2021). This causes oxidative stress altering enzyme production and microbial activity. The following describes how various types of NPs affect enzyme and microbial activity.

## 6.1 *Zinc Nanoparticles*

More than 300 enzymes contain zinc, which is essential as a catalyst (Long et al. 2024). The kind of soil and the length of incubation determine how NPs like CuO and ZnO affect dehydrogenase activity. Increased microbial and enzyme activity is observed in several soils with reduced NP concentrations. Additionally, the way that bacteria react to water-soluble copper and zinc is linked to the actions of these NPs (Dimkpa et al. 2012).

Therefore, the NPs that are mediated in compost at lower concentrations could be advantageous. Furthermore, a study found that ZnO NPs had a greater impact on enzyme activities than other metal oxide nanoparticles (NPs), including $TiO_2$, cerium dioxide ($CeO_2$), and $Fe_3O_4$ (Sharifan 2019). According to the study, introducing metal oxide nanoparticles into the soil may change the bacterial population and the activity of enzymes. Additionally, by lowering ROS, the ZnO NPs in co-composted biochar enhanced wheat growth, chlorophyll content, and antioxidant enzyme activity (Bashir et al. 2021).

## 6.2 *Silver Nanoparticles*

Rather than silver nanoparticles (AgNPs), soil management had an impact on the urease and β-glucosidase activities. In contrast, neither soil management nor AgNPs has an impact on phosphatase activity. The clay and organic matter content of the soil covering the NPs may be the cause. The particle size stabilizes as a result of electrostatic repulsion (Adetunji 2019).

As a result, it shields these extracellular enzymes, and the direct interaction with NPs is complex enough to prevent any potential toxicity (Abbas et al. 2020). But occasionally, due to the increased immobilization of these NPs, the same soil with a higher organic matter content might also hinder certain microbial and enzyme activities, lowering the bioavailability to the microorganisms and consequently lowering enzyme activities (Yamini et al. 2023).

Moreover, it is deduced that AgNPs negatively impact the activity of several enzymes, including β-D-glucosidase, leucine-aminopeptidase, phosphomonoesterase, and arylsulfatase, in soils. In soils supplemented with organic matter, it lessened the inhibition of these actions. Because of its higher organic content, NP-mediated composting can therefore produce potential results with decreased toxicity if mediated in the necessary concentrations (Abbas et al. 2020).

### 6.3 *Iron Nanoparticles*

According to one study, nanomagnetite is more advantageous to actinomycetes, while nanohematite increased bacterial and fungal abundances more effectively among $Fe_3O_4$ NPs (Saif et al. 2016). These authors came to the conclusion that inorganic NPs, with the exception of $Fe_3O_4$, might be more hazardous than organic NPs (Sajid and Płotka-Wasylka 2020; Klekotka et al. 2022; Rahman et al. 2022). According to a number of studies, $Fe_3O_4$ NPs had fewer adverse effects than silver, copper, titanium, zinc, and gold NPs.

Among the $Fe_3O_4$ NPs (i.e., $Fe_2O_3$ and $Fe_3O_4$), $Fe_2O_3$ NPs have demonstrated amazing results during the composting process, including increased seedling development, germination index, and total organic matter loss. Additionally, they enhanced the activity of enzymes, including urease and dehydrogenase (Soni et al. 2023). $Fe_3O_4$ NPs often emit Fe2+, which lessens the enhancement of microbial activity. The fully oxidized crystals seen in $Fe_2O_3$ NPs are extremely stable in the environment and have a reduced capacity to produce ROS. As a result, it increased microbial activity and produced a number of advantageous composting results (Rajput et al. 2022).

### 6.4 *Copper Nanoparticles*

The activity of enzymes like dehydrogenase, urease, acid phosphatase, and alkaline phosphatase is adversely affected by nano-CuO of less than 50 nm in the case of inorganic NPs because the smaller particles dissolve more quickly and produce more reactive oxygen species (ROS). ROS puts microbial cells under oxidative stress, which leads to their decomposition. In the end, this leads to the inhibition of the microbes' enzyme secretion (Gattupalli et al. 2023).

A lower dosage of nanomaterials (<20 ppm) showed antioxidant activity, according to the meta-analysis results, while a higher dosage (>200 ppm) showed oxidant activity due to the excess production of ROS, which causes oxidative stress (Zandi and Schnug 2022). This suggests that the ideal dosage of nanomaterials can prevent excess ROS generation and reduce toxicity.

## 7 Enhancing Organic Waste Decomposition with Nanoparticles

Aerobic degradation is the breakdown of organic materials when oxygen is present. This is regarded as aerobic composting as well. "Nanocomposting," or composting using NPs, is a novel agricultural method to feed this rapidly expanding population. Nanocomposting aids in soil remediation, improved nutrition, and precision farming, among other things.

During composting, a variety of materials, including biochar, silver nanoparticles (AgNPs), silver-doped titanium dioxide nanoparticles ($Ag + TiO_2$), and engineered nanoparticles (ENPs), can be used as additions. Because of its porous structure, cation exchange ability, and functional groups, biochar is regarded as one of the best compost additives. Composting improves humification, microbial development, nutrient retention, and the inhibition of the bioavailability of heavy metals like zinc and copper (Singh and Kalamdhad 2013). Because of its adsorption activity, it can enhance animal waste composting by lowering gaseous emissions (Osman et al. 2022). AgNPs are advantageous when composting is taking place. The amount of organic matter lost during the composting of sewage sludge was negligible in the case of polyvinylpyrrolidone AgNPs (Gitipour et al. 2013). AgNPs have no effect whatsoever on the aerobic composting process in municipal solid waste (Ganguly and Chakraborty 2023). Similar findings were made, emphasizing that exposure to AgNPs had no effect on enzyme activity (Bao et al. 2018). As a result, applying AgNPs to the compost had no effect on its humification or stabilization. The microbial degradation process was unaffected by the presence of silver-doped $TiO_2$ NPs; in fact, significant carbon mineralization was noted (Messaoud et al. 2024). Metal, nonmetal, and carbon nanoparticles are among the several kinds of ENPs. $Fe_3O_4$ NPs improved the quality of the finished compost by accelerating the breakdown of organic materials during the composting process (Gattupalli et al. 2023). Additionally, they encouraged the passivation of cadmium and the release of accessible phosphorus (P) during the composting of pig dung (Noor et al. 2024). Research on ENP-mediated composting is extremely limited. Consequently, further research on the effects of various ENPs on the composting process is needed.

Some ions, which may be stable or unstable, are released when the NPs first dissociate. ROS can be produced by the unstable ions, which result in decreased microbial activity. In this instance, the release of ions as a nutrient is offset by the production of ROS. Furthermore, the stable ions have a limited ability to produce

ROS (Gold et al. 2018). Therefore, these ions function as nutrients and increase the diversity or abundance of microbes, which in turn speeds up the pace of decomposition.

## 8 Bioprocess Dynamics in Composting with Nanoparticles

The process of creating value-added material from a live source is called bioprocessing. Throughout the process, the source organism needs to be living and reacting to its surroundings. The biostimulation procedure entails providing the contaminated locations with enough nutrients (N, P, K, and electron donors) to promote bacterial growth and break down organic matter, toxic, and hazardous pollutants (Chettri et al. 2024), as mentioned in Table 4.

**Table 4** Impact of NPs on microbial composting process

| Nanoparticle type | Impact on microbial activity | Effect on composting process | Potential benefits | Potential risks |
|---|---|---|---|---|
| Zinc oxide (ZnO) NPs | Can stimulate microbial growth at lower concentrations<br>Toxic at higher concentrations | Enhanced microbial activity at low concentrations<br>Higher concentrations may decrease composting efficiency | Improved decomposition at optimal concentration | Toxicity to certain microbial species at high concentrations |
| Copper (Cu) NPs | Can inhibit the growth of specific microorganisms, especially bacteria | Reduced microbial diversity<br>Potential reduction in composting efficiency | Potential control over harmful bacteria in compost | Toxic to beneficial microorganisms |
| Iron (Fe) NPs | Can enhance the activity of specific microorganisms involved in aerobic decomposition | May improve oxygen availability in the composting process<br>Promotes microbial growth | Accelerated organic matter decomposition | High concentrations may disrupt microbial balance |
| Titanium dioxide ($TiO_2$) NPs | Can enhance the growth of aerobic microorganisms<br>Can have photocatalytic effects under light | Potential increase in aerobic decomposition, especially in well-aerated compost | Potential to accelerate composting under sunlight or ultraviolet (UV) exposure | May inhibit microbial growth under dark conditions |

During the composting process, microbes require food and energy to carry out their tasks. Therefore, biostimulation involves supplying organic matter that serves as a nutrient source, thereby enhancing microbial activity and accelerating decomposition. After consuming the food (nutrients), the microbes begin to proliferate and carry out their functions through reclamation, degradation, and other processes. This procedure modifies the moisture content (MC), increases microbial density during composting, and promotes microbial activity, especially in the contaminated environment (Li et al. 2024). Furthermore, because microbes digest the pollutants under ideal conditions, it is essential to focus on oxygen and optimal MC to enhance the growth of the degrading microorganisms. Likewise, the majority of species need the first-row transition metals, which include copper, zinc, iron, nickel, manganese, and cobalt (Vishwakarma et al. 2021). Nanotechnology, or the mediation of metal nanoparticles, is the most efficient method of supplying these metals. In addition to microorganisms, the new era of nanotechnology might offer a workable solution for plants' nutrient availability (Singh et al. 2024).

The microorganisms hydrolyze sugars, amino acids, lipids, and refractory molecules like proteins, lignin, cellulose, and hemicellulose during the composting process, converting them into simpler materials (Xie et al. 2023). Microorganisms undertake bioprocess functions such as hydrolysis, oxidation, and reduction of complex compounds, releasing heat, $CO_2$, and $H_2O$. In the same vein, all living things require metals. Additionally, they play structural and catalytic roles in biological activities such as transcription, respiration, oxidative stress responses, DNA replication, and precursor biosynthesis (Chen et al. 2023). As a result, metal nanoparticle mediation throughout the composting process aids in the execution of several structural and catalytic biological processes, including hydrolysis, oxidation, and reduction, and leads to better bioprocess performance than control composting.

The antibacterial qualities of silver nanoparticles are widely recognized (Bruna et al. 2021). Prior studies have demonstrated that AgNPs are harmful to microorganisms in two ways:

1. By altering proteins and enzymes, the release and passage of Ag+ across membranes impacts cellular respiration, results in cell death.
2. The production of ROS on the surface of AgNPs results in oxidative stress. Thus, there was no discernible improvement in the organic matter's breakdown during the AgNP-mediated composting process.

Furthermore, AgNPs of varying sizes had distinct impacts on the composting process at the same concentration. This might be a result of the previously discussed size-based toxicity.

In terms of physicochemical properties, compost quality, and other aspects of the composting process, iron, a micronutrient, demonstrated a favorable effect (He et al. 2022). The microorganisms that are not suited to FeONPs are screened out, which results in the decreased diversity of microorganisms (Table 4). Of the FeONPs that have been studied in composting, $Fe_2O_3$NPs has a higher abundance of thermophilic microbes, or core microorganisms linked to the breakdown of organic matter, including *Bacillus, Thermobifida, Thermobacillus, Thermopolyspora, Ureibacillus,*

*Sinibacillus, Oceanobacillus, Symbiobacterium,* and *Ammoniibacillus*. These microorganisms may have higher degradation rates and enzyme activity because of their exceptional capacity to break down a variety of organic materials, including cellulose, lignin, labile organic matter, and carbohydrate organic matter (Gattupalli et al. 2023).

Additionally, they enhanced the microbial digestion of amino acids and carbohydrates in FeONP-mediated composting (Gattupalli et al. 2023). By composting cellulose and hemicellulose, the metabolism of carbohydrates can create a variety of chemicals that help microorganisms survive. Furthermore, by increasing the synthesis of amino acids and humic acids, amino acid metabolism can restore carbon supplies and energy for microbial activity and growth. Therefore, using various kinds of iron oxide nanoparticles (NPs) in composting, especially $Fe_2O_3$ NPs, is a practical method to boost microbial abundance, metabolism, and degradation rates during the composting process and enhance the overall effectiveness and caliber of the finished compost product. Additionally, it is necessary to investigate the various metal nanoparticles (NPs), including copper, manganese, and others, in the composting process.

## 9 Life Cycle Assessment of Composting Enhanced by Nanoparticles

The life cycle assessment (LCA), which takes into account all of the waste, assesses the potential environmental benefits and resources used in nanocompost. It examines every stage i.e. from waste generation to waste treatment and final application to plants, providing a comprehensive view of its overall impact.

There are four phases in the life cycle analysis. They are impact assessment, life cycle inventory, life cycle phases, and life cycle steps. From the cradle to the gate boundary, this life cycle assessment seeks to determine the environmental advantages of nanocompost (Bishop et al. 2021). Table 5 outlines the specific procedures from the cradle to the gate boundary along with the advantages of inputs and outputs.

Based on several life cycle assessment (LCA) studies, we infer that all composting systems have a lower environmental impact. Reduced environmental dangers are the outcome of applying compost to crops (Hiranmai et al. 2024). If administered in greater amounts, the NP-mediated compost may have negative effects on plants in addition to nutrition. To optimize the procedure and application amount, more research is required on nanocompost and its use on different crops. The LCA of NP-mediated composting has not been thoroughly studied. To learn more about comprehensive nanocomposts, it is advised to do a thorough life cycle assessment (LCA) of NP-mediated composting, especially from gate to grave.

**Table 5** Life cycle assessment of nanocompost

| Life cycle stage | Environmental benefits of nanocompost | Impact on environmental aspects | Potential advantages | Challenges or limitations |
|---|---|---|---|---|
| Raw material sourcing | Utilizes organic waste materials (e.g., agricultural residues)<br>Reduced need for synthetic fertilizers | Decreases the demand for chemical fertilizers and pesticides<br>Minimizes land degradation | Reduces reliance on nonrenewable resources<br>Promotes recycling of organic waste | Energy use in nanoparticle synthesis<br>Availability of raw materials |
| Nanofertilizer synthesis | Eco-friendly green synthesis methods reduce chemical use<br>Use of plant-based or biogenic materials | Lower energy consumption compared to traditional fertilizer production<br>Reduced $CO_2$ emissions | Sustainable synthesis methods lead to fewer emissions<br>Potential for low-cost production | Potential risk of using nonbiodegradable nanoparticles |
| Composting process | Accelerates decomposition of organic waste<br>Enhanced nutrient release for plant uptake | Faster nutrient cycling, leading to improved soil fertility<br>Reduced methane and $CO_2$ emissions | Reduced greenhouse gas emissions due to efficient composting<br>Increased carbon sequestration | Possible toxicity of nanoparticles at high concentrations |
| Product use (nanocompost) | Enhances soil health and microbial activity<br>Reduces soil erosion and degradation | Improves soil structure, water retention, and biodiversity<br>Reduces chemical runoff and leaching | Promotes sustainable agriculture and organic farming<br>Improves soil carbon storage | Long-term effects on soil microorganisms still unclear |
| End-of-life/ disposal | Biodegradable nature of nanocompost reduces environmental impact<br>Can be composted or used as a soil amendment | Less waste generated compared to chemical fertilizers<br>Reduced landfill burden | Reduced landfill contribution<br>Promotes circular economy through composting | Potential for nanoparticle persistence in the environment |
| Overall environmental impact | Reduced chemical fertilizer use and associated environmental pollution<br>Potential for sustainable agricultural practices | Decreased soil, water, and air pollution from chemical fertilizers and pesticides | Eco-friendly solution for nutrient management in agriculture<br>Supports sustainable agricultural practices | Concerns over nanoparticle toxicity and persistence in the environment |

# 10 Nanozymes

By lowering the activation energy, enzymes speed up the chemical reaction's journey to equilibrium (Cheung et al. 2024). Microbial enzymes can speed up the rate of disintegration and transform even dangerous substances into innocuous, useful products through the biodegradation or bioconversion process (Doukani et al. 2022). Additionally, NPs increase the rate of breakdown by boosting thermophilic microbes. Therefore, the organic matter decomposes more quickly in both directions. While nanomaterials are complicated, natural enzymes are mostly proteins and are regarded as soft materials. Nonetheless, NPs and natural enzymes share several characteristics, such as size, structure, and surface charge. They imitate enzymes as a result. Hematin, cyclodextrin, polymers, hemin, porphyrin, metal complexes, and biomolecules created as catalytic antibodies, proteins, and nucleic acids have all been successfully used to replicate the catalytic actions of natural enzymes (Aziz et al. 2022).

## *10.1 The Connection between the Breakdown of Organic Materials and Enzymes*

Research indicates that the use of enzyme technology in agriculture has a beneficial effect on the environment (Kaushik et al. 2024). After catalyzing the transformation of the substrate (S) into the product, the enzyme (E) separates into the enzyme (E) and the product (P) after forming an enzyme-substrate complex (ES) (Zong et al. 2024).

$$\text{Enzyme} + \text{Substrate} = \text{Enzyme-substrate complex} \rightarrow \text{Enzyme} + \text{Product}$$

There are not many amino acids in the active sites. For substrate and enzyme binding, a variety of noncovalent bonds—such as hydrogen bonds, hydrophobic bonds, and van der Waals contacts—are necessary (Mujafarkani et al. 2024). These enzymes are also highly specialized; they look for their particular substrate and release the end products. In a similar manner, the enzymes break down the complex organic matter to release products like $CO_2$, $H_2O$, humic-like chemical compounds, etc.

## *10.2 Nanozyme Biocatalytic Characterization*

Thanks to developments in a number of fields, such as fermentation technology, nanotechnology, bioinformatics, advanced spectroscopy, material sciences, molecular biology, and genetics, biocatalysis is becoming more and more prominent in the present and future scenarios of white biotechnology in the framework of green chemistry (Gattupalli et al. 2023). Since nanozymes are more stable, resilient, cost-effective, and have higher catalytic activity and substrate selectivity than natural enzymes, they are becoming more and more popular in a variety of applications (Zhang and Liu 2020). When integrating nanomaterials with biological systems, including nanorods, nanoclusters, nanobelts, nanowires, and crystals, nanozymes are crucial. Nanozymes can be used in a variety of applications, including biosensors, immunoassays, cancer diagnostics, therapy, stem cell growth, and environmental engineering, because they can carry out natural enzyme-like functions (Munir et al. 2020). $Fe_3O_4$ is the first metal nanoparticle found to have inherent peroxidase activity. Interest in determining different classes of enzyme activity from nanomaterials increased as a result of this startling discovery. Consequently, NPs that demonstrate catalytic activity similar to that of enzymes are referred to as "nanozymes." Additionally, the basic theoretical framework in enzymology has been utilized to study the activities of oxidase, superoxide dismutase, hydrolase, and other nanozymes.

The enzymes' novel catalytic capabilities can be obtained through the use of chemicals, metals, and cofactor changes. The catalytic activity of surface area changes, atomic ratio, electrical structure, electron transfer, and crystal facet are some ways that nanomaterials are improved (Patel et al. 2023). Several techniques have been developed to produce the enzymes, including solid substrate fermentation, recombinant DNA technology, recombinant protein expression, and purification of the desired enzymes from various sources, including bacteria, fungi, and plants (Sharma and Upadhyay 2020). Protein engineering, on the other hand, makes use of direct evolution approaches, computational methodologies, etc. Additionally, efforts have been made to establish enzyme stability in organic media using chemical modification, directed evolution, immobilization, and protein engineering (Sharma et al. 2021).

Commercially accessible enzymes are not cost-effective; effective and economical bioseparation techniques can achieve successful enzyme technology, necessitating the development of novel techniques and technologies (Sinha et al. 2022). Another important effort that may be utilized for their efficient use in the composting process is the employment of "nanozymes." In addition to providing nutrition, this nanozyme-mediated composting process aims to significantly enhance the breakdown of pollutants, toxins, and organic materials. They are also faster and more selective than enzymatic therapy alone. The nanozymes can help with substrate adsorption on their surface and lengthen exposure time due to their smaller size and larger surface area. Additionally, there is more interaction between the substrate and the nanozyme's active regions (Wang et al. 2020).

Cobalt ferrite NPs demonstrated peroxidase activity in soybean plants, boosting nitrogenase activity in soybean nodules and leghemoglobin buildup, according to research on the effects of nanozymes on soybeans (Ma et al. 2022a). Furthermore, ROS can be broken down by these nanozymes. Thus, NP-mediated composting may also be beneficial for the breakdown of contaminants and organic waste. When given as fertilizer, these NPs can also function as nanozymes, which are enzymes that nourish plants (Rajput et al. 2024).

## 11 Economic Implications of Nanocompost Technology

Compared to the conventional composting method, the nanomediated microbial composting process offers a few benefits. They are: (a) Metals are necessary for all living things, including microbes. For instance, it was demonstrated that $Fe_3O_4$ NPs enhanced the number of thermophilic bacteria, causing higher organic matter breakdown than the control (Galal et al. 2024). Additionally, it was asserted that the composting process's bacterial diversity was enhanced by silver nanoparticles (Gattupalli et al. 2023). The metal NPs, which are necessary for several bioprocesses like DNA replication, precursor biosynthesis, respiration, etc., may be the source of this. (b) In addition to providing nutrients, the NPs in compost help restore soils by eliminating heavy metals.

1. The commercial NPs can be harmful and are costly. The NPs produced naturally from microorganisms, plant extracts, etc., are cost-effective. Additionally, they offer capping agents that may lessen toxicity and bring potential economic results (Ahmed et al. 2017).
2. Because NP-mediated compost are already high in macronutrients and also deliver extra micronutrients in the form of NPs, using them can save money on commercial micronutrient fertilizers (Pradhan and Mailapalli 2017).
3. Larger-scale planning can shorten the composting process's duration because of the higher rates of degradation. It lowers the cost of labor, aeration, electricity, land use, etc. (Keng et al. 2020).

Additionally, it is necessary to look into the specific economic perspective of nanocomposting.

## 12 Understanding the Role of Nanoparticles in Soil Dynamics as Nanocompost

The soils received further applications of the mature compost that contained the NPs. Furthermore, these NPs permeate the various soil matrixes as a result of a variety of transport modes, including rain, floods, and cyclones. This section discusses their mode of action inside the soil matrix.

The NPs were either mobilized or retained as a result of the soil reactions occurring within the soil solution or soil pores. ENP transit and retention are significantly influenced by the soil's pH, organic matter, ionic strength (IS), clay concentration, and particle characteristics (Gattupalli et al. 2023). All of the nanoparticles (NPs) exhibit negative zeta potential regardless of pH when they are distributed in soil solutions (Strekalovskaya et al. 2024). The NPs' surface characteristics change as a result of the interaction between the organic matter and NPs, which intensifies the steric repulsions (Ali et al. 2022). By neutralizing the negative charge that the organic matter has transferred to their surface, these NPs will raise the zeta potential as cation concentration increases. The stability of the NPs in the soil solution matrix is explained by this zeta potential. Furthermore, the impact of metal oxide nanoparticles on the makeup and size of the bacterial population was significantly influenced by the type of soil (Rajput et al. 2018). The heterogeneous environment, the onset of aggregation, deposition, oxidation and dissolution, and the transformation of different NPs are the foundations of the varied mechanisms of NPs in the soil matrix.

## 12.1 Aggregation

Through ligand exchange and electrostatic synergy, iron nanoparticles are crucial for the retention and adsorption of anion nutrients in soils. Additionally, they aid in the stability of soil aggregates and the flocculation of clay (Bao et al. 2024). Additionally, large heteroaggregates may reduce the biological availability of NPs by preventing their translocation and absorption through plant cells and membranes. Seven distinct soil types were examined in the study: (i) sandy loam soil with high organic matter (HOM), (ii) sandy loam soil with low organic matter (LOM), (iii) loam soil with HOM, (iv) loam soil with LOM, (v) silty clay soil with HOM, and (vi) silty clay soil with LOM. The researchers came to the conclusion that only silty clay with HOM (bark compost) showed toxicity, primarily as a result of particle aggregation. Because of their greater IS, lower dissolved organic carbon, and higher cation content, which decrease stability and produce aggregation, they exhibit higher aggregation than other soils.

## 12.2 Deposition

The deposition of NPs in soils is caused by surface bonding and collision. Deposition at a particular site can also be caused via Brownian diffusion, interception, or gravitational settling inside the soil pore walls. Because of aggregation, collision, and interaction energy, the repulsive van der Waals forces that occur between NPs and collection surfaces promote an undesired deposition in soil (Osterberg et al. 2023). Compared to other colloids or ENPs, CNTs have more substantial deposition because of their greater aspect ratios and coiling properties around soil particles (Mishra et al. 2021). Thus, the transportation and deposition of ENPs are intricate processes that are simultaneously impacted by a number of variables, including hydrodynamic action, soil, pore water solution, and NP properties.

## 12.3 Transportation

Zeta potential and ENP surface coating may regulate the movement of ENPs throughout the soil solution. On the other hand, capping agents prevent degradation and transformation, while emulsifiers help stabilize them throughout processing. The movement of nanoparticles to collection sites is largely caused by hydrodynamic forces (Angardi et al. 2022). Ma et al. (2022b) demonstrated that the vast surface area and retention sites of ultisols decreased their migration. Furthermore, transport has been facilitated by Ag NPs' increased concentration, reduced particle size, and surface coating. Transport barriers include fine-grained silty loam, higher IS near divalent cations, and reduced flow rates. AgNPs also demonstrated good mobility in loamy sand at high flow rates and low IS. Diffusion predominates at low flow rates, allowing for additional advantageous locations as a transport mechanism that impedes movement. The compression of the electrostatic double layer of AgNPs and soil surfaces causes deposition at higher IS (Kyziol-Komosinska et al. 2024). The interaction between AgNPs and colloids serves two purposes: it can either speed up or slow down AgNP mass transit. The mobility of Ag NPs is thought to be enhanced by humic acids, and this enhancement was more noticeable at increasing humic acid concentrations (Ding et al. 2019). When natural organic materials are present, the combined effects of hydrodynamic diameter and contact increase the transport capacity of CuO NPs (Aniagor et al. 2022). Additionally, at different pH and IS conditions, the CuO NPs react differently in terms of transit and retention. CuO NPs had the lowest transportability at pH 6 due to a lower primary energy barrier.

## 12.4 Oxidation

ENPs often undergo a number of oxidation reactions, complexation with organic materials and chelating agents, and adsorption on the colloidal surface. Mg, Ag, Zn, and Cu are examples of soft metal cations that are susceptible to these reactions. The type, texture, and source of ENPs can all affect how quickly they dissolve in soils (Abbas et al. 2020). Chemical properties and soil elements like pH, organic carbon, texture, and size affect the dissolution of metal-based NPs. As the particle size decreases below 10 nm, it is assumed that the mineral solubility will significantly rise (Suazo-Hernández et al. 2023). Because oxidized Au (Au3+) is unstable, AuNPs were resistant to oxidative dissolution and readily reduced in soil (Hussain et al. 2022). Complex formation in soil between chelating agents and Ce3+ accelerated the oxidation and dissolution of $CeO_2$ NPs, reducing their mobility and bioavailability (Kurvet et al. 2017). Consequently, the kinetics of NP dissolution is a very complicated process, and the variables affecting the dissolution need to be taken into account and thoroughly examined.

## 13 Conclusion

The role of NPs in the composting process is highlighted in this chapter. NPs typically interact with bacteria by damaging nucleic acids, causing membrane breakdown, etc. By offering nutrients with less toxicity, on the other hand, the mediation of NPs in the composting process and compost may provide some potential results. This is because the organic matter in the compost prevents bacteria from coming into direct contact with it, reducing the possibility of toxicity.

Because of the electrostatic repulsions, which fix the particle size and shield these extracellular enzymes, the potential toxicity of NPs throughout the composting process is reduced, and the direct interaction with NPs is complex. Numerous bioprocesses, including respiration, precursor biosynthesis, DNA replication, and others, depend on the metal nanoparticles. When compared to composting that is not NP-mediated, this enhances compost quality and bioprocess performance. Because smaller particles dissolve more quickly and produce more ROS, the size and concentration of NPs have a significant impact on toxicity. Nanomaterials with a lower dosage show antioxidant activity, whereas those with a greater dosage show oxidant activity because of the overproduction of ROS, which causes oxidative stress. Therefore, it is more advantageous to use green or biologically synthesized NPs in the composting process to avoid excessive ROS formation and toxicity. To maximize NP size and quantity during the composting process, more research is required.

## References

Abbas Q, Yousaf B, Ali MU, Munir MAM, El-Naggar A, Rinklebe J, Naushad M (2020) Transformation pathways and fate of engineered nanoparticles (ENPs) in distinct interactive environmental compartments: a review. Environ Int 138:105646

Abhiram G (2023) Contributions of nano-nitrogen fertilizers to sustainable development goals: a comprehensive review. Nitrogen 4(4):397–415

Adetunji CO (2019) Environmental impact and ecotoxicological influence of biofabricated and inorganic nanoparticle on soil activity. In: Nanotechnology for agriculture: advances for sustainable agriculture. Springer, Singapore, pp 221–239

Agrawal K, Gupta VK, Verma P (2022) Microbial cell factories a new dimension in bio-nanotechnology: exploring the robustness of nature. Crit Rev Microbiol 48(4):397–427

Ali I, Tan X, Li J, Peng C, Naz I, Duan Z, Ruan Y (2022) Interaction of microplastics and nanoplastics with natural organic matter (NOM) and the impact of NOM on the sorption behavior of anthropogenic contaminants–a critical review. J Clean Prod 376:134314

Andraos, J., & Matlack, A.S. (2022). Introduction to green chemistry (3). CRC Press. https://doi.org/10.1201/9781003033615

Angardi V, Ettehadi A, Yücel Ö (2022) Critical review of emulsion stability and characterization techniques in oil processing. J Energy Resour Technol 144(4):040801

Aniagor CO, Igwegbe CA, Iwuozor KO, Iwuchukwu FU, Eshiemogie S, Menkiti MC, Ighalo JO (2022) CuO nanoparticles as modifiers for membranes: a review of performance for water treatment. Mater Today Commun 32:103896

Aziz AHA, Engliman NS, Mansor MF, Abdul PM, Arisht SN, Jamali NS, Tiang MF (2022) Synergistic enhancement of biohydrogen production by supplementing with green synthesized magnetic iron nanoparticles using thermophilic mixed bacteria culture. Int J Hydrog Energy 47(96):40683–40695

Bao S, Huang J, Liu X, Tang W, Fang T (2018) Tissue distribution of Ag and oxidative stress responses in the freshwater snail Bellamya aeruginosa exposed to sediment-associated Ag nanoparticles. Sci Total Environ 644:736–746

Bao T, Damtie MM, Wang CY, Li CL, Chen Z, Kuk CHO et al (2024) Iron-containing nanominerals for sustainable phosphate management: a comprehensive review and future perspectives. Sci Total Environ:172025

Bashir A, ur Rehman MZ, Hussaini KM, Adrees M, Qayyum MF, Sayal AU et al (2021) Combined use of zinc nanoparticles and co-composted biochar enhanced wheat growth and decreased Cd concentration in grains under Cd and drought stress: a field study. Environ Technol Innov 23: 101518

Bishop G, Styles D, Lens PN (2021) Environmental performance comparison of bioplastics and petrochemical plastics: a review of life cycle assessment (LCA) methodological decisions. Resour Conserv Recycl 168:105451

Bruna T, Maldonado-Bravo F, Jara P, Caro N (2021) Silver nanoparticles and their antibacterial applications. Int J Mol Sci 22(13):7202

Cao R, Wang J, Ben W, Qiang Z (2020) The profile of antibiotic resistance genes in pig manure composting shaped by composting stage: mesophilic-thermophilic and cooling-maturation stages. Chemosphere 250:126181

Chaudhary P, Ahamad L, Chaudhary A, Kumar G, Chen WJ, Chen S (2023) Nanoparticle-mediated bioremediation as a powerful weapon in the removal of environmental pollutants. J Environ Chem Eng 11(2):109591

Chen S, Wang X, Cheng Y, Gao H, Chen X (2023) A review of classification, biosynthesis, biological activities and potential applications of flavonoids. Molecules 28(13):4982

Cheung LKY, Sanders AD, Houfani AA, Grahame DAS, Bryksa BC, Dee DR, Yada RY (2024) Factors affecting enzyme activity and design. In: Improving and tailoring enzymes for food quality and functionality. Woodhead Publishing, pp 17–57

Chettri D, Verma AK, Verma AK (2024) Bioaugmentation: an approach to biological treatment of pollutants. Biodegradation 35(2):117–135

Das RK, Pachapur VL, Lonappan L, Naghdi M, Pulicharla R, Maiti S et al (2017) Biological synthesis of metallic nanoparticles: plants, animals and microbial aspects. Nanotechnol Environ Eng 2:1–21

Dimkpa CO, McLean JE, Britt DW, Anderson AJ (2012) Bioactivity and biomodification of Ag, ZnO, and CuO nanoparticles with relevance to plant performance in agriculture. Ind Biotechnol 8(6):344–357

Ding Y, Bai X, Ye Z, Gong D, Cao J, Hua Z (2019) Humic acid regulation of the environmental behavior and phytotoxicity of silver nanoparticles to Lemna minor. Environ Sci Nano 6(12): 3712–3722

Doukani K, Boukirat D, Boumezrag A, Bouhenni H, Bounouira Y (2022) Fundamentals of biodegradation process. In: Handbook of biodegradable materials. Springer International Publishing, Cham, pp 1–27

Faisal HA (2025) Effect of Nano-chelated iron and Potgrond organic fertilizer on the growth characteristics of date palm (Phoenix dactylifera L.) cv. Khalas. Int J Hortic Sci Technol:85–94

Galal A, Elgarahy AM, Haroun B, Safwat SM, El-Qelish M, Fahmi AA (2024) Enhanced biohydrogen production from thermally hydrolysed pulp and paper sludge via Al2O3 and Fe3O4 nanoparticles. Chem Eng Res Des 210:82–96

Ganguly RK, Chakraborty SK (2023) Valorization of fish wastes for circular bioeconomy: a concern toward antibiotic resistance genes (ARGs) contaminants for environmental safety. In: Fate of biological contaminants during recycling of organic wastes. Elsevier, pp 71–100

Gattupalli M, Dashora K, Mishra M, Javed Z, Tripathi GD (2023) Microbial bioprocess performance in nanoparticle-mediated composting. Crit Rev Biotechnol 43(8):1193–1210

Gauba A, Hari SK, Ramamoorthy V, Vellasamy S, Govindan G, Arasu MV (2023) The versatility of green synthesized zinc oxide nanoparticles in sustainable agriculture: A review on metal-microbe interaction that rewards agriculture. Physiol Mol Plant Pathol 125:102023

Gitipour A, El Badawy A, Arambewela M, Miller B, Scheckel K, Elk M et al (2013) The impact of silver nanoparticles on the composting of municipal solid waste. Environ Sci Technol 47(24): 14385–14393

Godeto YG, Ayele A, Ahmed IN, Husen A, Bachheti RK (2023) Medicinal plant-based metabolites in nanoparticles synthesis and their cutting-edge applications: an overview. In: Secondary metabolites from medicinal plants. CRC Press, pp 1–34

Gold K, Slay B, Knackstedt M, Gaharwar AK (2018) Antimicrobial activity of metal and metal-oxide based nanoparticles. Adv Ther 1(3):1700033

He Y, Huang X, Zhang H, Li H, Zhang Y, Zheng X, Xie L (2022) Insights into the effect of iron-carbon particle amendment on food waste composting: physicochemical properties and the microbial community. Bioresour Technol 351:126939

Hiranmai RY, Neeraj A, Vats P (2024) Improvement of soil health and crop production through utilization of organic wastes: a sustainable approach. Int J Recycl Org Waste Agric 13(1)

Hussain FS, Abro NQ, Ahmed N, Memon SQ, Memon N (2022) Nano-antivirals: a comprehensive review. Front Nanotechnol 4:1064615

Ibrahim EAA, El-Sherbini MAAA (2024) Advances in soil systems remediation: lessons from the global north. In: Sustainable soil systems in global south. Springer Nature Singapore, Singapore, pp 573–607

Jadav K, Durgam M, Perli M, Mailapalli DR (2024) Phosphorus transport modeling in soils treated with nano-phosphorus fertilizers. Plant Nano Biol 10:100112

Kaur N, Prakash C, Bhalla A, Chaudhary GR (2020) Nanodiamonds and other organic nanoparticles: synthesis and surface modifications. In: Functionalized nanomaterials I. CRC Press, pp 135–160

Kaur H, Hussain SJ, Mir RA, Verma VC, Naik B, Kumar P, Dubey RC (2023) Nanofertilizers–emerging smart fertilizers for modern and sustainable agriculture. Biocatal Agric Biotechnol 54: 102921

Kaushik N, Gupta R, Negi M, Kaushik A, Kim JH, Choi EH, Kaushik NK (2024) Integrating cutting-edge plasma technology for environmentally friendly smart horticulture: A proteomics approach. Appl Mater Today 37:102142

Keng ZX, Chong S, Ng CG, Ridzuan NI, Hanson S, Pan GT et al (2020) Community-scale composting for food waste: a life-cycle assessment-supported case study. J Clean Prod 261: 121220

Khan M, Khan MSA, Borah KK, Goswami Y, Hakeem KR, Chakrabartty I (2021) The potential exposure and hazards of metal-based nanoparticles on plants and environment, with special emphasis on ZnO NPs, TiO2 NPs, and AgNPs: a review. Environ Adv 6:100128

Klekotka U, Rogacz D, Szymanek I, Malejko J, Rychter P, Kalska-Szostko B (2022) Ecotoxicological assessment of magnetite and magnetite/Ag nanoparticles on terrestrial and aquatic biota from different trophic levels. Chemosphere 308:136207

Kumar Y, Tiwari KN, Singh T, Raliya R (2021) Nanofertilizers and their role in sustainable agriculture. Ann Plant Soil Res 23(3):238–255

Kurvet I, Juganson K, Vija H, Sihtmäe M, Blinova I, Syvertsen-Wiig G, Kahru A (2017) Toxicity of nine (doped) rare earth metal oxides and respective individual metals to aquatic microorganisms Vibrio fischeri and Tetrahymena thermophila. Materials 10(7):754

Kyziol-Komosinska J, Dzieniszewska A, Czupioł J (2024) Behavior of silver species in soil: Ag nanoparticles vs. Ionic Ag. Molecules 29(23):5531

Latif A, Sheng D, Sun K, Si Y, Azeem M, Abbas A, Bilal M (2020) Remediation of heavy metals polluted environment using Fe-based nanoparticles: mechanisms, influencing factors, and environmental implications. Environ Pollut 264:114728

Li J, Liu S, Xu Y, Xu C, Deng B, Cao H, Yuan Q (2024) Optimizing biochar addition strategies in combined processes: comprehensive assessment of earthworm growth, lignocellulose degradation and vermicompost quality. Bioresour Technol 406:131031

Liu L, Li W, Song W, Guo M (2018) Remediation techniques for heavy metal-contaminated soils: principles and applicability. Sci Total Environ 633:206–219

Long Y, Jin H, Li H, Zhu N, Sun E, Shan C et al (2024) Trace MnFe2O4 boosts polyphenol-maillard reaction and humification process for value-added composting: integrated effect of chemical and enzymatic catalysis. ACS ES&T Eng 4:3067

Ma J, Zhou Y, Li J, Song Z, Han H (2022a) Novel approach to enhance Bradyrhizobium diazoefficiens nodulation through continuous induction of ROS by manganese ferrite nanomaterials in soybean. J Nanobiotechnol 20(1):168

Ma R, Hu F, Xu C, Liu J, Zhao S (2022b) Response of soil aggregate stability and splash erosion to different breakdown mechanisms along natural vegetation restoration. Catena 208:105775

Mathur S, Singh D, Ranjan R (2022) Remediation of heavy metal (loid) contaminated soil through green nanotechnology. Front Sustain Food Syst 6:932424

Messaoud M, Salhi R, Baazaoui N, Hammami SB, Ezzine A, Mosbeh R et al (2024) Bio-nanoremediation of olive oil mill wastewater by Alternaria alternata fungi coupled with Ag-doped TiO2 nanoparticles and its use as biofertilizer for cereal crops. Euro-Mediterr J Environ Integr 9(1):81–93

Mishra D, Pandey V, Khare P (2021) Engineered nanoparticles in agro-ecosystems: implications on the soil health. In: Plant-microbes-engineered nano-particles (PM-ENPs) nexus in agro-ecosystems: understanding the interaction of plant, microbes and engineered nano-particles (ENPS). Springer, Cham, pp 103–118

Mujafarkani N, Agurokpon DC, Hussien RA, Ayoola OV, Alshdoukhi IF, Sampathkumar G et al (2024) Antimicrobial activity of guanidine-based terpolymers: synthesis, spectroscopy (FT-IR, 1H, and 13C NMR), quantum chemical investigation, and molecular docking. J Mol Struct 1298:137005

Mukhopadhyay R, Sarkar B, Khan E, Alessi DS, Biswas JK, Manjaiah KM et al (2022) Nanomaterials for sustainable remediation of chemical contaminants in water and soil. Crit Rev Environ Sci Technol 52(15):2611–2660

Munir S, Shah AA, Rahman H, Bilal M, Rajoka MSR, Khan AA, Khurshid M (2020) Nanozymes for medical biotechnology and its potential applications in biosensing and nanotherapeutics. Biotechnol Lett 42:357–373

Noor RS, Shah AN, Tahir MB, Umair M, Nawaz M, Ali A et al (2024) Recent trends and advances in additive-mediated composting Technology for Agricultural Waste Resources: a comprehensive review. ACS omega 9(8):8632–8653

Osman AI, Fawzy S, Farghali M, El-Azazy M, Elgarahy AM, Fahim RA et al (2022) Biochar for agronomy, animal farming, anaerobic digestion, composting, water treatment, soil remediation, construction, energy storage, and carbon sequestration: a review. Environ Chem Lett 20(4): 2385–2485

Osterberg M, Henn KA, Farooq M, Valle-Delgado JJ (2023) Biobased nanomaterials— the role of interfacial interactions for advanced materials. Chem Rev 123(5):2200–2241

Pal A, Goswami R, Roy DN (2021) A critical assessment on biochemical and molecular mechanisms of toxicity developed by emerging nanomaterials on important microbes. Environ Nanotechnol Monit Manag 16:100485

Pal A, Bhattacharjee S, Saha J, Sarkar M, Mandal P (2022) Bacterial survival strategies and responses under heavy metal stress: a comprehensive overview. Crit Rev Microbiol 48(3): 327–355

Patel AK, Dong CD, Chen CW, Pandey A, Singhania RR (2023) Production, purification, and application of microbial enzymes. In: Biotechnology of microbial enzymes. Academic, pp 25–57

Pathan SI, Arfaioli P, Bardelli T, Ceccherini MT, Nannipieri P, Pietramellara G (2020) Soil pollution from micro-and nanoplastic debris: a hidden and unknown biohazard. Sustainability 12(18):7255

Poudel A, Singh SK, Jiménez-Ballesta R, Jatav SS, Patra A, Pandey A (2023) Effect of nano-phosphorus formulation on growth, yield and nutritional quality of wheat under semi-arid climate. Agronomy 13(3):768

Pradhan S, Mailapalli DR (2017) Interaction of engineered nanoparticles with the agri-environment. J Agric Food Chem 65(38):8279–8294

Rady MM, Mossa ATH, Youssof AM, Osman AS, Ahmed SM, Mohamed IA (2023) Exploring the reinforcing effect of nano-potassium on the antioxidant defense system reflecting the increased yield and quality of salt-stressed squash plants. Sci Hortic 308:111609

Rahman A, Kang S, McGinnis S, Vikesland PJ (2022) Life cycle impact assessment of iron oxide (Fe3O4/γ-Fe2O3) nanoparticle synthesis routes. ACS Sustain Chem Eng 10(10):3155–3165

Rajmohan KS, Chandrasekaran R, Varjani S (2020) A review on occurrence of pesticides in environment and current technologies for their remediation and management. Indian J Microbiol 60(2):125–138

Rajput VD, Minkina T, Sushkova S, Tsitsuashvili V, Mandzhieva S, Gorovtsov A et al (2018) Effect of nanoparticles on crops and soil microbial communities. J Soils Sediments 18:2179–2187

Rajput VD, Minkina T, Kumari A, Shende SS, Ranjan A, Faizan M et al (2022) A review on nanobioremediation approaches for restoration of contaminated soil. Eur J Soil Sci 11(1):43–60

Rajput, V. D., Singh, A., Ghazaryan, K., Alexiou, A., Al-Tawaha, S., & Mohammad, A. R. (Eds.). (2024). Harnessing NanoOmics and Nanozymes for sustainable agriculture. IGI Global

Roshan RR, Velan ST, Devi PB (2023) A complete review on anaerobes and nanoparticles in wastewater treatment. In: Metagenomics to Bioremediation, pp 477–509

Saif S, Tahir A, Chen Y (2016) Green synthesis of iron nanoparticles and their environmental applications and implications. Nano 6(11):209

Sajid M, Płotka-Wasylka J (2020) Nanoparticles: synthesis, characteristics, and applications in analytical and other sciences. Microchem J 154:104623

Shaik S, Franchetti MJ (2022) Recycling and sustainability of nanocomposites. In: Applications of nanocomposites. CRC Press, pp 236–259

Sharifan H (2019) Interaction of nanoparticles (ZnO & CeO2) and coexistence of heavy metals/metalloids (Cd, Pb & As) with dietary plants, a food safety perspective. Doctoral dissertation, Texas A&M University

Sharma H, Upadhyay SK (2020) Enzymes and their production strategies. In: Biomass, biofuels, biochemicals. Elsevier, pp 31–48

Sharma A, Gupta G, Ahmad T, Mansoor S, Kaur B (2021) Enzyme engineering: current trends and future perspectives. Food Rev Intl 37(2):121–154

Sharma K, Tyagi S, Vikal S, Devi A, Gautam YK, Singh BP (2023a) Sustainable nanomaterials for environmental remediation. In: Handbook of green and sustainable nanotechnology: fundamentals, developments and applications. Springer International Publishing, Cham, pp 933–971

Sharma ND, Dhiman S, Varma A, Goel A (2023b) Recent perspectives of nanoparticles in industrial waste management—an overview. In: Waste management and resource recycling in the developing world. Elsevier, Amsterdam, pp 365–384

Sharma P, Bano A, Singh SP, Srivastava SK, Singh SP, Iqbal HM, Varjani S (2023c) Different stages of microbial community during the anaerobic digestion of food waste. J Food Sci Technol 60(8):2079–2091

Singh J, Kalamdhad AS (2013) Bioavailability and leachability of heavy metals during composting—a review. Int Res J Environ Sci 2(4):59–64

Singh A, Rajput VD, Pandey D, Sharma R, Ghazaryan K, Minkina T (2023) Nano zinc-enabled strategies in crops for combatting zinc malnutrition in human health. Front Biosci Landmark 28(8):158

Singh A, Sharma A, Singh O, Rajput VD, Movsesyan HS, Minkina T et al (2024) In-depth exploration of nanoparticles for enhanced nutrient use efficiency and abiotic stresses management: present insights and future horizons. Plant Stress 14:100576

Sinha R, Goel N, Fatima H, Khare SK (2022) Advances in downstream processing techniques. In: Current developments in biotechnology and bioengineering. Elsevier, pp 353–371

Soni S, Jha AB, Dubey RS, Sharma P (2023) Mitigating cadmium accumulation and toxicity in plants: the promising role of nanoparticles. Sci Total Environ:168826

Strekalovskaya EI, Perfileva AI, Krutovsky KV (2024) Zinc oxide nanoparticles in the "soil–bacterial community–plant" system: impact on the stability of soil ecosystems. Agronomy 14(7):1588

Suazo-Hernández J, Arancibia-Miranda N, Mlih R, Cáceres-Jensen L, Bolan N, Mora MDLL (2023) Impact on some soil physical and chemical properties caused by metal and metallic oxide engineered nanoparticles: A review. Nano 13(3):572

Szuplewska A, Sikorski J, Matczuk M, Ruzik L, Keppler BK, Timerbaev AR, Jarosz M (2023) Enhanced edible plant production using nano-manganese and nano-iron fertilizers: current status, detection methods and risk assessment. Plant Physiol Biochem 199:107745

Tee JK, Ong CN, Bay BH, Ho HK, Leong DT (2016) Oxidative stress by inorganic nanoparticles. Wiley Interdiscip Rev Nanomed Nanobiotechnol 8(3):414–438

Vaseghi Z, Nematollahzadeh A, Tavakoli O (2018) Green methods for the synthesis of metal nanoparticles using biogenic reducing agents: a review. Rev Chem Eng 34(4):529–559

Verma KK, Song XP, Degu HD, Guo DJ, Joshi A, Huang HR et al (2023) Recent advances in nitrogen and nano-nitrogen fertilizers for sustainable crop production: a mini-review. Chem Biol Technol Agric 10(1):111

Vishwakarma D, Mishra P, Patel B, Mishra RP, Khan M (2021) Importance of some transition metals and their biological role: a review. Int Res J Pure Appl Chem 22(5):12–23

Wang Z, Zhang R, Yan X, Fan K (2020) Structure and activity of nanozymes: inspirations for de novo design of nanozymes. Mater Today 41:81–119

Wang L, Wu WM, Bolan NS, Tsang DC, Li Y, Qin M, Hou D (2021) Environmental fate, toxicity and risk management strategies of nanoplastics in the environment: current status and future perspectives. J Hazard Mater 401:123415

Wang Q, Shan C, Zhang P, Zhao W, Zhu G, Sun Y et al (2024a) The combination of nanotechnology and potassium: applications in agriculture. Environ Sci Pollut Res 31(2):1890–1906

Wang YL, Lee YH, Chou CL, Chang YS, Liu WC, Chiu HW (2024b) Oxidative stress and potential effects of metal nanoparticles: a review of biocompatibility and toxicity concerns. Environ Pollut 346:123617

Xie S, Tran HT, Pu M, Zhang T (2023) Transformation characteristics of organic matter and phosphorus in composting processes of agricultural organic waste: research trends. Mater Sci Energy Technol 6:331–342

Yamini V, Shanmugam V, Rameshpathy M, Venkatraman G, Ramanathan G, Garalleh HA et al (2023) Environmental effects and interaction of nanoparticles on beneficial soil and aquatic microorganisms. Environ Res 236:116776

Zandi P, Schnug E (2022) Reactive oxygen species, antioxidant responses and implications from a microbial modulation perspective. Biology 11(2):155

Zhang J, Liu J (2020) Light-activated nanozymes: catalytic mechanisms and applications. Nanoscale 12(5):2914–2923

Zong Z, Zhang X, Chen P, Fu Z, Zeng Y, Wang Q et al (2024) Elucidation of the noncovalent interactions driving enzyme activity guides branching enzyme engineering for α-glucan modification. Nat Commun 15(1):8760

# Composting: Future Perspectives in Sustainable Agricultural in Different Geographical and Climatic Conditions

**Arne Sæbø, Lidia Sas-Paszt, Luisella B. Celi, Henrique Trindade, Francisco Saraiva, Joao Ricardo Sousa, and Rupesh Kumar Singh**

## 1 Introduction

The world is facing some very large challenges, many of which can be effectively handled only at the local scale. In this context, the ancient practice of composting is poised to play a vital role in future green societies. Traditionally, composting has served to convert organic residuals into fertilizers and soil amendments for agricultural and horticultural use. This function will remain essential. However, the future of composting lies in enhancing all the ecosystem services provided by this truly nature-based solution (NBS). As we face urgent and complex global challenges—including transforming food systems, mitigating climate change, promoting resource circularity, and ensuring cleaner production—composting stands out as a key component. The emerging bioeconomy will require significant scaling up in the production, valuation, handling, and treatment of biomass. In the final stages of the cascading use of biomass, composting will—as it has for millennia in traditional agriculture—play a crucial role in converting residual materials into clean soil

A. Sæbø (✉)
Norwegian Institute of Bioeconomy Research, Ås, Norway
e-mail: arne.sabo@nibio.no

L. Sas-Paszt
The National Institute of Horticultural Research, Skierniewice, Poland
e-mail: lidia.sas@inhort.pl

L. B. Celi
Department of Agricultural, Forest and Food Sciences, Università degli Studi di Torino, Grugliasco (TORINO), Italy
e-mail: luisella.celi@unito.it

H. Trindade · F. Saraiva · J. R. Sousa · R. K. Singh
Centre for the Research and Technology of Agro-Environmental and Biological Sciences (CITAB), Inov4Agro, UTAD, Vila Real, Portugal

 
R. K. Singh et al. (eds.), *Composting of Farm Residues by Nature-Based Solutions*,
https://doi.org/10.1007/978-3-032-12255-1_13

amendments, fertilizers, and growing media. These compost products will be essential not only at the end of the biomass processing chain but also at the beginning of new cycles of production. They will support the regeneration of resources for food, fiber, energy, and both traditional and innovative bioproducts, which are necessary to sustain life on a densely populated planet. Healthy soils and closed recycling circles will be crucial for the sustainability of agriculture, horticulture, and forestry systems as integrated and holistic systems. And we need to look at the "treatment train" from the initial high value products to low value, or even cost-driving waste biomasses. In the basic strategy of the future sustainable societies, "waste" may not be a much-used term. There are new expectations of composting actors, including increasing professionalization, advanced technology, and more focus on cleaning for contaminants in composting materials. We shall recycle the nutrients and increase the plant-growth-stimulating substances in composts. In the same processes, we shall end the recycling of contaminants, either of organic or inorganic nature. The aim is high-quality products and that the use of composts only has positive effects on production and the environment.

The input materials to composting will be side- and rest-products from agriculture and urban greening, like seed husks, peels, manure, some products from animal production, and park and garden wastes. However, humans produce large quantities of organic wastes. With the growth in human population and the increasing urbanization, we cannot ignore cities as a source of huge quantities of organic materials that can be composted, returning resources to the soils. However, these wastes often contain pathogens, persistent organic pollutants, medicines, antibiotic-resistance genes, and high levels of toxic heavy metals that pose danger to human health and the environment. In the future sustainable "biocities" (Scarascia-Mugnozza et al. 2023), composting will be one of the tools of great importance for treating organic materials as resources for food, fiber and energy production. Thus, cities will not only be sources for large quantities of waste, but they will also be important markets for compost. Thus, we need to look at composting in the future both from the perspective of decreasing the footprint on human health and ecosystems, and from a value chain perspective, using most of the available input resources for new and profitable production.

These perspectives lay the foundation for the future development of composting, highlighting the key challenges that lie ahead for policymakers, urban planners, citizens, and researchers alike. The overarching goal must be to identify effective ways to treat residual products and waste streams in order to produce the clean and healthy compost that are essential for sustainable systems. Efforts should focus on improving the efficiency of composting technologies, removing organic pollutants and heavy metals from biomass, and developing innovative products derived from compost. All of this must be embedded within a holistic framework aimed at fostering healthy ecosystems and resilient human societies. While these challenges are complex and demanding, addressing them is critical if we are to achieve the vision of a greener, more sustainable future.

## *1.1 Composting Facilities for the Future*

Composting has been a simple and low-cost treatment method for organic waste, often at small scale. In rural areas, this may still be the situation in the future, where there is enough space to carry out composting. Small entities can use cold composting as a viable alternative, for example for plant residues from private gardens and farms. However, such facilities need to be constructed to prevent runoff and eutrophication. If compost from small-scale units is to be sold or used beyond the property where it is produced, regulations require documentation of both the treatment methods and the quality of the finished compost to ensure its safety. However, if clean input materials, free of human pathogens and other contaminants, are used, such small-scale composting should be a cheap and safe option. Small-scale composting still needs space, which can be a challenge in densely populated residential areas. Arrigoni et al. (2024) addressed small-scale composting as an alternative to centralized composting facilities in cool climates, decreasing transport and increasing user participation. Their research was on how the addition of animal products to small compost bins can affect the composting process. The results showed that implementing animal products improved the composting efficiency and rasied temperatures to a level that could remove potential pathogens even in small-scale composting.

When large quantities are to be composted, or when a mixture of plant residues, food waste and contaminated biomass is to be treated, a professional approach in larger facilities should be the prefrred solution. Such facilities require substantial investments, with the infrastructure adapted to the input materials. The treatments, which will be more targeted in the future, with respect to the aims for the composting, need high competence and advanced technology to meet the demands for methods and quality indicators to be implemented. Composting will still have the two main functions: (1) decomposing and transforming organic materials considered as wastes into safe and useful products, and (2) producing compost and derived products as inputs to soil improvement in agriculture, horticulture, and forestry.

Composting facilities should be located close to both the sources of waste biomass and the areas where the resulting compost will be applied. Scalable composting systems must meet strict standards for safety and quality at every stage of the process—from the collection of input materials through treatment, maturation, and final delivery for both traditional and innovative uses. Productive collaboration between stakeholders with diverse expertise and goals can be highly beneficial. However, it is essential to understand the value proposition of composting within a broader evaluation framework. Key incentives for composting facilities include low operational costs, strong revenue potential, and the production of high-quality compost (Su et al. 2024).

The knowledge required to design and operate effective composting facilities spans multiple scientific disciplines—including biology, hygiene, engineering, and social sciences—as well as various administrative sectors at the national, regional, and municipal levels. This interdisciplinary nature highlights the critical need for coordination among all relevant stakeholders, including governmental authorities and their respective departments. Governments are responsible for establishing regulations and guidelines for the construction and operation of composting facilities. However, these frameworks must evolve alongside advances in knowledge and technology to ensure the production of safe, high-quality compost.

Effective two-way communication among citizens, composting operators, and government agencies is essential to achieving the broader goals of composting as a nature-based solution (NBS). As noted by Du et al. (2024b), factors such as high oil and salt concentrations in input materials were found to correlate with low germination indices in finished compost—emphasizing the importance of continuous monitoring and quality control. Scalable composting systems, from the smallest units fitting for homes, to the small-scale composting on farms and medium-sized enterprises, to the large facilities that need good regulation possibilities for the process and good facilities to make profitable products from high-quality, clean compost. This approach may increase the quality and valuation of compost along the whole value chain, from waste treatment and cleaning of the input materials to high-value compost products tailored for the specific uses in the production of food, fibers and fuel.

**Conclusion 1: Future Prospects for Scale in Composting**

(a) *Scalable solutions should be developed for composting facilities, for treatments of small, medium, and large quantities of input materials, with standard descriptions for quality parameters.*
(b) *Composting facilities and their production lines should be adapted to handle specific wastes and fractions with specific problematic contents..*
(c) *Indexes of compost quality needs to be revisited in research to give a reliable tool for composting facilities to evaluate their products and for the compost users to plan their safe use of compost in food production.*

## 2 Feeding Materials to Composting

The choice of input materials affects the composting process and quality parameters in multiple ways in all four phases of composting: heating, high temperature, cooling, and maturation. Defining the conditions and feedstocks that optimize the composting process is a key issue addressed elsewhere in this book. While many types of biomass have been proposed for bioenergy production or as bulking agents for composting, not all of them are sustainable choices. For example, forestry residues left after logging may not be suitable bulking materials, as their removal can negatively affect greenhouse gas emissions and soil organic carbon (Lan et al. 2024). Deadwood, both above- and below-ground, is important for nutrient cycling in forest systems. Its removal may therefore harm biodiversity and ecological functions in forests (Ranius et al. 2017). Materials for composting should primarily consist of residual byproducts from human activities, often as side streams from sectors such as agriculture, forestry, fisheries, and other ocean-based industries. Importantly, composting inputs should be selected with the end use of the compost in mind. Compost intended for use as soil amendments or fertilizers should be rich in nutrients and growth-promoting substances while being free of contaminants such as heavy metals and organic pollutants.

Making high-quality compost, according to regulations and their limits on content, is the responsibility of the composting facility. Contaminants in organic matter and persistent organic pollutants, for example, some herbicides, pesticides, and per and polyfluoralkyl substances (PFAS) compounds, may increase in concentration after composting is finished, merely because of the concentration effect when organic matter is broken down. Plant residues from agriculture and horticulture may carry contaminants, for example, herbicides that may pop-up as serious problems in the compost.

In the composting process, intermediate organic compounds may be harmful to plants when the compost is used in soil for agricultural and horticultural production. Therefore, careful evaluations and relevant analyses must be conducted primarily by composting companies but also be enforced through regulations and oversight by authorities. The use of emergent materials, such as degradable plastics for food packaging and mulching materials in agriculture, is expected to increase. Composting may accelerate the degradation of these plastics. However, Bernardi et al. (2024) found that composting such materials affected the fungal community more significantly than bacteria consortia. Earthworms were also affected, increasing gut-inflammation-related bacteria and some DNA damage. Manures and municipal wastes along with composts derived from these feedstocks, may introduce plastic and microplastic residues into soils. Up to 2 tons of plastic per ha were observed after 23 years of biannual compost application (Colombini et al. 2024). This will, according to the authors, lead to a ban on the use of such composts in the EU. Long-term application of chicken manure compost increased microplastic in a greenhouse, even deep in the soil (Zhang et al. 2024c). The combination of adding biochar and bacteria accelerated the degradation of plastic compared to composting without these additives, suggesting potential environmental and economic benefits from this approach (Wang et al. 2024b).

**Conclusion 2: Future Perspectives for Feeding Materials**

*To produce high-quality compost and fully realize the potential of composting as a recycling tool, composting facilities must enforce high standards for their feeding materials. Implementing thorough cleaning of feeding materials may be necessary. Furthermore, the processes must be continuously monitored and adapted to ensure the production of certified compost according to the regulatory requirements and high-quality standards.*

# 3 Efficient and Targeted Composting

## *3.1 Additives and Pretreatment Technologies*

To preserve the nitrogen (N) resources, the knowledge on management of processes involved in the transformation from raw wastes to high-quality composts is of paramount importance. However, there is a need for more knowledge on nitrate reduction to ammonium to be able to better control the process (Hoang et al. 2022). The authors suggested several strategies to decrease N losses, for example, through control of moisture content, appropriate oxygen supply, pH, the use of targeted microbes, and addition of $NH_4^+$-absorbing and adsorbing materials (biochar, zeolite, etc.) in the composting phase. The pretreatment addition of clay minerals changed the microbial community structure to promote humic substance production and fixation during straw composting (Shen et al. 2024b), probably through Fenton-like reactions (involving Reactive Oxygen Species). At the end of the composting process of swine manure, with an electric field (EK) in the compost pile, removal of antibiotic-resistant bacteria and antibiotic-resistance genes increased by 33.52–90.66% and 68.60–85.24%, respectively (Li et al. 2024). The decrease in antibiotic-resistant bacteria and antibiotic-resistance genes was probably caused by the increase in oxidants in the EK treatment.

Using ultra-high temperatures and prolonging the duration of the high-temperature phase seems promising for improving the composting process. Input materials with organic contaminants and antibiotic-resistance genes are best degraded by fast heating and several days at high temperatures (preferably >70 ° C). It has been routine to make mixtures of the main composting materials, for example, food residues or manure, with materials that both affect the C/N-ratio in the composting mass and affect $O_2$ supply to the whole composting piles. A C/N -ratio of 20–30 is recommended in composting materials (Shen et al. 2024a), with ratios over 30 demanding longer time in composting (beyond 45 days). Zhu et al. (2024) used a "Puls alternating ventilation composting technology" and an additive with mature compost as preparation for composting food residues. The supply of 10% (wet weight) of mature compost increased the thermophilic phase of the composting by 15 days and thus significantly increased the efficiency of composting. The authors concluded that the addition of mature compost increased the dissolution of humus precursors that subsequently condensed faster with hydrolysis products of the microbes in the composting pile. Such additives may be a cheap and efficient way to spur composting. Bulking components are important for the physical properties of compost and thus the temperatures of different phases of the composting process. Wang et al. (2024c) showed that the bulking materials, through affecting temperature in the thermophilic and cooling phases, affected the degradation of antibiotic-resistance genes during swine manure composting.

Microbial inoculants are often included as a pretreatment in composting. Zhang et al. (2024d) promoted the use of microbial inoculants for enhancing compost quality and minimizing the potential human health risks associated with the use of composts. Recalcitrant lignin in dead plant tissues protects the cellulose and hemicellulose from microbial enzymes, and the lignocellulose is very slowly decomposed. However, pretreating materials rich in lignocellulose can increase contact points between microbes and their enzymes to degrade the plant residues (Wu et al. 2022). To increase composting efficiency and quality while reducing the time spent on composting, the authors suggest (i) screening microbes and consortia of bacteria and fungi for their ability to efficiently produce the enzymes (lignocelluloses) that facilitate the degradation of lignin, cellulose, and hemicellulose, and (ii) exploring methods in pretreatments of the feed materials to enhance the contact points between substrates and enzymes, including biological methods (fungi and bacteria), high-temperature treatments, the use of oxidants, acids, alkali, and surfactants. Wu et al. (2022) also highlighted the need to adopt methods that do not introduce or induce toxic materials in the compost. Zhang et al. (2024d) used thermophilic bacteria (*Bacillus paralicheniformis* and *Streptomyces thermoviolaceus*) to increase the degradation of lignocellulose and showed significant changes in the microbial consortia of the composts, an increase in the temperature and duration of the thermophilic phase of composting, and an increase in degradation of different fractions of the lignocellulose.

**Conclusion 3: Future Prospects for Additives to Composting**
*The research on additives with microbes, nanochar, targeted bulking materials,* etc. *should increase, and new knowledge should be implemented in composting facilities.*

## 4 Decreasing Pathogens, Antibiotic Content, and Resistance Genes in Compost

Antibiotics and resistance genes are already widespread, especially in areas with large human activities and animal production. The human use of antibiotics is increasing in developing countries and stable in developed ones (Browne et al. 2021). The authors pointed out that both inappropriate use and lack of access to antibiotics are important for public health. However, livestock and poultry farming also use large quantities of antibiotics (see Zhao et al. 2024 for references), not only for treatment of diseases, but also to increase growth and production in animals. The differences between countries in animal use of antibiotics are probably at least as large as for the human use, given the costs related to the use of medicine and due to differences in regulations of antibiotic use.

High levels of antibiotics were shown to affect the composting process (in the thermophilic phase) and the compost quality (Ezzariai et al. 2017). Both, resistant microbes and resistance genes may be spread to soils and plants, and even to humans through aerosols when working in the composting facilities (Ezugworie et al. 2021). Care should be taken to establish the composting conditions, including temperature and physicochemical parameters, that promote degradation of antibiotics and resistance genes (Ezzariai et al. 2018). The most efficient composting seems to give the highest degradation of resistance genes (Ezugworie et al. 2021; Wang et al. 2024b; Zhao et al. 2024). Optimization of the composting processes aiming at decreasing antibiotic resistance and its spread should be addressed in research on composting. However, there may also be several recalcitrant molecules related to degradation of antibiotics, molecules that are unknown or have little attention because of analytical challenges (Ezzariai et al. 2018). The presence of high concentrations of metals may increase the danger of transfer (Horizontal Gene Transfer) of resistance genes between bacterial taxa (Chen et al. 2024).

Composting of chicken manure decreased some and increased other antibiotic-resistance genes and pathogenic bacteria (Zhang et al. 2024a), and the authors emphasize that the compost quality with respect to this needs to be monitored and managed. The effect of composting on plant pathogens and nematodes in soils happened through multiple and complex processes, thus with a potential increase in plant growth and yields (Bouchtaoui et al. 2024). However, the involved processes are seldom well known. Different composts act differently on soil-borne pathogens, due to the different components in the compost, different treatments, and protocols for the use of mature composts. Metagenomics may be the best tool to characterize the involved microorganisms, transcriptomics can show which genes, and proteomics which proteins that are involved in inducing and responsive processes in microorganisms and plants (Bouchtaoui et al. 2024)..

**Conclusion 4: Future Prospects for Potential Pathogens in Composts**
*The potential content of pathogens in compost is one of the major obstacles in the future.*

*Several new pathogens, for example COVID-19 and zoonoses, should be eliminated before composts are distributed.*

*Ultra-high-temperature composting is an important step, which needs more research with respect to bacteria and other composting factors.*

## 5 Decreasing Organic and Inorganic Contaminants and Pollutants

Harmful metals, medicines, including antibiotics, pesticides, herbicides, PFAS, PCBs, polycyclic aromatic hydrocarbons (PAHs), and many other compounds can be found in wastes. To make safe composts, the best way will be to analyze the critical compounds and discharge fractions that may pose risks. Analytical tools are continually improving, and in the future, we will be able to detect even trace concentrations of an increasing number of compounds. In some cases, we still do not know all the compounds that warrant our attention, but identifying them will be an important area of scientific advancement.

Plastic is a problem in composts and remediation for this pollutant is not easy to obtain. However, Tran et al. (2022) reported that composting is effective in degrading toxic plastic softener phthalates. The degradation process was enhanced by free radical reactions (Fenton-like reactions) in co-composting of biodegradable plastic bags and food residues (Lu et al. 2024). Gonzalez-Victoriano et al. (2024) studied the compostability of plastics used for packaging (films), disks, and forks and found promising results only for plastic bags. However, even when composting is possible, potential toxicity of the compost products needs to be clarified (Wang et al. 2023).

Kou et al. (2024) report that kitchen compost–derived organic matter was the main component affecting the degradation of chrysene, a polyaromatic hydrocarbon compound, and bacteria from the genera *Bacillus*, *Arthrobacter*, *Pseudomonas*, *Lysinibacillus*, and *Acinetobacter* were related to positive effects by the compost.

Plans should be implemented to treat new pollutants as they emerge. Addressing pollutants in compost requires many approaches, including source reduction and physical and chemical treatments. How to retrieve metals from biomass will be a most wanted technology, both to be able to reuse the valuable metals and to decrease human exposure to metals when compost is used to support plant growth.

**Conclusion 5: Future Prospects for Contaminants and Pollutants**
*The contaminants should be removed in priority of: (1) Removal or remediation before composting, and (2) Establishing compost conditions that lead to degradation of the pollutants.*

## 6 Decreasing Climate Gas Emissions

Products of composting include nutrients, metals, $CO_2$, $H_2O$, and humic substances, like fulvic acid, humic acid, and humin (Wang et al. 2024a). Thus, although composting releases greenhouse gases (GHG), the extent of emissions and losses of nutrients should be restricted as much as possible. Lack of $O_2$ may lead to anaerobic processes, with unwanted methane ($CH_4$) and $N_2O$ release. The loss of N from composting is a great problem, from both the agronomic and environmental point of view. Therefore, measures to decrease N losses in composting should be an important target for research and development. Nitrogen accumulation and the inhibition of N losses is an important aim in composting (Hoang et al. 2022). The increase of $N_2$-fixing bacteria in compost can enhance the N content in the compost (Pepe et al. 2013), decreasing agricultural dependency on mineral N fertilizers. Li et al. (2024) ascribed the increase in N content in their compost to the increase of bacteria in *Chryseolinea*. In composting pig manure, a significant decrease in $NH_3$ and P losses was observed by Gu et al. (2024).

Different feeding materials gave different emissions of $CO_2$, $N_2O$ and $NH_3$ from windrow composting (Rose et al. 2024). Tian et al. (2024) observed smaller emissions of climate gases in composting as affected by $Fe^0$ nanoparticles, through effects on microbe consortia composition (decreasing methanogenic bacteria) and Reactive Oxygen Species (ROS)–mediated climate gas production. Xiao et al.

(2024) used the Germination Index (GI) with a limit of 80% as an indicator of (sufficient) compost quality, at which the losses of $CO_2$ and $NH_4^+$ were decreased by 21% and 22%, respectively, as compared to longer composting time.

Tao et al. (2024) showed that nanobiochar of corn and rice could contribute to a higher content of carbon in compost, higher $NH_4^+/NO_3^-$ ratio, more adsorbed N, and, in general, better compost quality. They attributed the positive effect of nanochar in the composting to the high surface area, functional groups on the particle surfaces, the pores of biochar, and more beneficial microbes as compared to composting without nanochar. Xu et al. (2023) observed a strong increase in ammonification and decreased $N_2O$ and $NH_3$ losses in composting cow manure by use of ammonifying bacteria.

The best way to mitigate GHG emissions from composting is to monitor and optimize the process by ensuring adequate conditions for oxygen supply, temperature, and moisture content. This helps maintain a stable aerobic environment and prevents the onset of anaerobic processes. Recommendations should distinguish between measures that provide efficient and lasting reductions in emissions, and those that offer only short-term benefits with limited long-term impact. Machine learning may become a tool for monitoring and finding the best targets for decreasing emissions from composting (Wang et al. 2024a).

Growing media have traditionally been produced using peat extracted from bogs. Peat is almost ideal for this use, due to its high porosity, excellent water-holding capacity, low density, and low nutrient content. It is a clean material that decomposes without generating problematic byproducts. However, peat bogs store large quantities of carbon and host unique ecological features that have been dramatically diminished. They are also habitats for species adapted to wetlands and may play an important role in water retention and regulation within landscapes. For these reasons, peatlands are increasingly being protected, and natural peat may soon be banned for use in growing media and other products. In the search for peat-free alternatives, a variety of raw materials have been tested (Barrett et al. 2016; Sitzmann et al. 2025) and compost has emerged as a promising option. However, not all compost fractions are suitable for use in growing media (Vandecasteele et al. 2021), or their use implies long-distance transport. The search for viable substitutes in growing media must continue, and compost remains one of the most promising candidates. By replacing peat with compost, we not only reduce carbon emissions but also contribute to the conservation of biodiversity through the protection of peat.

**Conclusion 6: Future Prospects for Climate Emissions Amelioration**

*Composting is producing large quantities of GHG and often implies too large losses of both carbon and nutrients, especially nitrogen. Research should be directed to find improvements in the composting technology that decrease losses and emissions. How to regulate the many processes in the nitrogen cycle may be the most important factor.*

## 7 Quality During and After the Composting Process

Increasing attention will be placed on the ecological quality of compost, particularly regarding the chemical content and benefits of compost uses. This quality depends primarily on the characteristics of the input materials used in the composting process. Thus, we must know the chemical status of the biomass and the composting facilities needed in order to be able to document the state of the materials they receive. Limit values for allowable metal contents are well established, but they may be revised in the future, as our understanding of the effects evolves and analytical techniques improve in accuracy and methodology. For most plant nutrients, analysis is relatively straightforward, although further advancements are expected. However, the plant availability of nutrients in compost needs to be better described and predicted, as this is essential for determining the appropriate timing and application rates for different crops.

**Conclusion 7: Prospects for Quality**
*Process quality may be improved through research on composting. Product quality can be improved through attention to feeding materials, cleaning for metals and organic contaminants, and optimizing degradation processes during composting. The aim should be that users of compost should never have a reason to fear that product quality will be an issue.*

## 8 Geographic and Climatic Challenges

Geography and climate are strongly coupled not only between regions in the world but also between regions within countries and even between sites separated by very small distances, where precipitation and temperatures can be very different. This is especially the case in transects from coastal areas to inland and mountainous landscapes. On the seaside of mountains, the precipitation is usually rich, whereas on the opposite side of mountain ridges, precipitation may be extremely low. Both temperature and humidity in such landscapes are very different, even over small distances. Lack of water and low ambient temperatures may pose problems for composting, decreasing its efficiency, which add costs to adjust for suboptimal conditions. In such areas, composting in sheltered and controlled conditions may be the best option, considering issues like efficiency of composting processes, economy, and environment.

In general, low temperatures decrease microbial life in soils, but the presence of compost and biochar in the frost-exposed soils changed the microbial composition (toward degradation of the soil chemicals), with the largest effect when the two amendments were applied together (Shi et al. 2024). The combined effect of compost and biochar supply to a soil in freeze-thaw cycles gave the best result with respect to degradation of glyphosate, imidacloprid, and pyraclostrobin, mediated by adaptation of soil bacteria and fungi (Shi et al. 2024). Leaching of chemicals (pesticides) is affected by soil freezing (Holten et al. 2019), probably through effects of freezing on sorption effects in the soil. Probably, multiple interactions can be found, between climate, soil quality parameters, and additives such as compost.

Low temperatures at high and low latitudes, as well as at high elevations, can significantly affect microbial activity. During certain periods of the year, harsh climatic conditions can slow down or even halt the degradation and mineralization of organic matter. The use of compost feed additives may increase efficiency during the first stages of chicken manure composting. Bacteria present in soldier fly larvae frass have been shown to induce the thermophilic phase of composting despite the low temperatures (Zhang et al. 2024b). When composting is conducted under controlled conditions, temperature-related challenges can be strongly reduced. In well-developed composting systems, the compost energy may be recirculated to keep the optimal temperatures in the different stages of composting, compensating for low ambient temperatures. Sensor technologies and smart solutions may find good uses here.

Composting demands large quantities of water, especially in the thermophilic phase. Thus, in dry regions water may be a crucial minimum factor. In the future, water will probably be an even more valuable resource, maybe to an extent that composting will be costly at some places. However, the optimization of temperature and humidity will be best monitored and adjusted in closed-compartment composting.

Climate is also an important factor for the use of compost. Mineralization of nutrients in the soil and plant availability of nutrients depend on soil temperatures. Thus, the climatic conditions give a premise for the timing of application of composts to soils, and also determine what quantities are optimal for plant growth and for the avoidance of nutrient runoff.

**Conclusion 8: Prospects for Adaptations to Cold and Dry Environments**
*Low temperatures may decrease composting during winter, and water may be a scarce resource. The composting conditions should be adapted to conserve energy and recycle water lost from the composting piles.*

*Climate and soils should be evaluated when deciding the best practice for compost use.*

## 9 Composting in the Future Urbanized Societies

Urbanization is progressing at dramatic rates in the whole world, but mainly in developing countries. Cities must be transformed into smart and sustainable "Biocities" (Scarascia-Mugnozza et al. 2023), with short transport distances of both commodities into and wastes out of the city and the region. To make this happen, both production and waste treatment must be on the site where the resources can be reused and recirculated and used in new products. Production of food may be in closed compartments, on roofs, or other available spaces, close to where people live, work, and thrive in green urban environments. Composted waste will be used as fertilizers and growing media for the plant production that nourish the people and increase the citizens well-being and health in urban green areas. The composting in cities, as an important NBS in waste treatment, must be coordinated so that collection of input materials to the composting facilities, the process, and the distribution of compost are securing quality and supply to where the compost is needed. Energy exchange can be established between composting facilities, urban greenhouses, other food production units, and housing of the people. Thus, a flow of renewable energy, water, nutrients, and other resources will benefit the future urban and green societies. Composting can start in the housing, as bokashi or in areas outside the residents, if there is space for this. However, in densely populated areas, food residues, brown water, and blackwater will probably still need to be transported to centralized renovation and composting facilities, although hybrid solutions be developed. A holistic approach is necessary to give the good NBS and to use resources in the best manner.

## 10 Sustainable Intensification in Food Production: Demanding Cooperation and Co-creation

There is great attention to using renewable resources in all sectors of our societies, and often agriculture is, rightfully, the main actor for use of compost. However, changes in societies call for the attention to new and sustainable food production systems, adapted to the small available spaces and closed production systems. Co-creation in this area includes actors working on collection of organic waste, actors treating waste, users of composts, governments, actors doing planning and establishments of urban green areas and food production, many different local enterprises, and, finally, the citizens. It will be difficult to find the best solutions without this broad cooperation and co-creation. Not only specialized competences are needed in the many links of the treatment chain, but also groups that need to adjust attitudes and conceptions toward composting and compost use. All must contribute to the processes so that composting is accepted and incorporated as the useful NBS it is.

In rural agriculture and horticulture, compost will be an important resource for the sustainable intensification of food production. The replacement of mineral fertilizers and potentials for remediating and sustaining healthy soils emphasizes the importance of this NBS as a contributor to the future food systems. The division between conventional and organic farming will diminish in the future and compost use will be adopted in all production systems. However, the conditions for this to happen will be high-quality and safe composts with competitive prices as compared to other fertilizers and soil amendments.

## 11 Needed Development in the Future

Composting is a natural process that has existed for as long as life has been on Earth. However, the modern use of composting technology addresses pressing environmental and agricultural challenges, including the need for resource recirculation, the degradation of organic contaminants, the mitigation of GHG, and the demand for high-quality products. Achieving these ambitious goals requires research and development. Composts that positively impact soil health and nutrient supply. will remain a central focus in the future. In addition, novel uses, like, for example, remediation of soil pollutants, sources of organic fertilizers, peat-free growing media, and the use of compost as soil amendments should be further explored and practically developed to maximize the delivery of ecosystem services (ESS).

### *11.1 Unlocking Potential for Sustainable Intensification of Food Systems*

To promote sustainable intensification in horticulture and agriculture, the remediation and utilization of marginal soils must be one of the key strategies (Schröder et al. 2017). Better soil health and stronger plants, more resistant to diseases, are important goals in agriculture and horticulture. The use of composts and organic materials contributes to these objectives by enhancing the soil water storage capacity and increasing the content of humic acids, which in turn improve soil structure and plant nutrient supply. As also discussed elsewhere in this book, composts also affect plant pathogens (Bouchtaoui et al. 2024), an effect that should be used more in both conventional and organic food systems (Schröder et al. 2017). Compost application supports microbial soil life and affects the balance between pathogenic and beneficial bacteria. Achieving the ambitious aims for composting and its use will require a deeper understanding of the underlying biological processes and the role of specific microbes that contribute to desired outcomes.

### *11.2 Pretreatments*

Pretreatments and additives to the compost feeding materials should be more developed to increase efficiency and increase ESS from composting. Additives may consist of microbial consortia or other materials that contribute to the development of composts with high quality. Biochar with a high proportion of nanoparticles may be a factor that increases the quality of the compost-based fertilizers by improving cation exchange capacity (CEC) of growing media and soils, ammonium, and water retention. Yang et al. (2024b) observed that application of 0.5 g $SiO_2$ nanoparticles could decrease N losses from the composting of swine manure by 72%

and copper resistance genes by 29%. A very interesting prospect would be if other nanoparticle materials could also affect N losses and contribute to the optimization of composting.

Other additives could be from the food industry, for example vinasse or whey (side products from the distilling industry and yeast production, respectively). In research conducted in Poland, one of the potential sources of organic matter to be used in soil fertilization is also brown coal (lignite), of which there are significant reserves. Composting of lignite with the addition of nitrogen-rich organic agricultural waste is a method for accelerating the decomposition of humic substances and producing products useful for improving the soil structure, while having good fertilizing properties. The experimental compost was made from lignite-rich brown coal (80%) with the addition of 10% vinasse. This mixture is inoculated with the mycelium of the oyster mushroom (*Pleurotus ostreatus*) or the shiitake mushroom (*Lentinus edodes*), which have the ability to quickly degrade lignin. The finished compost contain over 50% organic carbon and small amounts of N. A similar product is compost made from lignite-rich brown coal (80%) with the addition of 19% whey. The finished composts contain over 50% organic carbon and small amounts of N. The composts improve soils by increasing the retention of moisture and reducing evaporation, which contributes to the improvement of the 'crumbly' structure A-layer of soils and can decrease erosion. The increase in soils of organic matter and humus brings benefits for the development of microflora. The tannins contained in the compost form compounds that chelate heavy metal cations, thus reducing their mobility in the soil. They also bind N compounds, which prevents rapid leaching from the soil.

## *11.3 How Can Quality Be Predicted and Described?*

The review by Onwosi et al. (2017) showed that maturation indexes are not always informative, only describing a part of the maturity complex. Factors used as indexes are respiration capacity, odor, phytotoxicity as described by germination indexes, C/N-ratio, $NH_4/NH_3$, $NH_4$-N/$NO_3$-N, pH, electrical conductivity (EC), water-soluble-C, volatile-organic C, cation exchange capacity (CEC), dissolved organic matter, electron transfer capacity (ETC), and more (See Onwosi et al. 2017 and references therein). The authors concluded that a set of indicators should be examined for efficient reporting of compost maturity, rather than presenting single indicators. Germination index was found to be affected by the presence of acetic acid, formic acid, and ammonium (Du et al. 2024a). The authors showed that the composting process is important for many compost quality parameters. However, the composition of the feedstock is also responsible for crucial factors, such as the metal content, which can result in higher concentrations in the final compost. Simple, easy-to-use indicators, such as organic matter content, can now be measured with sensors, showing promising results (Carvalho et al. 2024).

Smart solutions, using sensor technologies and remote sensing, may increase data collection and precision in composting management, while also decreasing operational costs. New sensors and AI that are more specific, accurate, and cheaper than earlier, and the development and adaptation of them to support monitoring and information on smart composting represent a great opportunity to improve the technology and compost quality. Analysis of Big Data is also an opportunity to direct composting to support market demands and make strategies on how to decrease waste and support the transition toward a more circular economy.

## *11.4 Combining Compost Technologies*

The combination of different composting technologies may give added value to compost quality. Zhou et al. (2024) showed that a combination of hyperthermophilic composting as a pretreatment, followed by vermicomposting, can combine the advantages of the two methods for transforming waste into high-quality composts. Using different technologies and tools for composting may advance progress in the treatment of wastes in both agricultural and urban settings. Sun et al. (2024) observed that a semi-permeable cover on the compost pile could increase the composting temperature (maximum 82 °C) by 15 °C, compared to the control without a membrane. The elevated temperature contributed to accelerating the degradation of antibiotic-resistance genes and mobile genetic elements.

## *11.5 Decreasing Pollutants*

We know too little about organic pollutants that accumulate in biomass. Xenobiotics, including antibiotics, pesticides, herbicides, PFAS, PCBs, and PAHs, are only some of the compounds that can be found in waste (Rivera-jaimes et al. 2018). It is urgent to map the possibility of exposure to harmful pollutants that may affect humans and our surrounding environments. Isotope labeling (Wang et al. 2024a) and molecular methods (Bouchtaoui et al. 2024; Zhang et al. 2024) will be important tools to elucidate mechanisms for the transformation of organic materials during composting and to describe compost quality, including pathogens and antibiotic-resistance genes.

## *11.6 Valuation*

Once high-quality composts have been produced, the next crucial step is to utilize this valuable resource effectively. Composting systems vary widely—from small-scale operations run by garden-interested citizens using cold or warm composting methods for personal land use, to large composting facilities that produce biogas, digestate, and compost for use in commercial production. These larger facilities must focus on developing competitive products that generate revenue sufficient to cover operational costs.

On one hand, the primary products include soil amendments relatively low in nutrients, and fertilizers without any adjustments to nutrient content (organic fertilizers) or enriched with nutrients to produce balanced fertilizers tailored to plant needs. In organic agriculture, nutrient sources are derived from organic materials and rock materials (Panday et al. 2024), most of them with nutrients that either rely on mineralization or weathering before nutrients become plant-available. To meet the demands of modern agriculture and circular economy goals, compost quality must move beyond variability caused by inconsistent feedstock composition. Instead, production should be optimized and targeted to achieve specific nutrient profiles and functional qualities in the final compost products.

On the other hand, composts that are nutrient-poor and unsuitable for food production can be used for other purposes. They can be used as soil improvers for restoration of degraded areas such as mining or industrial areas, and places contaminated with organic or inorganic pollutants. These composts can switch soil properties and increase the survival rate of bacteria, fungi, and plants, which can in turn transform and adsorb pollutants. This action can reduce pollutant emissions and their transfer from soil to the surrounding surface and groundwater.

## *11.7 Organic and Organo-Mineral Fertilizers*

The transformation of digestates and compost into fertilizers has received great attention but will need further development in the future to meet the Regulation (EU) 2019/1009, which classifies the organic matrixes into Product Function Categories. Fertilizers derived from compost may be used in pelletized form and applied in their natural state—based solely on compost—or further enriched with nutrients. For instance, nitrogen levels can be increased by adding blood meal; potassium through potash; phosphorus through struvite; and calcium and pH through lime. Micronutrients can be supplemented using various types of rock meals, and in some cases, synthetic mineral fertilizers may be incorporated to enhance nutrient balance. Utilizing industrial and mining by-products in this way not only improves the nutrient profile of the compost but also contributes to reducing environmental pollution and resource waste. Additionally, N-rich materials can be co-composted with C-rich substrates, such as lignite (brown coal) or by-products from the wood industry, like sawdust, to create more balanced and stable compost products.

A key objective is to develop organic and mineral-organic fertilizers that are economically viable, practical, and safe for farmers and gardeners to use. One common issue with fertilizers based on composted manure is their unbalanced nutrient profile—specifically, a high phosphorus (P) to nitrogen (N) ratio. Fertilization planning in the EU is typically based on nitrogen content, as mandated by the

Nitrates Directive 91/676/CEE. As a result, using these compost-based fertilizers can lead to over-application of phosphorus, increasing the risk of eutrophication in nearby water bodies. A more sustainable approach would be to plan fertilization based on phosphorus content and then supplement with additional nitrogen. However, in practice, farmers often prefer broad-spectrum fertilizers that provide balanced nutrients in a single application to minimize labor and fuel costs. One promising solution is the development of "hybrid fertilizers"—organic fertilizers enriched with mineral nutrients—or the use of multifunctional machines capable of delivering several fertilizers simultaneously, ensuring nutrient balance tailored to crop needs. Additives such as bone meal, blood meal, and biochar can be included depending on the desired properties of the final product—whether used as a fertilizer, growing medium, or soil amendment.

Newly developed fertilizers can be supplemented with microorganisms as single strains or consortia which favor the reduction of N and P losses from soil by increasing nutrient bioavailability. This management method contributes to increased organic matter content and biodiversity of microorganisms in arable and other agricultural soils. There are many factors leading to increased soil fertility and productivity. The addition of selected strains can improve some properties of compost. For example, the plant rhizospheric bacterium *Bacillus velezensis* SQR9, originating from plant rhizospheres, can modify soil N transformation processes. It significantly enhances the soil $N_2O$ reduction rate, thereby mitigating the soil $N_2O$ emission flux. Furthermore, a significant positive correlation exists between its inoculation concentration and the magnitude of soil $N_2O$ emission mitigation. The *Bacillus velezensis* SQR9 can be applied to fields to mitigate soil $N_2O$ emissions.

The use of innovative fertilizer products based on compost enriched with beneficial bacteria instead of traditional mineral fertilizers to provide nitrogen nutrients for crops can also promote the growth of crop roots, improve crop yield, and greatly reduce $NH_3$ and $N_2O$ emissions generated by traditional nitrogen fertilizer application. This can contribute to decreasing the problem of excessive ammonia and greenhouse gas emissions in agriculture. The technology for recovering N and P from biomass waste and balancing nutrients in fertilizers will contribute to recycling widely available organic waste, and to reducing the losses of these nutrients to the environment.

## *11.8 Growing Media*

Plant production in greenhouses, on balconies, and in rooftop gardens relies heavily on lightweight growing media, traditionally based on peat. However, the use of peat is increasingly restricted due to the urgent need to protect peatlands for their vital ecosystem services: they serve as significant carbon sinks, help regulate water flow to mitigate flooding, and support unique biodiversity. As a result, there is a growing need to identify sustainable alternatives to peat.

Compost is emerging as one of the most promising components in the development of peat-free growing media. While this application may not absorb a large share of the total compost produced, it offers a valuable revenue stream for composting facilities. Moreover, replacing peat with compost in horticulture contributes meaningfully to climate change mitigation by reducing GHG emissions and preserving C-rich ecosystems. To meet market demands, compost must be produced with selected and consistent quality, in substantial volumes, and with stable supply chains aligned with industrial requirements (Barrett et al. 2016). However, certain compost characteristics can limit its suitability for specific applications. For example, high pH and electrical conductivity (EC) can reduce the value of compost as an ingredient in peat-free growing media. Onwosi et al. (2017) suggest an ideal EC range of 2.0–3.0 mS $m^{-1}$, a threshold that is often challenging to meet. Furthermore, a pH above 7 can inhibit the uptake of certain essential nutrients by plants.

Hashemi et al. (2024), through a life cycle assessment (LCA), demonstrated that replacing peat in growing media with compost and other sustainable components can significantly reduce greenhouse gas emissions. This makes the targeted production of compost—using only feedstocks that deliver desired properties—an attractive and environmentally beneficial option. Importantly, compost intended for horticultural substrates can generate strong revenues for composting facilities, adding economic value in addition to climate benefits.

Moreover, compost and its derived products contribute to improving and sustaining soil health and productivity. As reviewed by Xu et al. (2024), compost-based biochar and specific compost additives can promote the safe recirculation of agricultural waste materials. Biochar, in particular, facilitates the immobilization of metals by bonding them to negatively charged particles. The addition of phosphorus compounds may further help to bind complex potentially toxic elements (PTEs) such as copper (Cu), zinc (Zn), and heavy metals like cadmium (Cd), lead (Pb), cobalt (Co), and arsenic (As). However, the authors emphasize that the long-term effects of such practices must be thoroughly investigated. Enriching soils with metals poses a serious risk—if these accumulate to harmful levels, they could impair plant growth and compromise food safety. Thus, the primary objective should be to extract and remove PTEs from waste streams, thereby preventing their deposition in soils.

Maintaining soil organic matter is essential for productivity and resilience, traditionally achieved using manure and organic fertilizers. However, compost can serve as an alternative to raw manure and is widely used to improve soil quality. Composts based on plant materials and animal excreta from non-intensive systems are approved for use in organic farming. Yet, changes in livestock production technologies have reduced manure availability in many regions. Additionally, manure from intensive animal farming—such as large-scale pig, poultry, or cattle operations—often contains residues of medicines, antibiotics, and feed additives, rendering it unsuitable for organic agriculture. This challenge underscores the need to develop and implement alternative strategies for maintaining soil fertility that are safe for human health and the environment.

## 11.9 Need for Fundamental Knowledge

There is a growing need for fundamental knowledge to advance composting processes to the next level. Developing improved composting technologies and practices will require new scientific approaches and a deeper understanding of the biological and chemical processes involved. Key challenges—such as reducing nitrogen losses and managing contaminants—may depend on understanding in situ processes occurring within the various composting compartments, even at the microscale (Yang et al. 2024a). This also includes the potential use of targeted microbial inoculants to optimize decomposition and contaminant degradation (Ji et al. 2023; Zhu et al. 2023).

## 11.10 Stake Holder Engagement: Governments, Public Institutions, Enterprises, and Citizens

Relying solely on knowledge generation and dissemination to drive progress in composting is overly simplistic. Effective implementation requires active communication and collaboration with stakeholders—including governments, public institutions, businesses, and citizens. Countries with established waste management systems and strong economies bear a particular responsibility to lead by example. They must support the development and transfer of effective regulatory frameworks, technological innovations, and financing mechanisms.

Governments have a crucial role in establishing strict regulations for waste treatment and compost quality standards for different end uses. Such regulations can drive innovation in both composting technologies and applications. For instance, requiring a transition from non-recyclable to compostable products would pressure industries to innovate. However, it is equally important to assess how truly degradable these products are and to identify potentially toxic intermediates or end products that may form during composting.

## 11.11 Toward Holistic, Sustainable Solutions

A multidimensional approach is essential. Every actor in society must take responsibility for the full life cycle of the materials they produce or consume. Though this systemic transformation is challenging, holistic solutions represent the only viable path forward. The future of waste treatment must focus on minimizing waste generation, improving recycling efficiency, and creating innovative, sustainable uses for organic materials.

Achieving this vision will require coordinated action from governments, businesses, and local communities. Only through collective responsibility and integrated strategies can we ensure that waste management becomes truly sustainable and beneficial to both society and the environment.

**Conclusion 9: Prospects for Development of Methods and Compost Products** *Research and development must continue in all parts of the processes: (1) decreasing and adapting waste production, (2) collecting feeding materials and additives, of biotic and abiotic natures, (3) composting process adaptations, (4) development of compost products, (5) analysis of compost and product quality and, finally, (6) valuation and marketing of compost and compost-based products. Overarching work will be to include all stakeholders and specialists to "cultivate" the ground for the good solutions.*

## References


Arrigoni JP, Paladino G, Garibaldi LA, Hedenström E, Zhang W, Laos F (2024) Performance of small-scale composting in low ambient temperatures: effects of adding animal by-products and recycling leachates. Waste Manage Bull 2:309–317. Journal homepage: www.elsevier.com/locate/wmb

Barrett GE, Alexander PD, Robinson JS, Bragg NC (2016) Achieving environmentally sustainable growing media for soilless plant cultivation systems – a review. Sci Hortic 212(2016):220–234

Bernardi AD, Bandini F, Marini E, Tagliabue F, Casucci C, Brunetti G, Vaccari F, Bellotti G, Tabaglio V, Fiorini A, Ilari A, Gnoffo C, Frache A, Taskin E, Rossa UB, Silva E, Ricardo L, Martins MO, Duca D, Puglisi E, Pedretti EF, Vischetti C (2024) Integrated assessment of the chemical, microbiological and ecotoxicological effects of a bio-packaging end-of-life in compost. Sci Total Environ 951:175403

Bouchtaoui EM, Haouas A, Dababat AA, Lahlali R, Benali A, Fahr M, Smouni A, Azim K, Liu Z, Li J, Mokrini F (2024) Exploring mechanisms of compost-mediated suppression of plant pathogens: a critical review. Appl Soil Ecol 203:105644

Browne AJ, Chipeta MG, Haines-Woodhouse G, Kumaran EPA, Kashef Hamadani BH, Zaraa S, Henry NJ, Deshpande A, Reiner RC, Day NPJ, Lopez AD, Dunachie S, Moore CE, Stergachis A, Hay SI, Dolecek C (2021) Global antibiotic consumption and usage in humans, 2000–18: a spatial modelling study. Planet Health 5:e893

Carvalho GS, Weindorf DC, Sirbescu MC, Ribeiro BT, Chakraborty S, Li B, Weindorf WC, Acree A, Roberto L, Guilherme G (2024) Prediction of compost organic matter via color sensor. Waste Manag 185(2024):55–63

Chen X, Song X, Liang Y, Wang F, Pan C, Wei Z (2024) Evaluation of the potential horizontal gene transfer ability during chicken manure and pig manure composting. Environ Pollut 360:124621

Colombini G, Senouci F, Rumpel C, Houot S, Biron P, Felbacq A, Dignac M-F (2024) Coarse microplastic accumulation patterns in agricultural soils during two decades of different compost application. Environ Pollut 363:125076

Du S, Ding S, Wen X, Yu M, Zou X, Wu D (2024a) Investigating inhibiting factors affecting seed germination index in kitchen waste compost products: soluble carbon, nitrogen, and salt insights. Bioresour Technol 406:130995

Du S, Zhang M, Zhang S, Wen X, Wang Y, Wu D (2024b) Evaluation of the quality of products from multiple industrial-scale composting treatment facilities for kitchen waste and exploration of influencing factors. Environ Res 262:119899

Ezugworie FN, Igbokwe C, Onwosi CO (2021) Proliferation of antibiotic-resistant microorganisms and associated genes during composting: an overview of the potential impacts on public health management and future. Sci Total Environ 784:147191

Ezzariai A, Barret M, Merlina G, Pinelli E, Hafidi M (2017) Evaluation of the antibiotic's effects on the physical and chemical parameters during the co-composting of sewage sludge with palm wastes in a bioreactor. Waste Manag 68:388–397

Ezzariai A, Hafidi M, Khadra A, Aemig Q, El Fels L, Barret M, Merlina G, Patureau D, Pinelli E (2018) Human and veterinary antibiotics during composting of sludge or manure: global perspectives on persistence, degradation, and resistance genes (review). J Hazard Mater 359: 465–481

Gonzalez-Victoriano L, Camacho ES, Hernández-Varela JD, Arredondo-Tamayo B, Gallegos-Cerda SD, Chanona Pérez JJ, Cervantes-Sodi F, Martínez-Mercado E (2024) Single-use commercial bio-based plastics under environmental degradation conditions: is their biodegradability and compostability a fact? Sci Total Environ 955:176763

Gu X, Li H, Shi Y, Li J, Li S (2024) Regulating bacterial dynamics by Mg-modified biochar addition to mitigate gaseous emissions during pig manure composting. J Clean Prod 465: 142839

Hashemi F, Mogensen L, Smith AM, Larsen SU, Knudsen MT (2024) Greenhouse gas emissions from bio-based growing media: a life-cycle assessment. Sci Total Environ 907:167977

Hoang HG, Thuy BTB, Lin C, Vo D-VN, Tran HT, Bahari MB, Le VG, Vu CT (2022) The nitrogen cycle and mitigation strategies for nitrogen loss during organic waste composting: a review. Chemosphere 300(2022):134514

Holten R, Larsbo M, Nicholas J, Stenrød M, Almvik M, Eklo OM (2019) Leaching of five pesticides of contrasting mobility through frozen and unfrozen soil. Vadose Zone J 18: 180201. https://doi.org/10.2136/vzj2018.11.020

Ji Y, Cao Y, Wang Y, Wang C, Qin Z, Cai W, Yang Y, Yan S, Guo X (2023) Effects of adding lignocellulose-degrading microbial agents and biochar on nitrogen metabolism and microbial community succession during pig manure composting. Environ Res 239:117400

Kou B, Huo L, Cao M, Yu T, Wu Y, Hui K, Tan W, Yuan Y, Zhu X (2024) Applying kitchen compost promoted soil chrysene degradation by optimizing microbial community structure. J Environ Manag 370:122761

Lan K, Zhang B, Lee T, Yao Y (2024) Soil organic carbon change can reduce the climate benefits of biofuel produced from forest residues. Joule 8:430–449

Li B, Sardar MF, Zhang X, Ye J, Tian Y, Song T, Wang X, Zhu C, Li H (2024) Electrokinetic technology enhanced the control of antibiotic resistance and the quality of swine manure composting. Chem Eng J 484:149581

Lu J, Qiu Y, Zhang L, Wang J, Li C, Wang P, Ren L (2024) Effects of Fe3O4 NMs based Fenton-like reactions on biodegradable plastic bags in compost: new insight into plastisphere community succession, co-composting efficiency and free radical in situ aging theory. J Hazard Mater 467:133689

Onwosi CO, Igbokwe VC, Odimba JN, Eke IE, Nwankwoala MO, Iroh IN, Ezeogu LI (2017) Composting technology in waste stabilization: on the methods, challenges and prospects. J Environ Manag 190:140–157

Panday D, Bushal N, Das S, Ghalehgolabbehbahani A (2024) Rooted in nature: the rise, challenges, and potential of organic farming and fertilizers in agroecosystems. Sustainability 16:1530

Pepe O, Ventorino V, Blaiotta G (2013) Dynamic of functional microbial groups during mesophilic composting of agro-industrial wastes and free-living (N2)-fixing bacteria application. Waste Manag 33(2013):1616–1625

Ranius T, Hamäläinën A, Egnell G, Olsson B, Eklof K, Stendahl J, Rudolphi J, Stens A, Felton A (2017) The effects of logging residue extraction for energy on ecosystem services and biodiversity: a synthesis. J Environ Manage 209:409–442

Rivera-Jaimes JA, Postigo C, Melgoza-Alemán RM, Aceña J, Barceló D, López de Alda M (2018) Pharmaceutical residues and xenobiotic contaminants: occurrence, analytical techniques and sustainable alternatives for wastewater treatment. Sci Total Environ 639:1532–1553. https://doi.org/10.1016/j.scitotenv.2018.05.206

Rose HM, Jain N, Misra SD, Kumar O, Prasad K, Purakayastha TJ, Bhatia A, Kumar SN, Sethi S (2024) Open windrow composting of lignocellulosic crop residues and neem litter: accounting for reactive nitrogen and greenhouse gas emissions. J Clean Prod 478:143964

Scarascia-Mugnozza GE, Guallart V, Salbitano F, Ottaviani Aalmo G, Boeri S (2023) Transforming biocities designing urban spaces inspired by nature. Future City 20. ISSN 1876-0899 ISSN 1876-0880 (electronic) Future City ISBN 978-3-031-29465-5ISBN 978-3-031-29466-2 (e-Book). https://doi.org/10.1007/978-3-031-29466-2

Schröder P, Beckers B, Daniels S, Gnädinger F, Maestri E, Mench M, Gomez RM, Obermeier MM, Oustriere N, Persson T, Poschenrieder C, Rineau F, Rutkowska B, Schmidt-Sutter T, Szulc W, Witters N, Sæbø A (2017) Intensify production, transform biomass to energy and novel goods and protect soils in Europe – a vision how to mobilize marginal lands. Sci Total Environ. https://doi.org/10.1016/j.scitotenv.2017.10.209

Shen B, Zheng, Zheng X, Yang Y, Xiao D, Wang Y, Sheng Z, Ai B (2024a) Insights from *meta*-analysis on carbon to nitrogen ratios in aerobic composting of agricultural residues. Bioresour Technol 413:131416

Shen B, Zhang X, Zhao Y, Tao W, Wei Z, Song C (2024b) Investigating the effect of Fenton-like pretreatment-clay mineral addition on humic substance during straw composting. Chem Eng J 496:154199

Shi G, Hou R, Fu Q, Li T, Chen Q (2024) Effects of biochar and compost on microbial community assembly and metabolic processes in glyphosate, imidacloprid and pyraclostrobin polluted soil under freeze–thaw cycles. J Hazard Mater 471:134397

Sitzmann TJ, Celi L, Moretti B, Zavattaro L, Grignani C (2025) Suitability of renewable organic materials for the synthesis of organo-mineral fertilizers: driving factors and replacement of peat. Heliyon 11(4):e42529

Su Y, Zhou S, Tian P, Qi C, Xu Z, Zhang Y, Sung-Yoon Huh B, Wenhai Luo A, Guoxue Li A, Yangyang Li A (2024) Techno-economic assessment of industrial food waste composting facility: evaluating bulking agents, processing strategies, and market dynamics. Bioresour Technol 408:131210

Sun B, Bai Z, Li R, Song M, Zhang J, Wang J, Zhuang X (2024) Efficient elimination of antibiotic resistome in livestock manure by semi-permeable membrane covered hyperthermophilic composting. Bioresour Technol 407:131134

Tao R, Cui M, Li Y, Wang J, He W, Zhao Y, Xie W, Shen Y, Feng Y, White JC (2024) Nanoscale biochar for fertilizer quality optimization in waste composting: microbial community regulation. Bioresour Technol 414:131571

Tian P, Yang S, Yang M, Xie D, Yu H, Wang X (2024) Effect of iron-based nanomaterials on organic carbon dynamics and greenhouse gas emissions during composting process. Environ Res 263(2024):120281

Tran H-T, Nguyen M-K, Hoang H-G, Hutchison JM, Vu CT (2022) Composting and green technologies for remediation of phthalate (PAE)-contaminated soil: current status and future perspectives. Chemosphere 307:135989

Vandecasteele B, Pot S, Maenhout K, Delcour I, Vancampenhout K, Debode J (2021) Acidification of composts versus woody management residues: optimizing biological and chemical characteristics for a better fit in growing media. J Environ Manage 277(2021):111444

Wang T, Hosseinzadeh M, Cuccagna A, Alakenova R, Casademunt P, Rovatti AR, Lòpez-Rubio A, Porte C (2023) Comparative toxicity of conventional versus compostable plastic consumer products: an in-vitro assessment. J Hazard Mater 459:132123

Wang B, Zhang P, Qi X, Li G, Zhang J (2024a) Predicting ammonia emissions and global warming potential in composting by machine learning. Bioresour Technol 411:131335

Wang J, Gao X, Wang G, Liu Y, Chang J, Jiang T, Li G, Ma R, Yang Y, Yuan J (2024b) The enrichment of antibiotic resistance genes in swine manure compost was related to the bulking agent types. Environ Technol Innov 36:103765

Wang N, He Y, Zhao K, Lin X, He X, Chen A, Wu G, Zhang J, Yan B, Luo L, Xu D (2024c) Greenhouse gas emission characteristics and influencing factors of agricultural waste composting process: a review. J Environ Manag 354:120337

Wang L, Qi Y, Cao L, Song L, Hu R, Li Q, Zhao Y, Liu J, Zhang H (2024d) Promoting role of nitrogen-fixing bacteria and biochar on nitrogen retention and degradation of PBAT plastics during composting. Environ Pollut 363:125228

Wu D, Wei Z, Mohamed TA, Zheng G, Qu F, Wang F, Zhao Y, Song C (2022) Lignocellulose biomass bioconversion during composting: mechanism of action of lignocellulase, pretreatment methods and future perspectives. Chemosphere 286:131635

Xiao R, Li L, Zhang Y, Fang L, Li R, Song D, Liang T, Su X (2024) Reducing carbon and nitrogen loss by shortening the composting duration based on seed germination index (SCD@GI): feasibilities and challenges. Sci Total Environ 933:172883

Xu Z, Li R, Zhang X, Liu J, Xu X, Wang S, Lan T, Zhang K, Gao F, He Q, Pan J, Quan F, Zhang Z (2023) Mechanisms and effects of novel ammonifying microorganisms on nitrogen ammonification in cow manure waste composting. Waste Manag 169(2023):167–178

Xu Q, Zhang T, Niu Y, Mukherjee S, Abou-Elwafa SF, Nguyen NSH, Al Aboud NM, Wang Y, Pu M, Zhang Y, Tran HT, Almazroui M, Hooda PS, Bolan NS, Rinklebe, Shaheen SM (2024) A comprehensive review on agricultural waste utilization through sustainable conversion techniques, with a focus on the additives effect on the fate of phosphorus and toxic elements during composting process. Sci Total Environ 942:173567

Yang M, Zhang T, Zhou X, Jin C, You X, Zhang L, Yang Y, Kong Z, Chu H, Zhang Y (2024a) New insight into the spatio-temporal patterns of functional groups of hotspot inside the composting aggregates by synchrotron-based FTIR in hyperthermophilic composting. Sci Total Environ 945:174139

Yang W, Jiang H, Zhang L, Gu J, Wang X (2024b) $SiO_2$ nanoparticles can enhance nitrogen retention and reduce copper resistance genes during aerobic composting of swine manure. Bioresour Technol 414:131577

Zhang Y, Wang N, Wan J, Jousset A, Jiang G, Wang X, Wei Z, Xu Y, Shen Q (2024a) Exploring the antibiotic resistance genes removal dynamics in chicken manure by composting. Bioresour Technol 410:131309

Zhang R, Zhang Y, Xi Y, Zhou J, Han T, Ma Q, Wang C, Zhu F, Ye X (2024b) Effect of black soldier fly larvae frass addition on humus content during low temperature co-composting. Bioresour Technol 412:131379

Zhang J, Ding W, Wang S, Ha X, Zhang L, Zhao Y, Wu W, Zhao M, Zou G, Chen Y (2024c) Pollution characteristics of microplastics in greenhouse soil profiles with the long-term application of organic compost. Res Environ Sustainability 17:100165

Zhang J, Zou Y-J, Wang S-L, Zhang W-W, Chen Q-J, Wang Q-Y, Guan TK, Zhang J-Y, Zhao M-R, Zhang G-Q (2024d) The inoculation of *Bacillus paralicheniformis* and *Streptomyces thermoviolaceus* enhances the lignocellulose degradation and microbial communities during spent mushroom substrate composting. Environ Res 263(Part 2):120157

Zhao K, Yin X, Wang N, Chen N, Jiang Y, Deng L, Xiao W, Zhou K, He Y, Zhao X, Yang Y, Zhang J, Chen A, Wu Z, He L (2024) Optimizing the management of aerobic composting for antibiotic resistance genes elimination: A review of future strategy for livestock manure resource utilization. Journal of Environmental Management 370:122766

Zhou X, Yu Z, Deng W, Deng Z, Wang Y, Zhuang L, Zhou S (2024) Hyperthermophilic composting coupled with vermicomposting stimulates transformation of organic matter by altering bacterial community. Sci Total Environ 954:176676

Zhu L, Huang C, Li W, Wu W, Tang Z, Tian Y, Xi B (2023) Ammonia assimilation is key for the preservation of nitrogen during industrial-scale composting of chicken manure. Waste Manag 170(2023):50–61

Zhu L, Ji Y, Huang H, Liu L, Wang X, Yao X, Hu B (2024) Self-circulating pulse alternating ventilation composting technology: biotic and abiotic effects of mature compost on humification during food waste composting. Chem Eng J 498:155581

# Existing Challenges and Opportunities in Composting on Different Scales

**Zhuoya Lv and Tao Zhang**

## 1 Introduction

Composting, the process of biologically breaking down organic waste into a nutrient-rich soil amendment, has gained prominence in recent years as a sustainable waste management solution (Xu et al. 2024). This practice not only aids in the reduction of organic waste but also plays a crucial role in enhancing soil fertility, improving water retention, and promoting carbon sequestration in soils (Li et al. 2022). With the global increase in population, urbanization, and food waste generation, composting has emerged as a practical and cost-effective alternative to traditional waste disposal methods, such as landfilling and incineration. In fact, composting is increasingly viewed as a key component in sustainable agriculture and waste management, directly contributing to a circular economy model, where waste is viewed as a resource rather than a burden (Boldrin et al. 2011; Li et al. 2020).

The scale at which composting is implemented—ranging from small household systems to large industrial operations—greatly influences its effectiveness and impact. Each scale presents distinct challenges and opportunities shaped by a variety of economic, environmental, and social factors. For instance, household composting offers a decentralized, easy-to-implement approach that can reduce waste sent to landfills. However, it often faces barriers such as a lack of knowledge, space constraints, and inconsistent organic waste generation (Jara-Samaniego et al. 2017). At the community level, composting initiatives can address local waste management needs while fostering environmental awareness and collaboration.

Z. Lv · T. Zhang (✉)
Beijing Key Laboratory of Farmland Soil Pollution Prevention and Remediation, College of Resources and Environmental Sciences, China Agricultural University, Beijing, China
e-mail: taozhang@cau.edu.cn

 
R. K. Singh et al. (eds.), *Composting of Farm Residues by Nature-Based Solutions*,
https://doi.org/10.1007/978-3-032-12255-1_14

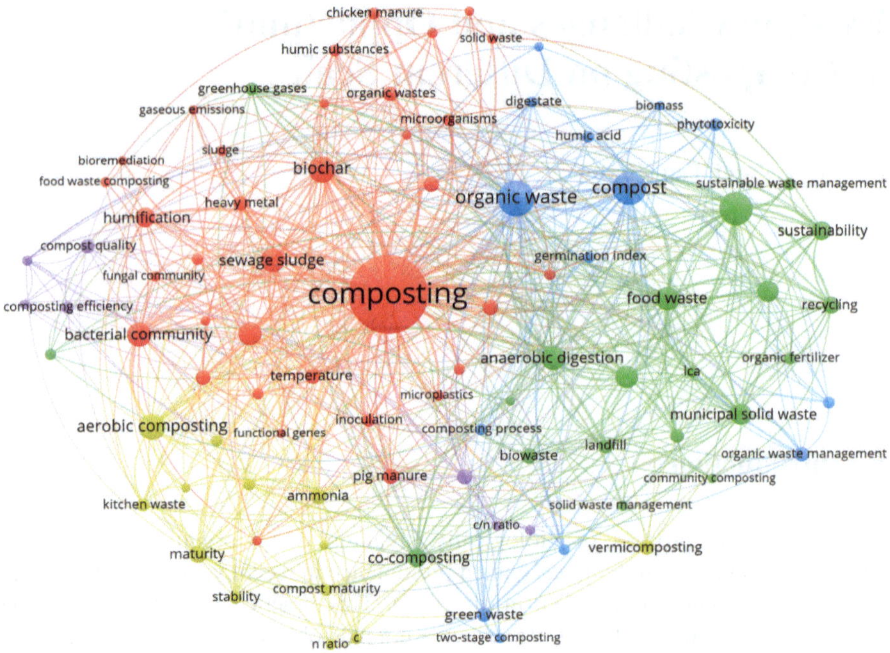


**Fig. 1** Keyword statistics of VOSviewer references on compost research

Nonetheless, these projects often encounter logistical complexities, including collection coordination, funding, and regulatory barriers (Al Mamun et al. 2020).

On an industrial scale, composting presents both challenges and significant opportunities for large-scale organic waste management. Industrial composting systems must deal with high operational costs, energy use, and managing emissions like methane and ammonia, but they also offer considerable potential for waste diversion, carbon credits, and the creation of high-value products, such as organic fertilizers and biochar (Awasthi et al. 2020). As the global community faces mounting pressure to mitigate climate change and reduce greenhouse gas emissions, composting at all scales represents a promising strategy for achieving environmental sustainability. Studies suggest that composting can reduce methane emissions from landfills, mitigate soil degradation, and enhance carbon sequestration in agricultural soils, thus contributing to climate change adaptation and mitigation efforts (Reyes-Torres et al. 2018).

Current research on composting processes has predominantly concentrated on organic waste composting and sustainable waste management. Therefore, it is crucial to comprehensively understand the challenges and opportunities associated with different composting scales in the context of sustainable development (Fig. 1). This chapter systematically explores the challenges and opportunities associated with composting at different operational scales. It highlights the technological advancements that have improved composting processes, such as smart monitoring systems and microbial inoculants, which have enhanced efficiency and product quality (Mejias et al. 2017). Furthermore, the chapter delves into the policy frameworks that support or hinder composting adoption, examining the role of regulations, subsidies, and public-private partnerships in scaling up composting efforts (Boldrin et al. 2011). Community engagement strategies, educational initiatives, and local government support are also discussed, as they play a crucial role in fostering composting practices at the household and community levels (Jara-Samaniego et al. 2017). Finally, the chapter provides insights into overcoming the various barriers to composting and leveraging this process to address pressing global issues, such as waste management, resource recovery, and climate change. Composting, despite the challenges it faces at different scales, holds substantial potential for creating more sustainable systems of waste management and resource recovery. As we move toward a more circular economy, the adoption of composting practices across various sectors is essential. Understanding the dynamics at each scale—household, community, and industrial—is critical for developing strategies that optimize composting systems and maximize their environmental, economic, and social benefits.

## 2 Challenges in Composting at Different Scales

Composting encounters distinct challenges depending on its scale. Small-scale household composting faces space limitations, odor, pests, and knowledge gaps, while community composting struggles with infrastructure maintenance and community involvement. Large-scale facilities require advanced technology, significant investment, regulatory compliance, and effective management of diverse organic materials. Table 1 shows the challenges of composting at different scales.

### *2.1 Household Composting*

Household composting, often practiced on a small scale, offers a decentralized approach to organic waste management. This method is particularly appealing as it provides an opportunity for individuals to contribute directly to waste reduction and soil enrichment. However, despite its potential to significantly reduce the amount of organic waste sent to landfills, several challenges hinder its adoption and effectiveness. At the same time, various opportunities exist to overcome these challenges and promote wider participation in household composting initiatives.

**Table 1** Challenges of composting at different scales

| Scales | Factors | Challenges | References |
|---|---|---|---|
| Household composting | Space constraints | Residents in cities often struggle to find adequate areas to implement traditional composting setups, such as open piles or large bins | De Boni et al. (2022) |
| Household composting | Time and effort | For individuals with busy lifestyles or competing responsibilities, home composting may seem burdensome, limiting their willingness to adopt composting practices | Manea et al. (2024) |
| Community-scale composting | Logistics | Inconsistent waste streams can disrupt the composting process, leading to inefficiencies and increased operational costs | Reyes-Torres et al. (2018) |
| Community-scale composting | Regulatory barriers | Regulatory frameworks often vary across regions, creating inconsistencies and administrative hurdles that discourage smaller initiatives from starting or expanding | Mejias et al. (2017) |
| Industrial-scale composting | High initial investment | The financial barrier is a significant deterrent to the expansion of industrial composting in developing countries, where waste management systems are already underfunded | Manea et al. (2024) |
| Industrial-scale composting | Energy and resource use | Balancing operational efficiency with resource conservation remains a pressing challenge for large-scale composting operations | Ayilara et al. (2020) |

### 2.1.1 Challenges

**Lack of Knowledge and Awareness** Many households lack an understanding of proper composting techniques, which often leads to common issues such as unpleasant odors, pest infestations, and incomplete decomposition. These problems can discourage individuals from continuing or attempting composting. For example, studies have shown that improper management, such as overloading compost piles or neglecting aeration, can result in anaerobic conditions that produce foul smells (Jara-Samaniego et al. 2017). Increased awareness and knowledge dissemination are essential to address these issues effectively.

**Space Constraints** Urban households, particularly those in high-density residential areas, frequently face space limitations that hinder their ability to set up composting systems. This challenge is particularly pronounced in apartment complexes or crowded neighborhoods where outdoor spaces are minimal or nonexistent. For instance, research highlights that residents in cities often struggle to find adequate areas to implement traditional composting setups, such as open piles or large bins (De Boni et al. 2022). These spatial constraints require innovative and compact solutions.

**Time and Effort** Maintaining a household composting system requires consistent effort, such as turning the pile to ensure proper aeration, monitoring moisture levels to avoid overly wet or dry conditions, and segregating waste to prevent contamination. For individuals with busy lifestyles or competing responsibilities, these tasks may seem burdensome, limiting their willingness to adopt composting practices (Manea et al. 2024). Simplifying maintenance processes is critical to making composting more accessible.

**Contamination** Improper segregation of organic and inorganic waste poses a significant challenge to household composting. Contamination by plastics, metals, or other noncompostable materials reduces the quality of the final compost and can introduce harmful substances into the soil. Moreover, contaminated compost can be less marketable or useful, which undermines its intended environmental benefits (Al Mamun et al. 2020). Enhanced waste sorting education and support mechanisms are needed to address this issue.

### 2.1.2 Opportunities

**Educational Campaigns** Educational initiatives can play a transformative role in empowering individuals with the knowledge and skills needed for effective composting. Organizing workshops, community events, and online tutorials can demystify the composting process and encourage broader participation. For instance, campaigns emphasizing the environmental and economic benefits of composting have successfully increased adoption rates in certain regions (Boldrin et al. 2011). Additionally, digital tools, such as mobile apps and video platforms, offer accessible and engaging ways to share composting knowledge.

**Compact Systems** Innovative solutions tailored to space-constrained environments have emerged to address the limitations of urban households. Compact composting systems, such as bokashi bins, vermicomposting kits, and countertop composters, provide practical options for individuals with limited outdoor space. These systems are designed to be odorless, efficient, and easy to use, making them particularly suitable for urban settings. Research indicates that these innovations have significantly increased the feasibility of composting in apartments and other small living spaces (Lou and Nair 2009).

**Community Initiatives** Shared composting facilities in neighborhoods or apartment complexes can help overcome individual household constraints by providing centralized locations for organic waste processing. These community-driven initiatives foster collaboration and allow residents to pool resources, making composting more convenient and cost-effective. Case studies from urban areas have demonstrated the success of such models, where local governments or community organizations manage shared composting hubs, ensuring proper waste handling and compost quality (De Boni et al. 2022). Additionally, these initiatives often serve as educational platforms, promoting sustainability within the community.

**Policy Support and Incentives** Government policies and incentives can further encourage household composting by reducing barriers to entry. Subsidies for purchasing composting equipment, tax deductions for waste reduction efforts, and mandates for source separation of organic waste are effective strategies. Policymakers can also collaborate with local organizations to provide technical support and resources for households, enabling more widespread adoption of composting practices. Such measures have been shown to increase participation rates and improve overall waste management efficiency (Jara-Samaniego et al. 2017).

In conclusion, household composting presents a viable solution for organic waste management, but its widespread adoption requires addressing the associated challenges through targeted strategies. Educational campaigns, innovative technologies, community-driven initiatives, and supportive policies collectively form a comprehensive approach to unlocking the full potential of household composting systems. By leveraging these opportunities, households can play an active role in reducing waste, enriching soils, and contributing to broader sustainability goals.

## *2.2 Community-Scale Composting*

Community-scale composting systems serve as a vital bridge between household and industrial composting, effectively addressing organic waste management at a local level. By engaging communities directly, these systems not only promote environmental sustainability but also foster social collaboration and awareness about waste reduction practices. Additionally, community-scale composting reduces the environmental impact associated with transporting organic waste to distant industrial facilities, thereby contributing to lower greenhouse gas emissions and more efficient waste management systems. Despite these benefits, several challenges must be overcome to ensure the success and scalability of community composting systems. At the same time, various opportunities exist to enhance their adoption and operational efficiency.

### 2.2.1 Challenges

**Logistics** Coordinating the various aspects of community-scale composting—including waste collection, transportation, and processing—can be highly complex. Variations in waste volume, composition, and timing from different households or institutions add to the logistical challenges. For example, a study by Reyes-Torres et al. (2018) highlights that inconsistent waste streams can disrupt the composting process, leading to inefficiencies and increased operational costs. Additionally, ensuring the timely and reliable transportation of organic waste to composting sites requires careful planning and significant resources.

**Funding** Securing adequate financial resources for establishing and maintaining community composting projects remains a significant hurdle. Many community initiatives rely on limited grants or subsidies, which are often insufficient to cover the costs of equipment, labor, and infrastructure. Wei et al. (2017) note that the lack of long-term funding sources frequently hampers the scalability and sustainability of these projects. Furthermore, small-scale operations often struggle to compete with larger industrial facilities in terms of cost efficiency.

**Regulatory Barriers** Compliance with local regulations and permitting requirements can be both time-consuming and costly for community composting projects. Regulatory frameworks often vary across regions, creating inconsistencies and administrative hurdles that discourage smaller initiatives from starting or expanding (Mejias et al. 2017). For instance, zoning laws and environmental impact assessments may impose restrictions that limit the location or capacity of composting facilities, further complicating project implementation.

**Quality Control** Managing diverse organic waste streams to ensure consistent compost quality presents a major challenge. Improperly sorted waste, often containing nonbiodegradable contaminants like plastics or metals, can significantly reduce the quality and marketability of the final compost product. According to Boldrin et al. (2011), contamination remains one of the most persistent issues in community composting, necessitating robust waste segregation and monitoring systems.

### 2.2.2 Opportunities

**Public-Private Partnerships** Collaboration between local governments, private companies, and nonprofit organizations can provide the financial and technical resources needed to support community composting initiatives. Such partnerships enable the pooling of expertise and resources, leading to more efficient and sustainable operations. Lou and Nair (2009) emphasize that public-private partnerships have successfully addressed funding gaps and operational challenges in various community composting projects worldwide. These partnerships can also foster innovation, such as the development of new composting technologies or waste collection systems.

**Technology Integration** The integration of digital tools and platforms into community-scale composting operations offers significant opportunities to optimize efficiency and improve outcomes. For example, mobile apps can be used to schedule waste collection, track participation, and provide real-time feedback to users. Additionally, sensors and monitoring systems can track key composting parameters, such as temperature, moisture, and oxygen levels, ensuring optimal conditions for decomposition. Shimizu (2017) demonstrate that technology-driven solutions have enhanced the operational efficiency of several community composting initiatives, making them more reliable and scalable.

**Policy Support** Government policies and incentives can play a crucial role in driving the adoption of community composting practices. Introducing tax benefits, grants, and subsidies for composting projects can reduce financial barriers and encourage more communities to participate. Policies that mandate source separation of organic waste or provide technical assistance to community composting initiatives further enhance their viability (Al Mamun et al. 2020). For instance, regions with well-established policy frameworks have seen higher rates of participation and success in community composting programs. Additionally, governments can support awareness campaigns to promote the benefits of community-scale composting and educate residents on proper waste segregation.

**Community Engagement and Education** Fostering active community participation is critical to the success of community-scale composting initiatives. Engaging residents through educational workshops, volunteer programs, and local events can help build a sense of ownership and responsibility toward waste management. Community engagement also ensures that residents understand the importance of proper waste segregation and their role in maintaining the composting system. Moreover, initiatives that involve schools, businesses, and other local institutions can create a collaborative network that supports the long-term sustainability of composting projects.

Community-scale composting offers a promising approach to organic waste management by addressing the gap between household and industrial systems. While challenges such as logistics, funding, regulatory barriers, and quality control persist, targeted strategies can mitigate these issues. By leveraging public-private partnerships, integrating advanced technologies, and fostering supportive policy environments, community composting can become a more accessible and effective solution. Furthermore, engaging in and educating communities will ensure the long-term success of these initiatives, paving the way for a more sustainable and collaborative approach to waste management.

## *2.3 Industrial-Scale Composting*

Industrial-scale composting represents the largest and most sophisticated approach to organic waste management. These systems are designed to efficiently process vast quantities of organic waste, often serving entire cities or regions. By handling large waste volumes, industrial composting facilities play a critical role in diverting organic matter from landfills, reducing greenhouse gas emissions, and contributing to sustainable resource management. However, despite their significant potential, industrial composting facilities face a range of challenges that must be addressed to ensure their long-term viability and effectiveness. At the same time, various opportunities exist to enhance the efficiency and economic sustainability of these systems.

### 2.3.1 Challenges

**High Initial Investment** Setting up industrial composting facilities requires substantial financial resources for infrastructure, specialized equipment, and permitting. The initial costs include the acquisition of land, construction of processing facilities, and installation of advanced composting technologies, such as in-vessel systems or aerated static piles. These high costs can deter municipalities and private investors, particularly in regions with limited access to capital or government subsidies. Manea et al. (2024) highlight that the financial barrier is a significant deterrent to the expansion of industrial composting in developing countries, where waste management systems are already underfunded.

**Energy and Resource Use** Industrial composting systems consume considerable amounts of energy and water, raising concerns about their environmental footprint. For instance, aeration systems, temperature control mechanisms, and leachate management require continuous energy input. Efficient resource management is therefore essential to ensure that industrial composting facilities operate sustainably. Ayilara et al. (2020) note that balancing operational efficiency with resource conservation remains a pressing challenge for large-scale composting operations.

**Odor and Emissions** Managing odors and emissions is one of the most significant operational challenges for industrial composting facilities. The decomposition of organic waste generates odors and greenhouse gases, including methane and ammonia, which can lead to public opposition if not properly controlled. Effective odor control measures, such as biofilters and odor-neutralizing agents, are essential to maintain public acceptance and comply with environmental regulations. De Guardia et al. (2008) emphasize that poorly managed emissions can undermine the environmental benefits of composting and result in regulatory penalties.

**Market Development** Developing and sustaining markets for compost products is another critical challenge. Although compost serves as a valuable organic fertilizer and soil conditioner, it often competes with synthetic fertilizers that are cheaper and more readily available. Market demand for compost products can also fluctuate based on seasonal agricultural needs, economic conditions, and consumer awareness. Boldrin et al. (2011) highlight that the lack of well-established markets can hinder the economic sustainability of industrial composting facilities.

### 2.3.2 Opportunities

**Advanced Technologies** Technological innovations offer significant opportunities to improve the efficiency and scalability of industrial composting systems. Advanced methods, such as in-vessel composting, aerated static piles, and the use of specialized microbial inoculants, can accelerate the decomposition process and enhance compost quality. Reyes-Torres et al. (2018) report that these technologies reduce processing times, minimize odor generation, and improve the overall reliability of composting systems.

**Carbon Credits** Industrial composting facilities can benefit from participating in carbon trading schemes. By diverting organic waste from landfills, these facilities contribute to significant reductions in methane emissions, a potent greenhouse gas. Lou and Nair (2009) highlight that the carbon credits generated through composting activities can provide an additional revenue stream, making large-scale composting more financially viable.

**Product Diversification** Expanding the range of products derived from compost offers new opportunities for profitability and market growth. In addition to traditional compost, facilities can produce high-value products such as biochar, organic fertilizers, soil conditioners, and microbial-rich amendments. Al-Nawaiseh et al. (2021) note that diversified product lines can enhance the appeal of composting products to a broader range of industries, including agriculture, horticulture, and landscaping.

**Circular Economy Integration** Industrial composting facilities can serve as central hubs for organic waste valorization, transforming waste into valuable resources and contributing to a circular economy. For instance, food waste, agricultural residues, and other organic by-products can be processed into compost or other bioproducts that reenter the production cycle. De Boni et al. (2022) highlight that integrating composting into a broader circular economy framework not only reduces environmental impacts but also strengthens local economies by creating jobs and supporting sustainable practices.

Industrial-scale composting is a cornerstone of sustainable organic waste management, offering significant environmental and economic benefits. While challenges such as high initial investment, energy consumption, odor management, and market development persist, targeted strategies and innovative technologies provide promising solutions. By leveraging advanced technologies, participating in carbon trading schemes, diversifying product offerings, and integrating into circular economy models, industrial composting facilities can overcome existing barriers and achieve long-term success. As the global demand for sustainable waste management solutions grows, industrial composting will continue to play a critical role in addressing environmental challenges and advancing sustainability goals.

## 3 Emerging Technologies in Composting

The integration of emerging technologies in composting has revolutionized the efficiency, scalability, and sustainability of organic waste management processes (Xu et al. 2024). These advancements address traditional challenges in composting, such as monitoring, labor intensity, and contamination, by leveraging innovative solutions (Zhang et al. 2023). This section explores four key technological trends—

**Table 2** Emerging technologies in composting

| Technologies | Advantages | Effect | References |
|---|---|---|---|
| The internet of things (IoT) | Detect anomalies early | By providing real-time feedback, these systems allow operators to maintain optimal conditions for microbial activity, ensuring efficient decomposition and high-quality compost production | Senadheera et al. (2024) |
| Microbial inoculants | Mitigate odors and emissions | By promoting aerobic decomposition, these microorganisms reduce the production of methane and ammonia | He et al. (2024) |
| Automation technologies | Greater precision | Automated machinery can perform labor-intensive tasks such as waste sorting, pile turning, and aeration with greater precision and consistency than manual methods | Olawade et al. (2024) |
| Dewatering technologies | Shorten the composting cycle | Dewatering technologies effectively manage wet organic waste, such as food scraps and industrial sludges, by reducing moisture content and preventing anaerobic conditions | Ayilara et al. (2020) |
| Pasteurization technologies | High security and high quality | The addition of pasteurized Agaricus mycelium colonized compost can successfully offset negative effects from high pH and EC, as well as limited nutrient supply (and nitrogen immobilization) in peat-free, compost-based growing media | Stoknes et al. (2019) |

IoT and smart monitoring, microbial inoculants, automation, and waste pre-treatment, which have transformed composting practices and expanded their applicability (Table 2).

## 3.1 *IoT and Smart Monitoring*

The Internet of Things (IoT) has brought significant advancements to composting systems by enabling real-time monitoring and data-driven decision-making. IoT devices equipped with sensors can continuously measure critical composting parameters such as temperature, moisture, oxygen levels, and pH. By providing real-time feedback, these systems allow operators to maintain optimal conditions for microbial activity, ensuring efficient decomposition and high-quality compost production (Senadheera et al. 2024).

One of the primary advantages of IoT-enabled monitoring is its ability to detect anomalies early, such as excessive moisture or temperature fluctuations that could disrupt the composting process. Automated alerts enable timely interventions, reducing the risk of incomplete composting or contamination. IoT systems are particularly valuable for large-scale and industrial composting operations, where manual monitoring would be labor-intensive and prone to errors (Salehi-Amiri et al. 2022).

Moreover, IoT integration supports data collection and analysis, facilitating research and optimization of composting practices. The data-driven insights obtained through IoT applications have been instrumental in improving energy efficiency, reducing greenhouse gas emissions, and enhancing the overall sustainability of composting systems.

## 3.2 Microbial Inoculants

Microbial inoculants are specialized strains of microorganisms introduced into composting systems to accelerate decomposition and improve the quality of the final product (Li et al. 2022). These inoculants are often tailored to target specific types of organic waste, enhancing the breakdown of complex compounds such as lignin, cellulose, and proteins (Mishra and Yadav 2021).

The use of microbial inoculants has shown promising results in reducing composting time and improving nutrient content (Zhang et al. 2022). For example, studies have demonstrated that inoculants containing ligninolytic fungi or thermophilic bacteria can significantly enhance the decomposition of agricultural residues and food waste. Additionally, inoculants contribute to pathogen suppression, reducing the risks associated with contaminated compost.

Another notable benefit of microbial inoculants is their ability to mitigate odors and emissions during the composting process. By promoting aerobic decomposition, these microorganisms reduce the production of methane and ammonia, which are common by-products of anaerobic conditions (He et al. 2024). The use of microbial inoculants aligns with the sustainability goals by minimizing environmental impacts and improving the marketability of compost products.

## 3.3 Automation

Automation technologies, including robotics and automated systems, have transformed the operational efficiency of composting facilities. Automated machinery can perform labor-intensive tasks such as waste sorting, pile turning, and aeration with greater precision and consistency than manual methods (Olawade et al. 2024).

One significant advantage of automation is its ability to reduce labor costs while maintaining high productivity. For instance, automated turning systems ensure uniform aeration of compost piles, optimizing microbial activity and reducing the time required for decomposition. Robotics equipped with AI-powered sorting capabilities can effectively separate organic waste from contaminants, improving the quality of input materials and the final compost product.

Furthermore, automation enhances workplace safety by minimizing human exposure to hazardous conditions, such as high temperatures and bioaerosols commonly associated with composting operations. As automation technologies continue to advance, their integration into composting systems is expected to further enhance scalability and cost-effectiveness (Cheah et al. 2022).

### 3.4 *Waste Pre-treatment*

Waste pre-treatment technologies, such as shredding, dewatering, and thermal processing, play a critical role in improving the efficiency of composting systems. Pre-treatment reduces the size and volume of organic waste, increasing the surface area available for microbial activity and accelerating the decomposition process (Reyes-Torres et al. 2018).

Shredding is particularly beneficial for processing bulky materials like yard waste, wood, and agricultural residues. By breaking these materials into smaller particles, shredding facilitates uniform composting and reduces processing time. Dewatering technologies, on the other hand, are effective in managing wet organic waste, such as food scraps and industrial sludges, by reducing moisture content and preventing anaerobic conditions (Ayilara et al. 2020).

Thermal pretreatment methods, such as pasteurization, can enhance the safety and quality of compost by eliminating pathogens and weed seeds. This approach is especially important for compost intended for agricultural applications, where contamination can have significant consequences. Studies have shown that combining pre-treatment technologies with advanced composting systems can achieve superior results in terms of efficiency, product quality, and environmental performance (Reyes-Torres et al. 2018).

Emerging technologies have brought transformative changes to composting practices, addressing traditional challenges and unlocking new opportunities for sustainable waste management. IoT and smart monitoring systems enable real-time optimization of composting conditions, while microbial inoculants enhance decomposition efficiency and compost quality. Automation reduces labor intensity and improves operational safety, and waste pre-treatment technologies streamline the composting process by preparing organic waste for efficient microbial breakdown. As these technologies continue to evolve, their integration into composting systems will play a pivotal role in meeting global sustainability goals and promoting a circular economy. By adopting these innovations, composting operations can achieve greater efficiency, scalability, and environmental benefits, contributing to a more sustainable future.

## 4 Policy and Regulatory Landscape

Effective policies and regulations play a pivotal role in promoting composting practices and ensuring their successful implementation across different scales. Governments and organizations worldwide have recognized the importance of composting in waste management systems and have introduced various measures to support its adoption. This section explores three critical aspects of the policy and regulatory landscape: incentives for composting, standardization, and waste segregation laws.

## 4.1 Incentives for Composting

Financial incentives, such as tax benefits, grants, and subsidies, are among the most effective tools to encourage the adoption of composting practices. These incentives reduce the financial burden on households, communities, and industrial facilities, making composting systems more accessible and economically viable. For example, government grants can support the establishment of community composting hubs, while subsidies on composting equipment can encourage households to adopt small-scale systems (Boldrin et al. 2011).

Tax benefits for businesses engaged in composting or using compost products also create a favorable economic environment. Such policies have been successfully implemented in several countries, leading to increased participation in organic waste recycling programs. Additionally, carbon credit schemes provide a significant financial advantage for industrial-scale composting facilities by monetizing the reduction of greenhouse gas emissions (Lou and Nair 2009). These incentives not only promote composting but also align with broader environmental sustainability goals.

## 4.2 Standardization

Establishing quality standards for compost products is crucial for building market trust and ensuring consistency in compost quality. Standardization helps consumers and businesses differentiate between high-quality compost and substandard products, promoting confidence in the use of compost for agricultural and landscaping purposes (Ayilara et al. 2020).

Regulations often define parameters such as nutrient content, pH levels, and the absence of contaminants, such as heavy metals or pathogens. For instance, in the European Union, stringent standards under the "EU Fertilizing Products Regulation" have significantly improved the marketability of compost products (Reyes-Torres et al. 2018). Similarly, in the United States, the US Composting Council's Seal of Testing Assurance (STA) program ensures that compost meets specified quality criteria, facilitating its acceptance by consumers and industries alike.

Standardization also benefits international trade by harmonizing compost quality requirements across regions. This allows producers to access broader markets and encourages innovation in composting technologies to meet global benchmarks. By implementing robust quality standards, policymakers can enhance the credibility and adoption of composting systems.

### *4.3 Waste Segregation Laws*

Mandating source separation of organic waste is a foundational policy measure that significantly enhances the effectiveness of composting systems. Waste segregation laws require individuals and organizations to separate organic waste from recyclables and general waste at the point of generation. This practice reduces contamination in organic waste streams and improves the efficiency and quality of composting operations (Manea et al. 2024).

In countries like South Korea and Germany, strict enforcement of waste segregation policies has led to remarkable success in organic waste recycling. For example, South Korea's "Volume-based Waste Fee System" incentivizes households to segregate food waste by imposing additional charges for mixed waste disposal (Al Mamun et al. 2020). Such policies not only promote composting but also reduce landfill reliance and associated greenhouse gas emissions.

Effective waste segregation laws also foster public awareness about the importance of sustainable waste management practices. Educational campaigns and community engagement programs can complement these laws by encouraging compliance and ensuring that waste segregation becomes a routine practice. Additionally, municipalities can invest in infrastructure such as separate collection bins and specialized waste collection vehicles to facilitate the implementation of these policies (Manea et al. 2024).

The policy and regulatory landscape for composting is integral to the success and scalability of organic waste management systems. Financial incentives reduce economic barriers and encourage participation, while standardization ensures the quality and marketability of compost products. Waste segregation laws enhance the efficiency of composting operations and promote environmental awareness among the public. By adopting and enforcing these measures, policymakers can drive the widespread adoption of composting practices, contributing to a more sustainable and circular economy.

## 5 Social and Environmental Impacts

Composting, as a sustainable waste management practice, has far-reaching social and environmental impacts. Beyond its role in diverting organic waste from landfills, composting contributes to climate change mitigation, community engagement, and soil health improvement. This section explores these key benefits in detail, highlighting their importance in achieving broader environmental and social sustainability goals.

## 5.1 Climate Change Mitigation

Composting significantly reduces greenhouse gas emissions, particularly methane, which is generated during the anaerobic decomposition of organic waste in landfills. By diverting organic waste to composting systems, the production of methane is minimized, contributing to climate change mitigation (Lou and Nair 2009). Additionally, the aerobic decomposition process in composting emits negligible amounts of greenhouse gases compared to landfilling, making it an environmentally friendly alternative (Boldrin et al. 2011).

Another crucial aspect of composting is its role in carbon sequestration. When compost is applied to soils, it increases the organic carbon content, enhancing the soil's capacity to store carbon over the longterm. This not only improves soil quality but also offsets carbon emissions from other sectors. Recent studies have shown that large-scale compost application in agricultural lands can play a significant role in global carbon management strategies (Oviedo-Ocaña et al. 2022).

## 5.2 Community Engagement

Composting fosters community participation and raises awareness about sustainable practices. Community composting programs, in particular, serve as platforms for individuals and groups to collaborate on waste reduction initiatives. These programs often involve educational campaigns, workshops, and volunteer opportunities that empower participants to adopt environmentally conscious behaviors (Senadheera et al. 2024).

Furthermore, community composting projects have been shown to strengthen social cohesion and foster a sense of shared responsibility for environmental stewardship. For example, initiatives in urban areas have successfully brought diverse groups together to address local waste management challenges while promoting sustainable lifestyles (Morrow and Davies 2022). By actively involving communities in composting efforts, these programs contribute to building environmentally aware and resilient societies.

## 5.3 Soil Health

The application of compost to soils has profound benefits for soil health. Compost improves soil structure by enhancing its aggregation, porosity, and water-holding capacity, which are critical for sustainable agricultural practices. Improved soil structure also reduces erosion and increases resilience to extreme weather events, such as heavy rainfall or droughts (Ayilara et al. 2020).

In addition to physical improvements, compost enriches soils with essential nutrients, such as nitrogen, phosphorus, and potassium, which are vital for plant growth. It also supports the proliferation of beneficial microorganisms, which enhance nutrient cycling and suppress soilborne diseases. As a result, compost application leads to increased crop yields and reduced reliance on chemical fertilizers, contributing to more sustainable farming practices (Sewak et al. 2021).

Moreover, compost acts as a natural buffer against soil contamination by immobilizing heavy metals and other pollutants. This makes it a valuable tool for land restoration and the remediation of degraded or contaminated soils (Reyes-Torres et al. 2018). The cumulative benefits of compost for soil health not only support agricultural productivity but also contribute to ecosystem services such as water purification and biodiversity conservation.

Composting offers substantial social and environmental benefits, making it an indispensable component of sustainable waste management and agricultural practices. By mitigating climate change, fostering community engagement, and enhancing soil health, composting addresses critical global challenges while promoting a circular economy. Continued investment in composting infrastructure, education, and research will be essential to maximizing these benefits and ensuring a more sustainable future.

## 6 Conclusion and Recommendations

Composting represents a versatile and highly effective solution for managing organic waste, offering significant environmental, economic, and social benefits. Its ability to divert waste from landfills, reduce greenhouse gas emissions, enrich soil health, and foster community engagement underscores its transformative potential. To fully harness the benefits of composting and address the challenges inherent at the household, community, and industrial scales, a comprehensive and collaborative approach is essential.

One of the most compelling aspects of composting is its alignment with global sustainability goals, including those related to climate change mitigation, circular economy frameworks, and food security. However, realizing its full potential requires targeted interventions to overcome existing barriers, such as lack of awareness, financial constraints, regulatory challenges, and technological gaps. By leveraging opportunities and fostering innovation, composting can serve as a cornerstone of sustainable waste management strategies worldwide.

**Key Recommendations**

- Invest in Education and Awareness: Promoting composting knowledge is crucial for increasing participation and improving practices. Targeted educational campaigns can demystify the composting process, address misconceptions, and empower individuals, communities, and businesses to adopt composting. Schools, community groups, and media platforms can be leveraged to create widespread awareness about the environmental and economic benefits of composting.

- Enhance Policy Frameworks: Governments play a pivotal role in facilitating composting initiatives through supportive policies and regulations. Financial incentives, such as grants, subsidies, and tax benefits, can lower entry barriers for households and businesses. Additionally, establishing quality standards for compost products ensures market trust and promotes the use of compost in agriculture and landscaping. Policies mandating source segregation of organic waste can further enhance the efficiency and effectiveness of composting systems.
- Adopt Innovative Technologies: The integration of advanced technologies, such as IoT-enabled monitoring, automation, and microbial inoculants, can optimize composting processes and improve outcomes. These innovations reduce labor costs, enhance operational efficiency, and ensure consistent compost quality. Investments in research and development of such technologies are essential to advancing composting systems and making them more accessible and scalable.
- Foster Partnerships: Collaboration among stakeholders, including governments, private companies, non-profit organizations, and local communities, is vital for the success of composting initiatives. Public-private partnerships can provide the necessary funding, expertise, and infrastructure to establish and maintain composting facilities. Engaging community members in decision-making processes and encouraging volunteer participation can also foster a sense of ownership and accountability, ensuring the long-term sustainability of composting projects.

By addressing the challenges and leveraging the opportunities at each scale, composting can play a transformative role in achieving sustainable waste management and contributing to global environmental goals. The key lies in adopting a holistic approach that combines education, policy support, technological innovation, and collaborative efforts. As the global community faces increasing pressure to manage waste sustainably and mitigate climate change, composting stands out as a practical and impactful solution that benefits both the people and the planet.

**Acknowledgments** The research was sustained by a grant from the National Key Research and Development Program of China, "Intergovernmental Cooperation in International Science and Technology Innovation" [Grant number 2023YFE0104700] and the National Natural Science Foundation of China [Grant Number 31401944].

## References

Al Mamun A, Hayat N, Malarvizhi CAN et al (2020) Economic and environmental sustainability through green composting: A study among low-income households. Sustainability 12(16):6488

Al-Nawaiseh AR, Aljbour SH, Al-Hamaiedeh H et al (2021) Composting of organic waste: A sustainable alternative solution for solid waste management in Jordan. Jordan J Civ Eng 15(3)

Awasthi SK, Sarsaiya S, Awasthi MK et al (2020) Changes in global trends in food waste composting: Research challenges and opportunities. Bioresour Technol 299:122555

Ayilara MS, Olanrewaju OS, Babalola OO et al (2020) Waste management through composting: Challenges and potentials. Sustainability 12(11):4456

Boldrin A, Andersen JK, Christensen TH (2011) Environmental assessment of garden waste management in the municipality of Aarhus, Denmark. Waste Manag 31(7):1560–1569

Cheah CG, Chia WY, Lai SF et al (2022) Innovation designs of industry 4.0 based solid waste management: Machinery and digital circular economy. Environ Res 213:113619

De Boni A, Melucci FM, Acciani C et al (2022) Community composting: A multidisciplinary evaluation of an inclusive, participative, and eco-friendly approach to biowaste management. Cleaner Environmental Systems 6:100092

De Guardia A, Petiot C, Rogeau D et al (2008) Influence of aeration rate on nitrogen dynamics during composting. Waste Manag 28(3):575–587

He Y, He Y, Al-Dhabi NA et al (2024) Effects of exogenous thermophilic bacteria and ripening agent on greenhouse gas emissions, enzyme activity and microbial community during straw composting. Bioresour Technol 407:131114

Jara-Samaniego J, Pérez-Murcia MD, Bustamante MA et al (2017) Composting as sustainable strategy for municipal solid waste management in the Chimborazo region, Ecuador: suitability of the obtained composts for seedling production. J Clean Prod 141:1349–1358

Li H, Zhang T, Tsang DCW et al (2020) Effects of external additives: Biochar, bentonite, phosphate, on co-composting for swine manure and corn straw. Chemosphere 248:125927

Li H, Zhang T, Shaheen SM et al (2022) Microbial inoculants and struvite improved organic matter humification and stabilized phosphorus during swine manure composting: Multivariate and multiscale investigations. Bioresour Technol 351:126976

Lou XF, Nair J (2009) The impact of landfilling and composting on greenhouse gas emissions–a review. Bioresour Technol 100(16):3792–3798

Manea EE, Bumbac C, Dinu LR et al (2024) Composting as a sustainable solution for organic solid waste management: current practices and potential improvements. Sustainability 16(15):6329

Mejias L, Komilis D, Gea T et al (2017) The effect of airflow rates and aeration mode on the respiration activity of four organic wastes: Implications on the composting process. Waste Manag 65:22–28

Mishra SK, Yadav KD (2021) Application of locally available microbial inoculant to accelerate green waste composting at a community level. Bioresour Technol Rep 16:100859

Morrow O, Davies A (2022) Creating careful circularities: community composting in New York City. Trans Inst Br Geogr 47(2):529–546

Olawade DB, Fapohunda O, Wada OZ et al (2024) Smart waste management: A paradigm shift enabled by artificial intelligence. Waste Manag Bull 2:244

Oviedo-Ocaña ER, Soto-Paz J, Domínguez I et al (2022) A systematic review on the application of bacterial inoculants and microbial consortia during green waste composting. Waste Biomass Valorization 13(8):3423–3444

Reyes-Torres M, Oviedo-Ocaña ER, Dominguez I et al (2018) A systematic review on the composting of green waste: feedstock quality and optimization strategies. Waste Manag 77:486–499

Salehi-Amiri A, Akbapour N, Hajiaghaei-Keshteli M et al (2022) Designing an effective two-stage, sustainable, and IoT based waste management system. Renew Sust Energ Rev 157:112031

Senadheera UE, Jayasanka J, Udayanga D et al (2024) Beyond composting basics: A sustainable development goals—oriented strategic guidance to IoT integration for composting in modern urban ecosystems. Sustainability 16(23):10332

Sewak A, Kim J, Rundle-Thiele S et al (2021) Influencing household-level waste-sorting and composting behaviour: What works? A systematic review (1995–2020) of waste management interventions. Waste Manag Res 39(7):892–909

Shimizu N (2017) Process optimization of composting systems. In: Robotics and mechatronics for agriculture. CRC Press, pp 1–22

Stoknes K, Wojciechowska E, Jasinska A et al (2019) Amelioration of composts for greenhouse vegetable plants using pasteurised agaricus mushroom substrate. Sustainability 11(23):6779

Wei Y, Li J, Shi D et al (2017) Environmental challenges impeding the composting of biodegradable municipal solid waste: A critical review. Resour Conserv Recycl 122:51–65

Xu Q, Zhang T, Niu Y et al (2024) A comprehensive review on agricultural waste utilization through sustainable conversion techniques, with a focus on the additives effect on the fate of phosphorus and toxic elements during composting process. Sci Total Environ 942:173567

Zhang T, Wu X, Shaheen SM et al (2022) Improving the humification and phosphorus flow during swine manure composting: a trial for enhancing the beneficial applications of hazardous biowastes. J Hazard Mater 425:127906

Zhang T, Li H, Yan T et al (2023) Organic matter stabilization and phosphorus activation during vegetable waste composting: multivariate and multiscale investigation. Sci Total Environ 891: 164608

# Composting: Safety and Regulatory Considerations in Sustainable Agriculture

**Sudhir Kumar Upadhyay**

## 1 Introduction

Composting is a well-known, eco-friendly biotechnological process that converts organic waste into stable, nutrient-rich soil amendments by means of microbial decomposition (Ho et al. 2022). Composting help to mitigate the consequences of human climate change are the biologically driven mechanisms increasing soil physicochemical characteristics, nitrogen cycling, and carbon sequestration. With increasing worldwide population and increased agricultural needs, which also lessen the reliance on synthetic fertilizers, composting is a sustainable option for traditional waste disposal (Tran et al. 2024). Although composting has agronomic and environmental benefits, it raises major biological and chemical safety issues, and therefore it requires careful management (Tran et al. 2024). Strong quality control policies in compost are needed to protect soil health, food safety, and ecological stability against the presence of viruses, heavy metals, and developing pollutants. Furthermore, influencing their general acceptance and commercial feasibility are the differences in composting rules across different geographical locations (Wani et al. 2024). Apart from managing organic waste, composting is essential to boost soil fertility and agronomic output. Mostly nitrogen (N), phosphorous (P), and potassium (K), along with humic compounds and bioavailable macronutrients produced by microbial catabolic activities, provide what plants require (Ahmed et al. 2023). Reducing the excessive use of synthetic fertilizer helps to lower the danger of eutrophication, soil acidity, and groundwater pollution. Compost also naturally conditions soil by improving cation exchange capacity (CEC), water retention, and soil aggregation (De la Rosa et al. 2022). Reducing soilborne illnesses helps the beneficial microbial consortia inside compost establish to improve plant disease resistance (De Corato

S. K. Upadhyay (✉)
Research and Development Cell, Lovely Professional University, Phagwara, Punjab, India
e-mail: sudhir.33064@lpu.co.in



R. K. Singh et al. (eds.), *Composting of Farm Residues by Nature-Based Solutions*,
https://doi.org/10.1007/978-3-032-12255-1_15

2020). Specifically, SDG 15 (Life on Land), SDG 12 (Responsible Consumption and Production), and SDG 2 (Zero Hunger) attributes make composting appropriate for the United Nations Sustainable Development Goals (SDGs). Composting also greatly reduces greenhouse gas (GHG) emissions, thereby helping to slow down global warming. Generated in anaerobic landfills, methane ($CH_4$), with 25 times the global warming potential of $CO_2$, is a powerful greenhouse gas. In particular, aerobic and regulated composting methods support oxidative decomposition, thereby reducing $CH_4$ emissions (Nordahl et al. 2023).

Including composting in circular bioeconomy models helps assign value to food and agricultural waste, thus improving long-term waste-to-energy projects. Although composting helps the environment, several safety concerns call for meticulous techniques. Microbial contamination is still a major problem, especially the explosion of dangerous bacteria such as *Salmonella enterica, Escherichia coli,* and *Listeria monocytogenes* (Lemunier et al. 2005). These bacteria may survive in poorly handled compost, seriously compromising human and animal health. To adequately inactivate pathogens, composting conditions must be maximized, that is, with appropriate aeration, regulated moisture levels, and extended exposure to thermophilic temperatures ($> 55$ °C). Another major safety concern is heavy metal buildup, that is, lead (Pb), cadmium (Cd), and mercury (Hg) which can occur from pesticide contamination, sewage sludge, or industrial wastes. The strong environmental persistence and bioaccumulating capability of heavy metals expose human consumers, plants, and animals to toxicity. Phytoremediation, charcoal additives, and nanotechnology-assisted immobilization techniques may help to lower heavy metal bioavailability in compost. Furthermore, attracting great attention are new pollutants, such as microplastics, endocrine-disruption chemicals (EDCs), and pharmaceutical wastes, due to their prospective ecotoxicological impacts. Advanced treatment options, including bioaugmentation, enzymatic biocatalysis, and tailored microbial consortia, are needed for low-efficiency traditional composting processes for breaking down persistent organic pollutants (POPs).

To ensure the safe application of compost in agricultural ecosystems, several international regulatory bodies including the European Union (EU), the United States Environmental Protection Agency (EPA), and the Food and Agriculture Organization (FAO) have set tight quality criteria for compost composition, microbial safety, and heavy metal levels. These recommendations define suitable pollution limits and best management practices (BMPs) to reduce environmental and health hazards. Still, the variations in composting laws among countries hinder global norms and cross-border trading. Major challenges still remain to compost quality assurance in poor nations, including waste management capacity, infrastructure monitoring, and regulatory enforcement. Dealing with these regulatory loopholes would necessitate legislative reforms, investment in composting technology, and significant stakeholder education, if we were to raise conformance with worldwide safety standards. Promising for increasing compost efficiency and safety are developing technologies such as artificial intelligence-driven waste analytics, sensor-based compost monitoring, and compost additives enhanced by nanotechnology. Ecologically benign and scientifically verified, composting improves soil condition by means of organic waste. Maintaining its safe use in agriculture, nonetheless, relies on resolving safety concerns regarding heavy metals, new pollutants, and microbial infections. Strict laws, technical developments, and worldwide policy harmonization will help to maximize the advantages of composting and reduce the related hazards. Future research should concentrate on improving composting methods, creating bio-based remedial remedies, and adding digital technology for better process management, thus strengthening the function of composting as a pillar of sustainable agricultural and circular bioeconomy models.

# 2 Composting Process and Microbial Dynamics

## *2.1 Biological Decomposition Process*

Composting is a biological process wherein organic waste is broken down to produce stable humic compounds and nutrient-dense compost (Ho et al. 2022). The degree of composting effectiveness depends on the metabolic activities of many microbial populations, including bacteria, fungi, and actinomycetes, all of which have special but interdependent functions in the biochemical breakdown of organic waste (Chandna et al. 2013). By helping complex organic macromolecules like cellulose, hemicellulose, lignin, and proteins mineralize, these bacteria transform them into easily available nutrients needed for plant growth (Izydorczyk et al. 2024). The succession of microbial communities throughout the composting process depends on temperature, moisture, oxygen availability, and carbon-to-nitrogen (C/N) ratio (Aguilar-Paredes et al. 2023). The three stages—mesophilic, thermophilic, and maturation phases—with their unique set of microbial populations and metabolic pathways, characterize composting. At 25–40 °C, mesophilic bacteria like *Pseudomonas, Bacillus,* and *Flavobacterium* predominate in the early phase of the breakdown of easily degradable organic substrates, including sugars, proteins, and lipids (Insam and Bertoldi 2007). These bacteria provide extracellular enzymes to break down challenging molecules into metabolic heat, hence increasing the compost temperature. Increased microbial respiration causes the compost temperature to rise from 40 °C to 70 °C, thereby favoring thermophilic bacteria such *Bacillus, Clostridium,* and *Geobacillus*. In pathogen inactivation as well as in the breakdown of refractory materials such as cellulose and hemicellulose, this phase is crucial (Finore et al. 2023). Thermophilic fungus called *Aspergillus* and *Thermomyces* help break down lignocellulose, therefore enhancing the durability of compost (Lopez et al. 2021). Actinomycetes, fungus, and mesophilic bacteria recolonize the compost in the final step to help humification, hence producing humic compounds that increase soil fertility. Lignin and chitin breakdown depends on actinomycetes, especially species of *Streptomyces* and *Micromonospora*, which helps to control microbial diseases and stabilize compost. Among the most important microbes used in composting include actinomycetes, fungus, and bacteria (Aguilar-Paredes et al. 2023). The main breakers in composting are bacteria, which also cause fast mineralization of organic waste. Specifically, the genera Actinobacteria, *Bacillus,* and *Pseudomonas* provide hydrolytic enzymes that hasten protein breakdown and cellulose digestion correspondingly (Jurado et al. 2019). Thermophilic bacteria essentially control illnesses by flourishing under heated conditions. Key breakdowners of lignocellulosic materials include fungus, especially *Aspergillus, Penicillium,* and *Trichoderma.* Their vast hyphal networks improve aeration and compost structure by letting plant wastes be enzymatically digested. Mature filamentous bacteria such as *Thermomonospora, Nocardia,* and *Streptomyces* break down complex polymers, including lignin, and generate secondary compounds having antibacterial action. Their impacts lower phytopathogen levels and improve compost stability. Including actinomycetes, bacteria, and fungi, the composting process generates a very active microbial environment where organic waste is effectively broken down. Compost quality may be enhanced by knowledge of microbial succession and management of environmental conditions, thereby guaranteeing its usage as a soil addition in sustainable development (Table 1). Future developments in microbial biotechnology, like bioaugmentation and metagenomic analysis, might help to enhance composting and prevent disease.

## 2.2 Factors Affecting Compost Quality

Through microbial activity, composting, a biologically driven breakdown process, turns organic waste into stable, nutrient-rich compost. Many basic physicochemical factors, including the carbon-to-nitrogen (C:N) ratio, temperature, moisture content, and aeration, influence compost efficiency and quality, and hence impact microbial dynamics and biochemical transformation. Optimizing these factors results in effective organic matter decomposition, pathogen control, and premium compost appropriate for agricultural application. The C:N ratio is one of the factors influencing microbial activity during composting. Although microbial protein synthesis requires nitrogen, carbon supplies energy. A C:N ratio of 25:1 to 35:1 helps to facilitate effective breakdown (Melis et al. 2023). A high C:N ratio (> 35:1) causes delayed breakdown, resulting in a carbon surplus, thereby limiting microbial development. Materials heavy in carbon, including straw or sawdust, require nitrogen supplements, such as manure or green waste, to maintain microbial equilibrium (Rajlakshmi et al. 2023). Low C:N ratios (< 20:1) increase nitrogen volatilization as ammonia ($NH_3$), hence producing odor and nitrogen loss. When composting high-nitrogen substrates such as food waste or manure, this is typical and requires the addition of carbonaceous bulking agents to stabilize the process (Barik 2019). Temperature swings affect microbial succession and metabolic activity throughout the composting process (Han et al. 2022). During the mesophilic phase (25 °C–40 °C), *Pseudomonas* and *Bacillus* attack and break down basic organic molecules. During the thermophilic phase (40 °C–70 °C), heat-tolerant bacteria (*Bacillus, Clostridium*) and fungi (*Aspergillus, Thermomyces*) break down complex macromolecules like cellulose and lignin. This period determines both the inactivity of

**Table 1** Microorganisms, compost constituents, enzymes, functions, mechanisms, and environmental factors in the composting process

| Microorganisms | Compost material | Primary enzymes produced | Role in composting phase | Mechanism & function | Composting condition |
|---|---|---|---|---|---|
| *Bacillus* spp. | Agricultural waste, manure | Protease, amylase, cellulase | Thermophilic phase | Degrades proteins, starches, and cellulose to simpler molecules; increases temperature and contributes to pathogen reduction | Thermophilic (45 °C–70 °C), pH 6.5–8.0 |
| *Pseudomonas* spp. | Food and garden waste | Lipase, esterase | Mesophilic and maturation | Degrades lipids and hydrocarbons; aids in odor control and secondary metabolism | Mesophilic (25 °C–45 °C), moisture ~55% |
| *Aspergillus* spp. | Lignocellulosic biomass | Cellulase, ligninase, xylanase | Thermophilic and maturation | Breaks down cellulose, hemicellulose, and lignin; enhances compost maturity and structure | Thermophilic, aerobic |
| *Trichoderma* spp. | Agro-industrial residues | Cellulase, chitinase | Maturation | Promotes humification, suppresses pathogens, and enhances decomposition of cellulose and chitin | Aerobic, pH 5.5–7.5, moisture 50%–60% |
| Actinomycetes | Lignocellulosic organic waste | Cellulase, xylanase, ligninase | Late thermophilic & maturation | Decomposes complex organics like lignin and cellulose; stabilizes compost and reduces odors | Thermophilic to maturation, aerobic |
| *Streptomyces* spp. | Sawdust, plant residues | Lignocellulolytic enzymes | Maturation | Degrades aromatic compounds; contributes to pathogen resistance and compost stabilization | Mesophilic to thermophilic, moisture ~55% |
| *Lactobacillus* spp. | Kitchen waste, food scraps | Lactic acid, protease | Early mesophilic phase | Produces lactic acid, lowers pH to control pathogens; enhances early fermentation | Anaerobic to facultative aerobic |
| *Penicillium* spp. | Fruit and vegetable waste | Pectinase, cellulase | Maturation | Degrades pectin and softens organic structure; assists in final compost refinement | Aerobic, temperature 30 °C–50 °C |

infections and the destruction of weed seeds. The mature phase below 40 °C) results from cooling following stable degrees of microbial activity. Mostly functioning as *Streptomyces* and *Micromonospora,* actinomycetes help to humify and stabilize compost. Composter safety and pathogen removal rely on temperatures maintained for at least three days above 55 °C. Extreme heat more than 70 °C may, however, trigger microbial death and reduce enzyme activity, hence preventing breakdown. Water also influences microbial metabolism and enzymatic activities as well. The perfect moisture level for composting is between 50% and 60% (El Fels et al. 2024). Too much moisture ($> 65\%$) results in anaerobic conditions that cause incomplete decomposition, bad smells, and $CH_4$ emissions. Less than 40% moisture reduces microbial activity, therefore reducing breakdown efficiency and generating worthless, dry composting (Kim et al. 2016).

Regular moisture monitoring and frequent watering or aeration adjustments help to sustain microbial activity to some degree. Mostly an aerobic process, composting relies on enough oxygen ($O_2$) for oxidative breakdown and bacterial respiration. Good composting requires more than 10% oxygen content (Wang et al. 2007). Insufficient aeration generates anaerobic conditions that promote the development of facultative anaerobes like *Methanobacterium* and *Clostridium* (Harirchi et al. 2022). Bad smells and worse compost quality are brought about by these anaerobes that generate methane ($CH_4$) and hydrogen sulfide ($H_2S$). Too much aeration may induce heat loss and moisture evaporation, thereby reducing microbial activity. In regulated systems, forced aeration is used or routinely switched on to improve oxygen diffusion and lower compaction. Maximizing the C:N ratio, temperature, moisture, and aeration will help to produce premium compost and preserve microbial efficiency (Azis et al. 2023). Pathogen-free, nutrient-rich compost may be generated by means of an understanding of microbial dynamics and factor control, thereby increasing its agronomic value in sustainable agriculture. Future innovations, such as sensor-based compost monitoring and bioaugmentation using specific microbial consortia, might significantly increase composting efficiency and quality.

## *2.3 Role of Nanotechnology and Microbial Inoculants in Enhancing Composting Efficiency*

A microbe-based bioconversion process, composting turns organic waste into nutrient-dense humic molecules. But a slowdown in decomposition rates, nutrient losses, and the persistence of contaminants like heavy metals and pathogens all afflict traditional composting methods. Recent developments in nanotechnology and microbial inoculants provide fresh approaches for enhancing composting efficiency by accelerating organic matter decomposition and raising microbial activity, thus minimizing environmental issues (Behl et al. 2024). By optimizing composting conditions, these growing techniques provide premium compost fit for sustainable agricultural applications. By incorporating nanoparticles that enhance microbial activity, control nutrient dynamics, and help to break down pollutants, nanotechnology presents a new approach to increasing composting efficiency. Important nanoparticles for composting are metal oxide nanoparticles (e.g., ZnO, $Fe_2O_3$, $TiO_2$), which increase microbial enzymatic activity, thereby enabling breakdown of complex organic polymers like lignin and cellulose (Zhang et al. 2019; Gattupalli et al. 2023; Wawrzynczak et al. 2024). Pyrolyzed organic materials produce nanobiochar that advances carbon sequestration, microbial colonization, and moisture retention (Shafiq et al. 2023). Strong metal immobilization and contaminant degradation iron nanoparticles and nanoclays effectively absorb heavy metals (e.g., Pb, Cd, Hg), thereby lowering their bioavailability in compost (Attia et al. 2020). By breaking down POPs and pharmaceutical wastes, photocatalytic nanoparticles such as $TiO_2$ and ZnO increase compost safety (Majumder et al. 2020). Nanofertilizers buried in compost provide delayed release of vital nutrients (N, P, K), thereby enhancing soil fertility and reducing leaching losses (Easwaran et al. 2024). Specific microbial strains called microbial inoculants, often referred to as bioenhancers, hasten the breakdown of organic matter, lower pathogen count, and improve compost stability (Fig. 1). Among the inoculants are bacteria, fungus, and actinomycetes,

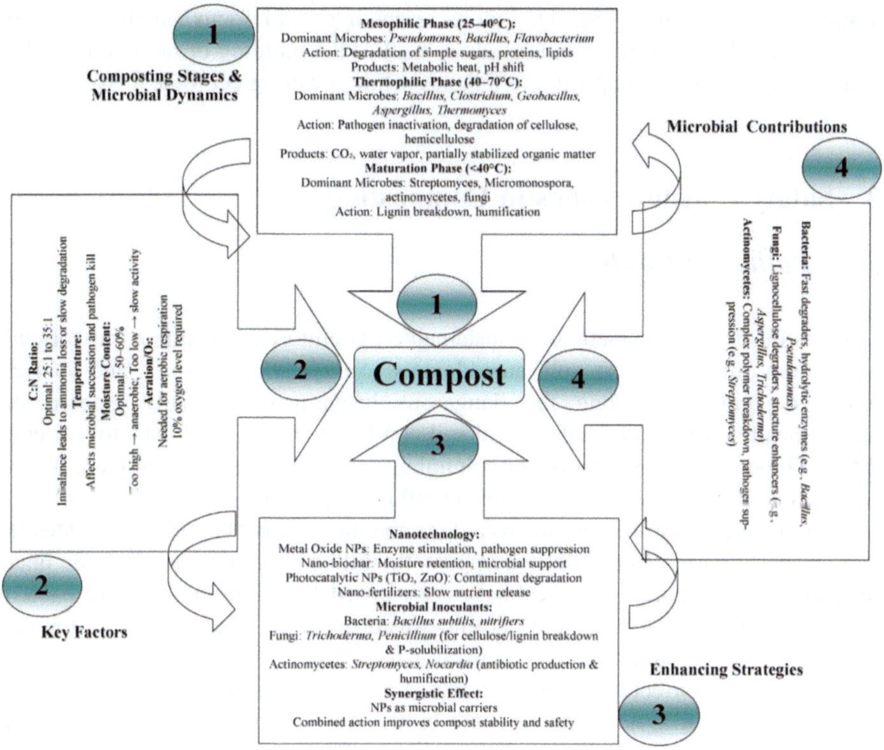


**Fig. 1** Diagrammatic illustration of microbial dynamics, composting stages, and efficiency-boosting measures

each of which aids in the composting process in unique manner. During the thermophilic phase, bacterial inoculants such as *Bacillus subtilis and Pseudomonas fluorescens* can promote decomposition of organic materials. By stabilizing nitrogen, nitrifying bacteria *Nitrosomonas* and *Nitrobacter* can lower ammonia loss. Lignocellulolytic fungi (*Aspergillus, Trichoderma, Penicillium*) can boost lignin and cellulose breakdown, thereby accelerating compost maturation. *Penicillium, Glomus,* and other phosphate-solubilizing fungus release bioavailable phosphorus, thus enhancing the nutritional value of compost (Timofeeva et al. 2022). *Streptomyces* and *Nocardia*, two actinomycetes, break down difficult chemical molecules and generate antibiotics to fight destructive bacteria (Alam et al. 2022).

Combining nanotechnology with microbial inoculants generates a synergistic approach to composting, thereby boosting microbial activity, breakdown of pollutants, and compost stability. Acting as microbial transporters, nanoparticles might boost inoculant bioavailability and viability. Moreover, metal oxide nanoparticles and nanobiochar might affect microbial metabolism, thereby improving the efficiency of composting and acting ecologically friendly. Composting science takes a major step forward when nanotechnology and microbial inoculants are combined. These techniques maximize nutrient dynamics, enhance decomposition kinetics, and lower compost contamination, thereby producing high-quality compost for use in sustainable agriculture (Fig. 1). Future research should focus on the long-term environmental impact of nanomaterials as well as the building of tailored microbial consortia to improve composting efficiency while lowering ecological issues.

## 3 Safety Considerations in Composting

### *3.1 Pathogen Control*

Composting is a main method in sustainable agriculture, as it transforms organic waste into nutrient-dense soil amendments. *Salmonella, E. coli*, and *Listeria monocytogenes* are among the dangerous bacteria substantially jeopardizing human health, agricultural safety, and soil condition (Jones 1999; Mantovam et al. 2025). Good use of compost in agricultural systems relies on efficient management of diseases. Among the many parameters influencing pathogen survival are temperature, moisture, aeration, and microbial competition. This chapter investigates the prevalence of harmful bacteria in compost, the influence of the thermophilic phase on pathogen reduction, and the likely risks posed by antibiotic-resistant bacteria (ARB) in compost. Among the organic waste materials used in composting are pathogenic microorganisms derived from manure, food waste, sewage sludge, and plant waste. If incorrect treatment of composting conditions is done, these pathogens may stay and poll agricultural soil and crops. Among the most harmful bacteria found in compost are *Listeria monocytogenes, Escherichia coli,* and *Salmonella* species. Human and animal food-related diseases are caused by the gram-negative enteric bacterium *Salmonella*. Under unfavorable circumstances, it may persist in compost and subsequently contaminate soils and crops. The main sources of *Salmonella* in compost are raw manure and badly broken down organic elements. Likewise, *E. coli*, especially Shiga toxin-producing strains (STEC) like *E. coli* O157:H7, may pose substantial health hazards when present in compost (Lim et al. 2010). Originally found in animal feces, this bacterium may cause digestive issues for humans. Unsuitable treatment of compost may allow pathogenic *E. coli* bacteria to persist, thereby increasing contamination risk. *Listeria monocytogenes*, a psychotropic bacterium capable of surviving at low temperatures and in damp surroundings, is another main disease one should be concerned about. It harms human health, especially in pregnant women and immunocompromised people. Common in animal manure and food processing waste, *Listeria* needs strict composting methods to stop its dissemination (Osek et al. 2022).

Composting consists of three distinct stages: mesophilic, thermophilic, and maturation. Higher temperatures and microbial antagonism make the thermophilic phase—which runs at 40 °C–70 °C—essential for pathogen inactivation. In the thermophilic phase, heat-resistant bacteria such as *Bacillus, Geobacillus, and Thermus* grab control and generate extracellular enzymes capable of breaking down intricate chemical compounds. Moreover, breaking down lignocellulose facilitates the stable compost matrix production by thermophilic fungi such as *Aspergillus* and *Thermomyces*. Thermal inactivation is among the finest techniques to completely eliminate bacteria in compost. Research shows that *Salmonella, E. coli,* and *Listeria* may be destroyed with at least three days at temperatures between 55 °C and 65 °C (Kim et al. 2019).

Although harmful proteins and nucleic acids are broken down, extended thermophiles support bacterial cell lysis. If one desires best pathogen reduction, several composting methods must be used. Ideal aeration is therefore required, as aerobic conditions promote thermophilic microbial metabolism and restrict the development of facultative anaerobic pathogens. Frequent rotating of compost heaps helps to distribute heat uniformly, thereby inactivating microorganisms. Especially crucial is moisture management, as a level of 50–60% boosts microbial enzymatic activity and heat production. However, moisture levels over 65% may produce anaerobic zones, allowing bacteria to proliferate. pH control is another very successful way of pathogen control. While acidic (< 5.5) and alkaline (> 8.5) circumstances limit their development, germs flourish at neutral pH levels (6.5–7.5) (Padan et al. 2005). Lime ($CaO$) or charcoal may affect pH and reduce *Salmonella* survivability (Placha et al. 2008). Moreover, competitive exclusion by beneficial bacteria considerably helps reduce infections. Actinomycetes like *Streptomyces* and *Nocardia,* as well as probiotic bacteria including *Lactobacillus* and *Bacillus subtilis*, generate antimicrobial chemicals that kill harmful bacteria (Al-Ansari et al. 2019).

## 3.2 *Potential Risks of Antibiotic-Resistant Bacteria (ARB) in Compost*

Growing global issues include antibiotic-resistant bacteria (ARB) and antibiotic resistance genes (ARGs) found in compost (Ondon et al. 2021; Li et al. 2023). Residual antibiotics from sewage sludge, manure, and hospital waste all drive selection pressure on resistant bacterial populations in compost. This is particularly concerning, as ARB-contaminated compost may expose people and other animals to health risks by spreading resistance genes to soil and plant microbiomes (Graham et al. 2019). Antibacterial-treated animal manure is one of the most frequently occurring factors contributing to antibiotic resistance in compost. Resilient bacteria and ARGs abound in this manure; if not properly handled, they might linger in compost. Moreover, food waste containing antibiotic residues might promote ARB formation (Lima et al. 2020).

Effective thermophilic composting might not kill ARGs, allowing them to spread in agricultural soils. The spread of ARB poses significant dangers. By means of plasmids and transposons, horizontal gene transfer (HGT) accelerates the spread of ARGs to dangerous bacteria, thereby raising the environmental incidence of antibiotic resistance (Bobate et al. 2023). Badly controlled composting operations' created bioaerosols might spread ARB into the atmosphere, aggravating the problem. Many reducing strategies have to be used to control ARB dangers in compost. Since sustained temperatures above 60 °C effectively degrade ARGs and ARB, extended thermophilic processing is needed. Microbial antagonism and biocontrol strategies, including the introduction of probiotic bacteria like *Lactobacillus* and *Bacillus subtilis,* might outcompete and prevent ARB growth. Additionally, possible solutions for ARB reduction include chemical and nanotechnology therapies. The effective adsorbability of ARGs and neutralizing effects of iron oxide nanoparticles, biochar, and nanosilver (AgNPs) have been shown. These substances might be included in composting systems to reduce contamination risk and increase microbiological safety. Reducing ARB contamination in compost depends heavily on monitoring and regulating guidelines. Strict antibiotic use limits in animal farms might assist in reducing the ARB spread into composting facilities. Moreover, frequent screening of compost for ARGs guarantees biosafety compliance and lowers public health risks. Ensuring agricultural and environmental safety depends heavily on pathogen control in composting. Pathogen inactivation depends on the thermophilic phase; therefore perfect aeration, moisture control, and microbial antagonism all help increase compost safety (El Hayany 2021).

However, the spread of antibiotic-resistant bacteria in compost generates fresh problems, requiring better mitigation strategies like nanotechnology applications, microbial biocontrol, and prolonged high-temperature composting. Future research should focus on developing microbial consortia capable of breaking down ARGs and improving compost safety by means of new biosanitization techniques. Strong regulations and scientifically based composting techniques might assist in lowering health risks and ensuring the safe use of compost in sustainable development.

### 3.3 Heavy Metals and Chemical Contaminants

One commonly used method in sustainable agriculture that increases soil fertility while lowering organic waste is composting. However, the heavy metal and chemical contaminants in compost seriously compromise human health, agricultural safety, and soil quality. Pollution from industrial waste, pesticides, and fertilizers may cause heavy metal accumulation in compost, including lead (Pb), cadmium (Cd), mercury (Hg), arsenic (As), and chromium (Cr) (Khatun et al. 2022). These dangerous compounds bioaccumulate in plants, are quite stable in the environment, and cause health problems via the food chain (Khatun et al. 2022). Understanding the causes, bioavailability, and accumulating dangers, as well as using advanced remedial solutions like charcoal application, nanotechnology treatments, and microbial-assisted detoxification helps one to solve heavy metal contamination in compost. Among the many causes of heavy metal contamination in compost are organic waste inputs, pesticide applications, and industrial waste. Significant amounts of heavy metals abound in industrial waste, including effluent from mining, smelting, electroplating, and textile sectors. Should they be disposed of unlawfully or mixed with biodegradable products, these poisons may stay in the composting process and find their way into agricultural soils. Heavy metals collected from household and industrial wastewater may also find their way into sewage sludge and municipal solid waste utilized in composting. The misuse of fertilizers and pesticides in conventional farming contributes to heavy metal contamination. Many contemporary herbicides include small amounts of metals, including arsenic (As) and copper (Cu), which might gather in composted agricultural waste (Alengebawy et al. 2021).

Likewise, as the phosphate rock used in fertilizer production naturally contains Cd pollutants, phosphate fertilizers are a major cause of cadmium toxicity (Liang et al. 2024). Extended application of contaminated compost in agricultural fields might lead to the accumulation of toxic metals in soils, thereby rendering them bioavailable for use by crops (Zheng et al. 2022; Priya et al. 2023). Often used as compost feedstocks, animal manure and food waste may also include heavy metals. Lead and mercury are among the heavy elements found in manure from cattle raised on contaminated food or water. Likewise, food waste gathered from municipal trash streams and commercial kitchens might contain metallic residues from cookware, processing tools, and packaging materials. These poisons linger in the composting process, and without proper treatment, might cause long-term problems. Many factors define the bioavailability of heavy metals in compost: pH, organic matter content, redox potential, and microbial activity. Human and animal health difficulties arise from bioavailable heavy metals being absorbed by plant roots and accumulated in edible plant tissues. Plant type, soil quality, and climatic conditions all influence the degree of metal accumulation in different crops. Therefore, heavy metal mobility is strongly influenced by soil pH. In an acidic environment ($pH < 5.5$), heavy metals are more soluble and hence simpler for plants to absorb.

By contrast, alkaline conditions (pH > 7.5) reduce bioavailability and increase metal precipitation (Zheng et al. 2024).

The organic componenets of compost may combine with heavy metal compounds to immobilize and limit their absorption (Wei et al. 2020). However, microbial breakdown of organic debris might finally release bound metals, thereby enhancing their mobility in the ground. Certain crops are unfit for use in contaminated soils, as their heavy metal accumulation potential is very strong. Leafy greens like spinach and lettuce, for instance, could absorb and store high amounts of cadmium and lead. Root crops, like potatoes and carrots, could potentially draw metals from contaminated compost-amended soil. Compost containing heavy metals used long-term might produce bioaccumulation in plants, thereby endangering the chronic exposure risk to consumers (Vasilachi 2023).

Furthermore, the leakage of heavy metals into groundwater systems could aggravate environmental harm. Many remedial techniques have been developed to lower heavy metal contamination in compost and agricultural soils, including the use of biochar, nanomaterials, and microbial treatments. For sustainable farming, these techniques raise metal immobilization, microbial detoxification, and soil quality. Increasingly favored as a soil addition for heavy metal cleaning is biochar, a carbon-rich compound produced from pyrolyzed organic sources. The strong cation exchange capacity (CEC) and porous structure of biochar help it absorb and immobilize heavy metals, thereby reducing their bioavailability in compost and soil (Viotti et al. 2024). Moreover, biochar raises microbial activity and reduces soil pH, thereby supporting heavy metal stability (Zheng et al. 2024). Research has shown that biochar created from agricultural waste, such as rice husks and wood chips, effectively lowered cadmium and lead mobility in soil. Metal binding and sequestration are facilitated by the surface functional groups of biochar—hydroxyl (-OH) and carboxyl (-COOH) (Dong et al. 2024). Moreover, biochar improves water retention and soil aeration, thereby supporting plant growth and lowering heavy metal absorption. New approaches for heavy metal remediation in compost come from nanomaterials like custom clays, carbon-based nanoparticles, and metal oxides. Excellent for adsorption and stabilization of dangerous metals, nanoscale materials have a great surface area and reactivity (Olawade et al. 2024). Research on iron oxide nanoparticles ($Fe_3O_4$) has focused especially on their ability to bind heavy elements like arsenic and lead, thereby preventing their uptake by plants (Khan et al. 2022). Likewise, by the use of electrostatic interactions and adsorption techniques, graphene-based nanomaterials encourage metal immobilization (Ahmad et al. 2020).

Using nanoparticles in composting systems could help improve organic pollution breakdown and heavy metal stability. To guarantee their safe usage in agricultural applications, questions about the environmental fate and toxicity of manufactured nanomaterials need greater research. Sustainable nanotechnology approaches provide good substitutes with less effect on the environment, as they are bionanocomposites created from natural polymers. Another effective method of heavy metal cleanup in compost and soil is microbie-assisted bioremediation. Among the microbial strains that transform and store heavy metals via biosorption, bioaccumulation, and biomineralization mechanisms are metal-resistant bacteria and fungi (Priya et al. 2022). By producing extracellular polysaccharides, bacteria like *Pseudomonas, Bacillus,* and *Rhizobium* have great metal-binding capacity and might immobilize heavy metals (Gupta and Diwan 2016). Moreover, sulfate-reducing bacteria (SRB) reduce bioavailability by encouraging the precipitation of heavy metals as insoluble sulfides (Ayangbenro et al. 2018).

By means of enzymatic reduction and cell wall binding routes, fungi like *Aspergillus* and *Trichoderma* have also demonstrated notable efficiency in metal absorption and detoxification (Dinakarkumar et al. 2024). By means of mutually beneficial interactions, the use of microbial consortia that combine multiple bacterial and fungal strains can improve bioremediation performance (Zhang et al. 2022). Moreover, better metal-binding properties of genetically engineered microbes (Pande et al. 2022) might provide choices for higher heavy metal detoxification in composting facilities. For human health, agricultural safety, and soil quality, heavy metals and chemical toxins in compost raise serious concerns. Agrochemicals, industrial waste, and organic waste all add to heavy metal contamination that results in bioaccumulation in crops and associated environmental hazards. Remedial therapies for heavy metal stabilization and detoxification include biochar application, nanotechnology treatments, and microbial-assisted detoxification, thereby helping to allay these worries. Future studies should focus on developing affordable and ecologically friendly solutions to allow the safe use of compost in sustainable development. Integrating these strategies into composting processes could help enhance soil quality, reduce pollution issues, and support environmentally friendly food production methods.

### 3.4 Emerging Contaminants

Composting is considered a commonly acknowledged approach for handling organic waste, enhancing soil fertility, and advancing sustainable development. Compost safety and quality are seriously threatened, nevertheless, by rising contaminants, including endocrine-disruption chemicals (EDCs), pharmaceutical wastes, and microplastics (Li et al. 2024). Often found in the environment and resulting from a range of human activities, these poisons pose long-term ecological and health hazards. Their existence in compost raises concerns about probable human and animal health risks, crop absorption, and soil contamination. New monitoring methods and innovative mitigating strategies must be used to guarantee the safe application of compost in agriculture and thereby solve these issues. Compost created from organic waste increasingly contains microplastics—plastic particles smaller than 5 mm. Plastic containers, agricultural mulching films, synthetic fibers from textiles, and sludge from wastewater treatment plants are among the many sources of these particles (Masciarelli et al. 2024).

Microplastics found in compost seriously compromise microbial communities and soil quality. Microplastics have been proven to compromise the nitrogen cycle, reduce water-holding capacity, and change soil structure (Zhou et al. 2024).

Moreover, exacerbating environmental contamination, microplastics may transport dangerous chemicals such as heavy metals and persistent organic pollutants (POPs) (Rafa et al. 2024). Plants' root systems seem to be able to absorb microplastics, which might help to cause bioaccumulation in the food chain (Hasan and Tarannum 2024). Human exposure to microplastics from contaminated agricultural food causes concerns over long-term health effects, including inflammatory responses and endocrine disturbance (Yuan et al. 2022). Pharmaceuticals, including hormones, analgesics, and antibiotics, are regularly detected in composted organic waste (Cucina et al. 2018). Usually, food waste, including expired or improperly managed drugs, municipal biosolids, and animal feces, brings these contaminants into the composting process (O'Connor et al. 2022).

Chemical stability, composting temperature, and microbial breakdown ability define the permanency of medicinal compounds in compost. Antibiotics in compost raise concerns about resistance genes in agricultural soils and the proliferation of antibiotic-resistant bacteria (ARB). Spraying antibiotic-containing compost on fields might raise the selection pressure for resistant bacterial strains, thus possibly helping to propagate antimicrobial resistance (AMR) (Haynes et al. 2020). This is a major public health issue, as AMR lowers the effectiveness of medications used to cure bacterial infections in humans and animals. Moreover, pharmaceutical leftovers might affect the microbial counts in soil, thereby affecting organic matter decomposition and nutrient cycling. Estrogens and progestins contained in human excrements and animal dung may survive in compost and pass through agricultural soils. These substances have endocrine-disrupting effects, thereby impairing animal reproductive health and possibly upsetting human hormonal balance by means of contaminated food and water supplies. Made either naturally occurring or synthetic, endocrine-disruptive chemicals (EDCs) disrupt the hormone balance in living things. Plastics, pesticides, industrial chemicals, and personal hygiene items all often include these molecules. The most commonly occurring EDCs in compost are bisphenol A (BPA), phthalates, dioxins, and organochlorines. Compost with EDCs has important consequences for human health, plant growth, and soil biodiversity. Developmental abnormalities, reproductive problems, immune system dysfunction, and an elevated risk of hormone-related cancer have been linked to EDCs from contaminated soil and food. In agricultural environments, EDCs may interfere with plant hormone transmission, thereby lowering crop productivity and quality. Moreover, these chemicals could leach into groundwater and contaminate sources of drinking water, thereby posing long-term environmental hazards. Agricultural sustainability and food safety depend on lowering EDC contamination in compost, as their persistence and bioaccumulative character call for it.Finding and eliminating emerging toxins in compost calls for advanced monitoring and mitigating strategies. Although nutritional content and pathogen levels define traditional compost quality assessments, analytical techniques for monitoring microplastics, pharmaceutical residues, and EDCs are becoming more and more important. Thanks to recent developments in analytical chemistry, new toxins in compost may be exactly identified. Microplastics and chemical pollutants are mostly found and quantified using spectroscopic and chromatographic methods, like Fourier-transform infrared spectroscopy (FTIR), Raman spectroscopy, and gas chromatography–mass spectrometry (GC–MS) (Primpke et al. 2020).

Low-level pharmaceutical residue and EDC identification is much aided by liquid chromatography–tandem mass spectrometry (LC-MS/MS). New technologies under investigation for real-time contamination detection include biosensors and machine-learning-based prediction models in order to increase monitoring efficiency. These techniques might help identify hotspots of contamination in composting processes and affect rules for compost safety criteria. Several remedies have been proposed to lower the risks posed by evolving contaminants in compost. By means of better trash-sorting methods and the development of biodegradable plastic replacements, microplastic contamination at the source might be somewhat lowered. Composting facilities may also remove plastic particles before they are applied by means of filtration and separation technologies. Thermophilic composting, bioaugmentation, and enzyme-based degradation include advanced composting techniques for pharmaceutical wastes and EDCs. Using high temperatures (50 °C–70 °C), thermophilic composting accelerates the breakdown of organic contaminants, including certain pharmaceutical compounds. Including specialized microbial consortia in bioaugmentation increases microbial enzyme activity, thereby improving the biodegradation of drugs and EDCs. Tailored biochar and nanocatalysts are among the nanotechnology-based treatments under testing for their ability to absorb and neutralize EDCs and pharmaceutical wastes. For organic pollutants, functionalized biochar treated with metal oxides or carbon-based nanomaterials showcases remarkable adsorption efficiency, thereby reducing their bioavailability in compost. Moreover, stricter waste management guidelines and government actions might assist in keeping emerging toxins out of the composting process. Public awareness campaigns and farmer education programs can potentially help reduce contamination by motivating appropriate waste disposal and sustainable composting methods. New contaminants, like endocrine-disrupting chemicals, pharmaceutical residues, and microplastics, found in compost cause significant environmental and health hazards. Plastic waste, drugs, and industrial chemicals are among the many sources of these toxins, which compromise human well-being, agricultural safety, and soil health. Spectroscopy and chromatography are among the advanced monitoring techniques needed to find and measure these contaminants in compost. Promising for reducing contamination issues are mitigating strategies, including improved waste management, thermophilic composting, microbial bioremediation, and technologies derived from nanotechnology. Compost safety calls for a multidisciplinary approach, combining public awareness, legislative rules, and scientific progress to support environmental health and sustainable agriculture.

# 4 Regulatory Frameworks for Composting

## *4.1 Global Regulations and Guidelines for Composting*

Composting is an essential component of sustainable waste management and improvement of soil condition. Its safety and effectiveness, however, depend on tight laws controlling compost production, quality, and application. Rules regulating heavy metal, pathogen, and organic matter concentration have been enacted in some places, including the European Union (EU), the United States (USA), India, and China, thereby ensuring compost safety (Table 2). These rules aim to prevent environmental contamination, protect human health, and promote long-term agricultural usage of organic compost. Moreover, particularly in organic farming uses, certification systems are crucial in ensuring compliance with these standards. Reflecting their objectives for environmental safety, agricultural output, and waste management, various countries approach compost legislation differently. The European Union's Fertilizing Products Regulation (EU 2019/1009) sets strict quality and safety standards for organic fertilizers and soil additions, thereby guiding composting (Athanasoulia et al. 2021; Huygens et al. 2019; Marini et al. 2020).

To ensure that compost products neither compromise human health nor the environment, the regulation sets reasonable levels of heavy metals, pathogens, and contaminants. The EU also promotes the use of End-of-Garbage Criteria, which determine when compost is no longer regarded as waste and starts to be a valued product. The Environmental Protection Agency (EPA) and the Department of Agriculture (USDA) handle United States composting guidelines. Pathogen control rules and heavy metal limits in biosolid composting are governed by EPA Part 503 Rule (EPA–United States, 1995). Moreover, the USDA National Organic Program (NOP) mandates that compost used in organic farming satisfy certain processing criteria, like keeping temperatures between 55 °C and 77 °C in windrow systems for at least 15 days to eradicate pathogens (Pittman et al. 2004; Manual 2023).

The 2016 Solid Waste Management (SWM) Rules– published by the Ministry of Environment, Forest, and Climate Change, control composting limits in India (GOI 2016). Establishing minimum requirements for organic matter content (at least 12%), carbon-to-nitrogen (C:N) ratios (between 20:1 and 30:1), and heavy metal limits such as lead (Pb), cadmium (Cd), and arsenic, the Fertilizer Control Order (FCO) of 1985 guides the sale of compost as an organic fertilizer (Biofertilizers and Organic Fertilizers, Order 1985). By means of municipal compost certification and incentives for the use of organic fertilizer, the Indian government also promotes composting. Thorough composting guidelines abound in China's Organic Fertilizer Standard (NY 525-2021) and the Solid Waste Pollution Prevention and Control Law (Cui et al. 2023). The criteria give compost maturity and organic matter concentration top priority, along with contaminant levels. China also has green labeling systems for organic compost in order to support environmentally friendly farming practices. Rising urbanization and industrial waste pollution still present problems

**Table 2** Key considerations and aspects of composting for safety and regulatory compliance in sustainable agriculture

| Aspect | Key considerations | Recent updates/limitations | References |
|---|---|---|---|
| Reducing pathogens | Eliminating pathogens like *Salmonella, Listeria*, and *E. coli* requires thermophilic composting (> 55 °C for three plus days) | Emerging infections and antibiotic-resistant bacteria (ARB) subvert conventional hygienic standards | El Hayany et al. (2021), Lima et al. (2020), Mantovam et al. (2025) |
| Heavy metal content | For Cd, Pb, Cr, and Hg there are set regulatory levels. Beyond these thresholds, phytotoxicity and health hazards might result | Lack of harmony in world standards; bioavailability and cumulative impact usually disregarded | Alengebawy et al. (2021), Vasilachi et al. (2023), Liang et al. (2024) |
| Organic contaminants | Covers drugs, pesticides, and newly discovered microplastics | Limited monitoring in impoverished nations; microplastic decomposition during composting is not well known | O'Connor et al. (2022), Rafa et al. (2024), Masciarelli et al. (2024) |
| Stability & maturity | Evaluating compost maturity using C/N ratio (< 20: 1), low respiration index, and seed germination index | Variability in feedstocks influences homogeneity; no worldwide agreement on ideal maturity indicators | El Fels et al. (2024), Azis et al. (2023), Kim et al. (2016) |
| Odor & air emissions | Release of $NH_3$, $CH_4$, and VOCs from little aeration or excessive moisture content | Still understated in national inventories are greenhouse gas (GHG) emissions from composting | Tran et al. (2024), Nordahl et al. (2023) |
| Antibiotic resistance | From manure and biosolids, compost may harbor antibiotic-resistant genes (ARGs) | Lack of regular ARG monitoring; danger of ARG movement to groundwater and agriculture remains high | Li et al. (2023), Graham et al. (2019), Bobate et al. (2023) |
| Application regulations | Application rules based on nutrients stop eutrophication and leaching | In areas without organizations for compost certification, enforcement is poor | EPA (1995), GOI (2016), Huygens et al. (2019) |
| Certification & labeling | Compost classification and sales are governed by standards such FCO India or EU Regulation 2019/1009 | Small producers might lack access to testing facilities; existing certifications do not cover newly discovered pollutants | Regulation (EU) (2018), Biofertilizers and Organic Fertilizers Order, (1985) |
| Technological limitations | Quality monitoring is using internet of things (IoT) tools and oxygen sensors among other technologies more frequently | Low-income areas' high tech adoption costs; lack of knowledge about digital technologies | Senadheera et al. 2024; Duguma and Bai 2025 |
| Legal and policy gaps | Circular economy ideas and the SDGs support composting | Many times, national policies are out of current or disjointed; poor connection with agriculture laws and waste rules | Athanasoulia et al. (2021), Pittman et al. (2004), Cui et al. (2023) |

with enforcement, however. Although geography affects the production quality standards for compost, they usually rely on three factors: heavy metals, pathogens, and quantity of organic matter.

(i) Limitations in Heavy Metal:

Compost derived from sources of industrial or sewage waste may accumulate heavy metals like lead (Pb), cadmium (Cd), mercury (Hg), and arsenic (As). In order to lower soil contamination and food chain bioaccumulation, rules in important locations set maximum allowed levels for certain elements, as follows: (a) EU: Pb $\leq$ 120 mg/kg, Cd $\leq$ 1.5 mg/kg, Hg $\leq$ 1 mg/kg; (b) USA (EPA Part 503 Rule); Pb $\leq$ 300 mg/kg, Cd $\leq$ 39 mg/kg, Hg $\leq$ 17 mg/kg, and (c) India (FCO, 1985); Pb $\leq$ 100 mg/kg, Cd $\leq$ 5 mg/kg, Hg.

(ii) Pathogen Thresholds:

Composting depends on the control of pathogens, especially in cases of involving organic fertilizers derived from municipal waste or animal manure. To kill dangerous bacteria like *Salmonella, E. coli,* and *Listeria monocytogenes*, composting facilities have to maintain high temperatures during the thermophilic phase—50–70 °C.

The EU requires compost to be free of *Salmonella* in 25 g samples and to have *E. coli* counts less than 1000 CFU/g. Class A compost must contain fewer than 1000 MPN (Most Probable Number) of fecal coliforms per gram, as per US EPA Part 503 Rule.

Though they do not specify any particular microbiological criteria, India's FCO guidelines imply complete pathogen elimination.

(iii) Organic Matter

Compost quality is also evaluated using organic matter content and the carbon-to-nitrogen (C:N) ratio, both of which affect soil fertility. For at least 30%–40% organic matter and a C:N ratio of 20:1 to 30:1 for best microbial activity and nutritional balance, most countries need compost. Commonly used certification and compliance processes help confirm the safety and efficacy of compost products. These progams help ensure that compost meets legal requirements and is fit for organic agriculture.

EU Organic Regulation (EU 2018/848) bans the use of synthetic additives and requires strict quality testing; therefore, compost producers in the EU seeking organic certification must abide by it (Regulation (EU) 2018). Under the USDA National Organic Program (NOP), compost used on certified organic farms in the United States must follow strict pathogen-reducing methods and include no prohibited chemicals. Third-party certification bodies assess conformity; OMRI (Organic Materials Review Institute) is one example (https://www.omri.org/faq).

Based on its commitment to FCO 1985 and National Program for Organic Production (NPOP) (https://apeda.gov.in/national-programme-for-organic-production-npop), India has established the Green and Organic Certification Program to certify compost goods Following the GB/T 23486-2009 guidelines, China's organic compost certification guarantees that composted materials satisfy soil fertility criteria and are free of hazardous contaminants, thus ensuring environmental safety, soil health, and agricultural output depends on global composting criteria (Ma et al. 2022). Although many countries have established varying criteria for heavy metal, virus, and organic matter content, the general goal is to promote safe and sustainable compost use. Compost makers, especially those aiming for organic certification, must comply with these standards. Composting will always be crucial for sustainable waste management; however, worldwide standards must be harmonized and enforcement strategies improved to guarantee high-quality compost generation and thereby minimize environmental issues.

### *4.2 Waste Management Policies and Sustainable Development Goals (SDGs)*

Increasing sustainability depends on waste management rules, particularly with regard to organic waste recycling and composting. Composting has become more important for achieving sustainable waste management, as environmental preservation and resource efficiency take center stage globally. To support circular economy ideas, enhance soil health, and lower environmental pollution, composting rules must be in line with the United Nations Sustainable Development Goals (SDGs), particularly SDGs 12 (Responsible Consumption and Production) and 15 (Life on Land). Broad acceptance of sustainable composting methods is still hampered, however, by legislative enforcement problems, especially in developing nations. By means of waste reduction, enhanced resource efficiency, and sustainable material management, the Sustainable Development Goal 12 (SDG 12) seeks to guarantee sustainable consumption and production patterns. By taking organic waste from landfills and reusing it as valuable soil nutrients, composting directly aids this goal. Effective composting promotes responsible organic waste management, reduces landfill greenhouse gas emissions, and lessens the demand for synthetic fertilizers. In a similar vein, SDG 15 (Life on Land) emphasizes the preservation, rehabilitation, and sustainable use of terrestrial environments. By raising soil fertility, enhancing biodiversity, and thereby lowering land degradation brought on by too much synthetic fertilizer use, composting contributes toward this goal. A key component of sustainable agriculture, high-quality compost boosts microbial diversity, carbon sequestration, and soil rehabilitation. Nations incorporating SDG 12 and SDG 15 into their waste management initiatives have seen significant benefits for both the environment and their economy. Composting is one way the European Green Deal and the EU Waste Framework Directive encourage a circular bioeconomy. Likewise, India's Swachh Bharat Mission has underlined composting as part of its national waste management strategy, which corresponds with SDG 12 objectives by distributing composting systems that manage organic waste at the source. By emphasizing resource efficiency, waste reduction, and ongoing use of resources, a circular economy model helps minimize environmental impact. With a focus on organic waste recovery, quality control, and financial incentives for compost use, composting laws all around the world are progressively including circular economy concepts.

The Circular Economy Action Plan drives biowaste recycling into premium compost or biogas, minimizes landfilling, and promotes organic waste separation in the European Union. To increase composting and anaerobic digesting techniques, the EU mandates that member states make biowaste collection mandatory by 2024. Composting is also supported by the US Environmental Protection Agency (EPA) as part of its circular economy strategy. Consistent with sensible waste management goals, the United States Food Recovery Hierarchy recommends composting over landfilling and incineration. Laws like California's SB 1383 demand organic waste disposal and promote municipal level composting (https://www.smcsustainability.org/waste-reduction/sb-1383/).

China has included composting in its national agricultural targets as part of circular bioeconomy strategies. The Chinese Circular Economy Promotion Law stresses sustainable agriculture, organic waste recycling, and compost use. Large-scale composting implementation is difficult, nevertheless, because of contamination problems and uneven regulatory control over different regions. Although composting laws and circular economy ideas are becoming more important globally, many developing countries lack strong regulations because of structural and socio-economic challenges. The lack of facilities for waste separation and composting is one of the most important challenges. Many low-income nations mix inorganic materials with municipal solid waste, making high-quality compost production challenging. Effective source separation limits the contamination of composting operations by plastics, heavy metals, and harmful chemicals, thereby reducing the usefulness of compost as an agricultural input. Inadequate regulatory enforcement is another key issue. Although waste management policies abound in many developing countries, institutional inefficiencies, lack of monitoring mechanisms, and inadequate funding cause inefficient enforcement measures. For instance, poor municipal execution of the Solid Waste Management Rules of 2016 has resulted in limited adoption of distributed composting, even in India (https://cpcb.nic.in/uploads/MSW/SWM_2016.pdf). Likewise, in African nations, limited market acceptance and varied compost production stem from the absence of rigorous compost quality criteria (Mmereki et al. 2023).

Composting projects in developing nations may find their efficacy hampered by economic constraints. Because subsidies often support chemical inputs, many farmers lack incentives to utilize compost over synthetic fertilizers. Moreover, inadequate composting tools and waste collection techniques can lead to excessive production expenses for compost. Smallholder farmers should be encouraged to utilize compost by governments via public awareness campaigns, training courses, and financial incentives. Moreover, in many places public knowledge and view of composting still remain limited. In some developing countries, composting is still seenas an informal or secondary waste disposal method rather than a necessary part of sustainable development. Along with capacity-building initiatives, increasing understanding of the environmental and financial benefits of composting might help to enhance policy execution and community involvement.

Composting, as a sustainable solution for organic waste recycling, calls for efficient waste management practices linked with SDGs 12 and 15. Including concepts from the circular economy into composting guidelines in developed nations has produced significant benefits, both economically and environmentally. Meanwhile, developing countries still fail to apply composting rules because of poor public understanding, inadequate infrastructure, weak enforcement systems, and financial limitations. Maximizing composting's potential as a key contribution to sustainable agriculture and environmental preservation depends on overcoming these obstacles via improved waste management systems, financial incentives, and public engagement.

# 5 Innovations in Safe Composting Practices

## *5.1 Innovations in Safe Composting Practices, Artificial Intelligence (AI) and IoT-Based Composting Monitoring Systems*

Given more attention to sustainable waste management, composting has emerged as a significant method for recycling organic waste and increasing soil fertility. Still, maintaining the efficiency and safety of composting calls for significant biological and technological innovations. Changing composting operations are emerging technologies like artificial intelligence and IoT-based composting monitoring systems, biochar and nanomaterials for pollution removal, and microbial consortia for enhanced biodegradation and disease suppression. These developments assist in building safe and efficient composting systems, lowering environmental concerns, and increasing compost quality. As adsorbents and stabilizers help remove contaminants from compost, biochar and nanoparticles have become somewhat well known. Useful for adsorbing heavy metals, organic pollutants, and surplus nutrients, biochar, a carbon-rich material produced from biomass pyrolysis, is very porous and has a large surface area. By providing a stable habitat for beneficial bacteria, biochar reduces hazardous chemical leaching, enhances compost structure, and increases microbial activity in composting systems. Moreover, biochar maximizes microbial decomposition and reduces greenhouse gas emissions by improving the carbon-to-nitrogen (C:N) ratio.

Both carbon-based nanomaterials and metal oxides (such as $TiO_2$, $Fe_3O_4$, and ZnO) exhibit antibacterial and pollution-degrading properties. These nanoparticles provide compost safety by accelerating the breakdown of refractory organic compounds, like drugs and pesticide residues. Furthermore, they are used to immobilize heavy metals like cadmium (Cd) and lead (Pb), nanoscale zero-valent iron (nZVI) reduces their bioavailability in compost. The requirement for environmentally benign nanotechnological applications in composting is underlined by questions about the environmental persistence and probable toxicity of artificial nanoparticles.

## 5.2 AI and IoT-Based Composting Monitoring Systems

Artificial intelligence (AI) and the Internet of Things (IoT) are changing composting methods by improving process management, optimizing decomposition conditions (Senadheera et al. 2024), and ensuring compost quality. Conventional composting requires continuous warmth, moisture, aeration, and microbial activity, all of which can take time and be variable. Combining IoT-based sensor networks with AI-driven algorithms helps provide real-time data collection and autonomous modifications, thus considerably enhancing composting efficiency (Choudhary et al. 2025). IoT sensors used in composting systems measure factors like oxygen levels, carbon dioxide emissions, and ammonia content, thereby enabling artificial intelligence algorithms to forecast optimal composting conditions (Navarro et al. 2020; Duguma and Bai 2025). These creative fixes reduce process uncertainty, save energy, and eliminate overly strong odor emissions. Early identification of pollution or microbial imbalances, made feasible by artificial intelligence-based predictive modeling, would also allow rapid treatments to stop pathogen proliferation. Still another amazing development is the evaluation of microbial succession during composting, using machine learning techniques. Understanding microbial dynamics enables artificial intelligence to choose the suitable additives or inoculants to increase decomposition rates and control illnesses. Integration of IoT and artificial intelligence into composting systems allows remote monitoring and autonomous aeration control, hence lowering labor costs and enhancing compost uniformity (Chen et al. 2024).

## 5.3 Microbial Consortia for Enhanced Biodegradation and Pathogen Suppression

Microbial populations contribute significantly to composting by breaking down organic materials, controlling pathogens, and stabilizing compost. Recent researches have focused on creating microbial consortia—synergistic groupings of bacteria, fungi, and actinomycetes—that speed breakdown and improve compost safety (Wu et al. 2023). These specifically microbiologically based treatments eliminate toxins, break down lignocellulosic materials, and outcompete harmful bacteria. Important microbiological actors that break down complex plant waste include lignocellulose-degrading fungi (*Trichoderma* spp. and *Aspergillus* spp.) (Shinde et al. 2022). Microorganisms like Bacillus and Pseudomonas flourish at high temperatures, thereby increasing compost stabilization (Shi et al. 2022); actinomycetes generate antibiotic compounds that inhibit harmful bacteria like *Salmonella* and *E. coli* (Kolawole et al. 2023). Furthermore, microbial consortia of nitrogen-fixing and phosphate-solubilizing bacteria increase the availability of compost nutrients, thereby boosting the value of the produced end product for soil fertility (Ahmed et al. 2023). In composting, using these microbial consortia reduces pathogen burden, improves organic matter breakdown, and promotes plant-beneficial bacteria. Developments in biochar and nanotechnology, artificial intelligence-driven compost monitoring, and microbial consortia all help dramatically increase composting safety and efficiency. While IoT-based technologies improve composting conditions in real time, charcoal and nanoparticles help remove contaminants. Excellent compost results from microbial consortia enhancing biodegradation and lowering pathogen count. These developments will help implement safe and sustainable composting techniques, which complement worldwide efforts aimed at improving waste management, soil condition, and environmental sustainability.

## 6 Future Prospects and Research Gaps

Although composting is becoming increasingly well-known as a sustainable waste management solution, major challenges still exist that call for changes in technology, risk assessment strategies, and legislative frameworks. Important problems requiring focused study and policy interventions include the harmonization of worldwide composting standards, the development of advanced risk assessment strategies for novel contaminants, and the prospect of smart composting technology. These components will define how well composting promotes agricultural productivity, environmental sustainability, and soil health over the long term.

Though most people agree that composting is a sustainable approach, national and regional regulations for pathogen limits, compost quality, and pollution levels vary widely. Different composting policies among the EU, EPA, China, India, and regulatory authorities of other nations produce differences in compost quality control and market acceptance. International trade, high-quality compost production, and the environmental and health risks connected with contaminated compost all depend on a uniform global composting system. Establishing common guidelines for heavy metal limitations, pathogen thresholds, organic matter concentration, and emerging contaminants would improve the credibility of composting as a sustainable method. Cross-border regulatory cooperation, certification systems, and standardized compost testing methods may all contribute to increasing consumer confidence and agricultural compost acceptance. Emerging pollutants like microplastics, pharmaceutical wastes, POPs, and EDCs compromise human well-being, plant growth, and soil quality. Usually neglecting the long-term consequences of these recently discovered pollutants, conventional compost quality assessments are mostly focused on heavy metals and microbial contamination. Sophisticated risk assessment methods, combining high-density mass spectrometry, bioinformatics, and predictive modeling, are required to tackle these challenges. Metagenomic research, biosensors, and quantitative structure-activity relationship (QSAR) modeling, among other approaches, might provide further details on pollution fate and transformation during composting. Moreover, studies on biodegradation, including microbial and enzymatic paths for the breakdown of microplastics and pharmaceutical wastes, might guide the development of efficient mitigating strategies. Smart composting technologies, such as AI, IoT, and automation, taken together, have great potential to raise compost efficiency and preserve continuous compost quality. Real-time monitoring of temperature, moisture, aeration, and gaseous emissions, made possible by IoT-enabled sensor networks might enable exact control of composting conditions. Prediction models driven by artificial intelligence might increase microbial activity, hasten the decomposition of organic waste, and find early signs of process failure or contamination. Blockchain technology might also increase traceability and openness in compost production, thereby ensuring regulatory compliance and enabling farmers to get premium, certified compost. By producing compost fit for specific soil and crop demands, smart composting systems will not only help in climate-resilient agriculture but also enhance waste reutilization. Future studies on composting should focus on establishing global rules, enhancing risk assessment strategies for new contaminants, and using smart technology to maximize processes. Dealing with these research gaps will ensure that composting is a scalable, safe, and effective substitute for sustainable agriculture and organic waste management.

## 7 Conclusion

Although composting is a vital method for sustainable waste management and increasing soil fertility, many safety concerns and legal obstacles have to be solved before it can be generally adopted. Major safety concerns include pathogenic germs (e.g., *Salmonella, E. coli, Listeria*), heavy metal contamination from industrial waste and fertilizers, and novel contaminants like microplastics and pharmaceutical residues. Furthermore, undermining quality assurance, international trade, and general acceptance of composting techniques are disparities in global composting standards. Compost quality and safety are erratic without clear criteria for pathogen eradication, heavy metal limits, and organic matter content, thereby compromising human health and the environment. Strategic improvements including politicians, scientists, and farmers will help improve compost safety and regulatory compliance. Policymakers should aim to standardize international compost laws, set high standards for quality, and support the use of smart monitoring technologies to instantly verify compost safety. Researchers should focus on developing advanced risk assessment strategies, enhancing microbial consortia for disease control, and investigating new remedial options like biochar, nanomaterials, and artificial intelligence-based monitoring systems. To generate high-quality compost that satisfies safety and sustainability criteria, farmers and compost producers must be trained in suitable composting techniques, including accurate feedstock selection, temperature management, and contaminant mitigating. Future composting systems must combine automation,

IoT-enabled monitoring, and AI-driven process management if we are to raise efficiency and lower risks. Moreover, maintaining compost safety will depend on improved waste segregation policies, biodegradable substitutes for plastics, and microbial-based degradation methods for newly introduced contaminants. Solving these problems and using innovative ideas will help composting become a very effective, safe, and globally regulated method that is crucial for environmental protection and sustainable development.

## References

Aguilar-Paredes A, Valdés G, Araneda N, Valdebenito E, Hansen F, Nuti M (2023) Microbial community in the composting process and its positive impact on the soil biota in sustainable agriculture. Agronomy 13:542. https://doi.org/10.3390/agronomy13020542

Ahmad SZN, Wan Salleh WN, Ismail AF, Yusof N, Mohd Yusop MZ, Aziz F (2020) Adsorptive removal of heavy metal ions using graphene-based nanomaterials: toxicity, roles of functional groups and mechanisms. Chemosphere 248:126008. https://doi.org/10.1016/j.chemosphere.2020.126008

Ahmed T, Noman M, Qi Y, Shahid M, Hussain S, Masood HA, Xu L, Ali HM, Negm S, El-Kott AF, Yao Y, Qi X, Li B (2023) Fertilization of microbial composts: a technology for improving stress resilience in plants. Plants 12:3550. https://doi.org/10.3390/plants12203550

Alam K, Mazumder A, Sikdar S, Zhao YM, Hao J, Song C, Wang Y, Sarkar R, Islam S, Zhang Y, Li A (2022) Streptomyces: the biofactory of secondary metabolites. Front Microbiol 13:968053. https://doi.org/10.3389/fmicb.2022.968053

Al-Ansari M, Alkubaisi N, Vijayaragavan P, Murugan K (2019) Antimicrobial potential of Streptomyces sp. to the Gram positive and Gram negative pathogens. J Infect Public Health 12:861–866. https://doi.org/10.1016/j.jiph.2019.05.016

Alengebawy A, Abdelkhalek ST, Qureshi SR, Wang MQ (2021) Heavy metals and pesticides toxicity in agricultural soil and plants: ecological risks and human health implications. Toxics 9:42. https://doi.org/10.3390/toxics9030042

Athanasoulia IG, Novakovic J, Kyriazi M (2021) Inventory and integrated assessment of the EU regulations related to the production and application of bio-based fertilisers. D6.1 Report,. EU Project Deliverable

Attia TS, Elsheery NI (2020) Nanomaterials: scope, applications, and challenges in agriculture and soil reclamation. In: Sustainable agriculture reviews 41: nanotechnology for plant growth and development. Springer, Cham, pp 1–39

Ayangbenro AS, Olanrewaju OS, Babalola OO (2018) Sulfate-reducing bacteria as an effective tool for sustainable acid mine bioremediation. Front Microbiol 9:1986. https://doi.org/10.3389/fmicb.2018.01986

Azis FA, Choo M, Suhaimi H, Abas PEA (2023) The effect of initial carbon to nitrogen ratio on kitchen waste composting maturity. Sustainability 15:6191. https://doi.org/10.3390/su15076191

Barik D (2019) Energy extraction from toxic waste originating from food processing industries. In: Barik D (ed) Energy from toxic organic waste for heat and power generation. Woodhead Publishing Series in Energy, pp 17–42. https://doi.org/10.1016/B978-0-08-102528-4.00003-1

Behl K, Jaiswal P, Pabbi S (2024) Recent advances in microbial and nano-formulations for effective delivery and agriculture sustainability. Biocatal Agric Biotechnol:103180. https://doi.org/10.1016/j.bcab.2024.103180

Biofertilizers and Organic Fertilizers, Order 1985—the Fertiliser (Inorganic, Organic or Mixed) (Control). https://nconf.dac.gov.in/uploads/quality_control/Green-Book-FCO-2023-Edition.pdf
Bobate S, Mahalle S, Dafale NA, Bajaj A (2023) Emergence of environmental antibiotic resistance: mechanism, monitoring and management. Environ Adv 13:100409. https://doi.org/10.1016/j.envadv.2023.100409
Chandna P, Nain L, Singh S, Kuhad RC (2013) Assessment of bacterial diversity during composting of agricultural byproducts. BMC Microbiol 13:99. https://doi.org/10.1186/1471-2180-13-99
Chen L, He P, Zhang H, Peng W, Qiu J, Lü F (2024) Applications of machine learning tools for biological treatment of organic wastes: perspectives and challenges. Circ Econ:100088. https://doi.org/10.1016/j.circeco.2024.100088
Choudhary V, Guha P, Pau G, Mishra S (2025) An overview of smart agriculture using internet of things (IoT) and web services. Environ Sustain Indicat:100607. https://doi.org/10.1016/j.indic.2024.100607
Cucina M, Tacconi C, Ricci A, Pezzolla D, Sordi S, Zadra C, Gigliotti G (2018) Evaluation of benefits and risks associated with the agricultural use of organic wastes of pharmaceutical origin. Sci Total Environ 613:773–782. https://doi.org/10.1016/j.scitotenv.2017.09.181
Cui D, Xi B, Tan W (2023) Composting industry under the Chinese municipal solid waste sorting policy: challenges, opportunities, and directions. Environ Sci Pollut Res 30:19513–19519. https://doi.org/10.1007/s11356-023-25367-0
De Corato U (2020) Disease-suppressive compost enhances natural soil suppressiveness against soil-borne plant pathogens: a critical review. Rhizosphere 13:100192. https://doi.org/10.1016/j.rhisph.2020.100192
De la Rosa JM, Santa-Olalla A, Campos P, López-Núñez R, González-Pérez JA, Almendros G, Knicker HE, Sánchez-Martín Á, Fernández-Boy E (2022) Impact of biochar amendment on soil properties and organic matter composition in trace element-contaminated soil. Int J Environ Res Public Health 19:2140. https://doi.org/10.3390/ijerph19042140
Dinakarkumar Y, Gnanasekaran R, Reddy GK, Vasu V, Balamurugan P, Murali G (2024) Fungal bioremediation: an overview of the mechanisms, applications and future perspectives. Environ Chem Ecotoxicol 6:293–302. https://doi.org/10.1016/j.enceco.2024.01.025
Dong X, Chu Y, Tong Z, Sun M, Meng D, Yi X, Gao T, Wang M, Duan J (2024) Mechanisms of adsorption and functionalization of biochar for pesticides: a review. Ecotoxicol Environ Saf 272:116019. https://doi.org/10.1016/j.ecoenv.2024.116019
Duguma AL, Bai X (2025) How the internet of things technology improves agricultural efficiency. Artif Intell Rev 58:63. https://doi.org/10.1007/s10462-024-11046-0
Easwaran C, Moorthy G, Christopher SR, Mohan P, Marimuthu R, Koothan V, Nallusamy S (2024) Nano hybrid fertilizers: a review on the state of the art in sustainable agriculture. Sci Total Environ:172533. https://doi.org/10.1016/j.scitotenv.2024.172533
El Fels L, Naylo A, Jemo M, Zrikam N, Boularbah A, Ouhdouch Y, Hafidi M (2024) Microbial enzymatic indices for predicting composting quality of recalcitrant lignocellulosic substrates. Front Microbiol 15:1423728. https://doi.org/10.3389/fmicb.2024.1423728
El Hayany B, El Fels L, Ouhdouch Y, Hafidi M (2021) Fate of pathogenic microorganisms during lagooning sludge composting and exploration of bacteriophages as indicator of hygienization. Environ Technol Innov 21:101268. https://doi.org/10.1016/j.eti.2020.101268
EPA—United States. Office of Wastewater Management (1995) A guide to the biosolids risk assessments for the EPA Part 503 rule. US Environmental Protection Agency, Office of Wastewater Management. https://www.epa.gov/biosolids/guide-biosolids-risk-assessments-epa-part-503-rule
Finore I, Feola A, Russo L et al (2023) Thermophilic bacteria and their thermozymes in composting processes: a review. Chem Biol Technol Agric 10:7. https://doi.org/10.1186/s40538-023-00381-z

Gattupalli M, Dashora K, Mishra M, Javed Z, Tripathi GD (2023) Microbial bioprocess performance in nanoparticle-mediated composting. Crit Rev Biotechnol 43(8):1193–1210. https://doi.org/10.1080/07388551.2023.2222764

GOI—Ministry of Environment, Forest and Climate Change (2016) The Solid Waste Management Rules, 2016. The Gazette of India, Part II, Section 3(ii), S.O. 1357(E), April 8, 2016. https://cpcb.nic.in/uploads/MSW/SWM_2016.pdf

Graham DW, Bergeron G, Bourassa MW et al (2019) Complexities in understanding antimicrobial resistance across domesticated animal, human, and environmental systems. Ann N Y Acad Sci 1441(1):17–30. https://doi.org/10.1111/nyas.14036

Gupta P, Diwan B (2016) Bacterial exopolysaccharide mediated heavy metal removal: a review on biosynthesis, mechanism and remediation strategies. Biotechnol Rep Amst 13:58–71. https://doi.org/10.1016/j.btre.2016.12.006

Han W, Chen S, Tan X et al (2022) Microbial community succession in response to sludge composting efficiency and heavy metal detoxification during municipal sludge composting. Front Microbiol 13:1015949. https://doi.org/10.3389/fmicb.2022.1015949

Harirchi S, Wainaina S, Sar T et al (2022) Microbiological insights into anaerobic digestion for biogas, hydrogen or volatile fatty acids (VFAs): a review. Bioengineered 13(3):6521–6557. https://doi.org/10.1080/21655979.2022.2035986

Hasan MM, Tarannum MN (2024) Adverse impacts of microplastics on soil physicochemical properties and crop health in agricultural systems. J Hazard Mater Adv:100528. https://doi.org/10.1016/j.hazadv.2024.100528

Haynes E, Ramwell C, Griffiths T, Walker D, Smith J (2020) Review of antibiotic use in crops, associated risk of antimicrobial resistance and research gaps. Report to Department for Environment, Food and Rural Affairs (Defra) & The Food Standards Agency (FSA), pp 1–83. https://www.food.gov.uk/sites/default/files/media/document/review-of-antibiotic-use-in-crops_1.pdf

Ho TTK, Le TH, Tran C-S, Nguyen P-T, Vo T-D-H, Thai V-N, Bui X-T (2022) Compost to improve sustainable soil cultivation and crop productivity. Case Stud Chem Environ Eng 6: 100211. https://doi.org/10.1016/j.cscee.2022.100211

Huygens D, Saveyn H, Tonini D, Eder PDSL, Delgado Sancho L (2019) Technical proposals for selected new fertilising materials under the Fertilising products regulation (regulation (EU) 2019/1009). Publications Office of the European Union, Luxembourg. https://data.europa.eu/doi/10.2760/186684

Insam H, Bertoldi M (2007) Microbiology of the composting process. In: Diaz LF, de Bertoldi M, Bidlingmaier W, Stentiford E (eds) Waste management series, vol 8. Elsevier, Amsterdam, pp 25–48. https://doi.org/10.1016/S1478-7482(07)80006-6

Izydorczyk G, Skrzypczak D, Mironiuk M et al (2024) Lignocellulosic biomass fertilizers: production, characterization, and agri-applications. Sci Total Environ 923:171343. https://doi.org/10.1016/j.scitotenv.2024.171343

Jones DL (1999) Potential health risks associated with the persistence of *Escherichia coli* O157 in agricultural environments. Soil Use Manag 15(2):76–83. https://doi.org/10.1111/j.1475-2743.1999.tb00069.x

Jurado MM, Suárez-Estrella F, López MJ, López-González JA, Moreno J (2019) Bioprospecting from plant waste composting: Actinobacteria against phytopathogens producing damping-off. Biotechnol Rep (Amst) 23:e00354. https://doi.org/10.1016/j.btre.2019.e00354

Khan S, Akhtar N, Rehman SU, Shujah S, Rha ES, Jamil M (2022) *Bacillus subtilis* synthesized iron oxide nanoparticles ($Fe_3O_4$ NPs) induced metabolic and anti-oxidative response in rice (*Oryza sativa* L.) under arsenic stress. Toxics 10(10):618. https://doi.org/10.3390/toxics10100618

Khatun J, Intekhab A, Dhak D (2022) Effect of uncontrolled fertilization and heavy metal toxicity associated with arsenic (As), lead (Pb) and cadmium (Cd), and possible remediation. Toxicology 477:153274. https://doi.org/10.1016/j.tox.2022.153274

Kim E, Lee DH, Won S, Ahn H (2016) Evaluation of optimum moisture content for composting of beef manure and bedding material mixtures using oxygen uptake measurement. Asian-Australas J Anim Sci 29(5):753–758. https://doi.org/10.5713/ajas.15.0875

Kim C, Alrefaei R, Bushlaibi M, Ndegwa E, Kaseloo P, Wynn C (2019) Influence of growth temperature on thermal tolerance of leading foodborne pathogens. Food Sci Nutr 7(12): 4027–4036. https://doi.org/10.1002/fsn3.1268

Kolawole AA, Adelegan AJ, Chukundah UD (2023) A review of the application of actinomycetes in the control of *Salmonella* species. BW Acad J 2023:7–7. https://bwacademicjournal.com/index.php/bwaj/article/view/94

Lemunier M, Francou C, Rousseaux S, Houot S, Dantigny P, Piveteau P, Guzzo J (2005) Long-term survival of pathogenic and sanitation indicator bacteria in experimental biowaste composts. Appl Environ Microbiol 71(10):5779–5786. https://doi.org/10.1128/AEM.71.10.5779-5786.2005

Li S, Ondon BS, Ho SH, Li F (2023) Emerging soil contamination of antibiotic resistance bacteria (ARB) carrying genes (ARGs): new challenges for soil remediation and conservation. Environ Res 219:115132. https://doi.org/10.1016/j.envres.2022.115132

Li X, Shen X, Jiang W, Xi Y, Li S (2024) Comprehensive review of emerging contaminants: detection technologies, environmental impact, and management strategies. Ecotoxicol Environ Saf 278:116420. https://doi.org/10.1016/j.ecoenv.2024.116420

Liang H, Tan Y, Yin J, Peng Y, Wei M, Chen H, Chen Q (2024) Phosphate fertilizers' dual role in cadmium-polluted acidic agricultural soils: dosage dependency and passivation potential. Agronomy 14(10):2201. https://doi.org/10.3390/agronomy14102201

Lim JY, Yoon J, Hovde CJ (2010) A brief overview of *Escherichia coli* O157:H7 and its plasmid O157. J Microbiol Biotechnol 20(1):5–14. https://doi.org/10.4014/jmb.0907.07007

Lima T, Domingues S, Da Silva GJ (2020) Manure as a potential hotspot for antibiotic resistance dissemination by horizontal gene transfer events. Vet Sci 7(3):110. https://doi.org/10.3390/vetsci7030110

Lopez MJ, Jurado MM, López-González JA, Estrella-González MJ, Martínez-Gallardo MR, Toribio A, Suárez-Estrella F (2021) Characterization of thermophilic lignocellulolytic microorganisms in composting. Front Microbiol 12:697480. https://doi.org/10.3389/fmicb.2021.697480

Ma D, Wang Y, Ye Y, Ge X, Lu X (2022) Effects of three sludge products from co-treatment of wastewater on the soil properties and plant growth of silty loam. Int J Environ Res Public Health 19(7):4385. https://doi.org/10.3390/ijerph19074385

Majumder S, Chatterjee S, Basnet P, Mukherjee J (2020) ZnO-based nanomaterials for photocatalytic degradation of aqueous pharmaceutical waste solutions–a contemporary review. Environ Nanotechnol Monit Manage 14:100386. https://doi.org/10.1016/j.enmm.2020.100386

Mantovam VB, Dos Santos DF, Giola Junior LC, Landgraf M, Pinto UM, Todorov SD (2025) *Listeria monocytogenes*, *Salmonella* spp., and *Staphylococcus aureus*: threats to the food industry and public health. *Foodborne Pathog Dis*. (in press)

Manual G (2023) Best practices and strategies for pathogen-free compost production. Self-published

Marini M, Caro D, Thomsen M (2020) The new fertilizer regulation: a starting point for cadmium control in European arable soils? Sci Total Environ 745:140876. https://doi.org/10.1016/j.scitotenv.2020.140876

Masciarelli E, Casorri L, Di Luigi M, Beni C, Valentini M, Costantini E, Aielli L, Reale M (2024) Microplastics in agricultural crops and their possible impact on farmers' health: a review. Int J Environ Res Public Health 22(1):45. https://doi.org/10.3390/ijerph22010045

Melis E, Asquer C, Carboni G, Scano EA (2023) Role of *Cannabis sativa* L. in energy production: residues as a potential lignocellulosic biomass in anaerobic digestion plants. In: García-Tejero IF, Durán-Zuazo VH (eds) Current applications, approaches, and potential perspectives for hemp. Academic Press, pp 111–199. https://doi.org/10.1016/B978-0-323-89867-6.00003-2

Mmereki D, David VE, Wreh Brownell AH (2023) The management and prevention of food losses and waste in low- and middle-income countries: a mini-review in the Africa region. Waste Manag Res 42(4):287–307. https://doi.org/10.1177/0734242X231184444

Navarro E, Costa N, Pereira A (2020) A systematic review of IoT solutions for smart farming. Sensors (Basel) 20(15):4231. https://doi.org/10.3390/s20154231

Nordahl SL, Preble CV, Kirchstetter TW, Scown CD (2023) Greenhouse gas and air pollutant emissions from composting. Environ Sci Technol 57(6):2235–2247. https://doi.org/10.1021/acs.est.2c05846

O'Connor J, Mickan BS, Siddique KH, Rinklebe J, Kirkham MB, Bolan NS (2022) Physical, chemical, and microbial contaminants in food waste management for soil application: a review. Environ Pollut 300:118860. https://doi.org/10.1016/j.envpol.2022.118860

Olawade DB, Wada OZ, Egbewole BI, Fapohunda O, Ige AO, Usman SO, Ajisafe O (2024) Metal and metal oxide nanomaterials for heavy metal remediation: novel approaches for selective, regenerative, and scalable water treatment. Front Nanotechnol 6:1466721. https://doi.org/10.3389/fnano.2024.1466721

Ondon BS, Li S, Zhou Q, Li F (2021) Sources of antibiotic resistant bacteria (ARB) and antibiotic resistance genes (ARGs) in the soil: a review of the spreading mechanism and human health risks. Rev Environ Contam Toxicol 256:121–153. https://doi.org/10.1007/398_2020_60

Osek J, Lachtara B, Wieczorek K (2022) *Listeria monocytogenes*—how this pathogen survives in food-production environments? Front Microbiol 13:866462. https://doi.org/10.3389/fmicb.2022.866462

Padan E, Bibi E, Ito M, Krulwich TA (2005) Alkaline pH homeostasis in bacteria: new insights. Biochim Biophys Acta 1717(2):67–88. https://doi.org/10.1016/j.bbamem.2005.09.010

Pande V, Pandey SC, Sati D, Bhatt P, Samant M (2022) Microbial interventions in bioremediation of heavy metal contaminants in agroecosystem. Front Microbiol 13:824084. https://doi.org/10.3389/fmicb.2022.824084

Pittman HM, Mirus S, Peters K, Reed R (2004) A legal guide to the National Organic Program. NationalAgLawCenter.com. https://nationalaglawcenter.org/publication/a-legal-guide-to-the-national-organic-program/

Placha I, Venglovský J, Maková Z, Martinéz J (2008) The elimination of *Salmonella typhimurium* in sewage sludge by aerobic mesophilic stabilization and lime hydrated stabilization. Bioresour Technol 99(10):4269–4274. https://doi.org/10.1016/j.biortech.2007.08.056

Primpke S, Fischer M, Lorenz C, Gerdts G, Scholz-Böttcher BM (2020) Comparison of pyrolysis gas chromatography/mass spectrometry and hyperspectral FTIR imaging spectroscopy for the analysis of microplastics. Anal Bioanal Chem 412:8283–8298. https://doi.org/10.1007/s00216-020-02878-1

Priya AK, Gnanasekaran L, Dutta K, Rajendran S, Balakrishnan D, Soto-Moscoso M (2022) Biosorption of heavy metals by microorganisms: evaluation of different underlying mechanisms. Chemosphere 307:135957. https://doi.org/10.1016/j.chemosphere.2022.135957

Priya AK, Muruganandam M, Ali SS, Kornaros M (2023) Clean-up of heavy metals from contaminated soil by phytoremediation: a multidisciplinary and eco-friendly approach. Toxics 11:422. https://doi.org/10.3390/toxics11050422

Rafa N, Ahmed B, Zohora F, Bakya J, Ahmed S, Ahmed SF, Mofijur M, Chowdhury AA, Almomani F (2024) Microplastics as carriers of toxic pollutants: source, transport, and toxicological effects. Environ Pollut 343:123190. https://doi.org/10.1016/j.envpol.2023.123190

Rajlakshmi, Jadhav DA, Dutta S, Sherpa KC, Komal J, Saravanabhupathy S, Mohanty KT, Banerjee R, Kumar J, Rajak RC (2023) Co-digestion processes of waste: status and perspective. In: Hussain CM, Kushwaha A, Bharagava RN, Goswami L (eds) Adv Zero Waste Tools Bio-Based Mater Waste Energy Gener Resour Manag, vol 5, pp 207–241. https://doi.org/10.1016/B978-0-323-91149-8.00010-7

Regulation (EU) (2018) Regulation (EU) 2018/848 on organic production and labelling of organic products. https://data.europa.eu/eli/reg/2018/848/oj

Senadheera UE, Jayasanka J, Udayanga D, Hewawasam C, Amila B, Takimoto Y, Hatamoto M, Tadachika N (2024) Beyond composting basics: a Sustainable Development Goals-oriented strategic guidance to IoT integration for composting in modern urban ecosystems. Sustainability 16:10332. https://doi.org/10.3390/su162310332

Shafiq F, Anwar S, Zhang L, Ashraf M (2023) Nano-biochar: properties and prospects for sustainable agriculture. Land Degrad Dev 34:2445–2463. https://doi.org/10.1002/ldr.4602

Shi W, Dong Q, Saleem M, Wu X, Wang N, Ding S, Huang J, Wang X, Zhou B, Gao Z (2022) Microbial-based detonation and processing of vegetable waste for high quality compost production at low temperatures. J Clean Prod 369:133276. https://doi.org/10.1016/j.jclepro.2022.133276

Shinde R, Shahi DK, Mahapatra P, Naik SK, Thombare N, Singh AK (2022) Potential of lignocellulose degrading microorganisms for agricultural residue decomposition in soil: A review. J Environ Manage 320:115843. https://doi.org/10.1016/j.jenvman.2022.115843

Timofeeva A, Galyamova M, Sedykh S (2022) Prospects for using phosphate-solubilizing microorganisms as natural fertilizers in agriculture. Plants 11:2119. https://doi.org/10.3390/plants11162119

Tran HT, Binh QA, Van Tung T, Pham DT, Hoang HG, Nguyen NSH, Xie S, Zhang T, Mukherjee S, Bolan NS (2024) A critical review on characterization, human health risk assessment and mitigation of malodorous gaseous emission during the composting process. Environ Pollut 351:124115. https://doi.org/10.1016/j.envpol.2024.124115

Vasilachi IC, Stoleru V, Gavrilescu M (2023) Analysis of heavy metal impacts on cereal crop growth and development in contaminated soils. Agriculture 13:1983. https://doi.org/10.3390/agriculture13101983

Viotti P, Marzeddu S, Antonucci A, Décima MA, Lovascio P, Tatti F, Boni MR (2024) Biochar as alternative material for heavy metal adsorption from groundwaters: lab-scale (column) experiment review. Materials 17:809. https://doi.org/10.3390/ma17040809

Wang W, Wang X, Liu J, Ishii M, Igarashi Y, Cui Z (2007) Effect of oxygen concentration on the composting process and maturity. Compost Sci Util 15:184–190. https://doi.org/10.1080/1065657X.2007.10702331

Wani NR, Rather RA, Farooq A, Padder SA, Baba TR, Sharma S, Mubarak NM, Khan AH, Singh P, Ara S (2024) New insights in food security and environmental sustainability through waste food management. Environ Sci Pollut Res 31:17835–17857. https://doi.org/10.1007/s11356-023-26462-y

Wawrzynczak A, Chudzińska J, Feliczak-Guzik A (2024) Metal and metal oxides nanoparticles as nanofillers for biodegradable polymers. ChemPhysChem 25:e202300823. https://doi.org/10.1002/cphc.202300823

Wei Y, Zhao Y, Zhao X, Gao X, Zheng Y, Zuo H, Wei Z (2020) Roles of different humin and heavy-metal resistant bacteria from composting on heavy metal removal. Bioresour Technol 296:122375. https://doi.org/10.1016/j.biortech.2019.122375

Wu D, Wang W, Yao Y, Li H, Wang Q, Niu B (2023) Microbial interactions within beneficial consortia promote soil health. Sci Total Environ 900:165801. https://doi.org/10.1016/j.scitotenv.2023.165801

Yuan Z, Nag R, Cummins E (2022) Human health concerns regarding microplastics in the aquatic environment—from marine to food systems. Sci Total Environ 823:153730. https://doi.org/10.1016/j.scitotenv.2022.153730

Zhang L, Zhu Y, Zhang J, Zeng G, Dong H, Cao W, Fang W, Cheng Y, Wang Y, Ning Q (2019) Impacts of iron oxide nanoparticles on organic matter degradation and microbial enzyme activities during agricultural waste composting. Waste Manag 95:289–297. https://doi.org/10.1016/j.wasman.2019.06.039

Zheng X, Zou D, Wu Q, Wang H, Li S, Liu F, Xiao Z (2022) Review on fate and bioavailability of heavy metals during anaerobic digestion and composting of animal manure. Waste Manag 150:75–89. https://doi.org/10.1016/j.wasman.2022.09.023

Zheng X, Lin H, Du D, Li G, Alam O, Cheng Z, Liu X, Jiang S, Li J (2024) Remediation of heavy metals polluted soil environment: a critical review on biological approaches. Ecotoxicol Environ Saf 284:116883. https://doi.org/10.1016/j.ecoenv.2024.116883
Zhou Z, Hua J, Xue J, Yu C (2024) Differential impacts of polyethylene microplastic and additives on soil nitrogen cycling: a deeper dive into microbial interactions and transformation mechanisms. Sci Total Environ 942:173771. https://doi.org/10.1016/j.scitotenv.2024.173771

# Index


**B**

Biocatalysis, 323, 380

**C**

Cattle industry waste, 138, 140, 141, 144–146
Chemical fertilizers, 4, 11, 22, 31, 36, 73–75, 108, 116, 124, 127, 137, 229–242, 250, 252, 254, 255, 259, 260, 263, 278, 279, 288, 294, 297, 322, 375
Circular economy, 30, 40, 75, 90, 102, 122, 127, 138, 147, 148, 179, 322, 350, 351, 359, 361, 368, 371, 373, 375, 395, 397–399
Compost, 1, 30, 65, 90, 122, 138, 153, 179, 230, 249, 270, 307, 336, 362, 379
Composting, 1, 30, 64, 89, 116, 138, 153, 177, 230, 249, 271, 307, 335, 359, 379
Crop productivity, 6, 37–39, 61, 125, 147, 148, 199, 200, 242, 254, 275, 278, 284–287, 290, 297, 392

**E**

Ecosystem, 30, 35, 48, 61, 74, 91, 92, 97, 119, 127, 138, 147, 148, 153, 155, 158–160, 169, 178, 179, 201–205, 233, 250–254, 271, 295, 314, 335, 336, 348, 352, 353, 375, 380
Environmental health, 89–110, 119, 291, 393

**F**

Feedstocks, 4, 22, 30–32, 35, 39, 41, 45, 92–95, 98–100, 102, 104, 156, 181, 184, 186, 187, 189, 190, 197, 249, 255, 258, 280, 281, 291, 294–296, 338, 339, 349, 351, 353, 389, 395, 402

**G**

Greenhouse gas mitigation, 95, 278
Green manures, 7, 8, 30, 36–37, 46, 230, 234, 240, 288

**H**

Hazards, 97, 116, 121, 234, 236, 294, 295, 380, 381, 387, 391–393, 395
Heavy metals, 22, 38, 40, 41, 43, 44, 47, 101, 102, 108, 121, 139, 145, 146, 184, 198, 203, 204, 239–242, 252, 262, 274, 288, 290, 295, 297, 307–309, 314, 318, 325, 336, 339, 349, 353, 372, 375, 379–381, 384, 385, 389–392, 394–399, 401, 402

**I**

Innovative fertilizer, 352




R. K. Singh et al. (eds.), *Composting of Farm Residues by Nature-Based Solutions*,
https://doi.org/10.1007/978-3-032-12255-1

**M**

Metagenomics, 154–165, 167, 168, 342, 382, 401
Metaproteomics, 154, 155, 159, 161–163, 167
Metatranscriptomics, 154, 155, 159–162, 167
Methane ($CH_4$), 2, 22, 62, 89, 91–95, 98–101, 106, 109, 118, 138–140, 144, 146–148, 154, 160, 163, 167, 168, 177, 181, 187, 188, 204, 236, 280, 322, 343, 360, 367–370, 374, 380, 384
Microbial dynamics, 43, 96, 100, 137–148, 162, 166, 381–386, 400

**N**

Nanocomposting, 309, 318, 321, 322, 325–327
Nanoparticles (NPs), 307–322, 324–328, 343, 348, 349, 384–386, 388, 390, 399–401
Nanotechnology, 269, 320, 323, 380, 384–386, 388–391, 393, 401
Nanozymes, 309, 323–325
Nitrous oxide emissions, 95–96, 101, 147
Nutrient bioavailability, 256, 263, 352
Nutrient management (NM), 7, 200–201, 205, 270, 271, 277–279, 285, 322
Nutrient recycling, 48, 89, 93, 107, 188, 250, 254, 271
Nutrient uptake, 186, 233, 269–297

**O**

Organic amendments, 34, 60, 61, 153, 156, 197, 198, 230, 238, 239, 242, 255, 258–261, 278, 286
Organic macromolecules, 381
Organic matter degradation, 41, 43, 100, 101, 124, 127, 141, 157, 158
Organic residues, 30, 35, 59–62, 64, 65, 70, 71, 75, 115–128, 140, 147, 148
Organic wastes, 2, 7, 9, 10, 12–17, 19, 22, 30, 33, 36, 40, 45, 46, 60, 89, 90, 94, 95, 97, 98, 100, 101, 107–110, 138, 140, 146, 148, 177, 179–181, 184–186, 189, 190, 195, 204, 205, 230, 231, 233, 237, 242, 249, 257, 270, 280, 288, 290, 294, 307, 308, 311, 318–319, 322, 324, 336, 337, 347, 352, 359–361, 363–376, 379, 381–384, 386, 389, 391, 397–399, 402

**P**

Pathogen control, 382, 386–388, 394
Productivity, 6, 7, 29, 31, 32, 37–39, 46, 47, 60, 61, 75, 90, 97, 108, 109, 116, 125, 138, 147, 148, 179, 185, 195, 199–202, 231, 233, 234, 237, 242, 250, 252, 254, 255, 258, 263, 270, 271, 275, 277, 278, 284–290, 294, 297, 312, 352, 353, 370, 375, 392, 401

**R**

Recycling, 7, 10–12, 30, 31, 34, 39, 40, 64, 108, 110, 139, 140, 147, 179–181, 184, 195, 205, 230, 257, 288, 322, 336, 339, 352, 354, 372, 373, 397–399
Renewable resources, 347
Resource efficiency, 137, 140, 179, 181, 397

**S**

Soil biodiversity, 182, 188, 201, 392
Soil fertility, 2, 3, 6–8, 15, 22, 29, 31–33, 35, 36, 38, 39, 60, 67, 73, 74, 108, 116, 117, 125, 127, 138, 140, 179, 195–198, 200–205, 230, 234, 238, 240, 241, 251, 253, 254, 256, 258, 263, 270, 275, 279, 284–288, 290, 296, 322, 352, 353, 359, 379, 381, 385, 389, 391, 396, 397, 399, 401, 402
Soil fertility enhancement, 180, 181
Soil health, 7, 22, 30–33, 36–38, 40, 41, 46, 47, 60, 61, 66, 67, 71–75, 90, 107–109, 118, 138, 140, 147, 148, 153, 181, 195, 197, 199–205, 231, 249–263, 270, 275, 276, 278, 285, 287, 288, 295–297, 322, 348, 353, 373–375, 379, 393, 397, 401
Soil organic carbon (SOC), 108, 109, 179, 196, 198, 200, 202, 234, 249, 250, 252, 253, 255, 256, 259, 261, 263, 274, 278, 286, 288, 294, 338
Sustainability, 9, 10, 29, 30, 36, 41, 45, 47, 48, 71, 73, 75, 90, 98, 102, 110, 125, 138, 140, 146–148, 178, 179, 201, 233, 250, 251, 254, 257, 278, 285, 287, 336, 360, 363–368, 370–373, 375, 376, 392, 397, 401, 402
Sustainable agriculture, 6, 10, 36, 59, 61, 74, 75, 127, 137, 138, 147, 177, 181, 197–205, 252, 275, 279, 286, 290, 297, 307, 322, 359, 379–403
Sustainable composting, 47, 109, 145, 393, 397, 401

Sustainable waste management, 13, 40, 90, 92, 122, 137, 140, 142, 147, 148, 178, 182, 188, 189, 205, 359, 360, 368, 371, 373, 375, 376, 394, 397, 399, 401, 402

**V**

Vermicomposting (VC), 14–17, 46, 59–73, 75, 102, 108, 119, 123–127, 184–186, 188, 230–233, 280, 350, 363

**W**

Waste management, 6, 7, 9–12, 20, 47, 59, 91, 94, 95, 97, 100, 106–110, 119, 120, 138–141, 146, 148, 179–181, 185, 187–190, 195, 205, 291, 354, 355, 359–362, 364, 366–368, 372–374, 380, 393, 394, 397–399, 401, 402

MIX
Papier aus verantwortungsvollen Quellen
Paper from responsible sources
FSC® C105338

If you have any concerns about our products,
you can contact us on
**ProductSafety@springernature.com**

In case Publisher is established outside the EU,
the EU authorized representative is:
**Springer Nature Customer Service Center GmbH**
**Europaplatz 3, 69115 Heidelberg, Germany**

Printed by Libri Plureos GmbH
in Hamburg, Germany